Low Carbon Materials
Paving a Sustainable Path for a Greener Future

Low Carbon Materials

Paving a Sustainable Path for a Greener Future

Edited by

Santanu Patra
Danmarks Tekniske Universitet, Denmark
Email: sanpa@dtu.dk

Sudheesh K. Shukla
Graphic Era (Deemed to be University), Dehradun, India
Email: sudheeshkshukla.bt@geu.ac.in

and

Mika Sillanpää
University of Johannesburg, South Africa
Email: mikaESillanpaa@gmail.com

ROYAL SOCIETY OF CHEMISTRY

Hardback ISBN: 978-1-83767-392-6
PDF ISBN: 978-1-83767-606-4
EPUB ISBN: 978-1-83767-605-7
A catalogue record for this book is available from the British Library

The Royal Society of Chemistry is a charity, registered in England and Wales, Number 207890, and a company incorporated in England by Royal Charter (Registered No. RC000524), registered office: Burlington House, Piccadilly, London W1J 0BA, UK, Telephone: +44 (0)20 7437 8656.

For further information see our website at www.rsc.org

For general enquiries, please contact books@rsc.org

For EU product safety enquiries, please email books@rsc.org or contact Royal Society of Chemistry Worldwide (Germany) GmbH, Römischer Hof, Unter den Linden 10, 10117 Berlin.

Printed in the United Kingdom by CPI Group (UK) Ltd, Croydon, CR0 4YY, UK

Preface

The challenges caused by climate change and environmental damage are getting more serious, so we need to rethink the way we develop, produce, and utilize materials. An essential component of sustainable development, low-carbon materials provide novel approaches to reducing the impact on the environment by lowering emissions of greenhouse gases and other harmful byproducts. These materials have the ability to transform various industries, including construction, manufacturing, energy, and healthcare, due to their reduced carbon intensity over their entire life cycle. This includes extraction, processing, use, and disposal. In addition to lowering emissions, low-carbon materials help in promoting ecological balance by using renewable energy, efficient production processes, and the concepts of the circular economy.

This book covers the development, current uses, and potential future of low-carbon materials, making it an excellent resource for anybody interested in this dynamic field. Case studies and state-of-the-art research are discussed to demonstrate their significant potential across multiple industries. In addition to the technological details, the economic and social consequences of widespread use of these materials are also included in the discussion. This book seeks to encourage researchers, industry experts, and policymakers to accelerate the shift toward a sustainable and greener future by increasing comprehension of the science and creativity underlying low-carbon materials.

Chapter 1 outlines the definition of low-carbon materials and their association with a reduced carbon footprint, along with their economic and environmental advantages. Chapters 2–6 discuss the application of low-carbon materials in energy systems like batteries, supercapacitor, fuel cells, energy harvesting, and water splitting. Chapters 7–9 present the importance of low-carbon materials in sensors, mostly for healthcare applications.

Low Carbon Materials: Paving a Sustainable Path for a Greener Future
Edited by Santanu Patra, Sudheesh K. Shukla and Mika Sillanpää

Published by the Royal Society of Chemistry, www.rsc.org

Chapters 10–13 summarize the application of low-carbon materials in construction and infrastructure, environmental monitoring, transportation and automobiles, and air pollution control, respectively. Chapter 14 highlights how we can recycle low-carbon materials for useful applications. The last chapter, Chapter 15, is dedicated to the concluding remarks on how these low-carbon materials can support the shift toward a Greener Future.

We are very thankful to all the authors of the chapters for their very informative and passionate contributions to this book. Special thanks are expressed to Helen Armes, Amina Headley, and the entire Royal Society of Chemistry Books team for their dedicated support and co-operation during the process of compiling and publishing this book.

Santanu Patra, *Lyngby, Denmark*
Sudheesh K. Shukla, *Dehradun, India*
Mika Sillanpää, *Johannesburg, South Africa*

Contents

Chapter 1 Low-carbon Materials: Unlocking Economic and Environmental Benefits 1
Jai Prakash Chaudhary, Mukesh Chandra Joshi and Kamla Rawat

1.1 Introduction 1
1.2 Understanding Low-carbon Materials 2
1.2.1 Criteria and Types of Low-carbon Materials 2
1.3 Economic Benefits of Low-carbon Materials 3
1.3.1 Cost Reduction 3
1.3.2 Innovation and Market Opportunities 4
1.3.3 Regulatory Compliance 5
1.4 Environmental Benefits of Low-carbon Materials 6
1.4.1 Reduction in Greenhouse Gas Emissions 6
1.4.2 Resource Efficiency 7
1.4.3 Impact on Ecosystems and Biodiversity 7
1.4.4 Enhancement of Air and Water Quality 8
1.5 Case Studies and Practical Applications 9
1.5.1 Successful Implementations in Construction 9
1.5.2 Industrial Applications 10
1.5.3 Community and Urban Planning 12
1.5.4 Global Examples and Best Practices 13
1.6 Challenges and Barriers 13
1.6.1 Technical and Material Limitations 13
1.6.2 Economic and Market Barriers 13

Low Carbon Materials: Paving a Sustainable Path for a Greener Future
Edited by Santanu Patra, Sudheesh K. Shukla and Mika Sillanpää

Published by the Royal Society of Chemistry, www.rsc.org

1.6.3 Regulatory and Policy Challenges 14
1.6.4 Social and Cultural Obstacles 14
1.7 Future Directions and Innovations 15
1.7.1 Emerging Technologies and Trends 15
1.7.2 Policy and Regulatory Innovations 15
1.7.3 Sustainability and Circular Economy 15
1.8 Sustainable Development Goals and Carbon Neutrality Approach 16
1.9 Conclusion 17
Acknowledgements 17
References 18

Chapter 2 Low-carbon Battery Materials 23
Felipe M. de Souza and Ram K. Gupta

2.1 Introduction 23
2.1.1 Importance of Energy-storage Devices 23
2.1.2 Role of Materials in Energy Devices 24
2.1.3 Importance of Low-carbon Materials 24
2.2 Batteries: Types and Characteristics 26
2.2.1 Types of Batteries 26
2.2.2 Working Principle of Batteries 27
2.2.3 Characteristics of Batteries 29
2.3 Low-carbon Materials in Batteries 29
2.3.1 Cathode Materials 29
2.3.2 Anode Materials 31
2.3.3 Electrolytes 32
2.4 Challenges and Opportunities 35
2.4.1 Recycling and Environmental Impact 35
2.4.2 Regulations 36
2.4.3 Sustainable Sources 37
2.4.4 Recent Advancements 38
2.4.5 Economic Benefits 39
2.5 Conclusion 40
References 41

Chapter 3 Low-carbon Supercapacitor Materials 46
A. Raouf Aliakbari, B. Elaheh Kowsari and C. Seeram Ramakrishna

3.1 Introduction 46
3.2 Low-carbon Electrode Materials 47
3.2.1 Biowaste to Low-carbon Electrode Materials 47

3.2.2 Convert Plastic Waste into Low-carbon Electrode Materials 51
3.2.3 Recycling Electronic Waste into Low-carbon Electrode Materials 53
3.2.4 Metal and Metal Oxides to Low-carbon Electrode Materials 55
3.3 Other Low-carbon and Green Components of Supercapacitors 57
3.4 Conclusion and Future Perspectives 59
Acknowledgements 60
References 60

Chapter 4 Low-carbon Fuel Cell Materials 65
Felipe M. de Souza and Ram K. Gupta

4.1 Introduction 65
4.1.1 Types of Energy Devices 65
4.1.2 Importance of Fuel Cells 66
4.1.3 Significance of Low-carbon Materials 67
4.2 Fuel Cell: Types and Technology 68
4.2.1 Types of Fuel Cells 68
4.2.2 Basics of Fuel Cells 68
4.2.3 Significance of Low-carbon Materials 72
4.3 Low-carbon Materials in Various Fuel Cells 72
4.3.1 Proton Exchange Membrane Fuel Cells (PEMFCs) 72
4.3.2 Solid Oxide Fuel Cells (SOFCs) 74
4.3.3 Molten Carbonate Fuel Cells (MCFCs) 76
4.3.4 Alkaline Fuel Cells (AFCs) 77
4.3.5 Biofuel Cells 79
4.3.6 Other Fuel Cell Technologies 82
4.4 Advancement and Challenges 83
4.4.1 Architectural Aspects of Fuel Cells 83
4.4.2 Efficiency and Performance 84
4.4.3 Environmental Impact 84
4.4.4 Economic Consideration 85
4.5 Future Trends and Opportunities 86
4.5.1 Emerging Technologies 86
4.5.2 Economic Benefits 86
4.5.3 Eco-friendly Materials 86
4.6 Conclusion 87
References 87

Chapter 5 Low-carbon Energy-harvesting Materials 93
Sinakshi Mehra, Shiwali Thakur, Neha Munjal, Jyoti Varma, Jagdeep Reddy, Phạm Thanh Đạng and Mukesh Kumar

5.1 Introduction 93
5.1.1 Overview of the Importance of Low-carbon Energy-harvesting Materials 93
5.1.2 Definition and Scope of Energy Harvesting 94
5.1.3 Energy-harvesting Materials and Their Importance for Sustainability 96
5.2 Fundamentals of Energy Harvesting 96
5.2.1 Explanation of the Ideas and Principles of Energy Harvesting 96
5.2.2 Types of Energy Sources for Harvesting 98
5.2.3 Role of Materials in Energy Conversion 98
5.3 Characteristics of Low-carbon Materials 99
5.3.1 Definition and Properties of Low-carbon Materials 99
5.3.2 Advantages of Using Low-carbon Materials in Energy Harvesting 100
5.3.3 Challenges and Limitations of Low-carbon Materials 101
5.4 Types of Low-carbon Energy-harvesting Materials 103
5.4.1 Photovoltaic Materials 103
5.4.2 Thermoelectric Materials 108
5.4.3 Piezoelectric Materials 114
5.4.4 Triboelectric Materials 118
5.5 Applications of Low-carbon Energy-harvesting Materials 120
5.5.1 Wearable Electronics 120
5.5.2 Internet of Things (IoT) Devices 121
5.5.3 Green Building Technologies 121
5.5.4 Sustainable Energy Solutions for Developing Regions 122
5.6 Challenges and Future Directions 123
5.6.1 Scalability and Commercialization Challenges 123
5.6.2 Integration with Existing Energy Infrastructure 124
5.6.3 Emerging Trends and Future Prospects 125

5.7 Conclusion 125
5.7.1 Importance of Continued Research and Development in the Field 125
5.7.2 Final Remarks on the Potential Impact of Low-carbon Energy-harvesting Materials on Global Sustainability 126
References 126

Chapter 6 Low-carbon Materials for Electrochemical Water Splitting 130
Nguyen Viet Khoa, Nguyen Minh Viet, Nguyen Minh Phuong, Phuong Thao, Nguyen Van Noi and Thanh-Dong Pham

6.1 Introduction 130
6.2 Electrochemical Water-splitting Process 132
6.2.1 Low-carbon Materials for Water-splitting Process 132
6.2.2 Electrocatalysts from Natural Minerals 138
6.2.3 Electrocatalyst from Biomass Waste 138
6.2.4 Electrocatalysts from Electronic Waste 140
6.3 Conclusions 141
References 142

Chapter 7 Low-carbon Materials for Healthcare Monitoring 144
M. A. Özbek, A. A. Topçu, F. Yilmaz and A. Denizli

7.1 Introduction 144
7.2 Nanocarbon-based Materials 146
7.2.1 Graphite 147
7.2.2 Carbon Nanotubes 148
7.2.3 Graphene 151
7.2.4 Carbon Black 153
7.2.5 Carbon Fibers and Their Properties 155
7.2.6 Fullerene 155
7.2.7 Carbon Dots and Fabrication 156
7.3 Sensor Types Used in Low-carbon Materials 157
7.3.1 Optical Biosensors 157
7.3.2 Electrochemical Biosensors 162
7.3.3 Thermal Sensors 164
7.3.4 Gravimetric/Piezoelectric/Mass-sensitive Biosensors 165
7.4 Applications of Nanocarbon-based Materials for Biosensing 166
7.5 Conclusions 170
References 172

Chapter 8 Low-carbon Materials in Biosensors Applications 177
Xiangyu Yao, Lanqing Yang, Baoqing Bai, Xiaomin Wang, Jinhua Zhang, Ligang Yu, Tao Bo, Ying Gu and Yukun Yang

8.1 Introduction 177
8.2 An Overview of Biosensors 178
8.2.1 Elements Contained Within a Biosensor 179
8.2.2 Characteristics of Biosensors 180
8.2.3 Different Types of Biosensors 180
8.3 Bio-based Biosensors 184
8.3.1 Enzyme-based Biosensors 184
8.3.2 Antibody-based Biosensors 187
8.3.3 Nucleic Acid Aptamer-based Biosensors 189
8.3.4 Cell-based Biosensors 196
8.3.5 Bacteriophage-based Biosensors 199
8.3.6 Antimicrobial Peptide-based Biosensors 199
8.3.7 Summary 200
8.4 Paper-based Biosensors 200
8.4.1 The Advantages of Paper as a Substrate Material for Biosensors 201
8.4.2 Applications of Paper-based Biosensors 201
8.4.3 Outlook and Conclusion 206
8.5 Silk-based Biosensors 207
8.5.1 Traditional Biosensors Based on Silk Substrates 207
8.5.2 Implantable Silk-based Biosensors 210
8.5.3 Conclusion and Future Perspectives 210
8.6 Carbon-based Biosensors 211
8.6.1 Types of Carbon Nanomaterials and Their Applications 211
8.6.2 Summary 214
8.7 Challenges and Future Trends 216
References 217

Chapter 9 Low-carbon Materials for Electrochemical Biomedical Sensors 225
Çiğdem Kanbeş Dindar, Selenay Sadak, Hülya Silah and Bengi Uslu

9.1 Introduction 225
9.2 Properties and Importance of Low-carbon Materials 227

9.3 Utilizing Low-carbon Material for the Development of an Electrochemical Biomedical Sensor 232
9.3.1 Chitosan and Its Utilization in Electrochemical Biomedical Sensors 232
9.3.2 Cellulose and Its Utilization in Electrochemical Biomedical Sensors 235
9.3.3 Silk and Its Applications in Electrochemical Biomedical Sensors 241
9.4 Conclusion and Future Perspectives 241
References 243

Chapter 10 Role of Low-carbon Materials in Construction and Infrastructure 254
Neetu Vishwakarma, Pooja Sharma, Upasana Bagri, Shalini Singh Chhetri, Monika Sindhu, Meenakshi Choudhary and Ajit Sharma

10.1 Introduction 254
10.2 Literature Review of Earlier Work 257
10.3 Innovation Challenges in the Construction Sector 258
10.4 Challenges Preventing the Building Sector from Using Substitute Materials 259
10.5 Study Goals and Limitations 261
10.6 Methods 261
10.7 Conclusions 262
References 263

Chapter 11 Low-carbon Materials for Environmental Monitoring 270
T. Tasić, V. Milanković and T. Lazarević-Pašti

11.1 Introduction 270
11.2 Principles of Low-carbon Materials 271
11.3 Low-carbon Materials for Environmental Sensors 275
11.4 Synthesis and Fabrication Methods of Low-carbon Materials 277
11.5 Applications of Low-carbon Materials in Environmental Monitoring 280
11.6 Challenges and Future Prospects 283
Declaration of Generative AI and AI-assisted Technologies in the Writing Process 288
Acknowledgements 289
References 289

Chapter 12 Low-carbon Materials: Transportation and Automotive Applications **295**
Atul Sharma, Rashmi Pundeer, Neha Dua, Harish Kumar, D. K. Sharma, Saif Ali Chaudhry and Bindu Mangla

12.1 Introduction 295
12.2 Low-carbon Materials in Vehicle Design 296
12.3 Advancements in Energy-storage Materials 297
12.3.1 Electric Vehicle Battery Materials 297
12.3.2 High-efficiency Engine Components 298
12.3.3 Fuel Cell and Hydrogen Storage Materials 299
12.4 Low-carbon Materials in Interior and Exterior Components 300
12.4.1 Sustainable Interior Materials 300
12.4.2 Low-emission Materials: Transportation Infrastructure Carbon Emission Reduction Strategies 302
12.4.3 Comparison of CO_2 Emissions between ICEVs and EVs for Better Insights into Low Carbon Emissions 303
12.5 Challenges of Utilizing Materials with a Lower CO_2 Footprint in the Automotive Industry 303
12.5.1 Cost and Regulatory Certification Challenges 303
12.5.2 Performance and Durability 304
12.5.3 Technological Barrier 305
12.6 Impact of Low-carbon Materials on the Transportation Sector 307
12.6.1 Reduction of Greenhouse Gas Emissions 307
12.6.2 Societal and Economic Benefits 309
12.7 Conclusion and Future Prospects 310
Acknowledgements 310
References 310

Chapter 13 Low-carbon Materials for Reducing Air Pollutants **314**
Abhishek Singh and Tanvir Arfin

13.1 Introduction 314
13.2 Low-carbon Materials 316
13.3 Essential Characteristics of Low-carbon Materials 317
13.4 Drive and Uncertainty 317
13.5 Manufacturing 318

13.6 Environmental Pollution 319
13.6.1 Particulate Matter 320
13.6.2 Volatile Organic Compounds 321
13.6.3 Carbon Dioxide 323
13.6.4 Carbon Monoxide 324
13.7 Low-carbon Materials and CO_2 Footprint 325
13.8 Future Perspectives 326
13.9 Conclusion 327
Acknowledgements 328
References 328

Chapter 14 Recycling of Low-carbon Materials **334**
Ahmed I. Osman, Chenhao Zhou, Zhonghao Chen, Mahmoud Nasr, Basem E. Keshta, David W. Rooney and Pow-Seng Yap

14.1 Introduction 335
14.2 Recycling Methods for Low-carbon Materials 336
14.2.1 Mechanical Recycling Techniques 337
14.2.2 Chemical Recycling Approaches 338
14.2.3 Biological Recycling Methods 339
14.3 Benefits and Impacts of Recycling Low-carbon Materials 340
14.3.1 Reduction of Greenhouse Gas Emissions 340
14.3.2 Conservation of Natural Resources 341
14.3.3 Energy Savings and Carbon Footprint Reduction 341
14.3.4 Circular Economy and Waste Management Solutions 342
14.4 Challenges in Recycling Low-carbon Materials 342
14.4.1 Collection and Sorting of Low-carbon Materials 343
14.4.2 Contamination Issues and Quality Control 344
14.4.3 Technological and Economic Barriers 344
14.4.4 Policy and Regulatory Considerations 345
14.5 Future Trends and Innovations in Low-carbon Material Recycling 345
14.5.1 Advancements in Recycling Technologies 345
14.5.2 Integration of Digitalization and Artificial Intelligence 346

14.5.3 Emerging Circular Business Models 347
14.5.4 Policy and Industry Initiatives 347
14.6 Conclusion 348
References 349

Chapter 15 Toward a Greener Future with Low-carbon Materials 359
Ashima Thakur, Saima Jan, Shreya Semwal, Nandini Dhupar, Phạm Thanh Đạng, Sudheesh Kumar Shukla and Ajit Sharma

15.1 Introduction 359
15.1.1 Low-carbon Materials for Greener Energy 360
15.1.2 Batteries with Low-carbon Materials 360
15.1.3 Low-carbon Fuel Cells (FCs) 361
15.2 Low-carbon Supercapacitors 362
15.2.1 Activated Carbon 362
15.2.2 Carbon Nanotubes 362
15.2.3 Low-carbon Energy-harvesting Materials 363
15.3 Low-carbon Materials for Greener Healthcare 364
15.3.1 Graphene as a Biosensor 364
15.4 Low-carbon Materials for a Greener Environment 364
15.4.1 Pollution Remediation 365
15.4.2 Water Treatment 365
15.4.3 Conversion of CO_2 into Fuel 365
15.5 Low-carbon Materials for Greener Construction 365
15.5.1 Addition of Recycled Aggregates 368
15.5.2 Adopting Recyclable or Low-embodied Carbon 368
15.5.3 Discovering New Concrete 368
15.5.4 Embracing Recycling 368
15.5.5 Considering Composite Window Frames 368
15.6 Low-carbon Materials in Transportation and the Automotive Industry 368
15.7 Steps Toward a Greener World with Carbon Neutrality 370
15.8 Conclusion 372
15.9 Future Prospective 372
References 373

Subject Index 381

CHAPTER 1

Low-carbon Materials: Unlocking Economic and Environmental Benefits

JAI PRAKASH CHAUDHARY,*[a] MUKESH CHANDRA JOSHI[a] AND KAMLA RAWAT*[b]

[a] Department of Chemistry, Kirori Mal College, University of Delhi, Delhi-110007, India; [b] Department of Chemistry, Jamia Hamdard, Delhi-110062, India
*Emails: jpsioasis@kmc.du.ac.in; kamlarawat@jamiahamdard.ac.in

1.1 Introduction

In recent years, the world has gained more awareness regarding the environmental issues associated with traditional manufacturing processes and materials. Thus, the world has had a greater emphasis on the development and usage of low-carbon materials. As environmental awareness grows, and as a result of the urgency in combating climate change, low-carbon materials have become an important solution for minimizing the negative impacts on the environment within different sectors. These materials represent a significant shift from conventional manufacturing practices, which often lead to high greenhouse gas emissions and resource depletion. The drive for sustainability has spurred the development and adoption of low-carbon materials in diverse sectors, including construction, agriculture, healthcare, waste management, transportation, and consumer goods.[1–5]

The low-carbon materials movement can be traced to the recognition of the environmental consequences associated with traditional manufacturing and material usage. Processes such as mining, refining, and manufacturing

Low Carbon Materials: Paving a Sustainable Path for a Greener Future
Edited by Santanu Patra, Sudheesh K. Shukla and Mika Sillanpää

Published by the Royal Society of Chemistry, www.rsc.org

contribute considerably to carbon emissions and resource depletion, calling for a review of industrial practices. With scientific evidence mounting on the increasing threats of climate change, governments, businesses, and consumers have started to look for alternatives that would minimize the ecological footprint of human activities.[1,4–6] Low-carbon materials' development is driven by technology, enhanced environmental awareness, and an increased demand for sustainable products. Advances in materials science, engineering, and manufacturing have opened up the availability of alternative products that have industry requirements but are also kept within environmental stewardship parameters. Lastly, carbon-reduction international agreements and the regulatory framework will make industries adjust their strategies and invest more in green alternatives.[4–6]

This chapter delves into the economic and environmental benefits of low-carbon materials, exploring their role in promoting innovation, creating new market opportunities, and supporting global efforts to address climate change. We examine how different industries are adopting these materials through case studies and analysis, paving the way for a more sustainable and resilient future.

1.2 Understanding Low-carbon Materials

Low-carbon material is defined as material having been engineered, produced, or extracted in ways that could reduce carbon dioxide emissions together with the overall carbon footprint throughout their life cycle. These materials are of extreme importance to sustainable development and the impacts from climate change because they release much less greenhouse gas emissions compared to traditional materials.

1.2.1 Criteria and Types of Low-carbon Materials

To be considered a low-carbon material, several criteria need to be met. First, it should have low embodied energy, that is, very little energy must be used in its extraction, processing, and manufacturing. Secondly, the material must be sourced sustainably, that is, with practices that do not cause much environmental degradation. Lastly, it should have a high recyclability potential so that it can be recycled and reused efficiently with minimum waste and demand for new materials. In addition, the material should be durable, long-lasting, and minimize frequent replacements, saving resources and energy. Lastly, some of the materials should be able to show carbon sequestration, that is, the ability to absorb and store carbon dioxide from the atmosphere, such as bio-based materials. These criteria set together the characteristics necessary for materials to qualify as low carbon, contributing to environmental sustainability and resource efficiency.

Types of low-carbon materials include natural low-carbon materials, engineered low-carbon materials, and innovative low-carbon materials. Natural low-carbon materials come directly from nature and are minimally processed. One of the best examples is bamboo, which grows faster than any other plant and can be harvested within 3–5 years—significantly faster than regular timber. Bamboo is renowned for sequestering substantial amounts of carbon dioxide and can be utilized in a wide range of applications such as construction, furniture, and textiles.[7] Other natural materials include straw, a by-product of cereal farming, which has been used in building construction for insulation purposes. Straw bale constructions offer excellent insulation, reducing energy consumption for heating and cooling interior spaces.[8]

Engineered low-carbon materials, on the other hand, are produced through industrial processes that aim to lower carbon emissions during production and use. For example, recycling steel significantly reduces the carbon footprint compared to producing new steel from iron ore, requiring less energy and generating fewer greenhouse gases.[9] Similarly, low-carbon concrete has been developed to mitigate emissions associated with traditional cement production. By substituting part of the cement with supplementary materials such as fly ash, slag, or calcined clays, the carbon emissions of concrete production can be significantly decreased.[10]

Innovative low-carbon materials leverage advanced technology and innovative approaches to achieve superior environmental performance. Examples include bioplastics derived from renewable biomass sources like vegetable fats, oils, and sugars from corn or sugarcane. These materials are designed to reduce carbon emissions and are often biodegradable.[11] Carbon fiber composites, despite being energy intensive to produce, offer lightweight properties that yield significant energy savings in applications like the automotive and aerospace industries. The reduced weight of vehicles and aircraft contributes to lower fuel consumption and emissions throughout their life cycles.[12]

1.3 Economic Benefits of Low-carbon Materials

1.3.1 Cost Reduction

1.3.1.1 *Initial Costs* versus *Lifecycle Costs*

Low-carbon materials are actually more expensive in securing at initial costs, although they are much cheaper than conventional materials at the lifecycle cost (see Figure 1.1). This is because they last longer, with reduced maintenance needs as well as having greater energy efficiency. Studies have shown that materials such as low-carbon concrete and recycled steel are more expensive at the beginning, but they have huge cost savings due to their durability and reduced maintenance costs.[13] Lifecycle analysis gives the most important analysis of all three phases of total cost, namely, production, in use, and disposal; therefore, it is considered a powerful tool in exhibiting the long-term economic advantage associated with these materials.

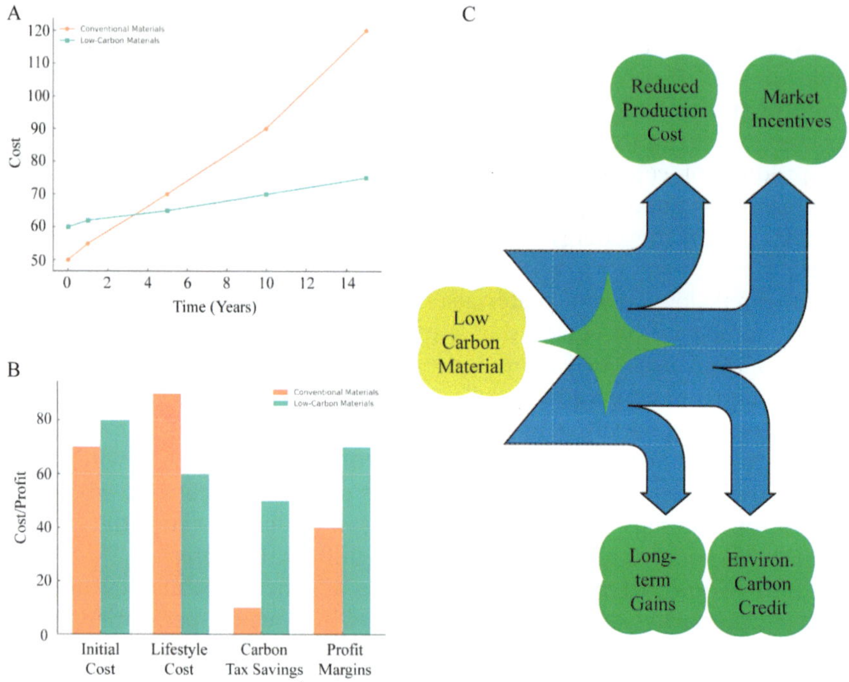

Figure 1.1 A schematic representation of costs and economic benefits of conventional and low-carbon materials over time. (A) Cost Comparison over Time Conventional *vs.* Low Carbon Materials. (B) Comparison of Economic Factors Conventional *vs.* Low Carbon Materials. (C) Economic benefits of Low Carbon Materials.

1.3.1.2 Energy Savings and Reduced Utility Bills

One of the easiest economic benefits connected with a low-carbon material is cutting down on energy consumption as an effect. Buildings constructed through low-carbon material, such as high-performance insulation and even energy-efficient windows, require less heating and cooling as these buildings consume much less energy, which directly and proportionately will reduce utility consumption for occupants and/or businesses. Buildings constructed with low-carbon materials save up to 50% energy when compared to traditional buildings, according to an International Energy Agency report.[14] This energy efficiency ensures cost minimization and lower emission of greenhouse gases to achieve the different goals set for global sustainability.

1.3.2 Innovation and Market Opportunities

1.3.2.1 Growth of Green Building Markets

The green building market is driven by increasing environmental sustainability. Builders and developers are more interested in building an environment-friendly building; hence, demand for low-carbon materials is increasing. The global market for green building materials is going to grow significantly and is going to be primarily driven by tighter environmental regulations and increasing preference for sustainable practices of construction.[15] This surge presents important opportunities for innovating and investing in the advancement and commercialization of new low-carbon materials.

1.3.2.2 Demand for Sustainable Products

The growing consumer demand for products that reduce carbon footprints has led to increased awareness and preference for eco-friendly options. This shift in consciousness is driving manufacturers and marketers to develop low-carbon materials. Such material are entering the market with other beneficial aspects such as excellent strength, bioplastics, carbon fiber composites, or recycled materials. Also, demand for sustainable products has grown due to new market opportunities or drives for more economic growth within the sustainable materials industry.[16]

1.3.3 Regulatory Compliance

Global regulations for environment-related and carbon-emission factors are now stringent. Governments are formulating policies and incentives that favor low-carbon materials. This could turn out to be financially profitable for business entities as compliance with regulations can benefit businesses. For example, companies adopting low-carbon materials can enjoy fiscal incentives, grants, and subsidies to encourage sustainability practices.[17] On the other hand, following the regulation standards will enhance the company's reputation and its competitiveness in the market and attract concerned consumers and investors.

The inset in Figure 1.1(A) illustrates a cost comparison over time between conventional materials and low-carbon materials. The graph shows that although low-carbon materials may have a marginally higher initial cost, their cost increase over time is much slower compared to conventional materials. This indicates a lower lifecycle cost, making them a financially sustainable option in the long run. These findings are consistent with previous studies that emphasize the long-term economic advantages of low-emission materials due to reduced operational and maintenance costs.[18,19]

The inset in Figure 1.1(B) compares several economic factors, namely initial cost, lifestyle cost, carbon tax savings, and profit margins, for both material types. While the initial costs are relatively comparable, low-carbon materials clearly outperform in lifestyle costs (associated with durability and energy efficiency), offer greater savings through carbon tax incentives, and yield higher profit margins. This analysis supports the growing body of evidence that transitioning to low-carbon materials can provide both direct economic savings and indirect financial benefits under current and evolving regulatory frameworks.[20,21]

The inset in Figure 1.1(C) presents a schematic overview of the economic benefits associated with the use of low-carbon materials. The diagram highlights four main advantages: reduced production costs, market incentives, long-term economic gains, and environmental carbon credits. These benefits arise from optimized resource usage, government subsidies or tax incentives, and growing consumer and investor preference for environmentally responsible practices. This multi-faceted economic viability aligns with global sustainability reports and lifecycle assessments of green materials.

In India, the government has also been proactive in encouraging the adoption of low-carbon materials through various initiatives and regulations. The Indian Green Building Council (IGBC) has established guidelines and rating systems to promote green buildings that utilize low-carbon materials. Additionally, the Indian government offers incentives such as the Perform, Achieve and Trade (PAT) scheme, which incentivizes industries to reduce their specific energy consumption through the use of energy-efficient and low-carbon technologies. The Smart Cities Mission also emphasizes sustainable urban development, encouraging the use of low-carbon materials in infrastructure projects. By aligning with these national policies and schemes, Indian companies can gain access to financial benefits, enhance their sustainability credentials, and contribute to the country's commitment to reducing greenhouse gas emissions.[22–25]

1.4 Environmental Benefits of Low-carbon Materials

1.4.1 Reduction in Greenhouse Gas Emissions

1.4.1.1 Carbon Footprint Analysis

A low-carbon material provides an opportunity to meet the objective of climate change mitigation, first of all, through less carbon intensity compared to traditional materials. The assessment of carbon footprints means the measurement of total CO_2 and other GHG emission over the entire life cycle of a material from extraction and manufacturing to the use phase and disposal. Research has revealed that products like low-carbon concrete and recycled steel reduce CO_2 emissions. For instance, remnant steel requires about 75% less energy in its production as compared to steel from iron ore, hence significantly reducing GHG emissions.[26]

1.4.1.2 Comparison with Traditional Materials

Low-carbon materials provide significantly greater environmental benefits than conventional construction products. For instance, traditional cement production—primarily from Portland cement—accounts for roughly 8% of global CO_2 emissions. In contrast, low-carbon concrete that incorporates industrial by-products such as fly ash or slag can reduce greenhouse gas (GHG) emissions by 30–50%.[27,28] Similarly, bioplastics derived from renewable resources generate far fewer GHGs during production compared to petrochemical-based plastics. Thus, the adoption of low-carbon materials plays a crucial role in mitigating climate change and promoting sustainable development.[11]

The application of low-carbon alternative materials is gradually being adopted in developing countries such as India. For instance, the utilization of fly ash as a constituent in concrete is helpful in reducing emissions and also in managing waste from coal-fired power plants.[29] With more research and development focused on superior materials such as geopolymer concrete that employs waste materials from industries, it is projected that future advancements in materials science are imperative for the sustainable development agenda in India, as well as for achieving its emission reduction targets under the Paris Agreement.[28]

1.4.2 Resource Efficiency

Low-carbon materials contribute to resource efficiency by reducing raw material consumption. For example, bamboo, a natural low-carbon material, grows quickly and can be harvested sustainably without depleting the soil. Its use in construction and manufacturing helps conserve forests and reduces deforestation by lowering the need for wood.[30]

Recycling and waste minimization are also essential aspects of low-carbon materials. The use of recycled materials, such as reclaimed wood, recycled steel, and post-consumer plastics, decreases the demand for virgin materials. This not only reduces the environmental impact of extraction and processing but also diverts waste from landfills. The recycling of construction and demolition waste into new building materials exemplifies a circular economy approach, further promoting sustainability and resource efficiency.[31]

1.4.3 Impact on Ecosystems and Biodiversity

1.4.3.1 Less Environmental Degradation

Low-carbon materials, when produced and utilized, lessen environmental degradation. Ecosystems can be highly disrupted because of the traditional extraction methods, such as mining for metal ores or quarrying for stone. Ecosystem disruption in these processes entails habitat destruction as well as soil erosion. Low carbon, in contrast, mostly means different, more sustainable uses, which minimize the damage from these practices. Agricultural residues such as straw are used in building construction, therefore avoiding the ecological losses of the typical materials.[32] Using industrial by-products, such as fly ash and slag, for construction reduces waste generation as well as reduces quarrying and mining activities, thereby protecting the surrounding ecosystem.[28]

In India, combined with the acceptance of fly ash use in construction materials from coal-using plants, waste generation and environmental protection solutions have gained ground. Such practices help curb the need for natural raw materials and offer better solutions for the eventual disposal of fly ash before it turns into an environmental hazard.[29] In addition, using low-carbon materials in the developing Indian construction industry has become important to prevent environmental degradation caused by rapid urbanization and infrastructure development.[33]

1.4.3.2 Preservation of Natural Habitats

As low-carbon materials do not require the extraction of raw materials, these products play a great role in the conservation of natural habitats. These practices serve to enhance the sustainability of products such as by sourcing bamboo and recycling materials to reduce harm to ecosystem diversity and wildlife. For instance, bamboo is characterized by its ability to regrow once cut and it has the elasticity of timber without having to fell trees for its use. Having applied this product in construction and production, there will be less strain on forests, hence helping in the conservation of forests.[7]

The effectiveness in the use of recycled materials is another factor that needs to be considered in the preservation of natural habitats. For example, recycling of steel prevents the utilization of iron ore, which leads to soil erosion and upset habitats. Other than saving natural resources, this practice is also useful in saving energy and the resultant greenhouse gases from the primary production processes. To achieve these goals in Italy, there have been the polices and incentives to support recycling activities in building practices.[34]

Also, low-carbon materials such as bioplastics, which are obtained from renewable resources such as biomass, are beneficial to habitat preservation since they are a better substitute for petroleum products. Bioplastics, because they are biodegradable, can disintegrate quicker within the environment, thus reducing plastic impacts on water and land animals and their habitats.[11] In the context of India, the usage of bioplastics is in line with Indian attempts to tackle the effects of plastic pollution on wildlife and ecosystems.[35]

1.4.4 Enhancement of Air and Water Quality

1.4.4.1 Reduced Pollution Levels

According to the study, with the production and utilization of low-carbon material, one is able to reduce the amount of airborne and water-soluble wastes as compared to conventional materials. Most conventional materials go through industrial processes whereby all sorts of pollutants are released into the atmosphere. For instance, it has been identified that the manufacturing of Portland cement releases high levels of CO_2, SO_2, and NO_x that are contaminants in the air and a factor in the production of acid rain.

On the other hand, low-carbon materials including geopolymers and other supplementary cementitious materials such as fly ash and slag incorporate far less emissions of these pollutants during their production.[35] Also, bioplastics and other materials derived from renewable sources reduce the emissions of bio-hazardous pollutants during their production cycle and in their processing.[28]

1.4.4.2 Improved Public Health Outcomes

In this way, reduced emissions of pollutants through the usage of low-carbon materials will guarantee healthier air and water quality to increase public health. The number of respiratory and cardiovascular diseases in the population is reduced due to the decrease in the level of dangerous emissions, for example, SO_2 and NO_x. Sustainable products release fewer pollutants into the environment, resulting in lower hospital admissions due to diseases caused by air pollution and thus reducing healthcare costs.[36]

Furthermore, using low-carbon materials that are free from harmful compounds such as volatile organic compounds (VOCs)—commonly found in paints and finishes—enhances indoor well-being. VOC emissions can cause a range of health problems, including headaches, dizziness, asthma, and even cancer. By choosing products that emit minimal VOCs, indoor air quality is significantly improved, leading to better health and comfort for occupants.[37] This is especially important in homes, schools, and workplaces, where people spend most of their time.

Apart from the reduction of indoor air pollution, low-carbon materials in structures can also safeguard water quality. For instance, pervious concrete and green roofing systems play a crucial role in stormwater management and minimize the pollution load in built-up areas and also the subsequent impact on water bodies. It was found that these systems filter and clean water naturally, and thereby help in preserving aquatic life and filter water that is safe for use by humans.[37]

1.5 Case Studies and Practical Applications

The use of low-carbon materials is the key to realizing sustainable development in different areas. This chapter is aimed at identifying some of the existing successful case studies and effective implementations of low-carbon materials in construction undertakings, industrial uses, community areas, and globally in general, with a view to showing their effectiveness and viability in other similar undertakings.[38–41]

1.5.1 Successful Implementations in Construction

1.5.1.1 Green Building Projects

The Bullitt Center in Seattle, USA, stands as a beacon of sustainable architecture. Constructed using primarily low-carbon materials such as Forest Stewardship Council (FSC)-certified wood, triple-pane windows, and photovoltaic panels, the Bullitt Center achieved net-zero energy and water use, demonstrating the feasibility of integrating sustainable practices in commercial buildings.[42]

1.5.1.2 Infrastructure Developments

The Oslo Fjord City project in Norway also demonstrates the proper utilization of low-carbon materials in an urban environment. It was proposed that Oslo Fjord City would actually decrease embodied carbon by focusing on the use of materials with a low-carbon content with "green" roofs and a sustainable water management system in mind.[43]

1.5.2 Industrial Applications

Examples of applications for low-carbon materials in different industrial sectors are summarised in Table 1.1.

1.5.2.1 Manufacturing Processes

Interface Inc., the world's largest modular flooring company, has taken low-carbon materials and sustainable manufacturing processes into its Mission Zero. Interface redesigned production lines to reduce waste and incorporate more recycled and bio-based materials. As a result, it has greatly reduced the company's environmental footprint while achieving similar product quality and performance.[44]

Table 1.1 Applications of low-carbon materials across key sectors.

Sector	Low-carbon material	Applications	Ref.
Agriculture	Bio-based polymers	Sustainable packaging, biodegradable films, and agricultural tools	57
Biomedical	Bio-based polymers, biocomposites	Biomedical devices, drug-delivery systems, prosthetics	57
Energy	Renewable materials (solar-grade silicon, recycled metals)	Solar panels, wind turbines, green energy infrastructure	59
Transportation	Recycled aluminum and plastics, biocomposites	Automotive parts, interior components, lightweight structures	57, 58
Electronics	Recycled rare-earth metals, biodegradable circuit substrates	Eco-friendly electronic components and PCBs	58
Construction	Recycled concrete, FSC-certified wood, green cement	Structural materials, insulation, interior finishing in sustainable buildings	58, 60

Other companies around the world are leaders in the production of low-carbon materials in manufacturing. Toyota has used bioplastics and recycled materials for most car interiors manufactured in Japan. Eco-Plastic, a plastic created from sugar cane and other renewable resources, makes the company's vehicles even greener.[45] Also in the construction industry, LafargeHolcim, a Swiss multinational, developed a low-carbon concrete portfolio branded as ECOPact. In their concrete portfolio, industrial by-products, such as fly ash and slag, were used to substitute materials that would otherwise lead to a 50% increase in CO_2 emissions in normal concrete.[46]

In India, Tata Steel has undertaken very significant progress in introducing sustainability in steel manufacturing processes. Utilizing state-of-the-art technology for the recycling of steel scrap and energy-saving processes has resulted in reducing Tata Steel's carbon emissions substantially. The company also has investigated the potential of using alternative raw materials and enhancing energy efficiency in its processes.[47] In addition, there is increasing interest in developing composite materials by using agricultural waste products, including rice husk and bagasse. These are renewable, have low embodied energy, and could generate more income streams for farmers and thus minimize the problems of waste disposal.[48]

1.5.2.2 Product Design and Innovation

In the field of product design and innovation, Novamont, an Italian company renowned for its expertise in bio-based plastics, represents a significant step toward sustainable material development. Its flagship product line, Mater-Bi, consists of bioplastics derived from renewable raw materials that are fully biodegradable, making them an effective alternative to traditional petrochemical-based plastics commonly used in packaging. The production of Mater-Bi bioplastics not only results in a substantially lower carbon footprint but also addresses the growing issue of plastic waste, as these materials decompose much faster after disposal.[49] The success of such innovations highlights how low-carbon materials can merge environmental responsibility with modern consumer demands, fostering sustainability-driven product design.

In India too, notable innovations with similar sustainable features are emerging across the industrial sector. Manufacturers are increasingly adopting low-carbon materials in response to both international and regional environmental regulations. For example, many Indian producers are utilizing agricultural residues such as rice husk and bagasse to manufacture composite materials. These renewable, resource-based materials not only have low embodied energy but also create additional income opportunities for farmers while reducing waste disposal issues.[48] Likewise, advancements in the use of natural fibers like jute and coir in the automotive and construction industries are showcasing durable and eco-friendly alternatives to synthetic materials.

1.5.3 Community and Urban Planning

1.5.3.1 Sustainable Cities and Communities

Low-carbon urban planning can be shown through an example like Vauban, Germany. This town incorporates low-carbon materials and places its infrastructures on ecological bases. In its construction, it has tried to reduce the use of cars to a minimum. For buildings, materials of low embodied energy were used, and in its construction, it has included renewable energy systems like solar panels and passive heating systems. This approach, while optimizing energy efficiency, also creates a feeling of community and well-being. The participatory nature of the residents in local decision-making processes makes the sustainability ethos of the neighborhood even stronger.[50] Such initiatives have proven that sustainable urban planning can create a vibrant, low-carbon community that serves as a model for future developments around the globe.

Similarly, near Pune in India, the development of Lavasa has taken sustainable practices as an integral part of its urban planning. This city uses rainwater harvesting, solar power, and dense green cover to reduce its environmental footprints. Sustainable development of the area has been made possible through the use of locally sourced low-carbon materials for construction in a geographically expanding region.[51]

1.5.3.2 Low-carbon Urban Infrastructure

The Marina Barrage in Singapore stands as a leading example of successfully implemented low-carbon urban infrastructure. Designed to serve multiple purposes—including flood control, freshwater storage, and hydropower generation—the project integrates sustainability at its core. During its construction, environmentally friendly building materials and low-carbon technologies were employed to minimize overall carbon emissions. These sustainable solutions not only address key urban challenges but also support Singapore's vision of becoming a fully sustainable urban metropolis.[52]

Other projects, aside from Marina Barrage, under the Singaporean HDB Greenprint program, include the retrofitting of existing older housing estates with green features. Works to be done include the installation of solar panels, green roofs, energy-efficient lighting systems, and the use of low-carbon materials in refurbishment. Existing urban infrastructure can easily be adapted to meet sustainability standards, and the carbon footprint can then be greatly reduced in urban areas.[53]

Copenhagen, Denmark, utilizes permeable pavements and green roofs to minimize the need for a massive drainage system that controls stormwater runoff. Such low-carbon solutions help minimize flood risks, while increasing the urban biodiversity and reducing the urban heat island effect.[54]

1.5.4 Global Examples and Best Practices

Integrating low-carbon materials into circular economy models is crucial for achieving long-term sustainability. Key principles of the circular economy include minimizing waste, extending product lifespans through reuse and recycling, and promoting sustainable consumption patterns. Low-carbon materials designed for easy disassembly, recyclability, and biodegradability play a significant role in closing the material loop, reducing the environmental impact, and contributing to a more sustainable system.[55]

Long-term sustainability strategies require the harmonization of economic, environmental, and social objectives through holistic approaches that align with the goals of corporations and governments. These strategies often involve cradle-to-grave design principles, lifecycle assessments, and eco-design approaches, which aim to enhance resource efficiency and minimize environmental footprints. By adopting such practices, businesses and governments can implement resilient, regenerative systems that support long-term sustainable development goals.[56]

1.6 Challenges and Barriers

The adoption of low-carbon materials is crucial for achieving sustainable development goals, however, various challenges and barriers hinder their widespread implementation across different sectors.

1.6.1 Technical and Material Limitations

One of the primary technical challenges faced by low-carbon materials is related to their performance and durability. Bio-based materials, for instance, may exhibit mechanical properties that differ from traditional materials, potentially affecting their suitability for various applications. Ensuring consistent performance and durability is essential to building trust and gaining acceptance among industries and consumers.[61]

Another significant limitation is the availability and scalability of low-carbon materials. Production processes for innovative materials, such as graphene or bio-based polymers, are often still in the developmental stages, leading to higher costs and limited supply chains. Scaling up production while maintaining quality standards presents a critical challenge for the widespread adoption of these materials in construction, manufacturing, and other sectors.[62]

1.6.2 Economic and Market Barriers

The higher initial costs associated with low-carbon materials present significant economic challenges. Investments in research and development, as well as sustainable production processes, often result in higher production costs compared to conventional materials. This cost differential can discourage potential adopters, particularly in price-sensitive markets where short-term economic considerations tend to dominate decision-making.[63]

Additionally, low-carbon materials face barriers related to market acceptance and demand uncertainty. Consumer preferences and industry standards play critical roles in determining the adoption of sustainable materials. A lack of awareness about the benefits of low-carbon materials, coupled with perceived risks associated with new technologies, can further hinder their market penetration and slow the pace of widespread adoption.[64]

1.6.3 Regulatory and Policy Challenges

Regulatory frameworks for low-carbon materials often suffer from inconsistency and a lack of coherence. Varying standards for environmental performance, certification requirements, and incentives across regions create significant barriers for manufacturers and can stifle innovation. Harmonizing policies and establishing clear, universally applicable guidelines are critical to fostering a regulatory environment that supports the adoption of sustainable materials.[65]

Additionally, the lack of robust regulatory support further complicates the adoption of low-carbon materials. Insufficient governmental subsidies, tax incentives, or procurement policies favoring sustainability hinder progress in this area. Strengthened regulatory support, including financial incentives and targeted policies, could serve as a powerful stimulus for industry players to embrace sustainable practices and integrate low-carbon materials into their operations.[66]

1.6.4 Social and Cultural Obstacles

One of the most significant social barriers to adopting low-carbon materials is resistance to change. Industries and consumers may be hesitant to transition from familiar, well-established materials to new sustainable alternatives due to perceived risks, compatibility concerns, or inertia associated with existing practices. This reluctance can slow the adoption of innovative materials, even when they offer clear environmental benefits.[67]

Another major challenge is poor public awareness and education regarding the advantages of low-carbon materials. As the global community works toward climate change mitigation and achieving sustainable development goals, low-carbon materials play an increasingly critical role. However, a lack of understanding among consumers, industry professionals, and policymakers about the environmental, economic, and social benefits of these materials can inhibit their uptake. Educating stakeholders about these advantages is essential for building confidence and encouraging widespread adoption.[68]

1.7 Future Directions and Innovations

As the global community steps up efforts to combat climate change and achieve sustainable development goals, the role of low-carbon materials is becoming even more important.

1.7.1 Emerging Technologies and Trends

Rapid advancements in materials science are driving innovations in low-carbon materials. Researchers are developing new bio-based polymers, lightweight composites, and sustainable alternatives to traditional construction materials like concrete and steel. These innovations do not only offer enhanced performance characteristics but also reduce the environmental impact throughout their life cycle.[69]

Smart materials and nanotechnology are further enhancing the functionality and sustainability of low-carbon materials. Smart materials, which possess self-repairing, adaptive properties and energy-harvesting capabilities, are being explored for applications in the construction, automotive, and electronics sectors. Nanotechnology, focusing on engineering materials at the molecular level, improves strength, durability, and environmental compatibility, contributing to the development of more sustainable and high-performance materials.[70]

1.7.2 Policy and Regulatory Innovations

One of the most significant social barriers to adopting low-carbon materials is resistance to change. Industries and consumers may be hesitant to transition from familiar, well-established materials to new sustainable alternatives due to perceived risks, compatibility concerns, or inertia associated with existing practices. This reluctance can slow the adoption of innovative materials, even when they offer clear environmental benefits.[71]

Another major challenge is poor public awareness and education regarding the advantages of low-carbon materials. As the global community works toward climate change mitigation and achieving sustainable development goals, low-carbon materials play an increasingly critical role. However, a lack of understanding among consumers, industry professionals, and policymakers about the environmental, economic, and social benefits of these materials can inhibit their uptake. Educating stakeholders about these advantages is essential for building confidence and encouraging widespread adoption.[72]

1.7.3 Sustainability and Circular Economy

Integrating low-carbon materials into circular economy models is crucial for achieving long-term sustainability. Key principles of the circular economy include minimizing waste, extending product lifespans through reuse and recycling, and promoting sustainable consumption patterns. Low-carbon materials designed for easy disassembly, recyclability, and biodegradability play a significant role in closing the material loop, reducing the environmental impact, and contributing to a more sustainable system.[73]

Long-term sustainability strategies require the harmonization of economic, environmental, and social objectives through holistic approaches that align the goals of corporations and governments. These strategies often involve cradle-to-grave design principles, lifecycle assessments, and eco-design approaches, which aim to enhance resource efficiency and minimize environmental footprints. By adopting such practices, businesses and governments can implement resilient, regenerative systems that support long-term sustainable development goals.

1.8 Sustainable Development Goals and Carbon Neutrality Approach

Closely knit with the sustainable development goals (SDGs) of the United Nations, the theme in this chapter backs a global approach toward carbon neutrality. We have almost balanced economic development by promoting and integrating low-carbon materials across different sectors. First, low-carbon materials contribute to SDG 7: affordable and clean energy through improved efficiency and implementation of renewable energy technologies like solar panels and wind turbines. The use of these materials decreases dependence on fossil fuels and makes clean energy more widely available and affordable. Besides, this contributes to sustainable industrialization as per SDG 9: industry, innovation, and infrastructure. Low-carbon materials are essential in developing sustainable industries, through innovation in materials science and resilient and low-impact infrastructure. For urban planning and construction, using low-carbon materials supports the aim of SDG 11: sustainable cities and communities, as a result of which they reduce the environmental footprint caused by urbanization and enhance quality living in cities. Moreover, they promote SDG 12: responsible consumption and production by promoting sustainable resource use, reducing waste, and even promoting a circular economy through recycled and bio-based materials. Climate action (SDG 13) would, therefore, take a step away if low-carbon materials did not reduce greenhouse gas emissions. The product life cycle is referred to when discussing the impacts of lowering emissions at the production, use, and disposal stages of its lifetime on climate change, and on mitigating to a high degree the significant contributions in an effort to hold global temperature increases to within safe limits. Furthermore, sustainable sourcing of materials, such as FSC-certified wood, contributes to SDG 15: life on land in terms of safeguarding ecosystems, lowering deforestation, and creating conditions for biodiversity conservation.

In line with the drive toward global carbon neutrality, low-carbon materials are an important part of reducing industries' carbon footprint and improving energy efficiency. Low energy consumption and emissions naturally follow the use of materials in insulation for buildings, lightweight transportation components, and renewable energy systems, which makes them central to the low-carbon economy. Moreover, bio-based materials can act as carbon sinks, capturing and storing carbon dioxide throughout their life cycle, thus helping to offset emissions and achieve carbon neutrality. Altogether, the incorporation of low-carbon materials into economic and environmental strategies helps the leap toward a sustainable future around sustainable development goals and enhances the global engagement toward carbon neutrality.

1.9 Conclusion

To summarize, this adoption of low-carbon materials is characterized by technical cost as well as environmental benefits. From the economic perspective, these materials offer a reduction in operational costs by improving energy efficiency and lowering costs incurred in sourcing raw materials as well as waste management during their lifetime. They also enhance market competitiveness and innovations, which promote the growth of a sustainable economy in the long term. From the environmental perspective, low-carbon materials have an effective role to play in reducing climate change due to their geo-engineering potential, minimizing GHG emissions and preserving resources. Their application into the circular economy submodels enables responsible consumption and production patterns, which are necessary for ecosystem balance and resilience to supply-side shocks.

In the future, and in order to be proactive, all stakeholders must respond to the call to action and expedite the adoption of use of low-carbon materials across the globe. It is worth noting that sustainability practices across sectors are strengthened by government policies that are supported by necessary laws and regulations. Stakeholders are encouraged to conduct cost-efficient research in new technologies and join forces in developing eco-friendly initiatives as well as operate in an open supply chain. Overall, these practices can help create a world where low-carbon materials are instrumental in building robust economies, and core principles of equity and environmental sustainability are maintained throughout the generations.

Acknowledgements

The authors, Jai Prakash Chaudhary and Mukesh Chandra Joshi, wish to convey their profuse thanks to Kirori Mal College, Delhi University, for the infrastructure and assistance provided during this research. In addition, Kamla Rawat is grateful to Jamia Hamdard, New Delhi, for the materials and facilities which played a major role in this work being completed.

References

1. A. M. Omer, Focus on low carbon technologies: The positive solution, *Renewable Sustainable Energy Rev.*, 2008, **12**(9), 2331–2357.
2. F. Orsini and P. Marrone, Approaches for a low-carbon production of building materials: A review, *J. Cleaner Prod.*, 2019, **241**, 118380.
3. C. Korra, The Essential Role of Low-Carbon Building Materials in Achieving Sustainable Development, *Int. J. Enhanced Res. Sci. Technol. Eng.*, 2021, **10**, 44–51.
4. M. da Graça Carvalho, M. Bonifacio and P. Dechamps, Building a low carbon society, *Energy*, 2011, **36**(4), 1842–1847.
5. M. C. McManus, Environmental consequences of the use of batteries in low carbon systems: The impact of battery production, *Appl. Energy*, 2012, **93**, 288–295.
6. H. Liu, X. Wang, J. Yang, X. Zhou and Y. Liu, The ecological footprint evaluation of low carbon campuses based on life cycle assessment: a case study of Tianjin, China, *J. Cleaner Prod.*, 2017, **144**, 266–278.
7. J. M. O. Scurlock, D. C. Dayton and B. Hames, Bamboo: an overlooked biomass resource?, *Biomass Bioenergy*, 2000, **19**(4), 229–244.
8. M. Lawrence, A. Heath and P. Walker, Determining moisture levels in straw bale construction, *Constr. Build. Mater.*, 2009, **23**(8), 2763–2768.
9. S. Yao, H. Zhu, S. Zhang, H. Chang and H. Wang, Green steel: The future path towards sustainable automotive manufacturing, *Resour., Conserv. Recycl.*, 2024, **200**, 107319.
10. G. Habert, J. B. d'Espinose de Lacaillerie and N. Roussel, An environmental evaluation of geopolymer based concrete production: reviewing current research trends, *J. Cleaner Prod.*, 2011, **19**(11), 1229–1238.
11. L. Shen, J. Haufe and M. K. Patel, *Product overview and market projection of emerging bio-based plastics*, Utrecht University, 2010.
12. S. Das, Life cycle assessment of carbon fiber-reinforced polymer composites, *Int. J. Life Cycle Assess.*, 2011, **16**, 268–282.
13. J. Smith, Lifecycle Cost Analysis of Sustainable Building Materials, *J. Environ. Econ.*, 2021, **45**(3), 123–134.
14. International Energy Agency, Energy Efficiency in Buildings: The Role of Low Carbon Materials, 2020, Retrieved from IEA.
15. Global Market Insights, Green Building Materials Market Report, 2022, Retrieved from Global Market Insights.
16. R. Doe and A. Lee, Consumer Demand for Sustainable Products: Trends and Implications, *Sustainability*, 2019, **11**(4), 567–580.
17. U.S. Green Building Council, Regulatory Incentives for Sustainable Building Practices, 2020, Retrieved from USGBC.
18. J. Giesekam, J. R. Barrett, P. Taylor and A. Owen, The greenhouse gas emissions and mitigation options for materials used in UK construction, *Energy Build.*, 2014, **78**, 202–214.

19. F. Pomponi and A. Moncaster, Embodied carbon mitigation and reduction in the built environment – What does the evidence say?, *J. Environ. Manage.*, 2017, **181**, 687–700.
20. E. Worrell and M. A. Reuter, *Handbook of Recycling: State-of-the-art for Practitioners, Analysts, and Scientists*, Elsevier, 2014.
21. J. M. Allwood, M. F. Ashby, T. G. Gutowski and E. Worrell, Material efficiency: A white paper, *Resour., Conserv. Recycl.*, 2011, **55**(3), 362–381.
22. K. Iyer, M. K. Hagnell and M. Åkermo, The multi-objective optimization framework: A step towards minimizing life-cycle costs and energy consumption of carbon fibre automotive structures, *Composites, Part B*, 2024, **271**, 111158.
23. M. J. Bataineh, P. Sánchez-Sellero and F. Ayad, Green is the new black: How research and development and green innovation provide businesses a competitive edge, *Bus. Strateg. Environ.*, 2024, **33**(2), 1004–1023.
24. R. Hanna, P. Heptonstall and R. Gross, Job creation in a low carbon transition to renewables and energy efficiency: a review of international evidence, *Sustainability Sci.*, 2024, 1–26.
25. S. H. Peng and S. L. Lo, An Economic Analysis of Energy Saving and Carbon Mitigation by the Use of Phase Change Materials for Cool Energy Storage for an Air Conditioning System–A Case Study, *Energies*, 2024, **17**(4), 912.
26. World Steel Association, *Steel Statistical Yearbook 2019*, World Steel Association, Brussels, 2019.
27. S. Bhuvaneshwari, H. Hettiarachchi and J. N. Meegoda, Recycled fly ash for sustainable construction, *Int. J. Environ. Sci. Dev.*, 2015, **6**(9), 676–680.
28. G. Habert, C. Billard, P. Rossi, C. Chen and N. Roussel, Cement production technology improvement compared to factor 4 objectives, *Cem. Concr. Res.*, 2013, **43**, 124–130.
29. V. S. Ramachandran, Fly Ash in Concrete, in *Waste Materials Used in Concrete Manufacturing*, Noyes Publications, 2018, pp. 603–662
30. P. van der Lugt, *et al.*, Bamboo as a building material: A critical review of its potential for sustainable construction, *Renewable Sustainable Energy Rev.*, 2018, **87**, 126–137.
31. F. Pacheco-Torgal, *Handbook of recycled concrete and demolition waste*, Woodhead Publishing, 2013.
32. W. Steffen, K. Richardson, J. Rockström, S. E. Cornell, I. Fetzer, E. M. Bennett, R. Biggs, S. R. Carpenter, W. De Vries, C. A. De Wit and C. Folke, Planetary boundaries: Guiding human development on a changing planet, *Science*, 2015, **347**(6223), 1259855.
33. S. Ghosh and T. Biswas, Sustainable Construction in India: A Perspective, *J. Cleaner Prod.*, 2018, **198**, 1498–1509.
34. J. Wang and W. Azam, Natural resource scarcity, fossil fuel energy consumption, and total greenhouse gas emissions in top emitting countries, *Geosci. Front.*, 2024, **15**(2), 101757.

35. R. P. Babu, K. O'Connor and R. Seeram, Current progress on bio-based polymers and their future trends, *Prog. Biomater.*, 2013, **2**(1), 1–16.
36. M. J. Mendell, A. G. Mirer, K. Cheung, M. Tong and J. Douwes, Respiratory and allergic health effects of dampness, mold, and dampness-related agents: a review of the epidemiologic evidence, *Environ. Health Perspect.*, 2011, **119**(6), 748–756.
37. M. L. Berndt, Sustainable concrete for low-carbon construction, *Proc. Inst. Civ. Eng. – Eng. Sustainability*, 2008, **161**(4), 161–172.
38. Q. Wang, X. Zhang, C. Guo, H. Zhou, Y. Zhao and B. Lin, Analysis of carbon emission differences between Chinese and Japanese buildings based on qualitative and quantitative comparisons, *J. Build. Eng.*, 2024, **89**, 109260.
39. M. M. Berlik, D. B. Kittredge and D. R. Foster, The illusion of preservation: a global environmental argument for the local production of natural resources, *J. Biogeography*, 2002, **29**(10–11), 1557–1568.
40. E. Massetti, M. A. Brown, M. Lapsa, I. Sharma, J. Bradbury, C. Cunliff and Y. Li, Environmental quality and the US power sector: air quality, water quality, land use and environmental justice, *ORNL*, 2017, **772**, 1–169.
41. M. A. Huston and G. Marland, Carbon management and biodiversity, *J. Environ. Manage.*, 2003, **67**(1), 77–86.
42. Bullitt Center, About the Bullitt Center, n.d., Retrieved from https://bullittcenter.org/about/.
43. City of Oslo, Oslo Fjord City, 2020, Retrieved from https://www.oslo.kommune.no/english/urban-development/projects-in-oslo/oslo-fjord-city/.
44. Interface, Mission Zero, n.d., Retrieved from https://www.interface.com/US/en-US/sustainability/mission-zero.
45. Toyota, Toyota Environmental Challenge 2050, 2018, Retrieved from Toyota.
46. LafargeHolcim, ECOPact: The Green Concrete, 2020, Retrieved from LafargeHolcim.
47. Tata Steel, Sustainability Report, 2021, Retrieved from Tata Steel.
48. I. Kaur, J. Singh and S. Arora, Utilization of agricultural waste products in composite material development: A review, *J. Cleaner Prod.*, 2019, **219**, 123–133.
49. Novamont, Mater-Bi, n.d., Retrieved from https://www.novamont.com/eng/mater-bi.
50. H. Diefenbacher and H. Sander, Sustainable urban development: The case of Vauban, Germany, *J. Environ. Planning Manage.*, 2017, **60**(8), 1431–1445.
51. S. Kumar, S. Jain and J. Mathur, Sustainable urban planning in India: The case of Lavasa, *Int. J. Urban Sustainable Dev.*, 2016, **8**(2), 187–204.
52. Public Utilities Board Singapore, Marina Barrage: A Model of Sustainable Development, 2020, Retrieved from PUB Singapore.
53. Housing & Development Board, HDB Greenprint, 2019, Retrieved from HDB.

54. Copenhagen Solutions Lab, Climate Adaptation Plan, 2015, Retrieved from Copenhagen.
55. Greater London Authority, London Climate Change Adaptation Strategy, 2018, Retrieved from https://www.london.gov.uk/sites/default/files/gla_migrate_files_destination/London-Climate-Change-Adaptation-Strategy.pdf.
56. Masdar, Masdar City, n.d., Retrieved from https://www.masdar.ae/en/masdar-city.
57. C. L. Reichert, E. Bugnicourt, M. B. Coltelli, P. Cinelli, A. Lazzeri, I. Canesi, F. Braca, B. M. Martínez, R. Alonso, L. Agostinis and S. Verstichel, Bio-based packaging: Materials, modifications, industrial applications and sustainability, *Polymers*, 2020, **12**(7), 1558.
58. P. Sormunen and T. Kärki, Recycled construction and demolition waste as a possible source of materials for composite manufacturing, *J. Build. Eng.*, 2019, **24**, 100742.
59. M. Z. Jacobson and M. A. Delucchi, Providing all global energy with wind, water, and solar power, Part I: Technologies, energy resources, quantities and areas of infrastructure, and materials, *Energy Policy*, 2011, **39**(3), 1154–1169.
60. S. Vierra, *Green building standards and certification systems*, National Institute of Building Sciences, Washington, DC, 2016.
61. J. H. Schwarz, C. M. Richardson and J. C. Castillo, *Bio-based materials for energy and environmental applications*, John Wiley & Sons, 2020.
62. S. Mishra, A. Das and J. K. Parikh, Challenges and prospects in sustainable construction materials: A critical review, *Constr. Build. Mater.*, 2021, **276**, 122015.
63. H. Rohracher and P. Späth, *Technology and innovation for sustainable development*, Springer, 2014.
64. A. R. Davies and M. Hillman, *Sustainable materials: With both eyes open*, UIT Cambridge, 2010.
65. J. Van Hoof, C. Egenhofer and J. Kern, *Policies and measures to decarbonise EU industry*, Centre for European Policy Studies, 2019.
66. European Commission, European Green Deal, 2020, Retrieved from https://ec.europa.eu/info/strategy/priorities-2019-2024/european-green-deal_en.
67. F. Berkhout and J. Muskens, *Sustainability transitions in the making: A closer look at actors, strategies and resources*, Edward Elgar Publishing, 2017.
68. A. S. Milani, M. Suardi and M. Rizzati, *Sustainability and innovation*, Routledge, 2018.
69. R. Phiri, S. M. Rangappa, S. Siengchin, O. P. Oladijo and H. N. Dhakal, Development of sustainable biopolymer-based composites for lightweight applications from agricultural waste biomass: A review, *Adv. Ind. Eng. Polym. Res.*, 2023, **6**(4), 436–450.
70. J. Sainz de Aja, M. Gracia and J. Terrados, *Nanotechnology and sustainability: The role of nanomaterials in achieving the sustainable development goals*, Springer, 2020.

71. European Commission, EU climate action: Sustainable Industry, 2021, Retrieved from https://ec.europa.eu/clima/policies/industry_en.
72. United Nations, The Paris Agreement, 2015, Retrieved from https://unfccc.int/process-and-meetings/the-paris-agreement/the-paris-agreement.
73. Ellen MacArthur Foundation, Circular economy, 2019, Retrieved from https://www.ellenmacarthurfoundation.org/explore/circular-economy.

CHAPTER 2

Low-carbon Battery Materials

FELIPE M. DE SOUZA[a] AND RAM K. GUPTA*[a,b]

[a] National Institute for Materials Advancement, Pittsburg State University, 1701 S Broadway Street, KS 66762, USA; [b] Department of Chemistry, Pittsburg State University, 1701 S Broadway Street, KS 66762, USA
*Email: ramguptamsu@gmail.com

2.1 Introduction

2.1.1 Importance of Energy-storage Devices

Energy-storage devices occupy an important place in modern society as they allowed the development of portable electronic products and addressed some of the challenges of intermittent energy which enabled efficient energy management. Such processes were possible through the introduction of devices like supercapacitors and batteries. Supercapacitors are devices that display high power density along with low energy density, which results in fast charge/discharge processes.[1] These electrochemical properties prompt their use when high-power bursts are needed. Some applications of supercapacitors can be found in regenerative braking in electric vehicles, backup power systems, power grid stabilization, and hybrid energy-storage systems to name a few. Batteries, on the other hand, can provide energy for extended periods with relatively lower power, hence they present a high energy density along with a lower power density. Some of their uses are in electric vehicles to provide the needed propulsion, portable electronics, grid energy storage, medical devices, stationary energy storage, aerospace, and aviation, among others. Many applications of these energy-storage devices emphasize their necessity in society as a core technology.

Low Carbon Materials: Paving a Sustainable Path for a Greener Future
Edited by Santanu Patra, Sudheesh K. Shukla and Mika Sillanpää

Published by the Royal Society of Chemistry, www.rsc.org

2.1.2 Role of Materials in Energy Devices

Accompanied by the myriad of applications in several technology sectors, there is a relatively modest number of materials that are currently utilized in the industry including graphite or other carbon-based materials such as carbon black or porous nanocarbon frameworks as anodes[2–6] and lithium-based compounds used in the cathodes which include $LiFePO_4$, $LiMn_2O_4$, $LiNi_{0.33}Mn_{0.33}Co_{0.33}O_2$, $LiNi_{0.8}Mn_{0.1}Co_{0.1}O_2$, and $LiNi_{0.8}Co_{0.15}Al_{0.05}O_2$. These materials are becoming increasingly important to the economy and energy management as there is a growing demand for energy-storage devices and it is forecast that by 2040 the stationary storage capacity may reach almost 3000 GW h.[7] The increase in demand for lithium-based batteries is due to their introduction in the transportation sector since the electroactive components in the electric engine consist of the most expensive parts of an electric vehicle. Therefore, greater effort is expected to be applied to the extraction of metallic resources such as Li, Co, and Mn to produce electroactive materials.

2.1.3 Importance of Low-carbon Materials

The alternative from the combustion engine to the electric engine provided using electroactive materials for the development of high-performance batteries led to several changes in the economy and shifts in demand. It is worth noting that, first, it can decrease the burden on non-renewable petrochemical resources. Yet, promoting this technology sustainably is a crucial aspect of making it a stable market that can function as a viable option. Achieving sustainability is related to CO_2 and greenhouse gases that are directly or indirectly emitted during the extraction of the metal resources (*i.e.*, Li, Ni, Co, and Mn), commonly referred to as the carbon footprint. Hence, developing ways to decrease the carbon footprint is an important factor in making the process eco-friendlier. However, it is also worth noting that the prices of metals such as Li, Co, Ni, and Mn are continuously increasing. Thus, further increasing their extraction can appear to be a short-term approach. To avoid that, there should be an investment in alternative materials that can deliver a similar electrochemical performance to those of noble metals. One of these options has been directed at an increase in the use of $LiFePO_4$ which is composed of relatively cheaper and more abundant materials as compared to Ni, Co, and Mn, while offering good cycle life.[8,9] However, $LiFePO_4$ presents a lower gravimetric density when compared to Ni–Co–Mn-based materials.[10] Because of that, batteries based on $LiFePO_4$ would be relatively heavier than Ni–Co–Mn-based ones to achieve the same energy density. One option to improve that consists of the addition of Mn which results in an improvement in the electrochemical properties. Also, $LiFePO_4$ is a safer material which makes it an attractive option.[11] Aside from utilizing abundant sources, it is also important to introduce a circular economy that can promote an everlasting system. To achieve that, recycling comes as an important aspect as it can reintroduce used electroactive materials in the market without the need for the extraction of non-renewable resources. Through that, the overall carbon footprint can be reduced while simultaneously making production more sustainable. Addressing such aspects is a priority at this point, given the importance of this technology for modern society. Through that, some of the main goals are to diminish the overall amount of greenhouse gases as well as finding sustainable and renewable sources for the manufacture of energy-storage systems that can be used in conjunction with renewable energy sources. Part of these objectives have been defined by international treaties such as the United Nations' Sustainable Development Goals (SDG), the Paris Agreement, and the Kyoto Protocol. From that, it is important to understand the life cycle analysis (LCA), which is a study that compiles the environmental impact of a material's life cycle that includes the raw material, processing to manufacture, application, discard process, and recycling. This analysis can provide information regarding the depletion of resources, greenhouse gas emissions, and the overall carbon footprint. A schematic that can potentially address such a situation is presented through a traditional battery recycling process that is presented in Figure 2.1. Such analysis allows for a clearer scope of the environmental impact of certain materials which can guide scientists to develop components for energy-storage devices that are eco-friendlier.

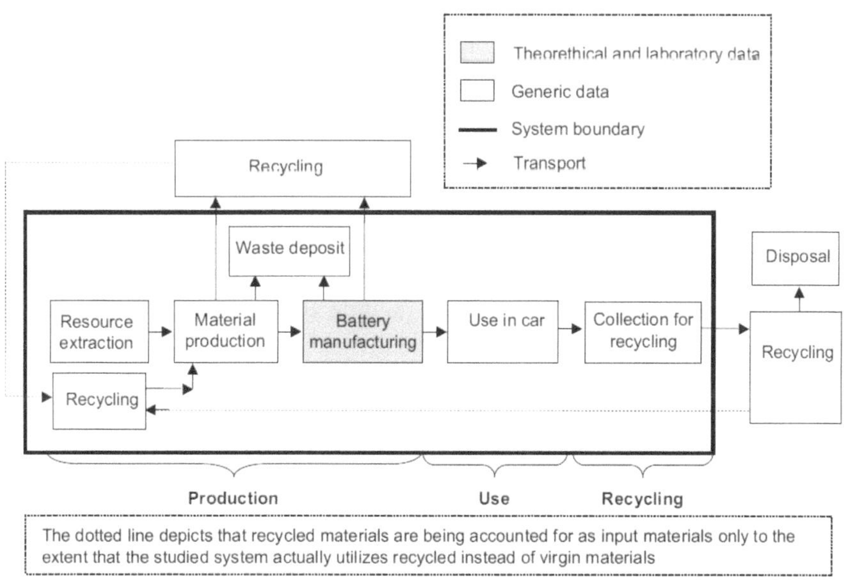

Figure 2.1 Schematic for a recycling system for batteries. Reproduced from ref. 12 with permission from Elsevier, Copyright 2010.

2.2 Batteries: Types and Characteristics

2.2.1 Types of Batteries

Energy-storage systems such as batteries are core aspects of the welfare of society, which pushes the need to develop new technologies within the energy-storage field that are required to be efficient as well as sustainable. Such a need has led to different types of batteries which can be roughly classified into (I) primary and (II) secondary batteries, (III) energy storage for grid systems, and (IV) electrochemical batteries.

Primary batteries are designated for one-time use as they cannot be recharged. They offer facile use and low maintenance. They can be defined through the nature of the electrolyte used in their system which can be aqueous or organic. Some examples are the Zn–C and alkaline batteries *i.e.*, non-rechargeable Li batteries, AAA, AA, C, and D cells. The Zn–C battery consists of a carbon-based cathode such as graphite that is in contact with a MnO_2 paste, the electrolyte is an acidic solution, and Zn serves as the anode. The conveniences of Zn–C batteries lie in their variety of voltages, capacities, sizes, and shapes along with an average shelf life. However, they also display relatively low energy density, capacity fading due to the formation of discharge products, and a tendency to leak. Also, the large amounts of Zn and Mg used in these batteries push the need for a proper discard process and recovery of metals. The alkaline batteries are like Zn–C batteries as they can be composed of the same materials with the difference being an alkaline electrolyte such as KOH is used. Their capacitance can be 10 times higher than conventional Zn–C batteries, while presenting good durability and low cost.[13] Further development of batteries led to the non-rechargeable lithium primary cells which present high cell voltage, wide working temperature, long durability, and appreciable power density.[14] Despite their dominance in the market, lithium batteries have some safety issues related to leakage that can lead to fire hazards due to the presence of flammable and toxic solvents along with metallic Li that reacts exothermically with O_2.

Secondary batteries are rechargeable cells often used in domestic applications such as electronics and power sources in hybrid and electric vehicles. Some of those rechargeable batteries include lithium-ion batteries (LIBs), Ni–metal hydride (Ni–MH), and Ni–Cd batteries. Among these, LIBs are widely used for micro-grid systems, small electronics, storage of renewable energy, and power sources for electric vehicles.[15,16] Some of their advantages include long electrochemical stability, wide working temperature, high energy density, and versatile design possibilities. One of the current goals is to develop electrolytes that can withstand higher operating voltages to improve the ionic transport between the electrodes. Commercially, one of the most common electrolytes used in LIBs is $LiPF_6$. However, its reactivity with water, relatively low thermal stability, and toxicity are pushing the development of solid-state electrolytes which are sought to solve these drawbacks.[17] The solid-state electrolytes can withstand even higher voltages compared to the current liquid ones, along with better safety credentials, leading to an overall improvement in capacity and performance. The main challenge remains in improving the contact interface between the solid electrolyte and the electrodes to decrease the system's internal resistance. Another technology that is receiving considerable attention is the Li–S batteries (LiSBs), which consist of S and Li as cathode and anode, respectively. Theoretically, S can deliver a specific capacity of around 1675 $mA\,h\,g^{-1}$, which is considerably higher than LIBs, which are around 150–250 $mA\,h\,g^{-1}$. Such high value is attributed to the multiple electrons involved during the redox process of S as well as the larger accommodation of Li atoms performed by S per unit of weight when compared to conventional cathode materials used in LIBs. Despite this attractive factor, several challenges need to be solved, such as the formation of lithium polysulfide species that can lead to permanent loss in capacity, high resistance of S, and self-discharge. To solve such issues the use of carbon-based materials that can act as an encapsulating agent that can prevent the shuttling effect and improve the overall conductivity of the system, enabling better capacity and electrochemical stability, has been proposed.[18]

In terms of batteries for grid-scale energy, there are two main technologies under development: redox flow batteries (RFBs) and Na–S batteries (NaSBs).[19] The RFBs are a system in which the electroactive components are dissolved in a liquid electrolyte to generate energy through redox processes that take place in separated half-cells. The dissolved electroactive material is reduced by removing electrons from the electrolyte, whereas in oxidation the electroactive material loses electrons to the electrolyte. The half-cells are connected through an external tank. The design of RFBs allows separate control of energy and power as the energy can be stored on the separated electroactive materials and the power can be determined by the cell stack. The main disadvantages of RFBs are related to the need for pumps to increase the power and energy density to acceptable levels and the use of V, which is an expensive metal. On the other hand, RFBs present a tunable design, no self-discharge, scalability, and electrochemical stability.[13] The NaSBs consist of a system based on liquid Na and S that can be a possible candidate for grid-scale use. One of the drawbacks for its applications is the working temperature of around 300 °C. The advantages are related to the wide availability of both Na and S, high electrochemical stability, energy density, and robusticity.[20] Based on this discussion the variety of battery-related technologies is notable. For that, a relation between the discharge time, size, and power efficiency of some of these energy-storage devices is presented in Figure 2.2.

2.2.2 Working Principle of Batteries

The working principle of batteries is based on the conversion of potential chemical energy into electricity through redox processes. Their design consists of two electrodes, *i.e.*, an anode (negative electrode) and a cathode

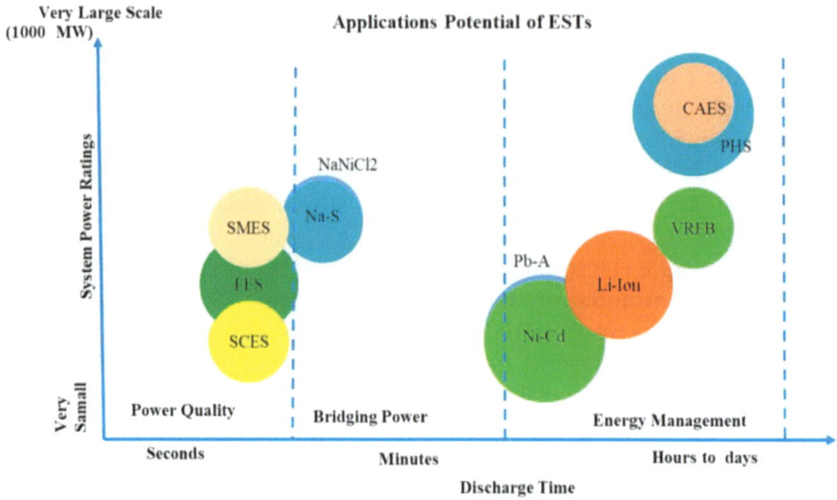

Figure 2.2 Comparison between energy-storage devices based on their discharge time and power density such as superconducting magnetic energy storage (SMES), flywheel energy storage (FES), supercapacitor energy storage (SCES), pumped hydro storage (PHS), compressed air energy storage (CAES), Pb-acid battery (Pb-A), and V redox flow battery (VRFB). Reproduced from ref. 21, https://doi.org/10.3390/su122410511, under the terms of the CC BY 4.0 license, https://creativecommons.org/licenses/by/4.0/.

(positive electrode), an electrolyte, a separator, and current collectors. During the discharge process the ions present in the electrolyte travel from the anode to the cathode, which generates an ionic current. During charging the ions in the electrolyte leave the cathode and return to the anode. Based on that, the anode undergoes an oxidation process, hence releasing electrons to the external circuit. In a conventional LIB, for instance, the anode is made of graphite and the oxidation reaction consists of the release of Li^+ that is intercalated in the graphite layered structure. Simultaneously, the cathode undergoes a reduction process, hence receiving electrons from the external circuit. In the case of LIBs, the cathode materials are $LiCoO_2$, $LiMn_2O_4$, $LiFePO_4$, combinations of materials based on Li–Ni–Mn–Co–O (NMC), and Li–Ni–Co–Al–O (NCA). The electrolyte is a material that connects the electrodes but allows the conduction of ions only that flow from the anode to the cathode (discharging) or from the cathode to the anode (charging). In LIBs, the electrolyte usually presents a salt containing Li that can be intercalated in the electrodes during the charging and discharging processes. The separator is a material that physically separates the electrodes to prevent the flux of electrons and short circuits, allowing the flow of only ions between the electrodes. Lastly, the current collectors transport the electrons generated through the electrochemical reactions to power an external circuit such as a device or electrical component. Simultaneously, the ions flow through

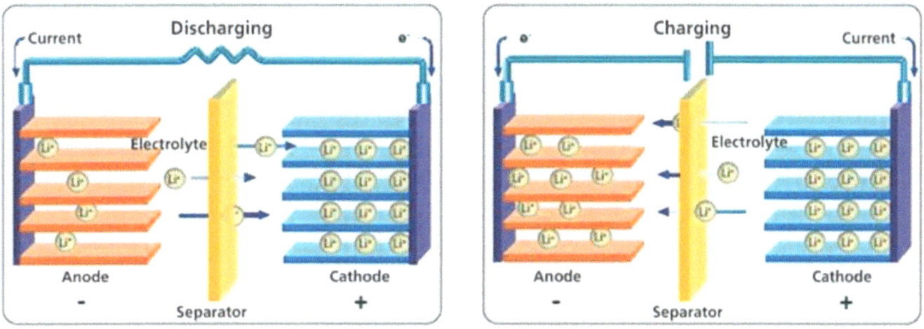

Figure 2.3 Schematic for the working principle of a typical LIB. Reproduced from ref. 22, https://doi.org/10.3390/batteries5010019, under the terms of the CC BY 4.0 license, https://creativecommons.org/licenses/by/4.0/.

the electrolyte to reach the electrodes. The reactions of oxidation and reduction can be described by reactions (2.1) and (2.2), respectively. A schematic of a typical LIB is presented in Figure 2.3.

$$\text{Anode:}\ \ A \rightarrow A^{+} + e^{-}\ (\text{Oxidation}) \tag{2.1}$$

$$\text{Cathode:}\ \ C^{+} + e^{-} \rightarrow C\ (\text{Reduction}) \tag{2.2}$$

2.2.3 Characteristics of Batteries

It is worth noting that batteries display a substantial difference from other energy-storage devices such as supercapacitors. In this sense, some of the characteristics of batteries consist of storing energy through electrochemical reactions that induce a phase change in the materials that compose the electrodes promoting the transport of electrons. Since these redox reactions are complex, they take longer to occur, which is associated with the discharge time of the battery. This aspect also promotes a higher energy density, meaning that batteries can store more energy per unit of weight or volume. Because of that, batteries find applications in long-term discharge devices. On the other hand, batteries present a lower power density due to the relatively slower kinetics involved in the electrochemical reactions, meaning that they provide a lower amount of energy in a given time.

2.3 Low-carbon Materials in Batteries

2.3.1 Cathode Materials

There is a plethora of materials that can be utilized for the fabrication of batteries. However, the current shift to the use of components with more sustainable, safer, and lower carbon emissions during production pushes the need for the development of materials that can meet such requirements while providing an appreciable electrochemical performance. Among the components used to fabricate batteries, the electrodes account for around 44% of the cost, as around 30% is attributed to the cathode which consists of materials based on Li, Ni, Mn, and Co. In the case of anodes, around 14% of the cost is attributed to the use of graphite.[23] Despite the dominance of LIBs, there is a limited number of materials that are used for the manufacture of such batteries which include $LiMn_{1.5}Ni_{0.5}O_4$, $Li_3V_2PO_4$, $LiMSiO_4$, and $LiMPO_4$, where M can be Fe, Mn, Co, or different ratios of these metals. A more detailed schematic of the materials utilized as anodes and cathodes in LIBs is presented in Figure 2.4.

One of the materials used to fabricate cathodes for portable devices was $LiCoO_2$. However, it had some drawbacks such as toxicity, high cost, chemical instability, and a capacity of around 135 A h kg^{-1} which limited its applications in larger energy storage systems. An alternative to that is $LiMn_{1.5}Ni_{0.5}O_4$ which displayed a relatively higher working voltage of around 4.7 V when compared to 4.1 V for $LiCoO_2$. It also displayed a more tunable structure that could be modified with other metals such as Fe, Zn, Co, Cr, or Ga or through surface modification with metal oxides that include ZnO, Al_2O_3, $AlPO_4$, and Bi_2O_3.[25,26] However, that process can be difficult to perform and, along with that, the commonly used electrolyte which is $LiFP_6$ can be dissolved in several organic solvents (*i.e.*, ethylene carbonate [EC], dimethyl carbonate [DMC], diethyl carbonate [DEC], propylene carbonate [PC], ethyl methyl carbonate [EMC], and methyl propyl carbonate [MPC]) that are not highly compatible with $LiMn_{1.5}Ni_{0.5}O_4$. Other materials that are feasible as cathodes are Li-excess layered oxides with the general formula of $Li_xM_yO_z$ where M can be Ni, Mn, or Co. Among these, $(Li_2MnO_3)_x(LiMO_2)_{1-x}$ can

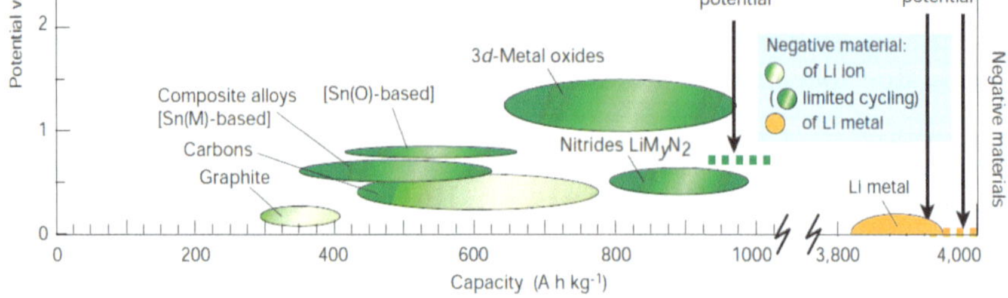

Figure 2.4 Plot of the potential against the capacity for the most utilized materials in the fabrication of LIBs. Reproduced from ref. 24 with permission from Springer Nature, Copyright 2001.

operate at around 4.0 V along with a capacity of 250 A h kg^{-1}.[27] Likewise, $LiMn_{1.5}Ni_{0.5}O_4$, coating with insulating metal oxides is an important process to reduce the capacity loss.[23] Yet, the relatively high reactivity of $LiMO_2$ can lead to side reactions between the electrolyte and the electrode, causing a decrease in the overall capacity due to passivation of the electrode's surface. One of the ways to address this is by improving the electrochemical stability of electrolytes that can withstand a higher working potential to avoid side reactions with the electroactive materials on the electrode while allowing them to perform lithiation/delithiation. Further, among the $LiMPO_4$ materials, $LiFePO_4$ is a convenient option as it is more accessible, less toxic, safer, and displays high electrochemical stability. One of the more recent studies has proposed the use of Li_2MSiO_4, which presents two Li atoms that can be removed and re-inserted in the host structure which could promote better electrochemical lithiation/delithiation without deteriorating the electrode's structure.[28]

2.3.2 Anode Materials

Aside from the cathode, the other components such as the anode, electrolyte, and separator account for a considerable mass percentage of a battery. Taking into consideration batteries composed of $LiFePO_4$ and $Ni(OH)_2$, Table 2.1 describes the main components of these types of batteries along with their mass percentages.

Another important component in batteries is the anodes which are usually made of carbonaceous materials that display convenient properties such as a high specific capacitance of around 372 mA h g^{-1}, and high electrochemical stability that allows them to withstand many cycles of intercalation with Li^+ ions. Given the diverse allotropic forms of carbon, several materials can be utilized as anodes. In that regard, there is some consistent research on carbon-based nanomaterials such as carbon quantum dots (0D), carbon nanotubes (CNTs) (1D), graphene (2D), and graphene framework (3D), along

Table 2.1 Components and their mass percentages in lithium–iron–phosphate and Ni–MH. Adapted with permission.[29] Copyrights 2016, Elsevier.[a]

Battery	Cathode	Anode	Electrolyte	Separator	Shell
Lithium iron phosphate	$LiFePO_4$ AC Al	MCMB AC Cu	$LiPF_6$ EC EMC	PE membrane	Steel
Mass percentage (%)	36.8	37.12	16.11	0.05	9.92
Ni–MH	$Ni(OH)_2$ Ni foam	$LaNi_5$ Ni foam	KOH LiOH 255 carbonyl Ni powder PTFE	PP membrane	Steel
Mass percentage (%)	14.89	16.42	7.12	2.33	59.24

[a] Acetylene black (AC), mesocarbon microbeads (MCMB), polytetrafluoroethylene (PTFE), polyethylene (PE), and polypropylene (PP).

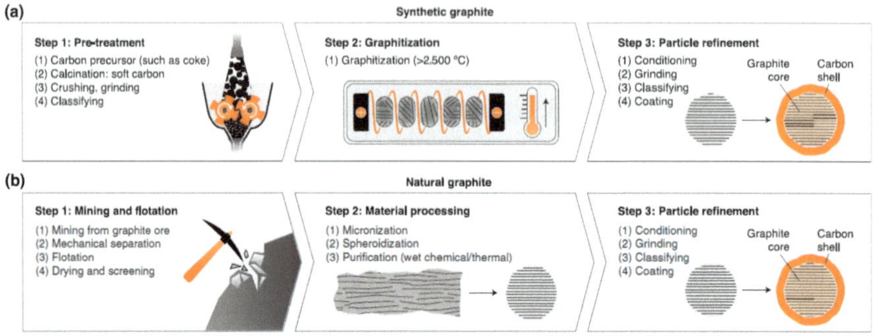

Figure 2.5 Schematics for the manufacturing of carbon-based anodes for LIBs are divided into (a) SG and (b) NG. Reproduced from ref. 39 with permission from Springer Nature, Copyright 2018.

with their composites that can be utilized as electroactive materials to further improve their electrochemical properties.[30–33] Yet, even though carbon is an abundant element, commercially, the carbon-based materials utilized as anodes require high purity and specific techniques for their synthesis which leads to an increase in cost.[34] Related to the carbon-based anodes, state-of-the-art materials can be divided into synthetic graphite (SG) and natural graphite (NG) as both can be either hard or soft.[35,36] The processing of these materials is quite complex. In the case of SG, the process takes place in three steps. First, the precursor carbon such as coke undergoes a calcination process that is around 800–1200 °C. The obtained soft graphite is ground into particles with different sizes within the micron scale. Then, a further graphitization process is performed at temperatures above 2500 °C which can be performed for days or weeks. Lastly, mechanical refining is performed through grinding to obtain particles with the desirable size.[37,38] The NG is obtained from the mining of graphite ores followed by several separation steps to remove Si, S, Na, and K, among others. Then, the carbon particles are ground to form spherical particles. Furthermore, leaching processes with HF or high thermal annealing (>2000 °C) are performed to further increase the purity which is one of the most important factors for high electrochemical performance.[38] The complexity of such processes leads to prices of 13 and 8 US $\$ kg^{-1}$ for SG and NG, respectively, as reported in 2016.[39] The processing schematics of SG and NG are presented in Figure 2.5.

2.3.3 Electrolytes

The commercially available electrolytes for status quo LIBs are usually $LiPF_6$ dissolved in organic solvents. Despite the conveniences of these electrolytes, there are several challenges related to their inherent toxicity, formation of lithium dendrites, and harm that can be caused to the environment. To avoid that, some studies have been made to obtain electrolytes that emphasize improving the environmental aspects as well as meeting the requirements for ionic transport. One example was performed by Liu *et al.*[40] who fabricated a quasi-solid polymer electrolyte membrane (QPE) composed of a complex structure based on polyether-*block*-polyamide (PEO-*b*-PA_6)/ lithium bis(trifluoromethanesulfonyl)imide (LiTFSI)/tetraethylene glycol dimethyl ether. According to the authors, the presence of hydrogen bonds between PEO and PA_6 is one of the factors that added robusticity for the transport of Li^+. In addition, the presence of PE separators can avoid short circuiting of the battery and the QPE allowed for ionic conduction of around 1.02×10^{-4} S cm^{-1}. In another study, Huang *et al.*[41] presented siloxane-based electrolytes that presented high compatibility with Li metal anodes as well as improved the chemical stability against oxidation to the cathodes. However, there are still efforts being made to introduce solid-state electrolytes into the market as one of the main challenges is associated with the high resistance between the solid interphase with the electrodes that hinder some of the ion flux. Yet, the benefits, such as higher working voltage, safety credentials, capability to be used as both electrolyte and separator, and application on stationary or electric vehicles, have prompted research in this area. A study performed by Larrabide *et al.*[42] showed that PEO/LiTFSI presented the lowest environmental impact due to the lower CO_2 footprint in comparison to other solid polymer electrolytes (SPEs) as it displayed a value of 0.37 kg of CO_{2eq} g^{-1} of material. A comparison with other SPEs is illustrated in Figure 2.6. Yet, balancing performance with eco-friendliness is an important aspect of implementation in the market. Based on that, ionic conductivity is one of the main factors related to the improvement of

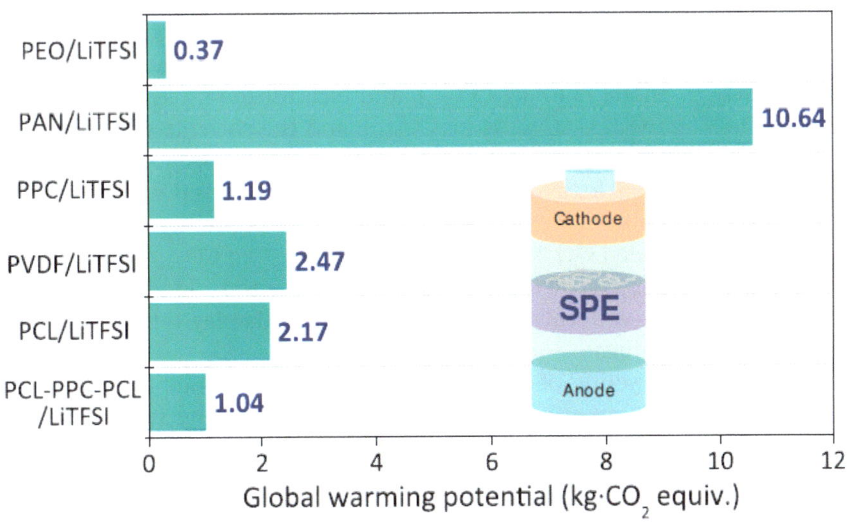

Figure 2.6 The carbon footprint for the main state-of-the-art SPEs per 1 g of material. Reproduced from ref. 42, doi.org/10.1002/aesr.202200079, under the terms of the CC BY 4.0 license, https://creativecommons.org/licenses/by/4.0/.

electrochemical performance. In the case of SPEs, the ionic conductivity is related to the motion of the polymeric segments that create free volume domains that allow Li^+ hopping.[43] Among the mentioned SPEs, PCL/LiTFSI displays the highest ionic conductivity of around 0.117 mS cm^{-1} at 20 °C which is attributed to the presence of ester O atoms that can interact with cations even at relatively lower temperatures.[44] In the case of PAN/LiTFSI, at 20 and 80 °C, it can reach an ionic conductivity of around 0.048 and 0.710 mS cm^{-1}, respectively. In this regard, SPEs should be able to create paths for the flux of Li^+ close to room temperature, which is strongly related to their glass transition temperature (T_g) which is the state in which polymer displays a higher degree of chain mobility that allows the movement of ions through its polymeric chain. It is also worth noting that, even though polymers tend to prevent the formation of Li dendrites that can cause short circuits, they cannot avoid the thermal decomposition of the electrolyte which can lead to side reactions and permanent loss of capacity. To avoid that some of the strategies include the introduction of thermally responsive polymer electrolytes that can block ions from the interface at higher temperatures and the incorporation of inorganic phases that can generally withstand higher temperatures.[45–47]

The need for the decrease of greenhouse gas emissions is a requirement in every industrial sector that can be partially measured through the carbon footprint which is the amount of greenhouse gases, such as CO_2, that are directly or indirectly emitted during the manufacturing of a given material. Based on that, simulated studies have proposed the carbon footprint of components present in $LiFePO_4$ and Ni–MH batteries.[29] Through that, it was proposed that a $LiFePO_4$ battery that produced 1000 kW h of electricity has a carbon footprint of around 12.7 kg of CO_{2eq} which is the sum of the other components that included the materials used to make the cathode (4.46 kg CO_{2eq}), anode (3.67 kg CO_{2eq}), and electrolyte (4.4 kg CO_{2eq}). Also, $LiFePO_4$ (3.85 kg CO_{2eq}), AC (3.04 kg CO_{2eq}), and the solvents analyzed such as $LiPF_6$ and EMC were 2.38 and 1.92 kg CO_{2eq}, respectively. A similar analysis was performed for Ni–MH batteries which displayed a considerably higher total carbon footprint of 124 kg CO_{2eq}. The higher kg CO_{2eq} was attributed to the anode (57.2 kg CO_{2eq}), cathode (49.7 kg CO_{2eq}), and electrolyte (10.3 kg CO_{2eq}), respectively. Furthermore, NH_3 (11.7 kg CO_{2eq}), Ni (66.8 kg CO_{2eq}), and LaO_2 (16.4 kg CO_{2eq}) also display relatively higher carbon footprint values.

The chemical recycling of battery materials can be a potential way to reduce the amount of greenhouse gases released during the manufacturing process as well as promote a circular economy for the non-renewable resources used in the manufacturing of energy-storage devices. A methodology to decrease the overall release of carbon and recycle the battery components was proposed by Zhang *et al.*[48] through shock-type heating which consisted of exposing the battery components to a high temperature of 1500 °C for 1 s. Such a process allowed a recovery rate of around 97%, with nearly full recovery of the metal foils used in the battery. By performing *in situ* spectroscopy, the authors observed that by applying an even distribution of heat through the material the organic binder can be quickly decomposed or form a thin carbonaceous layer on its surface. According to the authors, the short time, nearly zero generation of waste, and low carbon footprint for the process can make it cost-efficient and profitable, hence enabling viable recycling with low carbon generation.

The emphasis on Na-ion batteries (NIBs) can potentially be a valuable solution for the generation of energy that is closer to carbon neutrality along with more sustainable aspects due to the abundance of Na, less toxicity, and more convenient extraction when compared to other metals. Alongside these factors, Lai *et al.*[49] proposed a comparative study between Na-ion and Li-ion batteries to assess their differences in the amount of carbon generated for each process. Some of the findings in their work stated that the cathode materials utilized for LIB manufacture can represent more than half of the total amount of carbon emission, whereas for NIBs there was a considerably lower quantity.

2.4 Challenges and Opportunities

2.4.1 Recycling and Environmental Impact

The research and development teams in both industry and academia are focused on reducing the use of petrochemical-based materials due to the large amounts of greenhouse gases generated by them along with their non-renewability that can lead to future issues related to an energy crisis. The introduction of energy-storage systems offers an alternative to that. However, there is still a notable amount of greenhouse gases generated during the production of Li-based electroactive materials used in batteries. Li, as a non-renewable resource, is a scarce material that is becoming more expensive. Such situations make it necessary to develop a way to not only improve the overall efficiency and lifetime of current batteries but also to recycle the materials utilized in their fabrication as this could considerably reduce their environmental impact. Although this approach is not a zero-emission process it can lead to a considerably more sustainable path for the use of such resources as other alternatives to generate energy through solar, wind, tide, and biomass can potentially complement part of the energy demand. Hence, battery recycling is a crucial step for sustainability which can be performed through pyrometallurgy, hydrometallurgy, or a hybrid process between the two.

The pyrometallurgical process consists of crushing the materials and transferring them into a furnace at around 1260 °C without O_2 which results in metallic alloys. Yet, there is still the requirement to build recycling facilities that can be fully dedicated to Li. However, due to the high cost required for this technology, which is relatively recent in the case of Li, the tendency is to utilize recycling processes meant for other metals such as Fe, Ni, and Cr and incorporate the Li-based materials into the process. Despite

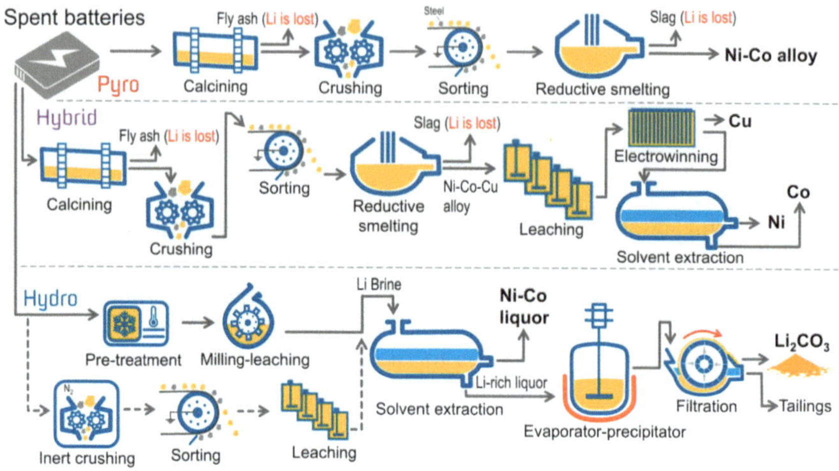

Figure 2.7 Scheme for the industrial-scale recycling process of spent LIB through pyrometallurgy, pyro-hydrometallurgy (hybrid process), and hydrometallurgy methods. Reproduced from ref. 50 with permission from Elsevier, Copyright 2021.

being more feasible economically, this process leads to considerable loss of Li into the slag and metal ashes.[50,51] In the case of the hydrometallurgical process, the spent LIBs are placed in a saline solution to discharge the batteries, which is followed by controlled crushing or sizing to avoid self-ignition or explosions which can be performed by freezing the materials under liquid N_2, direct milling under a CO_2 atmosphere, or utilizing dry organic solvents such as ethyl acetate.[52,53] Then, the metallic components can be separated by a magnetic process, whereas the graphite and polymers can be separated through the float–sink method. One of the notorious ways for Li recovery is based on freezing followed by milling and leaching performed in a hammer mill that contains Li brine. Then, the concentrated Li brine undergoes a process for the recovery of Li^+ through precipitation and evaporation.[54] A flowchart for the processes of pyrometallurgy, hydrometallurgy, and the hybrid process between the two is presented in Figure 2.7.

2.4.2 Regulations

Alongside the development of feasible recycling techniques, there are also some regulations in place for battery recycling placed for the European Union which requires that companies that place more than 1 ton of portable batteries in the market per year must pay for the collection, recycling, treatment, and disposal of the batteries' waste based on their market share proportion. Legislations such as these can lead to a long-term recycling approach that decreases the overall amount of virgin material extractions while enabling more sustainable production due to the reintroduction of materials used for the manufacturing of batteries.[55] In addition, investment in battery technologies can lead to a reduction of around 30% of overall greenhouse gas emissions in the transport sector and energy grid when accompanied by renewable intermittent sources.[56]

Another important aspect in terms of regulations is related to the United Nations' SDGs as it serves the purpose of directing the world toward a better future as it emphasizes nations address global issues such as climate change, environmental impacts, inequality, and lack of structure. For this case, the development of low-carbon battery materials can greatly aid in the reduction of carbon emissions, the development of sustainable technologies, and improve the availability of energy-storage devices. Through that, such technology is well aligned with several SDGs. Based on that the development of low-carbon battery materials follows SDG 7: affordable and clean energy; as this technology can be utilized in energy-storage devices with potential affordability while displaying a lower environmental impact. Also, for SDG 11: sustainable cities and communities; the development of more efficient and low-carbon emission energy storage devices can make communities more sustainable. For SDG 12: responsible consumption and production; the sustainable harvesting of carbon-based materials that can be recycled is a core aspect of responsible manufacturing and giving proper discard to the battery components. In regard to SDG 13: climate action; the use of low-carbon materials is a crucial factor in reducing the overall carbon footprint during the production of batteries by reducing the amount of greenhouse gases released into the atmosphere. In respect to SDG 15: life on land; one of the main concerns related to the use of some battery technologies is the presence of heavy metals that lead to bioaccumulation in the environment. The use of low-carbon batteries can avoid the use of such materials which can make the overall technology eco-friendlier.

2.4.3 Sustainable Sources

The pressing matters from international regulations that focus on the betterment of the global society have as some of the core aspects the use of important assets sustainably. In that regard, there is a need to further elaborate on recycling technologies to achieve more stable processing. The case of carbon-based materials utilized in batteries such as carbon black, carbon nanotubes, and graphene, among others, has sources that are widely abundant and renewable. However, the main challenge is related to the complex synthesis processes and the costly production of such materials which requires considerable investment.[57] On the other hand, to compensate for that, the carbon sources can be derived from biomass, which is inexpensive, non-toxic, and renewable. When processed, it can yield activated carbon materials with high porosity and different functionalities that can meet some of the main requirements for use in batteries. In this sense, biomass-derived carbon materials can present appreciable surface area, tailored pore structure (*i.e.*, micro- and/or meso-pores), high surface functionality that can potentially aid in electrochemical reactions, tunable conductivity, and hydrophilicity to a degree. The compilation of these factors can make them viable materials for LIBs and NIBs for example.[58]

The type of carbon utilized for such technologies is also a crucial factor. In this sense, the graphitizable carbon materials (soft carbons) can be converted into graphite when exposed to temperatures above 2000 °C, whereas non-graphitizable (hard carbons) do not convert into graphite even if exposed to temperatures as high as 3000 °C. Petrochemical-based sources such as pitch and coal tar have a high degree of aromatic compounds which leads to high graphitization. On the other hand, biomass-based sources have a lower level of compounds with aromatic rings which leads to the formation of hard carbons with high porosity and surface area.[59] In addition, such materials are reportedly better for battery-related applications due to good electrochemical stability, and they can reach high specific capacitances of around 1000 $mA\,h\,g^{-1}$ which are three times higher than for graphitic carbons, making such materials highly desirable for LIBs.[60,61] The myriad of sources available can include sisal fiber, banana peel, sugarcane bagasse, mustard seed waste, rice straws, jute fiber, and coffee waste grounds, among others.[62]

2.4.4 Recent Advancements

The search for new ways to improve the electrochemical properties of materials used in batteries is one of the main goals of the scientific community. In that sense, the thermal treatment of carbon-based sources such as pyrolysis plays a major role in their properties as the increase in the temperature treatment in carbon leads to an increase in pore volume and electrical conductivity while causing a decrease in surface functionalities and by default a decrease in O content. Hence, properly controlling the temperature is a valuable way to define the properties of carbon-based materials. Further, there are several advancements within the field of low carbon emission materials for batteries some of the previously mentioned examples include the uses of graphene and carbon nanotubes due to their high conductivity, and tunable structure in terms of morphology, functionalization, doping, and compositing. The solid-state batteries are being presented as a novel technology that can be applied in electric vehicles which are expected to offer higher energy density, better safety credentials, and lower carbon emissions when compared to traditional gel or liquid organic electrolytes used in LIBs. Another important research effort is being placed on Li–S batteries, which have the potential to deliver even higher energy densities due to the high theoretical capacitance of S which can surpass the status quo technology. The use of S, which is more abundant and less toxic to the environment, can be more feasible when compared to other resources. Despite the technical issues related to S such as high resistance, shuttling effect, and low cyclability there has been considerable progress to

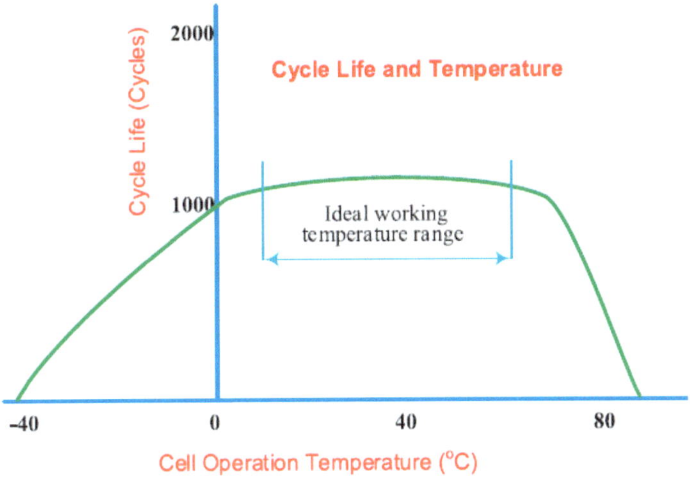

Figure 2.8 Relation between optimal working temperature and cycle life of a battery. Reproduced from ref. 56 with permission from Elsevier, Copyright 2020.

address these issues with the incorporation of carbon-based materials.[18] The development of technologies beyond LIBs such as NIBs, and K-ion and Mg-ion batteries can potentially serve as viable alternatives to the use of more abundant metals that can cause a reduced impact on the environment. In combination with all these advancements, there is also the need to introduce more reliable and stable recycling processes which can drastically improve the longevity of non-renewable resources with a more sustainable and eco-friendlier approach.

In a more recent aspect, one of the current challenges to optimize the performance of batteries in electric vehicles can be performed through the development of an efficient heat transfer process during its operation as it has been observed that heat generation can have a negative impact on the electrochemical stability of a battery. Based on that, it has been proposed that the optimal range of operation should be between 10 °C and 60 °C, as temperatures below 10 °C can lead to anode plating while temperatures above 60 °C can lead to chemical breakdown of electroactive components.[63] Figure 2.8 displays the optimal range for battery operation that contributes to a better efficiency and cycle life.

2.4.5 Economic Benefits

Investment in low carbon emission materials is a high-end technology that requires high investment. However, it tackles one of the core issues of modern society that is required in nearly every sector nowadays. Hence, it has great potential as a long-term and profitable investment. Based on that, technologies such as solid-state and Li–S batteries present a considerably higher energy density when compared to the status quo, which indirectly leads to reduced operational cost as it can lead to more durable electric vehicles, portable devices, and grid storage systems. The inherent electrochemical properties of carbon-based materials such as stability, good rate capability, and abundant sources make them an attractive type of investment. The introduction of novel technologies in the market can quickly lead to market opportunities and the generation of jobs, which introduces a new economic source within society that leads to a greater flux of capital within society. Finally, the establishment of recycling can lead to a virtuous circle of production, use, recycling, and reintroduction into the market that can be made profitable given that the amount and quality of the recycled material can meet the current necessities.

2.5 Conclusion

The plethora of applications that require the use of batteries and other energy-storage systems has turned this technology into a cornerstone for the functioning of society. However, the extraction of metallic resources for the production of electroactive materials utilized in batteries such as in the case of LIBs generates a considerable amount of CO_2-embedded emissions. Yet, given its importance and momentum in the market, the demand for such energy-storage systems is continuously increasing. Because of that, the focus has been shifting to further optimization of these batteries, resource management, a decrease in the carbon footprint, and the recyclability of the components. As discussed, there is a limited number of electroactive materials that can reach the desired performance for energy-storage systems in the same way that LIBs do since they have dominated the market for the last decade. Alongside that, the utilization of lithium-based batteries in electric vehicles applies a considerable strain on the overall demand for resources as the electric engine is the most expensive component. To address that, it is important to introduce a feasible recycling system that allows a circular economy to take place. The priority in that regard is due to the scarcity of metal resources such as Li, Co, and Mn which is leading to continuous increases in their price and the release of greenhouse gases during the extraction of virgin metallic resources. Yet, it is worth noting that technologies such as NIBs, and solid-state batteries, among other (*i.e.*, Li–S batteries, metal-air batteries, *etc.*) energy-storage systems can potentially serve as an alternative to LIBs to decrease the strain on their use. In addition, this chapter has displayed some of the recycling processes that can serve as a feasible and sustainable alternative to reintroduce Li-based devices into the market. Therefore, in the long run, recycling processes such as pyrometallurgy, hydrometallurgy, and hybrid approaches can aid in sustainably reintroducing Li-based devices which can decrease the strain on extraction of virgin metal resources that is damaging to the environment. Another important approach for the decrease of greenhouse gas emissions is the utilization of carbon sources derived from biomass for the development of carbon-based anodes. The reason for that is attributed to the high degree of graphitization and surface area that these carbon sources can achieve in comparison to petrochemical ones. Some of the new technologies that are gaining ground include solid-state batteries, Li–S batteries, and metal-ion batteries (*i.e.*, Na, K, and Mg). From that, solid-state batteries can offer higher energy densities, smaller designs, and improved safety aspects in terms of electrolyte flammability, thermal stability, longer electrochemical stability, and a reduction of hazardous materials usage. These advancements that can be offered by solid-state batteries can be nearly ideal in terms of performance and reduction of carbon emissions. Yet, there are still challenges related to the high production costs due to the complex manufacturing processes and expensive materials required. In addition, some inherent issues such as compatibility between the electrode/electrolyte interface and the temperature requirements impose some hindrances on the scalability of this technology. Another regarded technology is the Li–S battery which can also deliver high energy densities with potentially lower costs, and a wide range of operation temperatures. In addition, it embraces some advantageous environmental aspects such as avoidance of the use of Co and Ni, meanwhile S is a more abundant element that can be extracted without causing considerable harm to the environment. Despite these advantages, Li–S batteries face some issues related to their poor cycle stability, low coulombic efficiency, large size variation upon lithiation of S, and shuttling effect which can lead to permanent loss of capacitance due to the dissolution of S electrode into the electrolyte. Lastly, the development of metal-ion batteries such as Na, K, and Mg can potentially offer an alternative to the use of LIBs given the abundance of these former alkaline metals, safety, and environmental credentials, less hazardous waste, and simpler recycling. However, some challenges prevent their commercialization, such as lower energy densities than status quo LIBs, lack of a cost-effective manufacturing process, and the need for the development of compatible electrolytes. Through that, the scientific community must continue to place their efforts on the development of effective electroactive materials that display good environmental credentials and sustainable aspects as there are technologies that can lead to an appreciable sustainable process that allows for a circular economy. Yet, further research is required to optimize and tackle the inherent issues in these energy-storage technologies.

References

1. A. Gupta, C. A. Allison, M. E. Ellis, J. Choi, A. Davis, R. Srivastava, F. M. de Souza, D. Neupane, S. R. Mishra and F. Perez, Cobalt metal–organic framework derived cobalt–nitrogen–carbon material for overall water splitting and supercapacitor, *Int. J. Hydrogen Energy*, 2023, **48**, 9551–9564.
2. H. S. Using, P. Kahol and R. Gupta, Waste Coffee Management: Deriving Nitrogen-Doped Coffee-Derived Carbon, *C*, 2019, **5**, 44.
3. R. K. Gupta, Z. A. Alahmed and F. Yakuphanoglu, Graphene oxide based low cost battery, *Mater. Lett.*, 2013, **112**, 75–77.

4. C. Zequine, C. K. Ranaweera, Z. Wang, P. R. Dvornic, P. K. Kahol, S. Singh, P. Tripathi, O. N. Srivastava, S. Singh, B. K. Gupta, G. Gupta and R. K. Gupta, High-Performance Flexible Supercapacitors obtained *via* Recycled Jute: Bio-Waste to Energy Storage Approach, *Sci. Rep.*, 2017, 7, 1174.
5. S. Bhoyate, C. K. Ranaweera, C. Zhang, T. Morey, M. Hyatt, P. K. Kahol, M. Ghimire, S. R. Mishra and R. K. Gupta, Eco-Friendly and High Performance Supercapacitors for Elevated Temperature Applications Using Recycled Tea Leaves, *Glob. Challenges.*, 2017, **1**, 1700063.
6. C. Zequine, C. K. Ranaweera, Z. Wang, S. Singh, P. Tripathi, O. N. Srivastava, B. K. Gupta, K. Ramasamy, P. K. Kahol and P. R. Dvornic, High per formance and flexible supercapacitors based on carbonized bamboo fibers for wide temperature applications, *Sci. Rep.*, 2016, **6**, 1–10.
7. M. Mann, S. Babinec and V. Putsche, *Energy storage grand challenge: Energy storage market report*, National Renewable Energy Lab. (NREL), Golden, CO (United States), 2020.
8. J. Porzio and C. D. Scown, Life-Cycle Assessment Considerations for Batteries and Battery Materials, *Adv. Energy Mater.*, 2021, **11**, 2100771.
9. Y. Preger, H. M. Barkholtz, A. Fresquez, D. L. Campbell, B. W. Juba, J. Romàn-Kustas, S. R. Ferreira and B. Chalamala, Degradation of Commercial Lithium-Ion Cells as a Function of Chemistry and Cycling Conditions, *J. Electrochem. Soc.*, 2020, **167**, 120532.
10. L. Peiseler, V. Wood and T. S. Schmidt, Reducing the carbon footprint of lithium-ion batteries, what's next?, *Next Energy*, 2023, **1**, 100017.
11. Y.-S. Duh, Y. Sun, X. Lin, J. Zheng, M. Wang, Y. Wang, X. Lin, X. Jiang, Z. Zheng, S. Zheng and G. Yu, Characterization on thermal runaway of commercial 18650 lithium-ion batteries used in electric vehicles: A review, *J. Energy Storage.*, 2021, **41**, 102888.
12. M. Zackrisson, L. Avellán and J. Orlenius, Life cycle assessment of lithium-ion batteries for plug-in hybrid electric vehicles – Critical issues, *J. Clean. Prod.*, 2010, **18**, 1519–1529.
13. A. R. Dehghani-Sanij, E. Tharumalingam, M. B. Dusseault and R. Fraser, Study of energy storage systems and environmental challenges of batteries, *Renew. Sustain. Energy Rev.*, 2019, **104**, 192–208.
14. C. A. Vincent, Lithium batteries: a 50-year perspective, 1959–2009, *Solid State Ionics*, 2000, **134**, 159–167.
15. B. Scrosati and J. Garche, Lithium batteries: Status, prospects and future, *J. Power Sources*, 2010, **195**, 2419–2430.
16. C. Zhang, Y.-L. Wei, P.-F. Cao and M.-C. Lin, Energy storage system: Current studies on batteries and power condition system, *Renew. Sustain. Energy Rev.*, 2018, **82**, 3091–3106.
17. Z. Gao, H. Sun, L. Fu, F. Ye, Y. Zhang, W. Luo and Y. Huang, Promises, Challenges, and Recent Progress of Inorganic Solid-State Electrolytes for All-Solid-State Lithium Batteries, *Adv. Mater.*, 2018, **30**, 1705702.
18. S. Bhardwaj, F. Martins de Souza and R. K. Gupta, 4 - High-performance lithium–sulfur batteries: role of nanotechnology and nanoengineering,

in *Lithium-Sulfur Batteries*, ed. R. K. Gupta, T. A. Nguyen, H. Song and G. Yasin, Elsevier, 2022, pp. 57–73.
19. P. Taylor, R. Bolton, D. Stone, X.-P. Zhang, C. Martin and P. Upham, *Pathways for energy storage in the UK*, Report for the Centre for Low Carbon Futures, York. 2012.
20. D. Kumar, S. K. Rajouria, S. B. Kuhar and D. K. Kanchan, Progress and prospects of sodium-sulfur batteries: A review, *Solid State Ionics*, 2017, **312**, 8–16.
21. H. A. Behabtu, M. Messagie, T. Coosemans, M. Berecibar, K. Anlay Fante, A. A. Kebede and J. V. Mierlo, A Review of Energy Storage Technologies' Application Potentials in Renewable Energy Sources Grid Integration, *Sustainability*, 2020, **12**, 10511.
22. N. Chawla, N. Bharti and S. Singh, Recent Advances in Non-Flammable Electrolytes for Safer Lithium-Ion Batteries, *Batteries*, 2019, **5**, 19.
23. J. Xu, S. Dou, H. Liu and L. Dai, Cathode materials for next generation lithium ion batteries, *Nano Energy*, 2013, **2**, 439–442.
24. J.-M. Tarascon and M. Armand, Issues and challenges facing rechargeable lithium batteries, *Nature*, 2001, **414**, 359–367.
25. T.-F. Yi, Y. Xie, M.-F. Ye, L.-J. Jiang, R.-S. Zhu and Y.-R. Zhu, Recent developments in the doping of $LiNi_{0.5}Mn_{1.5}O_4$ cathode material for 5 V lithium-ion batteries, *Ionics*, 2011, **17**, 383–389.
26. J. Liu and A. Manthiram, Understanding the Improvement in the Electrochemical Properties of Surface Modified 5 V $LiMn_{1.42}Ni_{0.42}Co_{0.16}O_4$ Spinel Cathodes in Lithium-ion Cells, *Chem. Mater.*, 2009, **21**, 1695–1707.
27. Z. Lu, L. Y. Beaulieu, R. A. Donaberger, C. L. Thomas and J. R. Dahn, *J. Electrochem. Soc.*, 2002, **149**, A778.
28. D. Rangappa, K. D. Murukanahally, T. Tomai, A. Unemoto and I. Honma, Ultrathin Nanosheets of Li_2MSiO_4 (M = Fe, Mn) as High-Capacity Li-Ion Battery Electrode, *Nano Lett.*, 2012, **12**, 1146–1151.
29. Y. Liang, J. Su, B. Xi, Y. Yu, D. Ji, Y. Sun, C. Cui and J. Zhu, Life cycle assessment of lithium-ion batteries for greenhouse gas emissions-*Resour., Conserv. Recycl.*, 2017, **117**, 285–293.
30. L. L. Zhang and X. S. Zhao, Carbon-based materials as supercapacitor electrodes, *Chem. Soc. Rev.*, 2009, **38**, 2520–2531.
31. F. M. de Souza, R. K. Gupta, G. Yasin and T. A. Nguyen, Chapter 2 – Plasma produced by carbon nanotube-generated electron beam, in *Plasma at the Nanoscale*, ed. H. Song, T. A. Nguyen, A. Amrane, A. Amine Assadi and G. Yasin, Micro Nano Technol., Elsevier, 2022, pp. 21–35.
32. F. de Souza and R. K. Gupta, in *Bottom-Up Synthesis of 2D Nanomaterials for Energy Applications*, 2D Nanomaterials, CRC Press, 2022, pp. 53–70.
33. X. Zhang, J. Gao, Y. Xiao, J. Wang, G. Sun, Y. Zhao and L. Qu, Regulation of 2D Graphene Materials for Electrocatalysis, *Chem. – Asian J.*, 2020, **15**, 2271–2281.
34. S. Dühnen, J. Betz, M. Kolek, R. Schmuch, M. Winter and T. Placke, Toward Green Battery Cells: Perspective on Materials and Technologies, *Small Methods*, 2020, **4**, 2000039.

35. D. Andre, S.-J. Kim, P. Lamp, S. F. Lux, F. Maglia, O. Paschos and B. Stiaszny, Future generations of cathode materials: an automotive industry perspective, *J. Mater. Chem. A*, 2015, **3**, 6709–6732.
36. G. E. Blomgren, The Development and Future of Lithium Ion Batteries, *J. Electrochem. Soc.*, 2017, **164**, A5019.
37. R. Korthauer, *Handbuch lithium-ionen-batterien*, Springer, 2013.
38. X. Yuan, H. Liu and J. Zhang, *Lithium-ion batteries: advanced materials and technologies*, CRC Press, 2011.
39. R. Schmuch, R. Wagner, G. Hörpel, T. Placke and M. Winter, Performance and cost of materials for lithium-based rechargeable automotive batteries, *Nat. Energy*, 2018, **3**, 267–278.
40. P. Liu, L. Zhong, S. Wang, S. Huang, D. Han, Z. Huang, M. Xiao and Y. Meng, Ultra-long-life and ultrathin quasi-solid electrolytes fabricated by solvent-free technology for safe lithium metal batteries, *Energy Storage Mater.*, 2023, **58**, 132–141.
41. Y. Huang, R. Li, S. Weng, H. Zhang, C. Zhu, D. Lu, C. Sun, X. Huang, T. Deng, L. Fan, L. Chen, X. Wang and X. Fan, Eco-friendly electrolytes *via* a robust bond design for high-energy Li metal batteries, *Energy Environ. Sci.*, 2022, **15**, 4349–4361.
42. A. Larrabide, I. Rey and E. Lizundia, Environmental Impact Assessment of Solid Polymer Electrolytes for Solid-State Lithium Batteries, *Adv. Energy Sustain. Res.*, 2022, **3**, 2200079.
43. Z. Xue, D. He and X. Xie, Poly(ethylene oxide)-based electrolytes for lithium-ion batteries, *J. Mater. Chem. A*, 2015, **3**, 19218–19253.
44. H. J. Woo, S. R. Majid and A. K. Arof, Effect of ethylene carbonate on proton conducting polymer electrolyte based on poly(ε-caprolactone) (PCL), *Solid State Ionics*, 2013, **252**, 102–108.
45. T. Yu, P. Xue, S. Ma, Y. Gu, Y. Wang and X. Xu, Thermal Self-Protection Behavior of Energy Storage Devices Using a Thermally Responsive Smart Polymer Electrolyte, *ChemistrySelect*, 2022, **7**, e202104499.
46. J. Cui, Z. Zhou, M. Jia, X. Chen, C. Shi, N. Zhao and X. Guo, Solid Polymer Electrolytes with Flexible Framework of SiO2 Nanofibers for Highly Safe Solid Lithium Batteries, *Polymers*, 2020, **12**, 1324.
47. H. Zhang, L. Huang, H. Xu, X. Zhang, Z. Chen, C. Gao, C. Lu, Z. Liu, M. Jiang and G. Cui, A polymer electrolyte with a thermally induced interfacial ion-blocking function enables safety-enhanced lithium metal batteries, *eScience*, 2022, **2**, 201–208.
48. H. Zhang, Y. Ji, Y. Yao, L. Qie, Z. Cheng, Z. Ma, X. Qian, R. Yang, C. Li, Y. Guo, Y. Yuan, H. Xiao, H. Yang, J. Ma, J. Lu and Y. Huang, Transient and dry recycling of battery materials with negligible carbon footprint and roll-to-roll scalability, *Energy Environ. Sci.*, 2023, **16**, 2561–2571.
49. X. Lai, J. Chen, Q. Chen, X. Han, L. Lu, H. Dai and Y. Zheng, Comprehensive assessment of carbon emissions and environmental impacts of sodium-ion batteries and lithium-ion batteries at the manufacturing stage, *J. Clean. Prod.*, 2023, **423**, 138674.

50. C. B. Tabelin, J. Dallas, S. Casanova, T. Pelech, G. Bournival, S. Saydam and I. Canbulat, Towards a low-carbon society: A review of lithium resource availability, challenges and innovations in mining, extraction and recycling, and future perspectives, *Miner. Eng.*, 2021, **163**, 106743.
51. S. Al-Thyabat, T. Nakamura, E. Shibata and A. Iizuka, Adaptation of minerals processing operations for lithium-ion (LiBs) and nickel metal hydride (NiMH) batteries recycling: Critical review, *Miner. Eng.*, 2013, **45**, 4–17.
52. X. Zeng, J. Li and N. Singh, Recycling of Spent Lithium-Ion Battery: A Critical Review, *Crit. Rev. Environ. Sci. Technol.*, 2014, **44**, 1129–1165.
53. L. Gaines, J. Sullivan, A. Burnham and I. Belharouak, Life-Cycle Analysis of Production and Recycling of Lithium Ion Batteries, *Transp. Res. Rec.*, 2011, **2252**, 57–65.
54. B. Swain, Recovery and recycling of lithium: A review, *Sep. Purif. Technol.*, 2017, **172**, 388–403.
55. M. C. McManus, Environmental consequences of the use of batteries in low carbon systems: The impact of battery production, *Appl. Energy*, 2012, **93**, 288–295.
56. M. A. Hannan, A. Al-Shetwi, R. A. Begum, S. E. Young, M. M. Hoque, P. Ker, M. Mansur and K. Alzaareer, The value of thermal management control strategies for battery energy storage in grid decarbonization: Issues and recommendations, *J. Clean. Prod.*, 2020, **276**, 124223.
57. S. Iqbal, H. Khatoon, A. Hussain Pandit and S. Ahmad, Recent development of carbon based materials for energy storage devices, *Mater. Sci. Energy Technol.*, 2019, **2**, 417–428.
58. D. Alvira, D. Antorán and J. J. Manyà, Plant-derived hard carbon as anode for sodium-ion batteries: A comprehensive review to guide interdisciplinary research, *Chem. Eng. J.*, 2022, **447**, 137468.
59. R. E. Franklin and J. T. Randall, Crystallite growth in graphitizing and non-graphitizing carbons, *Proc. R. Soc. London. Ser. A. Math. Phys. Sci.*, 1997, **209**, 196–218.
60. S. Huang, Z. Li, B. Wang, J. Zhang, Z. Peng, R. Qi, J. Wang and Y. Zhao, N-Doping and Defective Nanographitic Domain Coupled Hard Carbon Nanoshells for High Performance Lithium/Sodium Storage, *Adv. Funct. Mater.*, 2018, **28**, 1706294.
61. W. Tian, L. Wang, K. Huo and X. He, Red phosphorus filled biomass carbon as high-capacity and long-life anode for sodium-ion batteries, *J. Power Sources*, 2019, **430**, 60–66.
62. P. Molaiyan, G. S. Dos Reis, D. Karuppiah, C. M. Subramaniyam, F. García-Alvarado and U. Lassi, Recent Progress in Biomass-Derived Carbon Materials for Li-Ion and Na-Ion Batteries—A Review, *Batteries*, 2023, **9**, 116.
63. E. Kallitsis, A. Korre, G. Kelsall, M. Kupfersberger and Z. Nie, Environmental life cycle assessment of the production in China of lithium-ion batteries with nickel-cobalt-manganese cathodes utilising novel electrode chemistries, *J. Clean. Prod.*, 2020, **254**, 120067.

CHAPTER 3

Low-carbon Supercapacitor Materials

A. RAOUF ALIAKBARI,[a] B. ELAHEH KOWSARI*[a] AND
C. SEERAM RAMAKRISHNA*[b]

[a] Department of Chemistry, Amirkabir University of Technology (Tehran Polytechnic), Hafez St., Tehran 15875-4413, Iran; [b] Department of Mechanical Engineering, Center for Nanofibers and Nanotechnology, National University of Singapore, 119260, Singapore
*Emails: Kowsarie@aut.ac.ir; seeram@nus.edu.sg

3.1 Introduction

Along with growing pollution and the consequences of climate change, the steep increase in gasoline prices has become one of the world's most visible challenges. One factor that has caused these increasing energy demands is the world's growing human population. Most of the world's fuels have been derived from petroleum up until this point, but alternative energy sources are required because petroleum is in short supply.[1] The creation of novel technological solutions to energy problems has garnered considerable attention in recent years. Research into low-carbon, ecologically friendly, lightweight, and conductive materials that can be used in cost-effective energy-storage and conversion systems is now critical for keeping up with contemporary society's demands and resolving new environmental issues. More efficient energy-storage devices, including batteries, capacitors, and supercapacitors, have been the subject of much research and development. One promising device for storing electrochemical energy is the supercapacitor. Its reliability and outstanding performance have garnered a lot of praise.[2]

Low Carbon Materials: Paving a Sustainable Path for a Greener Future
Edited by Santanu Patra, Sudheesh K. Shukla and Mika Sillanpää

Published by the Royal Society of Chemistry, www.rsc.org

Supercapacitors provide cyclic stability and a higher power density compared to batteries. The process by which they store charges, the adsorption, and subsequent release of ions from the electrode surface during charging and discharging are largely responsible for this. This distinguishes supercapacitors from batteries, which suffer from low power density and poor stability due to recurrent redox reactions. The principle of a supercapacitor is the same as that of a regular capacitor: a dielectric material separates two conductors. The separation of charges at the surface of the conductor is the source of the capacitance. The porous electrode, which offers effective surface areas for ion adsorption, has replaced the conductors in supercapacitors. A larger capacitance than a standard capacitor is also a result of the nanometer-scale separation of the two opposing charges.[3] Electric double-layer capacitors (EDLCs) are a kind of supercapacitor that stores charge solely by electrostatic forces and does not undergo Faradaic processes. Another one is the Faradaic process, which uses a pseudocapacitor and involves a reversible redox reaction. EDLC electrodes are carbon-based, whereas pseudocapacitor electrodes are made of conducting polymers and metal oxides. A hybrid supercapacitor, which combines the two kinds of charge storage mechanisms, has even better performance than EDLCs and pseudocapacitors.[4] Figure 3.1 shows different capacitive mechanisms in various supercapacitors.

Studies have recently brought up a novel method wherein "green supercapacitors" are being created by developing low-carbon and environmentally friendly materials for energy storage using green synthesis techniques.[5] The introduction of this method also allows for environmental concerns to be resolved. Materials are prepared using a variety of bio-based, green, or recycled materials and employ greener approaches for use in energy-storage systems.[6] The benefits of developing green and low-carbon supercapacitor materials from bio-based and end-of-life raw materials are that they aid in waste disposal by converting waste into high-value-added materials, and another is that sustainable supercapacitors provide a solid economic point for the need for energy-storage technology.[7] This chapter discusses the class of materials used in supercapacitors. Generally, this chapter considers biowaste, metal and metal oxides, electronic waste, and polymeric waste (plastic waste) as low-carbon electrode materials. This chapter also covers electrolytes, binders, separators, and current collectors.

3.2 Low-carbon Electrode Materials

3.2.1 Biowaste to Low-carbon Electrode Materials

Agricultural waste is recognized as a resource with high added value.[7] Agricultural waste consists of various organic materials, such as wood, grass, starch, shells, and husks, either individually or in combination. According to Karim *et al.*,[8] starch accounts for approximately 38% of the agricultural waste produced, while shells and husks contribute about 23%.

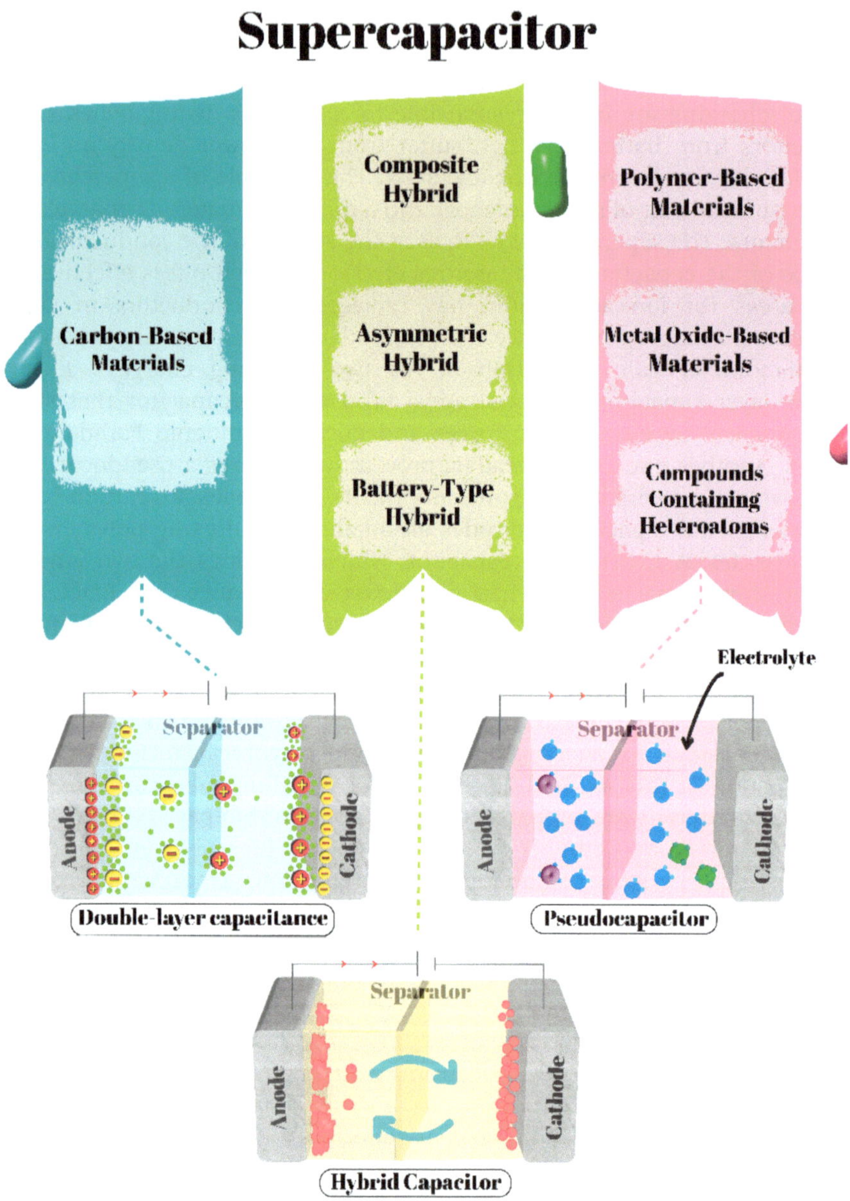

Figure 3.1 Different capacitive mechanisms in various supercapacitors.

Wood accounts for 18% of the waste, grass accounts for 3%, and combinations of these materials make up 6%. Other forms of agricultural waste make up the remaining 12%. Plants and their remains are the most plentiful sources of biomass, mostly consisting of forests, crops, and their by-products, which make up the majority of agricultural waste. Documents report an annual production of 998 million metric tons of agricultural residue waste.[101] By harnessing significant amounts of energy through thermochemical conversion, we can transform this waste into valuable products.[9] Given its ample availability, renewable nature, and many applications in energy storage and environmental preservation, the utilization of agricultural waste has garnered significant attention.

Converting agricultural waste into biochar can improve the circular economy by creating high-value products and reducing costs.[10] Via carbonization and treating methods, agricultural waste can be converted into high-value materials with a high surface area and large pore volume and utilized in energy-storage applications.[11] Researchers have developed several methods to enhance the electrochemical behavior of these materials in energy-storage applications, including combining these materials with synergistic materials in composites. For instance, electrode materials made of metal oxides or materials derived from agricultural waste have gotten a lot of attention because they have a high specific capacitance, are easy to get, are cheap, and are good for the environment. Olmos-Moya *et al.*[12] synthesized an electrode composed of banana peels and MnO_2. The synergistic effects of composite materials provided a specific capacity of 637 $F\,g^{-1}$. In research by Pradiprao Khedulkar *et al.*,[13] a nickel hydroxide/biochar composite derived from tea leaves was prepared for high-efficiency supercapacitor application. By recording a specific capacity of 945 $F\,g^{-1}$ and excellent stability of 95% after 10 000 charge–discharge cycles, the biomass-containing electrode gave extraordinary performance. For some other electrodes that employed metal/biomass composites, the performance was as follows: GO/biomass-derived activated carbon, 90.23 $F\,g^{-1}$ at 10 $mV\,s^{-1}$;[14] porous carbon based on Salvinia adnate/MnO, 436.8 $F\,g^{-1}$ at 1 $A\,g^{-1}$;[15] so on.

Sometimes, five different methods are used to make metal oxide/biochar composites: sol–gel, hydrothermal, *in situ* synthesis, co-precipitation, and impregnation. Figure 3.2 illustrates the sol–gel method. The biochar/metal

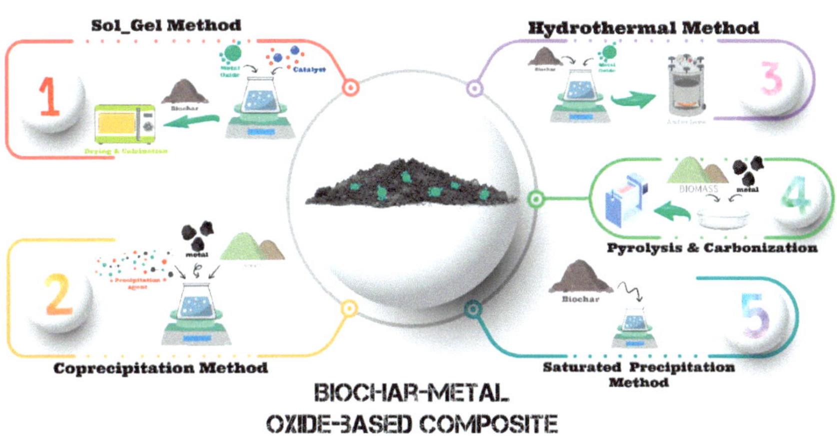

Figure 3.2 Metal oxide/biochar-based nanocomposites preparation routes.

oxide composite has a biochar–metal oxide synergistic effect, which makes the electrochemical behavior better and the charge transfer better. Ongoing research based on the sol–gel method focuses on optimizing synthesis conditions and exploring alternative greener catalysts and solvents.[16] Figure 3.2 illustrates the hydrothermal method. The unique circumstances offered by the hydrothermal method facilitate the formation and solidification of metal oxide species, resulting in improved storage behavior. The biochar matrix provides both structural support and enables electron transmission, improving electrode performance.[17] In Figure 3.2, the *in situ* synthesis method is shown. In this method, the simultaneous synthesis of metal oxide and biochar guarantees efficient charge transfer pathways. As a result of the synergistic interactions between metal oxides and biochar,[18] the composite demonstrates enhanced electrochemical behavior. In the co-precipitation method (Figure 3.2), composite materials take advantage of the high surface area of the carbon matrix, which facilitates the stabilization and adsorption of metal particles, resulting in an excellent electrochemical nature for energy storage.[19] Figure 3.2 also illustrates the impregnation method. The composites obtained from this method show improved electrochemical behavior due to the synergistic effects between the biochar matrix and metal oxide particles, which is critical for energy-storage applications.[20] Table 3.1 provides a summary of the advantages and disadvantages of each method.

Supercapacitors based on biowaste-derived electrode materials fulfill multiple sustainable development goals through the minimization of waste, resource efficiency, and reduction of carbon footprints. These supercapacitors provide energy-storage solutions, support the clean energy

Table 3.1 Advantages and disadvantages of sol–gel, hydrothermal, *in situ* synthesis, co-precipitation, and impregnation processes

Method	Advantages	Disadvantages	Ref.
Sol–gel	Controllable particle distribution and size	Time-consuming, as well as utilization of catalysts and organic solvents	21
Hydrothermal	Uniform particles	Difficult parameters control, and need for devices and equipment, as well as not scalable	22
In situ synthesis	One-step method No need for post-treatment	Prudent choice of precursors Optimizing conditions for carbonization	23
Co-precipitation	Economic Simple method Controlled particles size	Optimizing conditions	24
Impregnation	Scalable Simple method	Need for additional steps Limited dispersion of particles	25

transition, reduce greenhouse gas emissions, and stimulate innovation in materials science and engineering. In addition, biowaste-derived electrode materials help to maximize terrestrial ecosystems by identifying agricultural and natural waste as the main resource. Biowaste-based electrode materials for supercapacitor applications are playing a significant role in an efficient energy-storage system, along with generating awareness of sustainable development.[26]

3.2.2 Convert Plastic Waste into Low-carbon Electrode Materials

Two of the most urgent global concerns are the requirement for sustainable energy sources and the effective handling of plastic waste. Plastic waste-based activated carbon for energy-storage applications is a significant method for simultaneously achieving sustainable green energy and waste management. Using plastic waste to produce valuable carbon resources for supercapacitors is a practical approach. Additionally, it offers a solution for reducing the problem of plastic waste pollution. We can functionalize carbon-based compounds to make them usable in hybrid electrochemical electrodes for added convenience. The transformation of plastic waste into valuable resources, such as carbon-based materials, is an emerging area of study. In this specific scenario, we can thermally treat plastic waste to produce carbon-based materials. This can be achieved using many methods, such as molten salt methods,[27] hydrothermal methods,[28] catalytic carbonization,[29] microwave treatment,[30] anoxic pyrolysis,[31] flash Joule heating,[32] and pressure carbonization.[33] These procedures are frequently used to produce specific products, such as carbon nanosheets,[34] hierarchically porous carbon,[35] carbon nanotubes (CNTs),[36] graphene, and graphene-derived material.[37] In these processes, a variety of plastics have been used as raw materials, such as polytetrafluoroethylene,[38] polyethylene,[39] polypropylene,[40] polyimide, polystyrene,[41] and so on. Poly(ethylene terephthalate) (PET) is another plastic that has received a lot of attention.

PET is extensively used in several industries worldwide, resulting in significant global waste production.[42] Therefore, we are discussing it here as an important case study. Jung *et al.*[43] prepared activated carbon derived from PET through carbonization and then activation *via* CO_2 and KOH at 600–1000 °C and used it as a supercapacitor electrode. Activated carbon prepared under optimal conditions showed a specific surface area of 2076.7 $m^2 g^{-1}$, which is excellent for energy-storage applications. Kigozi *et al.*[28] prepared PET-derived activated carbon *via* the thermal-hydrothermal method, which showed high capacitance and stability in the role of symmetric supercapacitor electrodes. This demonstrates the potential for using activated carbon from plastic recycling in actual supercapacitor devices. Recycling PET extends beyond preparing activated carbon; it can also facilitate the preparation of materials with a high surface area. Recycling plastic waste into valuable metal–organic framework (MOF) materials is currently receiving significant attention as a crucial research field for addressing economic and environmental concerns. In comprehensive research, Dubey *et al.*[44] successfully prepared Zr-, Cu-, and Ti-containing MOFs in which recycled PET was used. In research by Li *et al.*,[45] waste PET bottles were used to prepare MIL-53 (Al) with 100% conversion under a hydrothermal process. Ni-MOF[46] and other PET-based MOFs have been developed by researchers.

Electrodes derived from recycled plastics must also fulfill the requirements related to energy storage; that is, in addition to being low-carbon, they must also be practical. Table 3.2 shows the electrochemical performance of PET-derived electrode materials. The data in Table 3.2 clearly confirm the acceptable performance of these electrodes.

The use of electrode materials derived from plastic waste for supercapacitor applications can help solve the problems of both waste disposal and energy storage. The production of high-performance electrodes from plastic waste contributes to sustainable development goals. This idea will not only reduce plastic pollution but also stimulate innovative thought in material science, eco-friendly technologies, and job creation within recycling industries. The application of plastic waste in supercapacitors is a response to several sustainable development goals for a better future.[54]

Table 3.2 Various electrodes obtained from plastic waste, sources of waste, methods of synthesis, and capacitive performance.

Active material	Plastic	Preparation route	Capacitive performance	Ref.
Activated carbon	Acrylonitrile butadiene styrene, polystyrene, polyethylene, and polystyrene	Pyrolysis, followed by activation	Specific capacitance of 220 $F g^{-1}$	47
Titanium carbide-MXene	PET	Thermal treatment	Specific capacitance of 556 $F g^{-1}$	48
Co_3O_4/carbon composite	PET	Carbonization and solvothermal synthesis	Specific capacitance of 110.8 $F g^{-1}$	49
Zr-MOF	PET	Solvothermal method	Specific capacitance of 822 $F g^{-1}$	50
Spherical nanoporous carbon	Polytetrafluoroethylene	Pyrolysis	Specific capacitance of 822 $F g^{-1}$	51
Carbon nanosheets	Polypropylene	Catalytic pyrolysis	Specific capacitance of 349 $F g^{-1}$	52
$NiCo_2O_4$/nitrogen carbon	PET	Pyrolysis	Specific capacitance of 913 $F g^{-1}$	53

3.2.3 Recycling Electronic Waste into Low-carbon Electrode Materials

Electronic waste, with an estimated annual production of between 20 and 50 MT, is a rapidly growing source of waste, consistently expanding at a rate of approximately 4–5% per year. In addition to increasing the commercial value, using electronic waste and secondary sources in supercapacitors reduces the need for oil-based fuels.[55] Pyrolysis is a highly effective approach for recovering valuable compounds from electronic trash, especially compared to other physicochemical methods. Non-metallic constituents of electronic waste are expected to produce activated carbon for use as electroactive material in supercapacitors, hence mitigating environmental concerns associated with electronic waste.[11] Chodankar *et al.*[56] recycled activated carbon from end-of-life supercapacitors using simple thermal activation. Recycled carbon materials showed a high specific surface area (about 1716 $m^2 g^{-1}$) and excellent morphology for supercapacitor application. The electrodes prepared through recycling ensured the specific capacity (57 $F g^{-1}$) and excellent stability (99% after 300 000 charging and discharging loops) with a wide potential range (2 V), which indicated the applicability of the method of these researchers. In published research by Nagaraju *et al.*,[57] waste cable wires were operated to obtain precious copper fibers as a cost-effective current collector for fabricating fiber-based hybrid supercapacitors through simple wet chemical approaches, and their electrochemical properties were investigated. In addition to the remarkable specific capacity (230.48 $mA h g^{-1}$), the electrode containing recycled materials presented acceptable cycling stability (about 83%).

Sathish *et al.*[58] presented an idea on how to convert e-waste into activated carbon using an activation process. Then, electrochemical behaviors were investigated. The specific capacity of 308 $F g^{-1}$ for prepared activated carbon accepts that activated carbon prepared from electronic waste is a suitable material for assembling supercapacitors. Printer devices, which release about 6000 T of carbon powder into the atmosphere every year, were recycled by Kaipannan *et al.*[59] through a one-step thermal conversion into carbon/Fe_3O_4 composites. The recycled carbon was calcined after thermal conversion, was finally applied as an active material in a supercapacitor electrode, and showed a high specific capacitance of 536 $F g^{-1}$, which is six times higher than commercial graphite carbon black. Excellent electrochemical stability of 97% after 5000 charging and discharging cycles, along with high specific capacity, promises the future of this type of recycling. An experiment designed by Rajkumar *et al.*[60] recycles copper oxide in end-of-life printed circuit boards with a simple and efficient approach and a one-step recovery process into electrode material for supercapacitors. The prepared electrode delivered a specific capacity of 408 $F g^{-1}$ and a retention capacity of 93.1% after 3000 charging and discharging cycles. A high-value recycling strategy for using non-metals in electronic waste to produce hierarchical porous carbon was proposed by Xi *et al.*[61] This study used non-metallic materials from the printed circuit board to prepare a porous supercapacitor electrode through carbonization and activation. High specific surface area (589 $m^2 g^{-1}$), temperature adaptability (−50 to 100 °C), and good electrochemical behavior were the properties of the synthesized carbon. The recycling of printer components has received extensive attention in research.[62]

Some other electronic wastes, such as keyboards, desoldering copper wicks, SIM cards, magnetic materials, *etc.*, have also been recycled and used as active material in supercapacitors. Through a one-pot green method, Rajkumar *et al.*[63] successfully recycled copper oxide nanoparticles from waste SIM cards and used them as supercapacitor electrodes. The electrode recorded typical pseudocapacitive behavior with a specific capacitance of 542 $F g^{-1}$. Ojeda *et al.*[64] recycled magnetite- and barium hexafrite-based electrodes from expired credit cards and speakers, respectively. These researchers may apply their proposed method to recycle other electronic waste that contains magnetic materials. Wasted desoldering copper wicks were proposed for synthesizing nickel selenide/copper oxide by Arbaz *et al.*[65] The recycled material after thermal modification was tested as electroactive material and recorded a specific capacity of 138.27 $\mu A h cm^{-2}$ with 98.7% stability after 50 000 charge/discharge cycles. In another piece of published research, porous activated carbon was recycled from an end-of-life keyboard and used as the positive and negative electrodes in an all-solid-state symmetric supercapacitor. The assembled supercapacitor showed excellent performance parameters with convincing cycle stability, indicating that supercapacitors designed through recycling have a promising future.[66]

Recycling battery waste for supercapacitor applications and *vice versa* has been suggested by researchers. One of the most important energy-storage systems utilized today is lithium-ion batteries (LIBs). However, because LIBs are mostly made of organic electrolytes and heavy metals, a large quantity of wasted batteries poses risks to the environment and resources.[67] Researchers are currently developing methods to extract metals and components from wasted LIBs. For example, Li–Ni–Mn–Co oxide was successfully extracted from end-of-life lithium batteries by a simple microwave/deposition method and composited with fullerene; it was then introduced as a supercapacitor electrode.[68] Recycling of cobalt in the form of CoS[69] and Co_3O_4,[70] graphite,[71] and $MnCO_3$[72] from LIBs is among other efforts made in this field. We also recycled other classes of batteries for use as supercapacitor electrodes. Recycling of Mn_3O_4 from Zn–C batteries through a thermal route,[73] graphene nanosheets from Zn–C batteries *via* a green electrochemical exfoliation process,[74] unpurified spent Zn–C battery powder by a ball milling route,[75] and chemically activated carbon from dry-cell batteries[76] are examples. It remains true that the methods used for recycling are not completely green and the recycled materials are not completely low-carbon. Still, it should be considered that recycling technology is an almost new topic and requires extensive research. For example, ionic liquids can be employed to recycle metals, which are greener materials than strong acids in the leaching process.[55]

Supercapacitors guarantee essential attributes required for sustainable urban infrastructure. Electronic waste presents an attractive possibility for developing sustainable electrode materials for supercapacitor applications. The production of electrode materials from electronic waste solves the environmental issues related to electronic waste disposal. Companies and researchers have the opportunity to address environmental pollution by repurposing waste electrode materials from electronics. This concept corresponds with the principles of a circular economy. By using electronic wastes, less reliance on virgin materials is achieved, thereby reducing the energy used and greenhouse gases emitted during extraction and processing activities. The use of electronic waste can be a viable option for making supercapacitor electrodes more sustainable and achieving several sustainable development goals.[77]

3.2.4 Metal and Metal Oxides to Low-carbon Electrode Materials

Metal oxides are considered the best electrode materials due to their inherent qualities, economic appeal, environmental friendliness, and abundant availability. Several metal and metal oxide particles (Ag, Au, Mn, Pt, Fe, Cr, Zn, Ti, Cu, Co, V, and Ni) have been synthesized using green methods and with a low carbon footprint.[78] The use of transitional metal oxides as electroactive materials has several advantages, one of which is that these oxides occur naturally on Earth. For example, ZnO and MoO are quite inexpensive. Despite the lower theoretical capacitance of transitional metal oxides, there are methods to enhance their performance as electrochemical supercapacitors. Metal oxide functionalization with less expensive carbonaceous materials, synthesizing mixed oxides and composites, and doping with electrochemically active compounds are all examples of these approaches. Nanoscale metal oxide preparation is another important and common technique. Because of their larger surface area, nanoparticles improve the efficiency of electrochemical processes for charge storage.[79] An electrode material needs several active sites to function as a supercapacitor, and Faradaic or non-Faradaic processes can achieve this. The morphology and structure of a material greatly influence its active surface. Metal oxide nanoparticles, with their high surface area and enhanced active sites, are perfect electrode materials for industrial-scale supercapacitors. Furthermore, nanoparticles provide reduced internal resistance and enhanced electrochemical conductivity by facilitating the rapid passage of ions and electrons *via* shorter diffusion pathways.[80] To that end, supercapacitor electrodes made of metal oxides are highly desirable. Nanoscale metal oxide production without agglomeration remains a challenge in both industrial and scientific domains, despite the availability of a more efficient and cost-effective synthesis method.

The literature is replete with methods for synthesizing nanomaterials; however, there is an immediate need to create inexpensive, efficient, and ecologically friendly methods that could enhance the device's performance while simultaneously reducing its total cost. There is a focus on sustainable metal oxide synthesis. Researchers have reported the synthesis of transitional metal oxides using a variety of processes, including co-precipitation synthesis, sol–gel, hydrothermal, solvothermal, irradiation-based approaches, and chemical precipitation. All of these investigations, however, make use of expensive and potentially dangerous chemical reagents.[81] Ultrasonication-assisted and microwave synthesis are two examples of processes that rely on cutting-edge methodology, even if they don't necessarily include dangerous or expensive substances. Physical, biological, and chemical methods are the three main approaches to nanomaterial synthesis. Nanomaterials are prepared using physical processes such as electrospraying, flame spray pyrolysis, melt mixing physical vapor deposition, laser pyrolysis, and inert gas condensation.[82] Plasma-enhanced chemical vapor deposition, chemical vapor hydrothermal synthesis, microemulsion technique, and polyol synthesis are examples of chemical processes that use expensive and potentially harmful chemical reagents. Enzymes and microbes are the biological agents used in the biological method, often called the "green synthesis approach," to prepare metal oxide. In biogenic synthesis, no chemicals, reagents, sophisticated procedures, or environmental factors are utilized. Furthermore, biogenic synthesis is an efficient and cost-effective method of synthesizing nanomaterials since it has lower cost, and uses less chemicals, energy, and other solvents, which in turn reduces the overall product cost.[83]

Since the principles of low-carbon materials must be accompanied by economics to be practical,[7] biogenic synthesis is considered here. Biogenic green synthesis synthesizes reagents from plants and microbes (*e.g.*, fungi, algae, bacteria). Biogenic synthesis makes use of microbes and plants as synthesis reagents. As an alternative to chemical reagents, active material synthesis uses compounds extracted from plants, leaves, fruits, various parts of fruits, and microbes. In contrast to conventional synthesis methods, green synthesis and plant-mediated synthesis, in particular, do not harm the environment. Instead, green synthesis stands out for its ease of use, low cost, scalability, and controllability over surface morphologies, among other advantages.[84] That is, electrode materials synthesized through the biogenic method follow the principles of low-carbon circular economy; therefore, in the following, we discuss the performance of metal ion-based electrodes that have been synthesized by this method. Azhar *et al.*[85] prepared CuO/Bi_2O_3 nanomaterials from *Amaranthus viridis* L. extract. The prepared electrode recorded a specific capacitance of 389 $F\,g^{-1}$. Anand *et al.*[86] prepared ZnO nanoparticles from leaves of the *Prunus dulcis* plant and reported the pseudocapacitive mechanism for the prepared ZnO nanoparticles. In Shaheen *et al.*'s research,[87] *E. cognata* extract was used to prepare functionalized MoO_3–$NiMoO_4$. The prepared electrode showed a specific capacity of 204 $F\,g^{-1}$. The provided examples demonstrate the wide applicability of this method.

The metal- and metal oxide-based electrode materials play an essential role in supporting sustainable development goals by enhancing the efficiency and performance of supercapacitors, which are necessary in energy-storage systems. These materials have all the desirable properties, such as high conductivity, a large specific surface area, and good electrochemical properties, essential for rapid charge and discharge processes, specifically for renewable energy applications, for instance solar and wind energy storage.[88] The metal- and metal oxide-based electrode materials aid in efficient energy storage. The metal- and metal oxide-based electrode is a necessity for many facets of current society and is important for global sustainability.[89]

3.3 Other Low-carbon and Green Components of Supercapacitors

This chapter will primarily focus on electrode materials, but also briefly discusses other components.

Designing electrolyte materials free of harmful F- and S-containing chemicals, as well as cyanide, is one of the primary goals of developing green supercapacitors. Therefore, creating environmentally friendly electrolytes necessitates utmost caution, cost-effectiveness, and proper disposal. In terms of environmentally friendly supercapacitor designs, the conventional aqueous electrolytes based on H_2SO_4 and KOH are not ideal due to their poisonous and corrosive properties. Other reported S-free electrolyte replacements include sodium formate and sodium acetate. These S-free substitutes produce a smaller diameter of dissolved anions. During full solvation, the majority of electrode material pores are able to accommodate the ions produced by these electrolytes. Nonetheless, the concentration of the solution has a significant impact on the ion mobility of these electrolytes.[90] Ionic liquids are another class of green materials that have been applied due to their unique properties.[91] Without a diluent, the ionic liquid at low temperatures shows little mobility. But when heated to higher degrees, it acts like organic solvents.[92] Types of supercapacitor electrolytes containing 1-ethyl-3-methylimidazolium tetrafluoroborate, *N*-propyl-*N*-methyl-pyrrolidinium bis(trifluoromethanesulfonyl)imide, 1-methyl-1-propylpiperidinium bis(fluorosulfonyl)imide, *etc.* have been introduced.[93]

Advances in binder technology highlight the importance of transitioning to biodegradable alternatives derived from renewable raw materials such as cellulose or lignin. Despite achieving performance standards comparable to traditional binders, these bio-derived polymers represent a departure from petroleum-derived materials and, as a result, meet greater sustainability requirements. Water-based adhesives further highlight the commitment to reducing environmental burdens associated with using solvents in electrode manufacturing processes. Carboxymethyl cellulose is the most advanced aqueous binder for supercapacitors. The coatings' mass loading is limited though, because they shrink when dried. Natural polymers such as guar gum, wheat starch, and potato starch were investigated by Ruschhaupt *et al.*[94] as potential substitutes. Combining guar gum and potato starch in a 75 : 25 (w/w) ratio yielded a promising result. Bauer *et al.*[95] reported a green approach for the production of a thin-film binder for electrical double-layer capacitors. The prepared binder, treated by a high-temperature route, presented a specific surface area of 527 $m^2 g^{-1}$ after an additional activation step. Jezowski *et al.*[96] used commercially available starch to prepare a conductive adhesive in a supercapacitor with a non-aqueous electrolyte. This study demonstrated the application of starch-based conductive adhesives in double-layer supercapacitor systems. Research results clearly confirm progress in this field.

When it comes to supercapacitors, the separator is a crucial part of their construction since it facilitates the movement of ions and their conveyance. Eliminating F and S elements is the main prerequisite for creating environmentally friendly supercapacitors with separators made of environmentally friendly materials. The chemicals used to construct the separators for green supercapacitors must be electrochemically stable. Porous membranes have been designed and manufactured for use in chemical filtering as green supercapacitors. These materials are affordable and easy to come by. They were chosen because they were electrochemically stable and had a low environmental impact.[97] For example, Rafique *et al.*[98] designed a scalable and facile blow-spinning route for the preparation of cellulose-containing separators. Good ashing stability and retention capacitance of 95% up to 1000 charge/discharge cycles were the characteristics of the supercapacitor assembled with this separator, which is very promising. In other research, a high-performance low-carbon cost-effective chitosan/poly(ethylene glycol)-*ran*-poly(propylene glycol)-based polymer composite was utilized by Raja *et al.*[99] This separator showed excellent electrochemical stability and high porosity. There is a demand for more investigation into green and low-carbon separators.

The current collector is the last component discussed in this section. Current collectors are essential for the optimal operation of an energy-storage system. Supercapacitors can exhibit significant differences in charging/discharging nature, energy/power density, and other important performance factors due to variances in material design, manufacturing, and surface features of current collectors.[100] Supercapacitors' current collectors rely on carbonaceous materials, such as carbon nanofiber mats. However, despite their graphite origin, these materials fail to provide sufficient conductivity. According to impedance spectroscopy, these materials show an equivalent series resistance higher than 200 ohms, which limits the values of specific capacitance. These considerations, together with the materials' complex and labor-intensive production processes, have hindered their investigation and potential use as environmentally friendly current collectors. Using the identical setup, the specific capacitance was 26% lower than that of metallic current collectors.[97] There is a need for more research and solutions to overcome the shortcomings of this type of green material.

3.4 Conclusion and Future Perspectives

The identification of low-carbon supercapacitor materials represents a significant advancement in the pursuit of sustainable energy solutions and environmental mitigation. This chapter discusses a range of electrode materials with low carbon content, including biowaste, e-waste, and the utilization of metal and metal oxides. Researchers have shown that waste materials can be converted into valuable resources for energy-storage applications using new techniques, including the carbonization of agricultural waste and the thermal treatment of plastic trash. Utilizing agricultural waste, plastic trash, and e-waste as electrode materials not only provides a sustainable alternative to conventional options but also aids in waste management and environmental preservation. The researchers showcased the exceptional electrochemical performance of these materials when utilized in supercapacitor electrodes, underscoring the potential of low-carbon materials in energy-storage applications.

The continuous transition from conventional energy to green energy, such as solar and wind energy, has increased the need to implement high-performance storage methods based on green materials. This is where low-carbon supercapacitors come in, as they offer fast charge–discharge rates, high energy density, and other significant properties. With the future improvements of hybrid supercapacitors based on low-carbon materials and waste materials that guarantee excellent electrochemical features, their energy storage can also be boosted without having to lose the benefits of supercapacitors' efficiency. So, in the process of evolving techno-politics, the introduction of low-carbon supercapacitors will undoubtedly play a significant role in advancing general energy security.

Scholars are expected to develop low-carbon and sustainable systems, such as low-carbon supercapacitors. Climate change creates demand for sustainable energy solutions, leading to the development of both low-cost and low-carbon alternatives. Through the development of low-cost methods for the preparation of low-carbon electrode materials by researchers, economies of scale can reduce production costs, allowing mass production for the consumer market. Effective collaboration between researchers, R&D units, entrepreneurs, and public authorities is crucial for industrial-scale production. Effective collaboration can help accelerate low-carbon supercapacitor technology development based on a low-carbon and circular economy.

Furthermore, the utilization of synthesis techniques such as sol–gel, hydrothermal, and biogenic procedures highlights the significance of implementing environmentally sustainable and economically efficient approaches to material synthesis. Biogenic synthesis is notable for its strict adherence to the principles of the circular economy, as it utilizes natural resources without causing harm to the environment. With the growing demand for energy storage, the advancement of low-carbon supercapacitor materials presents a hopeful path toward a sustainable future. Nevertheless, it is imperative to tackle obstacles related to scalability, affordability, and recycling procedures to fully use the capabilities of these materials in real-world scenarios. Overall, the investigation of low-carbon supercapacitor materials is a crucial advancement in the pursuit of sustainable energy-storage solutions. This research aims to harness the vast potential of waste materials while avoiding negative effects on the environment. By persistently conducting research and fostering innovation, the incorporation of low-carbon materials into supercapacitors holds the potential to completely transform the field of energy storage and propel us toward a more environmentally friendly and sustainable future.

Acknowledgements

The authors are grateful for the financial support provided by Amir Kabir University of Technology (AUT) in Tehran, Iran. The authors also acknowledge (Sustainable Tropical Data Centre Test Bed: A-0009465-05-00) awarded by the National Research Foundation of Singapore.

References

1. C. Liu, F. Li, L. P. Ma and H. M. Cheng, *Adv. Mater.*, 2010, **22**(8), E28–E62.
2. S. Sharma, P. Chand and S. Khokhar, *Introduction to low-carbon supercapacitors: new prospects*, 2023, pp. 34–62.
3. P. Sinha and K. K. Kar, *Handbook of nanocomposite supercapacitor materials II: Performance*, 2020, pp. 1–28.
4. R. Aliakbari, E. Kowsari, Y. Marfavi, S. Ramakrishna, A. Chinnappan and Z. A. Cheshmeh, *J. Energy Storage*, 2021, **44**, 103365.
5. A. Muzaffar, M. B. Ahamed and C. M. Hussain, *Renewable Sustainable Energy Rev.*, 2024, **195**, 114324.
6. H. T. Das, S. Dutta, T. E. Balaji and N. Das, in *Low-Carbon Supercapacitors: Towards Sustainability in Energy Storage Devices*, ed. M. B. Ahamed, C. M. Hussain and K. Deshmukh, The Royal Society of Chemistry, Cambridge, 2023, ch. 3, pp. 63–83.
7. R. AliAkbari, M. H. Ghasemi, N. Neekzad, E. Kowsari, S. Ramakrishna, M. Mehrali and Y. Marfavi, *J. Cleaner Prod.*, 2021, **293**, 126101.
8. A. A. Karim, M. Kumar, E. Singh, A. Kumar, S. Kumar, A. Ray and N. K. Dhal, *Crit. Rev. Environ. Sci. Technol.*, 2022, **52**(9), 1449–1490.
9. E. S. Goda, A. ur Rehman, B. Pandit, A. Al-Shahat Eissa, S. Eun Hong and K. Ro Yoon, *Chem. Eng. J.*, 2022, **428**, 132470.
10. B. Song, X. Cao, W. Gao, S. Aziz, S. Gao, C.-H. Lam and R. Lin, *Renewable Sustainable Energy Rev.*, 2022, **157**, 112057.
11. S. A. Bhat, V. Kumar, S. Kumar, A. Atabani, I. A. Badruddin and K.-J. Chae, *Fuel*, 2023, **337**, 127125.
12. P. M. Olmos-Moya, S. Velazquez-Martinez, C. Pineda-Arellano, J. R. Rangel-Mendez and L. F. Chazaro-Ruiz, *Carbon*, 2022, **187**, 216–229.

13. A. P. Khedulkar, B. Pandit, T. A. N. Bui, H. L. Tran and R.-A. Doong, *J. Colloid Interface Sci.*, 2022, **623**, 845–855.
14. S. K. Kandasamy, C. Arumugam, A. Sajitha, S. P. Rao, S. Selvaraj, R. Vetrivel, R. Selvarajan, A. M. Alosaimi, A. Khan and M. A. Hussein, *J. New Mater. Electrochem. Syst.*, 2021, **24**(1), 21–28.
15. H. Wang, H. Chen, X. Hou, H. Ye, Z. Guo, Z. Chen, Y. Jin, Y. Du and P. Ren, *Diamond Relat. Mater.*, 2023, **136**, 109888.
16. N. Anjum, N. Joyal, J. Iroegbu, D. Li and C. Shen, *J. Power Sources*, 2021, **499**, 229962.
17. F. Cheng and X. Li, *Catalysts*, 2018, **8**(9), 346.
18. H. Luo, Y. Wan, H. Zhou, Y. Cai, M. Zhu, Z. Dang and H. Yin, *Hazardous Materials*, 2022, **438**, 129461.
19. Y. Duan, T. Hu, L. Yang, J. Gao, S. Guo, M. Hou and X. Ye, *J. Alloys Compd.*, 2019, **771**, 156–161.
20. Y. Li, H. Sun, Y. Yang, Y. Cao, W. Zhou and H. Chai, *J. Colloid Interface Sci.*, 2020, **580**, 298–307.
21. X. Guo, Q. Zhang, X. Ding, Q. Shen, C. Wu, L. Zhang and H. Yang, *J. Sol–Gel Sci. Technol.*, 2016, **79**, 328–358.
22. M. Shandilya, R. Rai and J. Singh, *Adv. Appl. Ceram.*, 2016, **115**(6), 354–376.
23. N. Ullal, K. Muthamma and D. Sunil, *Chem. Pap.*, 2022, **76**(10), 6097–6127.
24. C. Xiang, Y. Liu, Y. Yin, P. Huang, Y. Zou, M. Fehse, Z. She, F. Xu, D. Banerjee and D. Hermida Merino, *ACS Appl. Energy Mater.*, 2019, **2**(5), 3389–3399.
25. B. A. Mehrabadi, S. Eskandari, U. Khan, R. D. White and J. R. Regalbuto, *Adv. Catal.*, 2017, **61**, 1–35.
26. Y. Al Haj, S. Mousavihashemi, D. Robertson, M. Borghei, T. Pääkkönen, O. J. Rojas, E. Kontturi, T. Kallio and J. Vapaavuori, *Chem. Eng. J.*, 2022, **435**, 135058.
27. K. Zhang, Z. Huang, M. Yang, M. Liu, Y. Zhou, J. Zhan and Y. Zhou, *SusMat*, 2023, **3**(5), 558–580.
28. M. Kigozi, G. N. Kasozi, S. Balaso Mohite, S. Zamisa, R. Karpoormath, J. B. Kirabira and E. Tebandeke, *Green Chem. Lett. Rev.*, 2023, **16**(1), 2173025.
29. Z. Chen, W. Wei, B.-J. Ni and H. Chen, *Environ. Funct. Mater.*, 2022, **1**(1), 34–48.
30. M. A. A. M. Abdah, F. N. M. Azlan, W. P. Wong, M. N. Mustafa, R. Walvekar and M. Khalid, *Chemosphere*, 2024, **349**, 140973.
31. L. Dai, O. Karakas, Y. Cheng, K. Cobb, P. Chen and R. Ruan, *Chem. Eng. J.*, 2023, **453**, 139725.
32. K. M. Wyss, W. Chen, J. L. Beckham, P. E. Savas and J. M. Tour, *ACS Nano*, 2022, **16**(5), 7804–7815.
33. S. K. Nayak, S. Saurabh, A. Kar, B. B. Sahoo, N. K. Sahoo and P. K. Sahoo, *Mater. Today: Proc.*, 2022, **67**, 1024–1032.
34. J. Xu, S. Dou, W. Zhou, C. Yang, I. Manke, P. Zhang, Z. Yan, Y. Xu, Q. Yuan and Y. Zhang, *Nano Energy*, 2022, **95**, 107015.

35. S. Chen, S. Fang, A. I. Lim, J. Bao and Y. H. Hu, *Chemosphere*, 2023, **322**, 138174.
36. A. Shah, S. Senapati, H. A. Murthy, L. R. Singh and M. Mahato, *ACS Omega*, 2023, **8**(37), 33380–33391.
37. M. Karakoti, S. Pandey, G. Tatrari, P. S. Dhapola, R. Jangra, S. Dhali, M. Pathak, S. Mahendia and N. G. Sahoo, *Mater. Adv.*, 2022, **3**(4), 2146–2157.
38. J. Zhou, L. Xu, L. Li and X. Li, *J. Colloid Interface Sci.*, 2019, **545**, 25–34.
39. I. Yang, J. H. Mok, M. Jung, J. Yoo, M. S. Kim, D. Choi and J. C. Jung, *Macromol. Rapid Commun.*, 2022, **43**(10), 2200006.
40. Q. Liu, M. Yang, Y. Deng, M. Zhang, C. Liu, F. Hu, X. Jian and Y. Chen, *J. Power Sources*, 2024, **602**, 234336.
41. X. Weng, Z. Wang, B. Zhang, X. Yu, X. Liu and T. He, *Polym. Adv. Technol.*, 2023, **34**(6), 1990–2000.
42. B. Sadeghi, Y. Marfavi, R. AliAkbari, E. Kowsari, F. Borbor Ajdari and S. Ramakrishna, *Mater. Circ. Econ.*, 2021, **3**, 1–18.
43. M. Jung, I. Yang, D. Choi, J. Lee and J. C. Jung, *Korean J. Chem. Eng.*, 2023, **40**(10), 2442–2454.
44. P. Dubey, V. Shrivastav, P. H. Maheshwari, M. Hołdyński, A. Krawczyńska and S. Sundriyal, *J. Energy Storage*, 2023, **68**, 107828.
45. J. Li, S. Zhang, Y. Hua, Y. Lin, X. Wen, E. Mijowska, T. Tang, X. Chen and R. S. Ruoff, *Green Energy Environ.*, 2024, **9**(7), 1138–1150.
46. D. Gangaraju, A. M. Shanmugharaj and V. Sridhar, *Polymers*, 2023, **15**(11), 2479.
47. I. Dědek, S. Bartusek, J. J. Dvořáček, J. Nečas, J. Petruš, P. Jakubec, V. Kupka and M. Otyepka, *J. Energy Storage*, 2023, **72**, 108660.
48. B. Padha, S. Verma, M. Prerna, A. Ahmed, S. P. Patole and S. Arya, *Appl. Energy*, 2024, **356**, 122402.
49. W. Gao, Y. Zhao, W. Chen, J. Zhuang, M. Shang, D. Sun and A. Xie, *Ceram. Int.*, 2023, **49**(5), 7266–7273.
50. A. M. Al-Enizi, A. Nafady, N. B. Alanazi, M. M. Abdulhameed and S. F. Shaikh, *Chemosphere*, 2024, **350**, 141080.
51. W. Jiang, X. Jia, Z. Luo and X. Wu, *Electrochim. Acta*, 2014, **147**, 183–191.
52. X. Liu, C. Ma, Y. Wen, X. Chen, X. Zhao, T. Tang, R. Holze and E. Mijowska, *Carbon*, 2021, **171**, 819–828.
53. N. Alhokbany, J. Ahmed, M. Ubaidullah, S. Mutehri, M. M. Khan, T. Ahamad and S. M. Alshehri, *J. Mater. Sci.: Mater. Electron.*, 2020, **31**, 16701–16707.
54. K. Shrivastava and A. Jain, *Wiley Interdiscip. Rev.: Energy Environ.*, 2024, **13**(2), e510.
55. R. AliAkbari, Y. Marfavi, E. Kowsari and S. Ramakrishna, *Mater. Circ. Econ.*, 2020, **2**, 1–27.
56. N. R. Chodankar, S. J. Patil, S.-K. Hwang, P. A. Shinde, S. V. Karekar, G. S. R. Raju, K. S. Ranjith, A. G. Olabi, D. P. Dubal, Y. S. Huh and Y.-K. Han, *Energy Storage Mater.*, 2022, **49**, 564–574.
57. G. Nagaraju, S. C. Sekhar and J. S. Yu, *Adv. Energy Mater.*, 2018, **8**(7), 1702201.

58. S. Sathish, P. Swaathi, S. B. Devi, R. Nirmala, P. Parthipan and R. Navamathavan, *ECS J. Solid State Sci. Technol.*, 2023, **12**(12), 121006.
59. S. Kaipannan, K. Govindarajan, S. Sundaramoorthy and S. Marappan, *ACS Omega*, 2019, **4**(14), 15798–15805.
60. S. Rajkumar, E. Elanthamilan, S.-F. Wang, H. Chryso, P. V. D. Balan and J. P. Merlin, *Resour., Conserv. Recycl.*, 2022, **180**, 106180.
61. Z. Xi, R. Gao, Z. Chen, H. Du and Z. Xu, *RSC Adv.*, 2024, **14**(2), 1386–1396.
62. R. R. Rajagopal, L. Aravinda, R. Rajarao, B. R. Bhat and V. Sahajwalla, *Electrochim. Acta*, 2016, **211**, 488–498.
63. S. Rajkumar, E. Elanthamilan, T. E. Balaji, A. Sathiyan, N. E. Jafneel and J. P. Merlin, *J. Alloys Compd.*, 2020, **849**, 156582.
64. L. Ojeda, A. Encinas, A. Zakhidov, A. I. Oliva, G. Gonzalez-Contreras, S. Diaz and J. Oliva, *J. Energy Storage*, 2023, **72**, 108733.
65. S. J. Arbaz, B. Ramulu and J. S. Yu, *Appl. Surf. Sci.*, 2023, **635**, 157696.
66. P. Dubey, S. Sundriyal and P. Maheshwari, in *Electrochemical Society Meeting Abstracts*, The Electrochemical Society, Inc., 244th edn, 2023, p. 992.
67. K. Du, E. H. Ang, X. Wu and Y. Liu, *Energy Environ. Mater.*, 2022, **5**(4), 1012–1036.
68. Y. I. Mesbah, N. Ahmed, M. M. Hasan and N. K. Allam, *Mater. Today Chem.*, 2023, **30**, 101521.
69. S. Yasa, B. Birol and M. Gencten, *New J. Chem.*, 2023, **47**(13), 6224–6234.
70. S. Yasa, B. Birol, K. B. Donmez and M. Gencten, *J. Energy Storage*, 2024, **81**, 110291.
71. S. Song, D. Deng, Y. Lu, W. Fan and W. Xu, in *Journal of Physics: Conference Seriesed*, IOP Publishing, 2024, vol. 2713, p. 012073.
72. S. Natarajan, M. Akshay and V. Aravindan, *Small*, 2023, **19**(17), 2206226.
73. R. Farzana, K. Hassan and V. Sahajwalla, *Sci. Rep.*, 2019, **9**(1), 8982.
74. P.-A. Le, P. L. Nguyen and T. V. Bac Phung, *ACS Sustainable Resour. Manage.*, 2024, **1**(2), 201–212.
75. P. Gao, B. Shen, P. Zhao, G. Shi and X. Zhao, *J. Power Sources*, 2023, **577**, 233231.
76. R. I. M. Vitto, M. T. Natividad and S. T. Palisoc, *J. Power Sources*, 2023, **582**, 233547.
77. A. Dutta, J. Mahanta and T. Banerjee, *Adv. Sustainable Syst.*, 2020, **4**(12), 2000182.
78. J. Meena, S. S. Sivasubramaniam, E. David and S. Kannappan, *RSC Sustainability*, 2024, **2**, 1224–1245.
79. Y. Ma, X. Xie, W. Yang, Z. Yu, X. Sun, Y. Zhang, X. Yang, H. Kimura, C. Hou and Z. Guo, *Advanced Composites and Hybrid Materials*, 2021, pp. 1–19.
80. I. Hussain, C. Lamiel, S. Sahoo, M. Ahmad, X. Chen, M. S. Javed, N. Qin, S. Gu, Y. Li and T. Nawaz, *Mater. Today Phys.*, 2022, **27**, 100844.
81. I. Shaheen, I. Hussain, T. Zahra, M. S. Javed, S. S. A. Shah, K. Khan, M. B. Hanif, M. A. Assiri, Z. Said and W. U. Arifeen, *J. Energy Storage*, 2023, **72**, 108719.

82. A. H. Phakatkar, M. T. Saray, M. G. Rasul, L. V. Sorokina, T. G. Ritter, T. Shokuhfar and R. Shahbazian-Yassar, *Langmuir*, 2021, **37**(30), 9059–9068.
83. V. T. Chebrolu, B. Balakrishnan, I. Cho, J.-S. Bak and H.-J. Kim, *Dalton Trans.*, 2020, **49**(41), 14432–14444.
84. D. Sharma, S. Kanchi and K. Bisetty, *Arabian J. Chem.*, 2019, **12**(8), 3576–3600.
85. S. Azhar, K. S. Ahmad, I. Abrahams, W. Lin, R. K. Gupta, M. Mazhar and D. Ali, *RSC Adv.*, 2021, **11**(49), 30510–30519.
86. G. T. Anand, D. Renuka, R. Ramesh, L. Anandaraj, S. J. Sundaram, G. Ramalingam, C. M. Magdalane, A. Bashir, M. Maaza and K. Kaviyarasu, *Surf. Interfaces*, 2019, **17**, 100376.
87. I. Shaheen, K. S. Ahmad, C. Zequine, R. K. Gupta, A. G. Thomas and M. A. Malik, *J. Energy Storage*, 2020, **29**, 101309.
88. R. Aliakbari, E. Kowsari, H. R. Naderi, S. Ramakrishna, A. Chinnappan and M. D. Najafi, *J. Alloys Compd.*, 2022, **914**, 165371.
89. S. K. Sahu, P. R. Sreekanth, N. D. Badgayan, T. J. Erinle, in *E3S Web of Conferencesed*, EDP Sciences, 2024, vol. 547, p. 03018.
90. R. Venkatkarthick and J. Qin, *Environ. Technol. Innovation* 2021, **22**, 101474.
91. Y. Marfavi, R. AliAkbari, E. Kowsari, B. Sadeghi and S. Ramakrishna, in *Ionic Liquid-Based Technologies for Environmental Sustainabilityed*, Elsevier, 2022, pp. 155–166.
92. E. Figueroa-Gonzalez, R. Mendoza, A. Oliva, C. Gomez-Solis, L. Valle-Garcia, E. Oliva, V. Rodriguez-Gonzalez, A. Encinas and J. Oliva, *J. Phys. D: Appl. Phys.*, 2021, **55**(3), 035501.
93. A. Bahaa, A. Alhammadi, K. C. Lethesh, R. A. Susantyoko and M. O. Bamgbopa, *Front. Chem.*, 2024, **12**, 1349864.
94. P. Ruschhaupt, A. Varzi and S. Passerini, *ChemSusChem*, 2020, **13**(4), 763–770.
95. C. Bauer, A. Bilican, S. Braxmeier, G. Reichenauer and A. Krueger, *Carbon*, 2022, **197**, 555–562.
96. P. Jeżowski and P. Ł Kowalczewski, *Polymers*, 2019, **11**(10), 1648.
97. A. Muzaffar and M. B. Ahamed, in *Low-Carbon Supercapacitors: Towards Sustainability in Energy Storage Devices*, ed. M. B. Ahamed, C. M. Hussain and K. Deshmukh, The Royal Society of Chemistry, Cambridge, 2023, ch. 4, pp. 84–116.
98. A. Rafique, I. Sequeira, A. S. Bento, M. P. Moniz, J. Carmo, E. Oliveira, J. P. Oliveira, A. Marques, I. Ferreira and A. C. Baptista, *Chem. Eng. J.*, 2023, **464**, 142515.
99. M. Raja, B. Sadhasivam, R. Dhamodharan and K. Ramanujam, *Sustainable Energy Fuels*, 2019, **3**(3), 760–773.
100. N. Kumar, L. Pradhan and B. K. Jena, *Wiley Interdiscip. Rev.:Energy Environ.*, 2022, **11**(1), e415.
101. O. Awogbemi and D. V. Von Kallon, *Heliyon*, 2022, **8**(10), e1227.

CHAPTER 4

Low-carbon Fuel Cell Materials

FELIPE M. DE SOUZA[a] AND RAM K. GUPTA*[a,b]

[a] National Institute for Materials Advancement, Pittsburg State University, 1701 S Broadway St, KS 66762, USA; [b] Department of Chemistry, Pittsburg State University, 1701 S Broadway St, KS 66762, USA
*Email: ramguptamsu@gmail.com

4.1 Introduction

4.1.1 Types of Energy Devices

The need for technologies that can generate and store energy in more sustainable and eco-friendly ways is one of the main objectives for the current society. There is a continuous strain on energy generation due to the high demands of modern civilization. The need to diminish the use of non-renewable resources has become a priority not only for the environment but also because it can lead to tensions across the globe due to its uneven distribution. One of the answers to alleviate these conditions can be presented based on further development and research of the different types of energy devices such as supercapacitors, batteries, electrolyzers, and fuel cells. Supercapacitors are energy-storage devices that function through fast electrostatic or redox reactions that allow for a quick burst of energy. Because of that, supercapacitors display high power density, meaning that they can release energy fast, while displaying a low energy density as they do not store a large quantity of energy and, therefore, discharge quickly. Batteries, on the other hand, store energy through complex intercalation reactions that result in higher energy densities. However, these reactions take longer to discharge, meaning that although the total stored energy is greater, the rate at which energy is delivered (power density) is lower. Electrolyzers are devices that can convert certain substances into fuels through an input of energy.

Low Carbon Materials: Paving a Sustainable Path for a Greener Future
Edited by Santanu Patra, Sudheesh K. Shukla and Mika Sillanpää

Published by the Royal Society of Chemistry, www.rsc.org

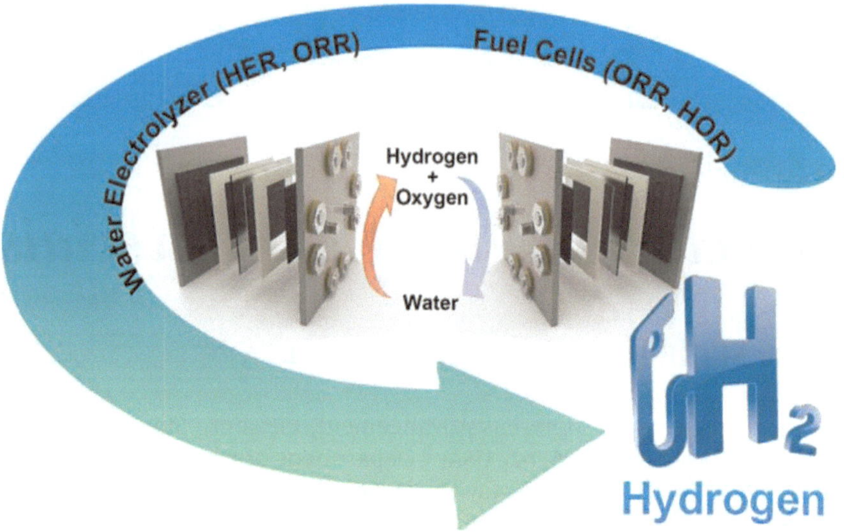

Figure 4.1 Schematic displaying the cyclic process between a water electrolyzer that produces H_2 and O_2 from H_2O that are utilized as fuel in a fuel cell that converts the gases into H_2O. Reproduced from ref. 1, https://doi.org/10.1002/aesr.202100019, under the terms of the CC BY 4.0 license, https://creativecommons.org/licenses/by/4.0/.

One of the most popular examples is water-splitting electrolyzers which break an H_2O molecule to generate H_2 and O_2 through a hydrogen evolution reaction (HER) and oxygen evolution reaction (OER), respectively. Such products from water-splitting electrocatalysts are the main reagents utilized in fuel cells that perform the combustion of H_2 through the two half-cells of hydrogen oxidation reaction (HOR) and oxygen reduction reaction (ORR) that leads to the release of H_2O and energy. Accordingly, the carbon emissions associated with fuel cells operating on hydrogen and oxygen primarily originate from the manufacturing of the materials used in their production. These two technologies hold great potential to move society in a sustainable direction for the generation of clean and sustainable energy. A schematic representing the cycling process between water electrolyzers and fuel cells is presented in Figure 4.1.

There are several fuel cell technologies such as proton exchange membrane fuel cells (PEMFCs), alkaline fuel cells (AFCs), phosphoric acid fuel cells (PAFCs), molten carbonate fuel cells (MCFCs), and solid oxide fuel cells (SOFCs). A summary of the main aspects of these fuel cells such as the fuels utilized, working temperatures, and type of electrolyte is presented in Figure 4.2.

4.1.2 Importance of Fuel Cells

Fuel cells are highly regarded energy generation devices due to their capabilities to provide clean energy efficiently which can potentially solve some

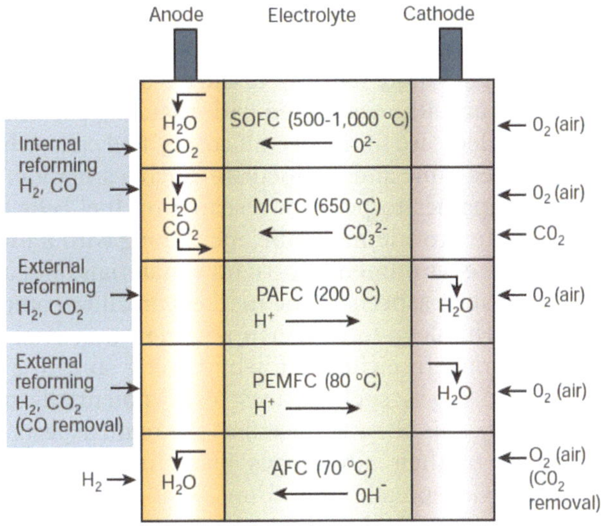

Figure 4.2 Schematic presenting the main fuel cell types in terms of fuels utilized, working temperature, and electrolytes. Reproduced from ref. 2 with permission from Springer Nature, Copyright 2001.

of the main issues related to the environment in relation to the emission of greenhouse gases that contribute to climate change. In addition, fuel cells display high efficiencies that can be greater than 80%, especially when used with configurations of combined heat and power (CHP). Another important factor is their flexibility in terms of fuels that can be utilized aside from H_2, including natural gas, biogas, methanol, and ethanol. Even though these fuels contribute to carbon emissions upon their combustion, excluding H_2, a device that can utilize different fuels is highly convenient and adaptable for different regions and occasions as it is more likely to prevent energy shortages. Fuel cells can be applied in urban regions such as residential and indoor areas due to their stationary and noise-free operation. In relation to that, fuel cells can work as decentralized energy production systems as they can be closer to the infrastructure to which they provide the energy. Such conditions can diminish energy losses due to long transmission distances and reduce the dependence on power plants, serving as a standalone energy provider system. By default, fuel cells can help stabilize the grid when in high demand as they can serve as a reliable and lasting system. Lastly, they can be integrated with renewable energy sources such as solar and wind power for instance, which can make them highly environment friendly.

4.1.3 Significance of Low-carbon Materials

The use of low-carbon materials in energy-related technologies is a highly desired condition that can play a major role in improving environmental aspects by decreasing the carbon footprint related to the manufacture of products utilized on fuel cells. In the long term, adopting such materials allows for better resource management and decreases the strain on the need for scarce materials that are more harmful to the environment such as noble metals, *i.e.*, Pt, Ru, Pd, and Rh. Associated with that, the use of low-carbon materials can lead to a decrease in the overall cost of fuel cells, which makes the technology more accessible and competitive. Such a situation serves as a good incentive for companies to perform research on fuel cells given their considerable potential to provide high efficiency, along with a lower cost of production, use of more environmentally friendly materials, and application versatility. Introducing such aspects in the market can allow companies that are pioneers in this technology to establish a system that is able to meet most the demands of modern civilization while doing so with a more sustainable approach that allows such devices to be potentially utilized for many years. In summary, this chapter describes the main aspects of fuel cell technologies in terms of their advantages and current bottlenecks that currently hinder their widespread applications. Along with that, the use of low-carbon materials is described in terms of the resources available and literature studies on the matter.

4.2 Fuel Cell: Types and Technology

4.2.1 Types of Fuel Cells

Fuel cells are devices that utilize the potential chemical energy of fuels and convert them directly into electricity. There are six main types of fuel cells: PEMFCs, AFCs, PAFCs, MCFCs, SOFCs, and DMFCs. They vary based on factors such as the type of electrolyte, working temperature, or the fuel utilized for the generation of electricity.

4.2.2 Basics of Fuel Cells

The PEMFC utilizes as an electrolyte a polymeric membrane able to conduct H^+ species while having an operational range from 85 to 105 °C. The first polymeric membrane used was made of polystyrene sulfonate (PSS) which demonstrated low stability after use. Later, there was the development of Nafion which is composed of polytetrafluoroethylene (PTFE) that can better withstand oxidizing or reducing environments, making it more chemically stable. The PEMFC utilizes an acidic electrolyte to provide protons. The fuel cell operates through two complementary reactions, which are the oxygen reduction reaction (ORR) that occurs at the cathode and the hydrogen oxidation reaction (HOR) that occurs at the anode. Both ORR and HOR are presented in reactions (4.1)–(4.3), respectively.

$$\text{Cathode: } \frac{1}{2}O_2 + 2H^+ + 2e^- \rightarrow H_2O \qquad (4.1)$$

$$\text{Anode: } H_2 \rightarrow 2H^+ + 2e^- \tag{4.2}$$

$$\text{Overall: } \frac{1}{2}O_2 + H_2 \rightarrow H_2O \tag{4.3}$$

One of the main factors to improve the PEMFC's efficiency is to increase the conductivity of H^+ of the polymeric membrane which is currently made of polymers based on perfluorocarbon sulfonic acid ionomers. For that, this polymer can be composited with PTFE or embedded with powdered additives to decrease the diffusion of gases through the polymeric membrane. One of the challenges in that regard is that thinner films tend to present higher conductivity while also increasing gas diffusion. In addition, the PEMFC often has to be filled with water to improve the conductivity. Yet, even though water itself is produced during the reaction at the cathode, it can be challenging to keep water in the membrane due to electro-osmotic factors that drag water through the cell. In this sense, a lower amount of water in the membrane leads to a decrease in conductivity. These issues can be partially solved by introducing humidified gases into the fuel cell. It has been observed that polysulfonic membranes can present better humidification when compared to Nafion 115, for example.[3]

The AFCs present some of the highest electrical efficiencies among the other fuel cells with the drawback of requiring highly pure gases for optimal operation, which is an economical burden for most applications. One of the reasons for its high efficiency is due to the use of high concentrations of KOH as an electrolyte which facilitates the ORR in the slow step of the half-cell reactions for the formation of H_2O due to the larger number of e^- and intermediate species involved. On the other hand, the HOR is slower on alkaline media. The reactions related to the O_2 reduction (cathode) and H_2 oxidation (anode) in alkaline media in an AFM are presented in reactions (4.4) and (4.5), respectively.

$$\text{Cathode: } \frac{1}{2}O_2 + H_2O + 2e^- \rightarrow 2OH^- \tag{4.4}$$

$$\text{Anode: } H_2 + 2OH^- \rightarrow 2H_2O + 2e^- \tag{4.5}$$

In the AFC, it has been observed that liquid KOH electrolyte displays more advantages than KOH stabilized in a gel-like matrix as the former can provide a circulating system that also works to cool down and clean the fuel cell.[4–6] Yet, one of the issues is related to the oxidation of carbon due to high open circuit voltages that can form carbonates either from the air or from the carbon support material. Such a process is presented in reaction (4.6). This process can clog the pores in the electrodes, preventing their contact with H_2 and causing the electrochemical performance to decrease. This drawback can be mostly avoided by utilizing a circulating electrolyte. In addition, a feasible way to remove CO_2 from the feed of H_2 can be performed by using an Fe-sponge system.[7–9]

$$CO_2 + 2OH^- \rightarrow {CO_3}^{2-} + H_2O \tag{4.6}$$

The PAFCs have gained more ground in comparison to the other fuel cells as they have been used in stationary power plants displaying power outputs of around 0.2–20 MW, which can be used to provide electricity and heat to hospitals, malls, and urban areas, for example. Such an important application was feasible due to the high chemical, electrochemical, and thermal stability of PAFCs along with the low volatility of its electrolyte that allows operating temperatures of between 150–200 °C. Such a temperature range results in faster processes when compared to PEMFCs that operate at around 100 °C since they follow the same electrochemical processes presented in reactions (4.1)–(4.3). Also, with the employment of materials that are more resistant to acid corrosion, the concentration of acids in the cell can be around 85% as higher concentrations of acid improve the electrolyte's conductivity which is stabilized in a SiC-based matrix. Yet, with the use of a liquid electrolyte it is important that the electrode presents a hydrophobic layer of PTFE as it can prevent the active pores from flooding. However, embedding with the optimal amount of PTFE is a challenge due to the need to find the proper amount of polymer that allows for good diffusion of gas due to low wettability instead of poor diffusion of gas due to high wettability.[10,11]

The MCFCs utilize carbonate ions obtained from molten salts of usually Li_2CO_3/Na_2CO_3 or Li_2CO_3/K_2CO_3 as a source of carbonates used as electrolytes as the former provides a relatively more alkaline system that helps to prevent the formation of Ni metal dendrimers from the electrode. Its high operating temperatures, that are in the range of 600–700 °C, melt the carbonate salts leading to the release of ion-conducting carbonates (CO_3^{2-}). The high temperature is required to maintain the CO_3^{2-} in its molten state as it allows the flux of ions and prevents the flux of e^- which passes through an external circuit to perform work. In addition, the high temperatures also increase the kinetics of the ORR processes. Based on that, MCFC functions in a cyclic system in which the HOR occurs at the anode and the ORR occurs at the cathode as shown in reactions (4.7) and (4.8), respectively. Then, the CO_3^{2-} from the molten electrode is converted into CO_2 at the anode reaction. Following that, the cathode consumes CO_2 and produces CO_3^{2-}. To stabilize the electrolyte either Al_2O_3- or Ce-based components are added as the latter is often more stable yet costlier.[12–14]

$$\text{Anode: } H_2 + CO_3{}^{2-} \rightarrow H_2O + CO_2 + 2e^- \tag{4.7}$$

$$\text{Cathode: } \frac{1}{2}O_2 + CO_2 + 2e^- \rightarrow CO_3{}^{2-} \tag{4.8}$$

SOFC is a gas–solid system in which a solid oxide is utilized as an electrolyte. Such a condition gives some advantages as it avoids problems with leakage as in the case of liquid electrolytes and it is relatively more stable than MCFCs. In addition, it does not present issues related to flooding of the catalyst's pores, water management requirements, or sluggish ORR kinetics. Yet, it is challenging to find materials that can withstand the thermal and chemical conditions at the relatively high operating temperatures required. It functions on a different mechanism when compared to the other fuel cells as the oxygen ions (O^{2-}), which are produced in the cathode (reaction (4.9)), migrate in the solid metal oxide to the anode (reaction (4.10)) through the oxygen vacancies in the crystal structure. The advantages of the SOFCs are related to the high stability of the solid oxide electrolyte which can prevent leaks and display high chemical stability.

$$\text{Anode: } H_2 + O^{2-} \rightarrow H_2O + 2e^- \tag{4.9}$$

$$\text{Cathode: } \frac{1}{2}O_2 + 2e^- \rightarrow O^{2-} \tag{4.10}$$

The DMFC is a system that operates similarly to that of a PEMFC with the difference that it can directly receive methanol as fuel in either liquid or gaseous form. The DMFC operates by direct injection of methanol to the anode that promotes its oxidation, which is a multi-step reaction that is simplified in reaction (4.11). The H^+ generated in the anode (reaction (4.11)) moves to the cathode through the electrolyte often consisting of an acid membrane. Then, the H^+ is utilized in the ORR reaction which leads to the formation of H_2O in the cathode as presented in reaction (4.12). The overall cell reaction is presented in reaction (4.13).

$$\text{Anode: } CH_3OH + H_2O \rightarrow CO_2 + 6H^+ + 6e^- \tag{4.11}$$

$$\text{Cathode: } \frac{3}{2}O_2 + 6H^+ + 6e^- \rightarrow 3H_2O \tag{4.12}$$

$$\text{Overall: } CH_3OH + \frac{3}{2}O_2 \rightarrow CO_2 + 2H_2O \tag{4.13}$$

When liquid methanol is used it is often diluted in water, which requires monitoring to maintain the feed concentration of fuel constant (around 1 to 2 mol L^{-1} of methanol). Such a process can be performed through the use of sensors that measure the current generated from the electro-oxidation processes of methanol.[15,16] In the case in which methanol is used in gaseous form it can lead to better performance of the fuel cell as it can decrease the cross-over effect with water. However, the fuel cell's design is more complex. In this regard, the polymeric membrane plays a major role as thinner membranes present higher permeability for liquid methanol, which is not desired since it can migrate to the cathode through electro-osmotic drag. To avoid that, thicker membranes like Nafion 120 are often utilized.[17] Despite the requirements for the use of methanol, this fuel has some advantages as it can be obtained from renewable sources and natural gas. Along with that, it provides high specific energy density. Yet, one of the main attractive factors is that the current infrastructure for transportation can be more easily adapted to the use of methanol. In summary, there are several types of fuel cells with distinct characteristics that perform the conversion of chemical energy into electricity through various routes. An overview of the main features of the fuel cells discussed is presented in Figure 4.3.

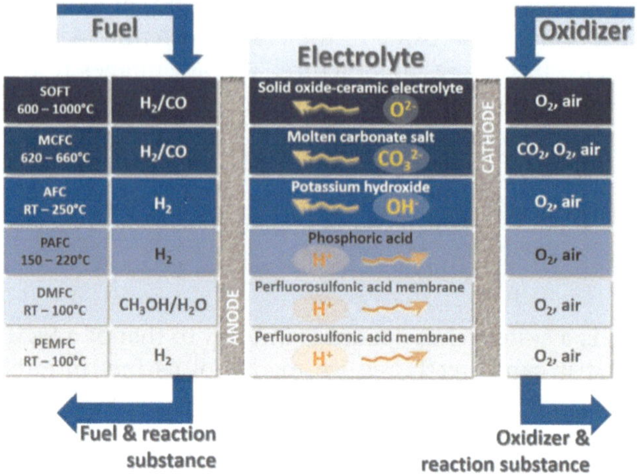

Figure 4.3 Schematics displaying the main characteristics of several types of fuel cells in terms of fuel, electrolyte, oxidizer, and working temperature. Reproduced from ref. 18, https://doi.org/10.3390/ijms22115430, under the terms of the CC BY 4.0 license, https://creativecommons.org/licenses/by/4.0/.

4.2.3 Significance of Low-carbon Materials

The development of technologies that can reduce the amount of greenhouse gases released into the atmosphere is one of the main concerns of modern society. Fortunately, fuel cells run on biofuels that can be obtained from renewable sources such as H_2 or methanol, for instance. Associated with that, energy production is inherently cleaner when compared to combustion processes of non-renewable sources that lead to the release of nitrogen and sulfur oxides, for example, which are responsible for environmental issues such as acid rain. The incorporation of low-carbon materials for fuel cells pushes the development of technologies that move toward a more sustainable future as research is focused on improvement of the cost-efficiency, durability, and the use of environmentally friendly materials. Further, fuel cells can find applications in portable devices, transportation, and stationary power generation, among others. A comparison between the main fuel cell technologies available regarding their applications, efficiency, power, and operating temperatures is presented in Table 4.1.

4.3 Low-carbon Materials in Various Fuel Cells

4.3.1 Proton Exchange Membrane Fuel Cells (PEMFCs)

PEMFCs can be composed of several low-carbon materials that contribute to the eco-friendly use of resources as well as potentially high-power efficiencies. The H_2 utilized as fuel for PEMFCs, and several other types of fuel

Table 4.1 Main aspects of different technologies for fuel cells. Reproduced from ref. 19, https://doi.org/10.3390/app6030075, under the terms of the CC BY 4.0 license, http://creativecommons.org/licenses/by/4.0/.

Fuel cell	Working temperature (°C)	Efficiency (%)	Power (kW)	Applications
PEM	60–110	40–55	0.01–250	Low power generation and portable devices
AFC	70–130	50–70	0.1–50	Portable devices, military, aerospace
PAFC	175–210	40–45	50–1000	Medium- and large-scale generation of power and combined heat and power
MCFC	550–650	50–60	200–100 000	Large-scale generation of power
SOFC	500–1000	40–72	0.5–2000	Medium- and large-scale generation of power and combined heat and power, auxiliary power units for vehicles, off-grid power
DMFC	70–130	40	0.001–100	Portable devices

cells, is a convenient source of energy when obtained from renewable sources as it leads to nearly zero carbon emissions when it undergoes electrochemical oxidation for the HOR that generates H^+ and e^- along with its complementary ORR process. For that, there is a considerable number of non-noble metals that can be used as catalysts for PEMFCs. However, Pt-based electrocatalysts have been more notorious for the ORR process. Some examples of these include carbon-based materials with Pt nanoparticles dispersed over their surface. Yet, Pt extraction leads to a considerable environmental impact along with its scarcity that imposes a hindrance to scalability.[20] In addition, when compared to other metals, plastics, and components utilized in PEMFCs, Pt and other noble metals presented the highest global warming potential over 100 years (GWP 100). Taking from that, Table 4.2 presents the GWP 100 (kg CO_2 eq.) for materials that can be utilized in PEMFCs. Through that, it is notable that graphite and other carbon-based materials may display the lowest GWP 100, followed by non-noble metals, and plastics.

The relatively lower operation temperature of PEMFCs leads to the need for higher loads of electrocatalysts due to slower kinetics. Therefore, it can induce the Ostwald ripening effect which consists of a decrease of active area due to the agglomeration of smaller particles into larger particles. Such an effect is often observed for Pt-based catalysts.[21] Given the drawbacks related to the use of Pt, the investigation of more available metals is necessary. In this sense, Fe-based electrocatalysts have demonstrated appreciable results as demonstrated in a previous study that described the pyrolysis (900 °C) of iron acetate that was adsorbed on perylene tetracarboxylic dianhydride in a system containing NH_3, H_2, and Ar. It was observed that a Fe content of around 0.2 wt% provided a comparable value of current density at

Table 4.2 GWP for materials per 1 g that can be utilized in the manufacturing of PEMFCs. Reproduced from ref. 20, https://doi.org/10.1002/ese3.441, under the terms of the CC BY 4.0 license, https://creativecommons.org/licenses/by/4.0/.

Material	GWP 100 (kg CO_2 eq.)
Nafion	8.31×10^{-1}
Glass fiber	1.73×10^{-3}
Polyether ether ketone	1.74×10^{-2}
PTFE	1.21×10^{-2}
Stainless steel	3.35×10^{-3}
Thermoplastics	4.28×10^{-3}
Al	8.26×10^{-3}
Si	5.82×10^{-3}
Carbon black	2.54×10^{-3}
Graphite	7.55×10^{-5}
Steel	2.19×10^{-3}
Pl	5.12
Pt	28.5
Polyvinylidene chloride	4.53×10^{-3}
Ru	6.43

0.5 V when compared to a 2 wt% state-of-the-art Pt/C.[22] Alongside the use of eco-friendly materials in the electrodes a report compared the use of three materials, *i.e.*, stainless steel, Al, and polymers, in the fabrication of the PEMFC tank.[23] The study results showed that stainless steel had the lowest CO_2 footprint of 72.1 kg per year when compared to Al (144 kg per year) and polymers (98 kg per year). Yet, stainless steel also presented the lowest energy production of 1.01 GJ per year when compared to Al (2.21 GJ per year), and polymers (1.76). Lastly, stainless steel had an average cost of 71.7 USD per year when compared to Al (40.7 USD per year) and polymers (143 USD per year).[3]

4.3.2 Solid Oxide Fuel Cells (SOFCs)

A SOFC is composed of an anode, a cathode, and a solid electrolyte that transports O^{2-} through its crystal structure. The anode (negative electrode) receives the H_2 that is dispersed over its surface generating H^+ and e^- that is used to power an external circuit. Simultaneously, the cathode (positive electrode) receives the O_2 and conducts the e^- from the external circuit to recombine them with the O^{2-} ions that are passed through the electrolyte along with H_2 to form H_2O and release energy. The electrolyte is the component that determines the working temperature of the cell and blocks the passage of electrons to the electrodes while allowing the flux of ions, such as O^{2-} or H^+, as the main difference is based on which side of the fuel cell the H_2O is produced. Schematics for a SOFC with the transport of O^{2-} and H^+ are presented in Figure 4.4a and b, respectively.

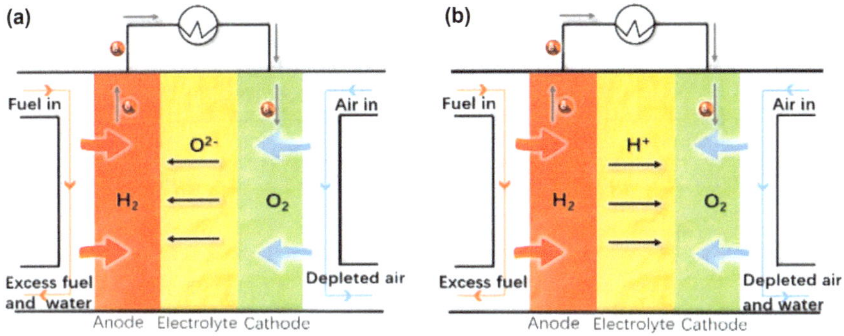

Figure 4.4 Schematics of a SOFC displaying a system of (a) O^{2-} conducting electrolyte and (b) H^+ conducting electrolyte. Reproduced from ref. 24, https://doi.org/10.1007/s41918-022-00168-0, under the terms of the CC BY 4.0 license, https://creativecommons.org/licenses/by/4.0/.

Following on from that, some of the most utilized materials as electrolytes for SOFCs are stabilized zirconia-based materials such as yttria-stabilized zirconia that includes Y_2O_3–ZrO_2 or $(Y_2O_3)_{0.08}(ZrO_2)_{0.92}$ (YSZ). The exclusive ionic conduction in YSZ functions is based on the crystal vacancies where O^{2-} are missing in the structure as the crystal array of ZrO_2 presents around two O^{2-} in proportion to a Zr ion. On the other hand, for Y_2O_3 there are 1.5 O^{2-} in proportion to a Y ion. This imbalance in the number of O^{2-} allows these ions to leap from vacancy to vacancy until they reach the anode. There is a plethora of metal oxide ceramics that can be used as an electrolyte for SOFCs such as cerium oxides with the general formula of $(Ce_xM_{x-1})O_y$ where M can be metals used as doping agents added in different proportions such as Ca, Sm, Gd, and Y. Also, ceramic metal oxides such as $Sr(Ce_{0.95}Yb_{0.05})O_3$, $(Ba_{0.75}Y_{0.25})_2O_3$, and $(Bi_{0.75}Y_{0.25})_2O_3$ are some examples of electrolytes that can be used in SOFCs. The anode usually consists of NiO composited with YSZ and $(Ce_xM_{x-1})O_y$, where M can be either Gd or Sm. The cathodes are almost exclusively composed of noble or electronic conducting metal oxides due to the high operating temperatures of SOFCs. Yet, given the unsuitability for large-scale and short-term stability of noble metals, other materials have been used as electrodes including $LaSrMnO_3$ (LSM) and $LaCaMnO_3$ (LCM). Such materials present a similar thermal expansion to those of Zr-based electrolyte which improves the fuel cell's durability along with appreciable performances at operating temperatures of around 600–800 °C. Following that, several other Perovskite-type La-based materials can be used as cathodes such as $(LaSr)(Fe)O_3$, $(LaSr)CoO_3$, $(LaSr)(CoFe)O_3$, $(LaSr)(MnFe)O_3$, $(SmSr)CoO_3$, and $(PrSr)(MnFe)O_3$, among others. As presented, several materials can be used in different components of SOFCs. Even though the electrolyte and cathode are usually made of materials that contain rare metals such as Zr, Y, Gd, and Sm, the advantage in their use lies in their high stability, which reduces the need for maintenance and improves the life cycle of the fuel cell. On the other hand, the anode usually consists of NiO

Table 4.3 GWP 100a materials utilized in SOFC. Reproduced from ref. 26, https://doi.org/10.1016/j.apenergy.2018.11.028, under the terms of the CC BY 4.0 license, http://creativecommons.org/licenses/BY/4.0/).

Component	Material	GWP 100a
Anode	NiO	133.04
	CeGdO	218.74
Electrolyte	Na_2CO_3	0.01
	$SrCO_3$	0.01
	Bi_2O_3	5.58
	TiO_2	1.47
Cathode	La_2O_3	60.49
	$Sr(NO_3)_2$	3.38
	Mn_3O_4	2.67
	YSZ	59.11

doped with other metals which are relatively abundant but require extensive mining processes that apply a strain on the environment. To reduce that, some alternative materials based on Sr, La, and Mn (SLM) or La, Sr, Co, and Fe (LSCF) which can be potentially more stable can serve as potential candidates.[25] In the context of low-carbon materials, Table 4.3 displays some of the main materials utilized in the fabrication of SOFCs and their GWP 100a, which is a similar measurement compared to GWP 100 with the difference being that it is calculated by neglecting the effect of short-lived pollutants such as CH_4.[26]

4.3.3 Molten Carbonate Fuel Cells (MCFCs)

The MCFCs are fuel cells that operate at high temperatures of around 650 °C to melt carbonate salts such as Li_2CO_3 and K_2CO_3 to form CO_3^{2-} ions that function as electrolytes. The most commonly used electrodes are Ni-based materials. The anodes are typically nickel alloys containing chromium or aluminum, which enhance the stability of their microstructure. The cathodes usually consist of lithiated nickel oxide formed *in situ*.[27] Yet, one of the recurrent issues is related to the solubility of the electrode active materials in the electrolyte given the high temperatures required for the formation of alkaline carbonate ions. One way that was found to address that issue consisted of introducing rare metal oxides such as Gd_2O_3, La_2O_3, or CeO_2. Through that, there was an increase in oxygen solubility and a decrease in corrosion of the electrolyte. Even though the improvement in stability and longevity was performed by using rare metal oxides, a previous report demonstrated a feasible approach to recycle such rare metals from scrap.[28] This work showed that a cathode fabricated with around 20% recycled metal scrap had a slightly lower performance when compared to the reference cell. Such a process suggested a possible way to improve the longevity of MCFCs while utilizing rare metal resources with an eco-friendlier approach. Yet, other reports presented feasible approaches that avoided the use of rare metal oxides. For example, in an NiO cathode, the use of some additives such as 20 mol% of Co led to a reduction of 66% in NiO solubility.[29] However, it also led to a decrease in conductivity when compared to bare NiO. Another study showed the use of different doping percentages of NiO with Ce and Co.[30] Their results showed the optimum values of 9.5% and 5% doping for Co and Ce, respectively, which led to a decrease of around 80% in NiO solubility while maintaining a similar porosity to the undoped material. Further, another report showed the effects of incorporating 2 mol% of ZnO into an NiO cathode which led to a reduction of electrode solubility of around 94%. However, it also led to an increase in charge transfer resistance when compared to bare NiO.[31] Aside from using additives, coating the surface of NiO has also been shown to be a viable approach to decrease its solubility. One case was performed through a mechanofusion approach to coat 5% of CoO particles over the surface of Ni which led to a decrease in NiO solubility of 90% without any apparent drawbacks.[32] In a similar case, another study showed the effect of coating NiO with 0.3 wt% of La which led to a 90% decrease in NiO solubility and a decrease in charge transfer resistance, providing an overall improvement in properties and performance of the cell.[33] Aside from the use of different methodologies to modify the properties of NiO there was an example from the literature presented by Tolkacheva *et al.*[34] who synthesized submicron particles of α-$LiAlO_2$ with a surface area of around 25–27 $m^2 g^{-1}$ through pyrolysis (650 °C) of $Al(OH)_3$ and Li formate. The eco-friendly factor of this approach lies in the presence of the formate which avoids the emission of NO_x along with better homogeneity of the product.

4.3.4 Alkaline Fuel Cells (AFCs)

The AFC is an attractive technology as it can display some of the highest electrical efficiencies when compared with other fuel cells. It functions through the half-cell reactions of HOR in the anode and ORR in the cathode. A general schematic for an AFC is presented in Figure 4.5, as a note on the anode there is an inlet for H_2, an outlet for H_2O, and an excess of fuel that did not react. Similarly, on the cathode, there is an inlet for O_2 and an outlet that releases unused O_2 gases from the system. Hence, considering an AFC with ideal efficiency there would be no release of leftover H_2 from the anode's outlet nor O_2 from the cathode's outlet.

Even though AFCs are regarded as the cheapest fuel cells and they can display some of the highest efficiencies, they have some challenges related to the need for high-purity gases to function properly. Also, the electrode materials utilized for their fabrication usually consist of porous carbon matrices embedded with noble metals. In addition, the high operating temperatures of around 200–230 °C improve the kinetics but also carries the requirement for KOH electrolyte management. Given such conditions, one of the main concerns is the presence of CO_2 in the atmosphere that can act as a poison. In addition, the high operating temperatures along with the carbon support

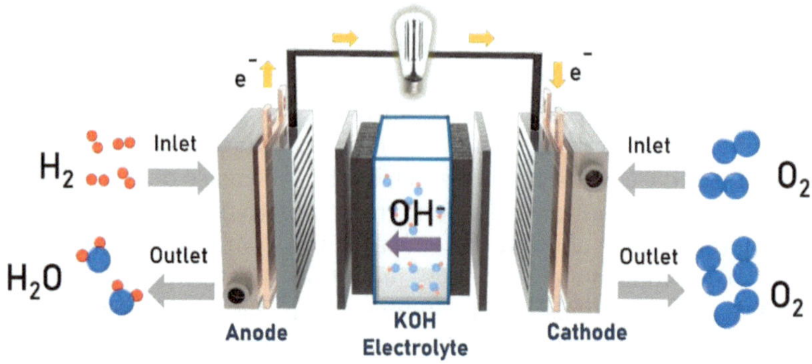

Figure 4.5 Schematic for the operation of an AFC. Reproduced from ref. 6, https://doi.org/10.3390/polym14061197, under the terms of the CC BY 4.0 license, https://creativecommons.org/licenses/by/4.0/

in the electrode lead to the formation of CO_2 within the cell which also leads to a performance decrease. It is proposed that reactions that take place are as presented in reactions (4.14) and (4.15). One of the issues related to these side reactions is the formation of K_2CO_3 which can potentially block the pores of the electrode, leading to a decrease in performance, whereas the CO_2 may decrease the diffusion of other gases into the system due to saturation.[35–37]

$$CO_2 + 2OH^- \rightarrow CO_3{}^{2-} + H_2O \tag{4.14}$$

$$CO_2 + 2KOH \rightarrow K_2CO_3 + H_2O \tag{4.15}$$

The effect of CO_2 poisoning could be mitigated based on the porous nature of the electrode as it was found that Ag-based electrodes could avoid the adsorption of CO_2 into its pores when compared to Raney-Ni. Such aspects are related to the surface area of the electrode that allow it to function for longer periods without a considerable decrease in performance. Yet, there is a focus on the use of non-noble metals such as Fe, Co, and Cu doped with N and dispersed in a carbon matrix. Such materials display appreciable electrocatalytic properties toward ORR in alkaline media. Such an aspect is believed to be potentially related to the presence of active sites that can have a facilitated reduction activity of $HO_2{}^-$ active species that are reduced to OH^-.[38] In regard to the HOR process, there is often a need for high loads of rare metals due to the sluggishness of the reaction. However, due to their inherently high release of carbon gases that contribute to the greenhouse effect and their scarcity, other metallic sources have been sought. One of the most researched electrocatalysts for HOR is Ni-based alloys with Cu, Co, and Mo, for instance.[39–41] Other composite materials such as Ni particles doped with N in a matrix of carbon nanotubes (Ni/N-CNTs) have also demonstrated appreciable electrocatalytic properties toward HOR in alkaline media while being relatively friendlier to the environment in comparison to other metals.[42] Yet, some of those that displayed the highest performances are Ni_3N and Ni/MoO_2-based materials.[43,44] It has been reported that Ni_3N particles displayed a current density exchange of around 3.08 mA cm^{-2}, which was 1.4 times higher than other Pt-based electrocatalysts.[43,45] Some of the reasons for the better performance observed for these materials were attributed to their capability to balance the binding of the hydrogen-containing intermediates to facilitate the HOR process. However, other factors such as conductivity, particle size, exposure of active sites, and electrochemical stability play a major role in the overall electrochemical performance.

4.3.5 Biofuel Cells

Biofuel cells consist of devices capable of converting chemical energy derived from biochemical processes into electricity. For that, biological components such as microorganisms or enzymes are used to catalyze and facilitate redox reactions in which the oxidation takes place at the anode, whereas the reduction takes place at the cathode. A schematic of a typical biofuel cell is presented in Figure 4.6. Several materials can be utilized to compose the electrodes such as carbon-based ones that include carbon nanotubes (CNTs), graphene, activated carbon, carbon paper, and carbon cloth. These materials often mostly display high conductivity and surface area, which are important aspects for the immobilization of enzymes or microorganisms that perform the biocatalytic processes. In addition, its high conductivity aids in the e^- transfer process which can occur with or without a mediator. Some of the options for low-carbon-footprint membranes are biopolymers such as cellulose and chitosan which are derived from biorenewable sources and are biodegradable. Also, they allow for ionic

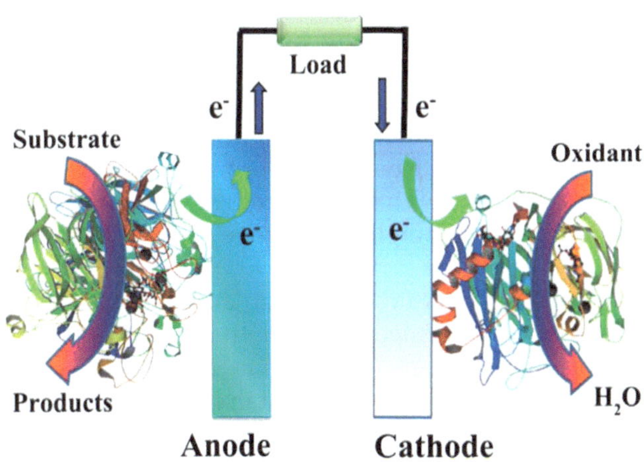

Figure 4.6 Schematic of a typical biofuel cell. Reproduced from ref. 46 with permission from the Royal Society of Chemistry.

conductivity based on their porosity. Other materials such as alginate-based hydrogels can be utilized to immobilize enzymes or microorganisms functioning as a support matrix as well as substrate which can increase the longevity and stability of a biofuel cell. Nanocellulose is another material that can provide mechanical support while adding eco-friendly credentials to the cell. Lastly, the enzymes and microorganisms that act as biocatalysts are among the core components for the fabrication of an electrode for biofuel cells. The immobilization of these bioactive components to perform an optimized redox process is crucial to improve the performance of biofuel cells, which can lead to a reduction in the overall carbon footprint of the system.

The substrates more often utilized in biofuel cells are compounds such as ethanol, glucose, and biowaste, among other organic compounds that can serve as a good source of chemical energy. These substrates provide advantages such as abundance and renewability, which aid in the overall decrease of carbon emissions. Also, most of these compounds such as carbohydrates are biocompatible, which allows them to be incorporated in applications related to biomedical and implantable devices. However, they present some disadvantages related to incomplete oxidation of the substrate which leads to a decrease in efficiency. Based on this issue, the use of an enzyme such as glucose oxidase in a biofuel cell can allow for the generation of 2 e^- even though the complete oxidation of glucose can yield up to 24 e^- until it is converted to CO_2. One of the most utilized approaches to harness these e^- is by developing a system with several enzymes that can efficiently perform a multi-step process of substrate oxidation. Such an approach was performed by Palmore *et al.*[47] who proposed the optimization of a biofuel cell using methanol as the substrate. The oxidation at the anode's surface was performed by introducing several enzymes (*i.e.*, enzymatic cascade) such as NAD-dependent alcohol dehydrogenase, aldehyde dehydrogenase, and formate dehydrogenase that could harness a large number of e^- throughout the complete conversion of methanol into CO_2. Along with the presence of different enzymes it is also important to introduce mediators in the system as they can improve the biofuel cell efficiency by shuttling e^- between the biological components to the electrode. For this case, the authors utilized quinone and diaphorase as mediators. In another study by Xu *et al.*,[48] a biofuel cell consisting of six enzymes was proposed to perform the oxidation of glucose into CO_2 with better harnessing of e^-. The enzymes utilized in the process required pyrroloquinoline quinone as a co-factor for proper functioning.

Another approach related to the use of an enzyme cascade on the surface of the electrode was performed through the use of a magnetic field.[49] For that, carbon-coated magnetic nanoparticles (ccMNPs) were used to promote the activation of the bioelectrocatalysis. The magnetically induced shuttling of the enzymes modified with the particles toward the surface of the electrode led to an improvement in the e^- transfer steps between the components. This approach allowed for simultaneous direct e^- transfer (DET) and mediated e^- transfer (MET), which led to multiple fuels being

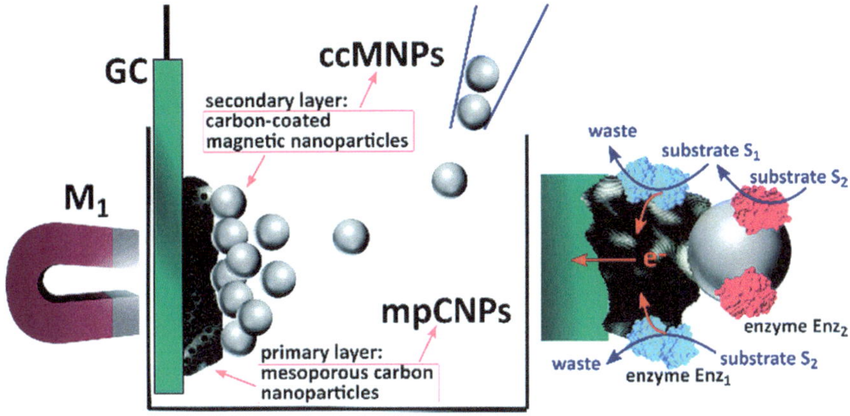

Figure 4.7 Schematic for the activation of an enzymatic cascade system through an applied magnetic field with enzymes modified with magnetic nanoparticles on the surface of mesoporous carbon nanoparticles. Reproduced from ref. 49 with permission from the Royal Society of Chemistry.

oxidized to generate electrical current. The advantage of this process was due to a high faradaic efficiency that was around 92.3–98.8%. However, some of the challenges faced were: (i) higher complexity of the system; (ii) the enzyme with the slowest activity governed the overall stability of the biofuel cell; (iii) the optimal temperature had a higher influence on the fuel cell performance given the different types of enzymes and substrates; (iv) control of pH became difficult as each enzyme may display a different optimum pH range of values to function properly; (v) the number of enzymes immobilized was limited and could present a heterogeneous distribution throughout the electrode, potentially leading to variations in performance; and (vi) the co-immobilization of two or three enzymes led to a decrease in catalysis efficiency because of the reduction of available surface area and steric hindrance of the enzymes for exposure of their active sites.[50] The optimization of e^- harnessing is a core step for efficiency enhancement which can indirectly reduce the overall amount of CO_2 generated. A schematic for the magnetic activation of the bio-electrocatalytic enzyme cascade is illustrated in Figure 4.7.

The improvement of the active life duration of the enzymes immobilized in the electrode is an important aspect of increasing their durability and efficiency over time. Generally, enzymes may last around 8 h to 2 days in a buffer solution. This time could be increased up to around 20 days through immobilization of the enzyme in the electrode's surface which can occur by physical adsorption, chemical bonding, entrapment, and cross-linking.[51] In addition, protein engineering is another strategy that aims to increase the electron transfer rate from the enzyme to the electrode's surface. Yet, it should be performed without compromising its active site area.[52–55] Protein engineering can also provide other advantages by improving the stability of the enzyme, optimizing its e^- transfer, and improving its adaptability toward other substrates (*i.e.*, enzymatic promiscuity), among other advantages. However, such a process is often complex and costly given it requires sophisticated techniques and laborious methods. Also, undesirable consequences may emerge upon modifying an enzyme structure.

Biofuel cells have great potential to integrate the future of energy generation as a sustainable and eco-friendly option. As it stands now, scientists are focusing on improving the biofuel cells' open circuit potentials and current densities for an overall increase in their efficiencies. Some of the approaches to improvement have been made through the incorporation of conducting polymers and nanomaterials that can result in an increment of current density due to better e^- transfer steps while providing a more stable cell structure.[56–58]

4.3.6 Other Fuel Cell Technologies

Some of the other fuel cell technologies include high-temperature PEMFC (HT-PEMFC), regenerative fuel cells (RFCs), microbial fuel cells (MFCs), metal–air fuel cells (MAFCs), and flow batteries. The HT-PEMFCs require materials that can withstand operating temperatures of around 100 °C which can be composed of sulfonated polyether ether ketones (SPEEK) and phosphoric acid-doped polybenzimidazole (PBI). In terms of catalysts for the ORR process in HT-PEMFC, there has been some research on the use of atomically dispersed FeN_4 over a carbon matrix.[59] Despite their good performance, some of the challenges are related to the low density of active sites, poor three-phase interface, and instability over long operation time. It has been observed that N-doped Fe over a carbon matrix (Fe–N–C) can be utilized in HT-PEMFCs as catalysts in the presence of phosphoric acid-doped PBI as the phosphate ions can avoid poisoning of the system, which is not the case with perchloric acid.[60] Graphite-encapsulated Fe (Fe_3C) is also a convenient material that can be utilized in HT-PEMFCs due to its tunable morphology that can be obtained by varying the pyrolysis temperature during the synthesis of Fe_3C using precursors such as cyanamide and ferrocene. The single-atom crystals on graphene (FeSA-G) are another type of catalyst that have demonstrated superior stability when compared to Pt/C for ORR processes at temperatures around 230 °C.[61] In addition, the uniform dispersion of Fe atoms throughout the graphene structure was a key factor in enhancing overall performance. This made it more stable and compatible than Pt/C, which tends to suffer from Ostwald ripening and particle agglomeration at high temperatures.[62]

RFCs are a type of device that can perform fuel combustion and electrolysis. In electrolysis mode, the RFC utilizes electricity to perform the water-splitting reaction to generate H_2 and O_2 that are utilized in a fuel cell to release energy upon the combustion of H_2 (reaction with O_2) to form H_2O. The device can be composed of materials that are utilized in PEMFCs and AFCs such as Nafion and anion exchange membranes that can conduct

OH^- species. In terms of catalysts, some of the materials utilized are Ni–Mo- and Fe-based for HER and ORR, respectively.

MFCs are devices that harness the energy from the metabolic activity of microorganisms and convert it into electricity. The anode of an MFC is composed of a microorganism, usually bacteria, that oxidizes a substrate which can be sugars, acetate, or wastewater while in the presence of an electron acceptor. The oxidation process leads to the release of e^-, which goes to the anode, and H^+ which flows through the electrolyte to the cathode. At the cathode, the e^- released at the anode combines with H^+ in the presence of O_2 to form H_2O. The use of microorganisms promotes an eco-friendlier aspect along with the possible consumption of biowastes and wastewater, which allows for the proper management of effluents.[63]

MAFCs generate electricity from the reaction between O_2 as a cathode and a metal anode. Through that, a metal anode (Al, Zn, or Li) undergoes an oxidation process releasing e^- and generating metal ions. At the cathode, O_2, H_2O, and e^- form OH^- species. The metal ions can then be reverted into their reduced state through an input of electricity. Some of the sought materials for MAFCs as anodes are Zn- and Al-based electrodes due to their relative abundance and relatively better environmental aspects when compared to other metals. In terms of cathodes, there is a continuous push for the development of porous carbon-based materials such as carbon nanotubes, graphene, and carbon black, among others that can facilitate the diffusion and reactions involving O_2. As electrolytes, some of the eco-friendly materials that can be employed are hydroxide-conducting polymers or alkaline solutions that display high ionic conductivity.

Flow batteries are a type of rechargeable electrochemical device that utilizes liquid electrolytes that are stored in external tanks which are designed for large-scale applications. Similarly to other fuel cells, flow battery components also consist of polymeric-based ion exchange membranes such as Nafion, and the use of non-noble metals is one of the aims of research to avoid the use of noble metals.

4.4 Advancement and Challenges

4.4.1 Architectural Aspects of Fuel Cells

As discussed, there is a plethora of fuel cells that convert different sources of chemical energy into electricity through different technological architectures. Yet, fuel cells display the general design of two electrodes. One is the anode at which the oxidation of the fuel (H_2, methanol, *etc.*) occurs. The cathode is the component at which the oxidant, O_2, is chemically reduced. The electrolyte is a component that allows only the flow of ions while preventing the flux of e^- which passes through the external circuit.

Yet, such components can be considerably different such as in terms of electrolytes which can be ionic solutions with aqueous or organic solvents or charged polymeric compounds that allow the flux of ions in between their chains. Yet, each fuel cell requires a different type of electrolyte for the flux of electrons along with their set of advantages and disadvantages. PEMs have the advantage of low operation temperatures of around 100 °C which demands less energy for its operation. Their high efficiency and power density, lightweight, compact, and low emission of carbon enable them to be integrated in portable and transportation technologies. Some of the disadvantages of PEMFCs include susceptibility to poisoning by fuel impurities, the current need for expensive and scarce electrocatalyst materials such as Pt, low stability at high temperatures, and maintenance of the PEM to avoid dehydration. AFCs are some of the cells with the highest efficiencies, which translates to high energy output, relatively lower cost, and high stability to be operated over long periods. However, they are more affected by gas poisoning, and have slower kinetics when compared to PEMFCs. PAFCs have the middle ground in terms of operating temperatures that are between 150–200 °C along with efficiency, stability, and good tolerance to impurities. Yet, they display a relatively lower power density in comparison to PEMFCs and have limited scalability. The MCFCs display high efficiency and power density which give them a potential use in power generation along with good tolerance of CO_2 in the system. However, they operate at considerably high temperatures that can go from 650 to 850 °C, which hinders their applications in portable devices, for example. On top of that, the corrosiveness of the electrolyte makes it necessary to find robust materials that can withstand such aggressive conditions for extended periods. Similarly, the SOFC also displays high operating temperatures of around 800 °C which improves the kinetics and allows for the use of different types of fuel and tolerance to CO_2. However, it also requires specific and robust materials that can withstand such aggressive conditions. Along with that the high weight and large size of these fuel cells, their applications are limited.

4.4.2 Efficiency and Performance

The different fuel cell technologies that are currently being developed present several variables in terms of power output, efficiency, working temperature, and application. Such conditions are important for expanding their realm of applications as each technology may be suitable in certain scenarios. Based on that, Table 4.4 displays some of the main aspects of different fuel cells in terms of efficiency, performance, operating temperatures, and applications.

4.4.3 Environmental Impact

Fuel cells can offer an alternative route for the generation of energy that can be friendlier to the environment when compared to conventional combustion systems of hydrocarbons derived from non-renewable sources. Yet, one of the main challenges lies in the decrease of CO_2 along with other

Table 4.4 Comparison between different fuel cells in terms of efficiency, power output, operating temperatures, and applications. Reproduced from ref. 19, https://doi.org/10.3390/app6030075, under the terms of the CC BY 4.0 license, http://creativecommons.org/licenses/by/4.0/.

Type of fuel cell	Efficiency (%)	Power output (kW)	Operating temperatures (°C)	Applications
PEM	40–55	0.01–250	60–110	Low power generation and portable
PAFC	40–45	50–1000	175–210	Generation of combined heat and power (CHP)
AFC	50–70	0.1–50	70–130	Mobile, military, space
MCDC	50–60	200–1 000 000	550–650	Large-scale generation of power
SOFC	40–72	0.5–2000	500–1000	Transportation, off-grid power, and micro-CHP
DMFC	40	0.001–100	70–130	Mobile and portable

gases that are notorious for contributing to the enhancement of the greenhouse effect. Some fuel cells such as PEMFCs can utilize H_2 or alcohols (*i.e.*, methanol and ethanol) as fuels. The former is the most convenient to the environment as the combustion of H_2 leads to the formation of only H_2O as a byproduct. However, there are some inherent challenges with the uses of H_2 such as its high flammability, instantaneous ignition, high permeation rate, and the fact that it can potentially lead to material embrittlement.[64–66] Despite such challenges, previous studies have described that when H_2 obtained from renewable sources is utilized in FCs most of the environmental impact is derived from the production of the materials used in the FC itself, which makes it a considerably friendlier technology to the environment in comparison to traditional combustion engines.[67–69]

4.4.4 Economic Consideration

There are promising economic aspects related to fuel cell technologies, such as the case of a previous report that suggested that vehicles operated through a fuel cell may display an efficiency that is 25–30% higher than those of combustion engine vehicles.[70] Despite the higher efficiency of fuel cells, other studies demonstrated that fuel cell-hybrid vehicles present considerably lower emission of greenhouse gases along with a higher cost per kilometer as it has been proposed that fuel cells, despite having lower gas emission and energy use, can roughly cost \$0.25 km^{-1}, whereas combustion engine vehicles cost around \$0.19 km^{-1}. Yet, future investments in technology and research on cheaper materials that can be used for their fabrication could potentially decrease their overall cost in the future.[71]

4.5 Future Trends and Opportunities

4.5.1 Emerging Technologies

There has been a continuous increase in the research into fuel cells due to their potential relevance and applicability. One of the main challenges is related to the use of Pt as a catalyst. To avoid that, there has been an increasing number of studies that have focused on the use of carbon-based materials such as graphene and carbon nanotubes doped with heteroatoms that can act as conducting support for Fe- or Co-based materials. Finding alternatives to the use of Pt is one of the main goals to promote the scalability of fuel cell technologies which have been directed to the use of SOFCs and PEMFCs.[72] For the case of SOFCs, one of the aims is to decrease their operation temperature to enable the use of materials such as stainless steel in the interconnections as well as to allow for quick starting and stopping of the fuel cell. For the case of PEMFCs, Nafion is still the main material utilized as an electrolyte due to its H^+ conductivity. Yet, their high cost and constant maintenance requirement for high relative humidity are some of the factors that increase their cost. As a means to address that, filler materials such as SPPEK, polyvinyl alcohol, and carbon-based and inorganic components are being researched to improve the durability and to ease the working conditions of bare Nafion.[73]

4.5.2 Economic Benefits

There has been increasing attention paid to the optimization and development of novel technologies related to fuel cells due to their highly desired aspects in terms of efficiency, eco-friendliness, and durability. Even though these technologies are currently facing some challenges due to the already established technologies, especially in transportation, such as combustion and electric engine vehicles, there has been considerable investment in niche markets such as the use of PEMFCs in buses for public transportation. Such investment can create feasible grounds for future investments while creating a stable source of income and profit. The continuous success of this technology can bring more investment from governmental funding agencies that can aid in the spreading of such technology in other fields. In addition, the portable electronics sector is adopting novel DMFCs that are directed at the semiconductor industry.[2] The expected performance of such technologies is expected to surpass those of lithium-ion batteries.

4.5.3 Eco-friendly Materials

One of the current goals within the fuel cell technologies sector is related to reducing the need for noble metals such as Pt, Rd, and Ru. For that, some of the metals that are less aggressive to the environment and more available are Fe-, Ni-, and Co-based materials which have demonstrated viable electrochemical properties that enable their use as electrocatalysts. In addition, carbon-based materials such as graphene, carbon nanotubes, and carbon black, among others hold great potential to be used along with transition metal materials. Furthermore, carbon-based materials can be utilized in gas diffusion layers and bipolar plates as their porosity can be controlled depending on the techniques used for their synthesis or the source of carbon. Their environmental aspects such as renewability, abundance, and ready availability make them highly desired for this type of application.

4.6 Conclusion

It is notable that fuel cell technologies are slowly progressing toward a more mature phase at which different designs and ways to convert chemical energy into electricity are being developed, as demonstrated by the different types of cells such as PEMFCs, AFCs, PAFCs, SOFCs, DMFCs, MCFCs, and so on. Some of their greatest points are related to their appreciable efficiency and power output along with their eco-friendlier credentials which can serve as a viable alternative to the current traditional combustion as well as electric engines. However, at this point, these technologies face different challenges related to durability, cost, use of expensive and scarce materials such as noble metals, constant maintenance, and high working temperatures that limit their use in large-scale applications. Despite that, some of the oldest fuel cells such as PEMFCs have found applications in public transportation which can serve as a good starting point for the dissipation of this technology. Also, continuous research into SOFCs can potentially reach lower operation temperatures that can facilitate their use in larger-scale technologies. In terms of eco-friendly aspects, fuel cells can function with nearly zero carbon emissions when utilizing H_2 as the fuel as the byproduct of its combustion is only H_2O. In this way, the only emissions generated by such fuel cells are related to the fabrication processes of their components, which is a feasible way to reduce the emission of greenhouse gases. Yet, to comply with that there is also a need to continue the development of feasible, cheap, and reproducible electrocatalysts for water splitting that can efficiently generate both H_2 and O_2 to serve as fuel in these fuel cells for the generation of electricity and heat. Investment in such a field is highly necessary considering the increasing demands on energy generation and the continuous strain on non-renewable resources.

References

1. D. Hyung Kweon, I.-Y. Jeon and J.-B. Baek, Electrochemical Catalysts for Green Hydrogen Energy, *Adv. Energy Sustain. Res.*, 2021, **2**, 2100019.
2. B. C. H. Steele and A. Heinzel, Materials for fuel-cell technologies, *Nature*, 2001, **414**, 345–352.

3. L. Carrette, K. A. Friedrich and U. Stimming, Fuel Cells: Principles, Types, Fuels, and Applications, *ChemPhysChem*, 2000, **1**, 162–193.
4. G. Couture, A. Alaaeddine, F. Boschet and B. Ameduri, Polymeric materials as anion-exchange membranes for alkaline fuel cells, *Prog. Polym. Sci.*, 2011, **36**, 1521–1557.
5. E. Han, İ Eroǧlu and L. Türker, Performance of an alkaline fuel cell with single or double layer electrodes, *Int. J. Hydrogen Energy*, 2000, **25**, 157–165.
6. G. Das, J.-H. Choi, P. K. Nguyen, D.-J. Kim and Y. S. Yoon, Anion Exchange Membranes for Fuel Cell Application: A Review, *Polymers*, 2022, **14**, 1197.
7. E. Gülzow, Alkaline Fuel Cells, *Fuel Cells*, 2004, **4**, 251–255.
8. K. Kordesch, V. Hacker, J. Gsellmann, M. Cifrain, G. Faleschini, P. Enzinger, R. Fankhauser, M. Ortner, M. Muhr and R. R. Aronson, Alkaline fuel cells applications, *J. Power Sources*, 2000, **86**, 162–165.
9. E. H. Yu, X. Wang, U. Krewer, L. Li and K. Scott, Direct oxidation alkaline fuel cells: from materials to systems, *Energy Environ. Sci.*, 2012, **5**, 5668–5680.
10. P. Stonehart and D. Wheeler, Phosphoric Acid Fuel Cells (PAFCs) for Utilities: Electrocatalyst Crystallite Design, Carbon Support, and Matrix Materials Challenges, in *Modern Aspects of Electrochemistry*, ed. B. E. Conway, C. G. Vayenas, R. E. White and M. E. Gamboa-Adelco, Springer US, Boston, MA, 2005, pp. 373–424.
11. P. Stonehart, Development of alloy electrocatalysts for phosphoric acid fuel cells (PAFC), *J. Appl. Electrochem.*, 1992, **22**, 995–1001.
12. C. Cui, S. Li, J. Gong, K. Wei, X. Hou, C. Jiang, Y. Yao and J. Ma, Review of molten carbonate-based direct carbon fuel cells, *Mater. Renew. Sustain. Energy.*, 2021, **10**, 12.
13. R. R. Contreras, J. Almarza and L. Rincón, Molten carbonate fuel cells: a technological perspective and review, *Energy Sources, Part A Recover. Util. Environ. Eff.*, 2021, **47**, 1–15.
14. J.-H. Wee, Carbon dioxide emission reduction using molten carbonate fuel cell systems, *Renew. Sustain. Energy Rev.*, 2014, **32**, 178–191.
15. S. A. Calabrese Barton, B. L. Murach, T. F. Fuller and A. C. West, A Methanol Sensor for Portable Direct Methanol Fuel Cells, *J. Electrochem. Soc.*, 1998, **145**, 3783.
16. W. Sun, G. Sun, W. Yang, S. Yang and Q. Xin, A methanol concentration sensor using twin membrane electrode assemblies for direct methanol fuel cells, *Int. J. Electrochem. Sci.*, 2006, **1**, 160–170.
17. X. Ren, T. E. Springer and S. Gottesfeld, Water and Methanol Uptakes in Nafion Membranes and Membrane Effects on Direct Methanol Cell Performance, *J. Electrochem. Soc.*, 2000, **147**, 92.
18. M. Ebrahimi, W. Kujawski, K. Fatyeyeva and J. Kujawa, A Review on Ionic Liquids-Based Membranes for Middle and High Temperature Polymer Electrolyte Membrane Fuel Cells (PEM FCs), *Int. J. Mol. Sci.*, 2021, **22**, 5430.

19. M. Irshad, K. Siraj, R. Raza, A. Ali, P. Tiwari, B. Zhu, A. Rafique, A. Ali, M. Kaleem Ullah and A. Usman, A Brief Description of High Temperature Solid Oxide Fuel Cell's Operation, Materials, Design, Fabrication Technologies and Performance, *Appl. Sci.*, 2016, **6**, 75.
20. R. Stropnik, A. Lotrič, A. Bernad Montenegro, M. Sekavčnik and M. Mori, Critical materials in PEMFC systems and a LCA analysis for the potential reduction of environmental impacts with EoL strategies, *Energy Sci. Eng.*, 2019, **7**, 2519–2539.
21. K. Kinoshita, *Electrochemical oxygen technology*, John Wiley & Sons, 1992.
22. S. Gottesfeld and T. F. Fuller, in *Proton Conducting Membrane Fuel Cells II: Proceedings of the Second International Symposium on Proton Conducting Membrane Fuel Cells II*, The Electrochemical Society, 1999.
23. T. Salameh, A. Al-Othman, A. G. Olabi, S. Issa, M. Tawalbeh and A. H. Alami, Comparative life cycle assessment for PEMFC stack including fuel storage materials in UAE, in *2020 Advances in Science and Engineering Technology International Conferences*, 2020, pp. 1–5.
24. F. Si, S. Liu, Y. Liang, X.-Z. Fu, J. Zhang and J.-L. Luo, Fuel Cell Reactors for the Clean Cogeneration of Electrical Energy and Value-Added Chemicals, *Electrochem. Energy Rev.*, 2022, **5**, 25.
25. A. J. Jacobson, Materials for Solid Oxide Fuel Cells, *Chem. Mater.*, 2010, **22**, 660–674.
26. L. Smith, T. Ibn-Mohammed, F. Yang, I. M. Reaney, D. C. Sinclair and S. C. L. Koh, Comparative environmental profile assessments of commercial and novel material structures for solid oxide fuel cells, *Appl. Energy*, 2019, **235**, 1300–1313.
27. A. Mehmeti, F. Santoni, M. Della Pietra and S. J. McPhail, Life cycle assessment of molten carbonate fuel cells: State of the art and strategies for the future, *J. Power Sources*, 2016, **308**, 97–108.
28. J. Milewski, K. Cwieka, A. Szczęśniak, Ł Szabłowski, T. Wejrzanowski, J. Skibinski, O. Dybiński, A. Lysik, A. Sienko and P. Stanger, Recycling electronic scrap to make molten carbonate fuel cell cathodes, *Int. J. Hydrogen Energy*, 2023, **48**, 11831–11843.
29. P. Ganesan, H. Colon, B. Haran, R. White and B. N. Popov, Study of cobalt-doped lithium–nickel oxides as cathodes for MCFC, *J. Power Sources*, 2002, **111**, 109–120.
30. M. H. Kim, M. Z. Hong, Y.-S. Kim, E. Park, H. Lee, H.-W. Ha and K. Kim, Cobalt and cerium coated Ni powder as a new candidate cathode material for MCFC, *Electrochim. Acta*, 2006, **51**, 6145–6151.
31. B. Huang, F. Li, Q. Yu, G. Chen, B. Zhao and K. Hu, Study of NiO cathode modified by ZnO additive for MCFC, *J. Power Sources*, 2004, **128**, 135–144.
32. T. Fukui, S. Ohara, H. Okawa, T. Hotta and M. Naito, Properties of NiO cathode coated with lithiated Co and Ni solid solution oxide for MCFCs, *J. Power Sources*, 2000, **86**, 340–346.
33. M. J. Escudero, X. R. Nóvoa, T. Rodrigo and L. Daza, Influence of lanthanum oxide as quality promoter on cathodes for MCFC, *J. Power Sources*, 2002, **106**, 196–205.

34. A. S. Tolkacheva, N. M. Porotnikova, A. S. Farlenkov, E. A. Il'ina and M. A. Konopelko, $LiAlO_2$ Prepared by Nitrates-Free Synthesis for Carbon Capture by MCFCs, *Refract. Ind. Ceram.*, 2021, **62**, 414–420.
35. M. A. Al-Saleh, S. Gültekin, A. S. Al-Zakri and H. Celiker, Effect of carbon dioxide on the performance of Ni/PTFE and Ag/PTFE electrodes in an alkaline fuel cell, *J. Appl. Electrochem.*, 1994, **24**, 575–580.
36. M. A. Al-Saleh, S. Gultekin, A. S. Al-Zakri and H. Celiker, Performance of porous nickel electrode for alkaline H_2/O_2 fuel cell, *Int. J. Hydrogen Energy*, 1994, **19**, 713–718.
37. K. V. Kordesch, Outlook for alkaline fuel cell batteries, *From Electrocatal. Fuel Cells.*, 1972, 157–164.
38. F. Xiao, Y.-C. Wang, Z.-P. Wu, G. Chen, F. Yang, S. Zhu, K. Siddharth, Z. Kong, A. Lu, J.-C. Li, C.-J. Zhong, Z.-Y. Zhou and M. Shao, Recent Advances in Electrocatalysts for Proton Exchange Membrane Fuel Cells and Alkaline Membrane Fuel Cells, *Adv. Mater.*, 2021, **33**, 2006292.
39. G. Wang, W. Li, B. Huang, L. Xiao, J. Lu and L. Zhuang, Exploring the Composition–Activity Relation of Ni–Cu Binary Alloy Electrocatalysts for Hydrogen Oxidation Reaction in Alkaline Media, *ACS Appl. Energy Mater.*, 2019, **2**, 3160–3165.
40. S. Porstmann, T. Wannemacher and W.-G. Drossel, A comprehensive comparison of state-of-the-art manufacturing methods for fuel cell bipolar plates including anticipated future industry trends, *J. Manuf. Process.*, 2020, **60**, 366–383.
41. Y. Gao, H. Peng, Y. Wang, G. Wang, L. Xiao, J. Lu and L. Zhuang, Improving the Antioxidation Capability of the Ni Catalyst by Carbon Shell Coating for Alkaline Hydrogen Oxidation Reaction, *ACS Appl. Mater. Interfaces*, 2020, **12**, 31575–31581.
42. Z. Zhuang, S. A. Giles, J. Zheng, G. R. Jenness, S. Caratzoulas, D. G. Vlachos and Y. Yan, Nickel supported on nitrogen-doped carbon nanotubes as hydrogen oxidation reaction catalyst in alkaline electrolyte, *Nat. Commun.*, 2016, 7, 10141.
43. W. Ni, A. Krammer, C.-S. Hsu, H. M. Chen, A. Schüler and X. Hu, Ni_3N as an Active Hydrogen Oxidation Reaction Catalyst in Alkaline Medium, *Angew. Chem., Int. Ed.*, 2019, **58**, 7445–7449.
44. S. Deng, X. Liu, X. Guo, T. Zhao, Y. Lu, J. Cheng, K. Chen, T. Shen, Y. Zhu and D. Wang, Insight into the hydrogen oxidation electrocatalytic performance enhancement on Ni *via* oxophilic regulation of MoO_2, *J. Energy Chem.*, 2021, **54**, 202–207.
45. F. Song, W. Li, J. Yang, G. Han, P. Liao and Y. Sun, Interfacing nickel nitride and nickel boosts both electrocatalytic hydrogen evolution and oxidation reactions, *Nat. Commun.*, 2018, **9**, 4531.
46. C. Zhao, P. Gai, R. Song, Y. Chen, J. Zhang and J.-J. Zhu, Nanostructured material-based biofuel cells: recent advances and future prospects, *Chem. Soc. Rev.*, 2017, **46**, 1545–1564.
47. G. T. R. Palmore, H. Bertschy, S. H. Bergens and G. M. Whitesides, A methanol/dioxygen biofuel cell that uses NAD+-dependent

dehydrogenases as catalysts: application of an electro-enzymatic method to regenerate nicotinamide adenine dinucleotide at low overpotentials, *J. Electroanal. Chem.*, 1998, **443**, 155–161.
48. S. Xu and S. D. Minteer, Characterizing Efficiency of Multi-Enzyme Cascade-Based Biofuel Cells by Product Analysis, *ECS Electrochem. Lett.*, 2014, **3**, H24.
49. K. Herkendell, A. Stemmer and R. Tel-Vered, Magnetically induced enzymatic cascades – advancing towards multi-fuel direct/mediated bioelectrocatalysis, *Nanoscale Adv.*, 2019, **1**, 1686–1692.
50. A. Szczupak, D. Aizik, S. Moraïs, Y. Vazana, Y. Barak, E. A. Bayer and L. Alfonta, The Electrosome: A Surface-Displayed Enzymatic Cascade in a Biofuel Cell's Anode and a High-Density Surface-Displayed Biocathodic Enzyme, *Nanomaterials*, 2017, **7**, 153.
51. J. Kim, H. Jia and P. Wang, Challenges in biocatalysis for enzyme-based biofuel cells, *Biotechnol. Adv.*, 2006, **24**, 296–308.
52. A. S. Campbell, H. Murata, S. Carmali, K. Matyjaszewski, M. F. Islam and A. J. Russell, Polymer-based protein engineering grown ferrocene-containing redox polymers improve current generation in an enzymatic biofuel cell, *Biosens. Bioelectron.*, 2016, **86**, 446–453.
53. R. C. Rodrigues, C. Ortiz, Á. Berenguer-Murcia, R. Torres and R. Fernández-Lafuente, Modifying enzyme activity and selectivity by immobilization, *Chem. Soc. Rev.*, 2013, **42**, 6290–6307.
54. G. Güven, R. Prodanovic and U. Schwaneberg, Protein Engineering – An Option for Enzymatic Biofuel Cell Design, *Electroanalysis*, 2010, **22**, 765–775.
55. T. S. Wong and U. Schwaneberg, Protein engineering in bioelectrocatalysis, *Curr. Opin. Biotechnol*, 2003, **14**, 590–596.
56. S. ul Haque, N. Duteanu, S. Ciocan and A. Nasar, Inamuddin, A review: Evolution of enzymatic biofuel cells, *J. Environ. Manage.*, 2021, **298**, 113483.
57. A. Nasar and R. Perveen, Applications of enzymatic biofuel cells in bioelectronic devices – A review, *Int. J. Hydrogen Energy*, 2019, **44**, 15287–15312.
58. A. Zebda, J.-P. Alcaraz, P. Vadgama, S. Shleev, S. D. Minteer, F. Boucher, P. Cinquin and D. K. Martin, Challenges for successful implantation of biofuel cells, *Bioelectrochemistry*, 2018, **124**, 57–72.
59. X. X. Wang, M. T. Swihart and G. Wu, Achievements, challenges and perspectives on cathode catalysts in proton exchange membrane fuel cells for transportation, *Nat. Catal.*, 2019, **2**, 578–589.
60. R. Haider, Y. Wen, Z.-F. Ma, D. P. Wilkinson, L. Zhang, X. Yuan, S. Song and J. Zhang, High temperature proton exchange membrane fuel cells: progress in advanced materials and key technologies, *Chem. Soc. Rev.*, 2021, **50**, 1138–1187.
61. Y. Cheng, S. He, S. Lu, J.-P. Veder, B. Johannessen, L. Thomsen, M. Saunders, T. Becker, R. D. e Marco, Q. Li, S. Yang and S. P. Jiang, Iron Single Atoms on Graphene as Nonprecious Metal Catalysts for

High-Temperature Polymer Electrolyte Membrane Fuel Cells, *Adv. Sci.*, 2019, **6**, 1802066.
62. C. Wan, X. Duan and Y. Huang, Molecular Design of Single-Atom Catalysts for Oxygen Reduction Reaction, *Adv. Energy Mater.*, 2020, **10**, 1903815.
63. Z. Du, H. Li and T. Gu, A state of the art review on microbial fuel cells: A promising technology for wastewater treatment and bioenergy, *Biotechnol. Adv.*, 2007, **25**, 464–482.
64. M. A. Abdelkareem, K. Elsaid, T. Wilberforce, M. Kamil, E. T. Sayed and A. Olabi, Environmental aspects of fuel cells: A review, *Sci. Total Environ.*, 2021, **752**, 141803.
65. N. F. Barilo, S. C. Weiner and C. W. James, Overview of the DOE hydrogen safety, codes and standards program part 2: Hydrogen and fuel cells: Emphasizing safety to enable commercialization, *Int. J. Hydrogen Energy*, 2017, **42**, 7625–7632.
66. R. Wurster, 9 - Hydrogen safety: An overview, in *Compendium of Hydrogen Energy*, ed. M. Ball, A. Basile and T. N. Veziroğlu, Woodhead Publ. Ser. Energy, Woodhead Publishing, Oxford, 2016, pp. 195–213.
67. Y. Chen, X. Hu and J. Liu, Life Cycle Assessment of Fuel Cell Vehicles Considering the Detailed Vehicle Components: Comparison and Scenario Analysis in China Based on Different Hydrogen Production Schemes, *Energies*, 2019, **12**, 3031.
68. Z. Yang, B. Wang and K. Jiao, Life cycle assessment of fuel cell, electric and internal combustion engine vehicles under different fuel scenarios and driving mileages in China, *Energy*, 2020, **198**, 117365.
69. S. Evangelisti, C. Tagliaferri, D. J. L. Brett and P. Lettieri, Life cycle assessment of a polymer electrolyte membrane fuel cell system for passenger vehicles, *J. Clean. Prod.*, 2017, **142**, 4339–4355.
70. M. Granovskii, I. Dincer and M. A. Rosen, Life cycle assessment of hydrogen fuel cell and gasoline vehicles, *Int. J. Hydrogen Energy*, 2006, **31**, 337–352.
71. A. Schäfer, J. B. Heywood and M. A. Weiss, Future fuel cell and internal combustion engine automobile technologies: A 25-year life cycle and fleet impact assessment, *Energy*, 2006, **31**, 2064–2087.
72. T. Ogawa, M. Takeuchi and Y. Kajikawa, Comprehensive Analysis of Trends and Emerging Technologies in All Types of Fuel Cells Based on a Computational Method, *Sustainability*, 2018, **10**, 458.
73. M. M. Tellez-Cruz, J. Escorihuela, O. Solorza-Feria and V. Compañ, Proton Exchange Membrane Fuel Cells (PEMFCs): Advances and Challenges, *Polymers*, 2021, **13**, 3064.

CHAPTER 5

Low-carbon Energy-harvesting Materials

SINAKSHI MEHRA,[a] SHIWALI THAKUR,[a] NEHA MUNJAL,[a] JYOTI VARMA,[b] JAGDEEP REDDY,[a] PHẠM THANH ĐẠNG[c] AND MUKESH KUMAR*[a]

[a] Department of Physics, Lovely Professional University, Phagwara, Punjab-144411, India, Email: nehamunjalphy@gmail.com; [b] Department of Chemistry, Lovely Professional University, Phagwara, Punjab-144411, India; [c] VNU Key Laboratory of Advanced Material for Green Growth, Faculty of Chemistry, University of Science, Vietnam National University, Vietnam
*Email: moka.esh9@gmail.com

5.1 Introduction

5.1.1 Overview of the Importance of Low-carbon Energy-harvesting Materials

In our quest for sustainable energy solutions, the significance of low-carbon energy-harvesting materials cannot be overstated. As the world faces the dual challenges of climate change and dwindling fossil fuel reserves, the urgency to transition toward renewable and low-carbon energy sources has never been more pressing. Energy harvesting, the process of capturing and converting various forms of ambient energy into usable electrical power, offers a promising avenue for sustainable energy generation. At the heart of this endeavour lie the materials used for energy harvesting, which play a pivotal role in determining efficiency, reliability, and environmental impact.[1] The importance of low-carbon energy-harvesting materials stems from their potential to address multiple sustainability objectives simultaneously.

Low Carbon Materials: Paving a Sustainable Path for a Greener Future
Edited by Santanu Patra, Sudheesh K. Shukla and Mika Sillanpää

Published by the Royal Society of Chemistry, www.rsc.org

Firstly, by minimizing carbon emissions throughout their life cycle, these materials contribute to mitigating the adverse effects of climate change. Unlike conventional energy-generation methods reliant on fossil fuels, which release greenhouse gases into the atmosphere, low-carbon energy-harvesting materials offer a cleaner alternative, helping to reduce the overall carbon footprint.

Secondly, low-carbon energy-harvesting materials align with the principles of resource efficiency and conservation. By harnessing renewable energy sources such as sunlight, heat differentials, mechanical vibrations, and ambient motion, these materials enable the utilization of readily available natural resources without depleting finite reserves. This aspect is particularly crucial in the context of achieving energy security and resilience against geopolitical uncertainties associated with traditional energy sources.

Furthermore, the adoption of low-carbon energy-harvesting materials promotes technological innovation and economic growth. As demand for sustainable energy solutions continues to rise, investment in research and development of these materials drives technological advancements and creates new opportunities for green industries. By fostering a culture of innovation and entrepreneurship, low-carbon energy-harvesting materials contribute to job creation, market diversification, and economic competitiveness in the rapidly evolving landscape of clean energy technologies.[2] Beyond environmental and economic considerations, the deployment of low-carbon energy-harvesting materials holds immense potential for addressing social challenges, particularly in underserved communities and developing regions. By providing access to decentralized and off-grid energy solutions, these materials empower marginalized populations, improve living standards, and catalyze socioeconomic development. From powering remote healthcare facilities and schools to enabling clean water and sanitation systems, the transformative impact of sustainable energy extends far beyond mere electricity generation.

In essence, the importance of low-carbon energy-harvesting materials lies in their ability to drive the transition toward a more sustainable, equitable, and resilient energy future. By harnessing the power of nature while minimizing environmental impact, these materials offer a pathway toward achieving the twin goals of combating climate change and advancing human prosperity. As we embark on this journey toward a greener and more sustainable world, the role of low-carbon energy-harvesting materials will undoubtedly remain central to our collective efforts toward building a brighter tomorrow.[3]

5.1.2 Definition and Scope of Energy Harvesting

5.1.2.1 Introduction

In the realm of sustainable energy solutions, the concept of energy harvesting stands as a beacon of innovation and promise. At its core, energy harvesting represents a paradigm shift in the way we conceive and harness energy, offering a versatile approach to capturing and converting ambient sources of power into usable electrical energy. To embark on a comprehensive exploration of low-carbon energy-harvesting materials, it is imperative to first establish a clear understanding of the definition and scope of energy harvesting.

Energy harvesting, often referred to as energy scavenging or power harvesting, encompasses a diverse array of techniques and technologies aimed at extracting energy from the surrounding environment and converting it into electrical power. Unlike traditional energy generation methods that rely on centralized power plants and finite fuel reserves, energy harvesting operates on the principle of capturing small, yet abundant, sources of ambient energy that permeate our everyday surroundings.

The scope of energy harvesting is broad and multifaceted, encompassing various forms of renewable and non-renewable energy sources. These sources include, but are not limited to:

a. Solar Energy: harnessing sunlight through photovoltaic cells or solar thermal collectors to generate electricity or heat.
b. Thermal Energy: utilizing temperature differentials or waste heat from industrial processes to produce electrical power through thermoelectric or pyroelectric conversion.
c. Mechanical Energy: capturing kinetic energy from mechanical vibrations, motion, or oscillations using piezoelectric or triboelectric materials.
d. Electromagnetic Energy: harvesting electromagnetic radiation from radio waves, microwaves, or electromagnetic fields for wireless power transmission or communication devices.
e. Chemical Energy: extracting energy from chemical reactions, such as fuel cells or microbial fuel cells, to generate electricity.[1,4]

The overarching goal of energy harvesting is to unlock the latent energy present in our environment and convert it into a usable form of power, thereby reducing reliance on traditional energy sources and minimizing environmental impact. By tapping into renewable energy sources that are inexhaustible and widely distributed, energy harvesting offers a decentralized and sustainable alternative to conventional energy generation methods.

Moreover, the scope of energy harvesting extends beyond mere electricity generation, encompassing a wide range of applications across various sectors. From powering wireless sensors and IoT devices to providing energy for wearable electronics, medical implants, and remote monitoring systems, the versatility of energy harvesting enables innovative solutions to address pressing challenges in fields such as healthcare, transportation, infrastructure, and environmental monitoring. In essence, the definition and scope of energy harvesting underscore its transformative potential as a key enabler of sustainable energy solutions. By harnessing the power of nature and leveraging advances in materials science and engineering, energy harvesting opens doors to a cleaner, greener, and more resilient energy future. As we delve deeper into the realm of low-carbon energy-harvesting materials, it is essential to keep this overarching vision in mind, guiding our exploration toward innovative solutions that pave the way toward a more sustainable world.[5,6]

5.1.3 Energy-harvesting Materials and Their Importance for Sustainability

Energy-harvesting materials have numerous potentials for powering the future. Sustainability in energy-harvesting materials is crucial for both current technology and environmental management. Population expansion, industrialization, and the development of electronic gadgets all contribute to an increase in energy demand globally. Low-carbon energy-harvesting materials have the potential to provide the required sustainability for one of the following reasons.

1. Material sourcing: harvesting materials should be easily accessible and have little effect on the environment during extraction and processing.
2. Life cycle assessment: the environmental impact of the material must be assessed over its whole life cycle, including production, use, and disposal.
3. Toxicity: the materials should be safe and shouldn't endanger human health when used.

Setting sustainable materials as a top priority can guarantee that low-carbon energy-harvesting technologies contribute to a healthier and cleaner environment.[7]

5.2 Fundamentals of Energy Harvesting

The process of obtaining ambient energy from the surroundings and turning it into useful electrical power is known as energy harvesting. Depending on certain physical phenomena and the characteristics of the materials being used, this captured energy can take on different forms during the conversion process. A block diagram of an energy-harvesting system is presented in Figure 5.1.

5.2.1 Explanation of the Ideas and Principles of Energy Harvesting

The conversion of ambient energy into electrical energy using accepted scientific and engineering principles is the basic idea behind energy harvesting. The following are the fundamental ideas.

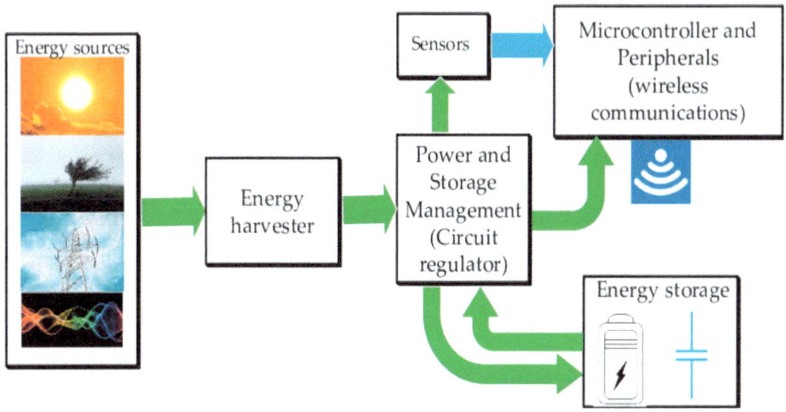

Figure 5.1 Block diagram of an energy-harvesting system. Reproduced from ref. 8, https://doi.org/10.3390/s20030647, under the terms of the CC BY 4.0 license, https://creativecommons.org/licenses/by/4.0/.

a. Energy transfer mechanisms: there are a number of ways that ambient energy can be transferred, such as:
   1. Radiation: the transfer of energy in the form of electromagnetic waves, such as sunlight and radio waves.
   2. Conduction: the direct physical contact of two objects at different temperatures that transfers heat.
   3. Convection: heat transfer caused by the motion of liquids or gases.
   4. Vibration: mechanical energy produced by an object moving quickly back and forth or up and down.

b. Transduction principles: certain principles of transduction are necessary for converting ambient energy that has been captured into electrical energy. The widely used principles are:
   1. Photovoltaic effect: light excites electrons in some materials when absorbed, such as silicon in solar cells, and this generates a potential barrier which produces an electric current.
   2. Piezoelectric effect: certain materials when subjected to mechanical stress or vibration produce electricity. Piezoelectric energy-harvesting devices use this principle to turn vibrations from machinery or footsteps into electrical power.
   3. Thermoelectric effect: a voltage differential that can drive electric current is produced when two dissimilar materials are joined by a temperature difference. The waste heat generated by industrial processes or even human body heat can be converted into energy using thermoelectric generators.
   4. Electromagnetic induction: a changing magnetic field can induce a voltage in a conductor according to Faraday's law of electromagnetic induction. To harvest energy from moving magnets or ambient radio waves, electromagnetic energy-harvesting devices use this principle.[6,9]

5.2.2 Types of Energy Sources for Harvesting

There are numerous potential sources of ambient energy within its broad range that could be used for energy-harvesting applications. A list of some typical kinds is provided in Table 5.1.[10]

5.2.3 Role of Materials in Energy Conversion

Materials play a crucial role in energy-harvesting systems because their configurations directly affect conversion efficiency. Here are some key considerations:

1. Material properties: the amount of energy converted is greatly impacted by the properties of the material, such as bandgap in solar cells or thermal conductivity in thermoelectric materials. For instance, a material used in a solar cell with a bandgap that maximizes solar absorption will produce more electricity than a material with a bandgap that is not well matched.
2. Material selection: to minimize environmental concerns with energy-harvesting solutions, it is essential to select materials that are non-toxic, sustainable, and abundant.
3. Material tailoring: to increase energy conversion efficiency, researchers are always working to create new materials or modify the characteristics of ones that already exist. Materials with nanostructures, for instance, may have improved qualities for energy-harvesting uses.

The advancement of materials science plays a pivotal role in pushing the boundaries of energy-harvesting technology. By developing new materials with

Table 5.1 Various energy sources and their applications in energy harvesting.

Energy source	Description	Examples of applications
Solar radiation	Energy from the sun in the form of electromagnetic waves	Solar cells, building-integrated photovoltaics (BIPV)
Mechanical vibrations	Kinetic energy from vibrations or movements	Piezoelectric generators in shoes or roads, industrial machinery vibration harvesting
Thermal gradients	Waste heat or temperature differences	Thermoelectric generators for industrial waste heat recovery, body heat-powered wearable electronics
Radio waves	Ambient radio waves or microwaves from radio frequencies (RF) signals	RF energy harvesting for powering low-power wireless sensors
Wind energy	Kinetic energy from moving air masses	Micro wind turbines for powering remote sensors or small devices

superior energy conversion properties, more efficient and cost-effective energy-harvesting devices can be created. Henceforth, energy harvesting offers a promising solution for a sustainable future by utilizing ambient energy through innovative materials and devices, reducing our reliance on fossil fuels.[11]

5.3 Characteristics of Low-carbon Materials

Materials with low environmental impact must be used in the transition to sustainable energy systems. In this context, low-carbon materials are crucial because they provide lower carbon footprints at every stage of their lives, from extraction and processing to use and disposal. The properties, benefits, and difficulties of low-carbon materials in energy harvesting are examined in this chapter.

5.3.1 Definition and Properties of Low-carbon Materials

Materials classified as low carbon emit few greenhouse gases (GHGs) over the course of their lifetime. This covers the stages of extraction, production, transportation, usage, and recycling or disposal.

These materials are essential for lowering the total carbon footprint of products and processes, especially those in the energy industry.[12]

5.3.1.1 Properties

1. Low Embodied Energy: this is the total energy needed to produce a material, including that of extraction, processing, and manufacturing. Materials with low carbon have substantially lower embodied energy than traditional materials. Materials like bio-based polymers and recycled aluminium, for instance, require less energy to produce than virgin aluminium or plastics derived from petroleum.
2. Renewability: renewable resources are the source of many low-carbon materials. The natural replenishment of bio-based materials, like those derived from plant fibres or biopolymers, contribute to their sustainability.
3. Recyclability and Reusability: one important characteristic that reduces the need for new raw materials and minimizes waste is the ability to recycle or reuse materials. Certain plastics and metals, such as steel and aluminium, can be recycled repeatedly without experiencing any major degradation of their properties.
4. Durability and Longevity: long-term resource consumption is decreased by durable materials because they require fewer replacements. For example, wind turbine blades made of advanced composites are intended to endure severe weather conditions over an extended period of time.

5. Non-Toxicity: the use of hazardous materials that can cause damage to the environment and people's health is frequently avoided in low-carbon materials. For instance, non-toxic flame retardants and lead-free solders are being utilized more often in construction materials and electronic equipment.
6. Energy Efficiency in Application: materials that contribute to a decrease in total energy consumption and improve the efficiency of energy-harvesting systems are regarded as low carbon. For instance, more sunlight is converted into electricity by highly efficient photovoltaic materials, which lowers the need for additional energy generation.[7,13]

5.3.2 Advantages of Using Low-carbon Materials in Energy Harvesting

5.3.2.1 Reduced Environmental Impact

Reduced environmental footprint is the main benefit of low-carbon materials. These materials reduce greenhouse gas emissions throughout their life cycle, which helps to mitigate climate change. For instance, the carbon emissions from the manufacture and disposal of photovoltaic panels can be greatly reduced by using bio-based materials.

5.3.2.2 Enhanced Sustainability of Energy Systems

Low-carbon materials support the use of renewable resources which reduce total reliance on fossil fuels and help to develop sustainable energy systems. Achieving long-term environmental and economic sustainability requires making this shift. For example, compared to conventional silicon-based cells, organic solar cells made of renewable materials may be a more sustainable option.

5.3.2.3 Improved Energy Efficiency

Energy-harvesting devices operate more efficiently using low-carbon materials. Perovskite solar cells, for instance, are an example of an advanced material that offers high efficiency at lower production costs, increasing accessibility to solar energy and decreasing the need for energy sources based on fossil fuels.

5.3.2.4 Economic Benefits

Using low-carbon materials can have positive economic effects, such as lower energy bills and the creation of green jobs. The development and production of sustainable materials foster innovation and have the potential to launch new businesses that boost the economy.[7,12]

5.3.2.5 Compliance with Regulations and Standards

The use of low-carbon materials aids businesses in adhering to increasingly stringent environmental regulations enforced by governments and international organisations. A company's reputation can be improved, legal risks can be decreased, and new markets that place a premium on sustainability can be opened up by this compliance.

5.3.2.6 Positive Social Impact

Low-carbon materials can improve living conditions, reduce exposure to harmful substances, and have a social impact. For instance, the environmental pollution and health risks are decreased when non-toxic, biodegradable materials are used in consumer products.

5.3.3 Challenges and Limitations of Low-carbon Materials

5.3.3.1 Technical Challenges

1. Performance Limitations: when compared to their conventional counterparts, low-carbon materials can have performance issues. Bio-based polymers, for example, might not always have the same mechanical strength or thermal stability as traditional plastics, which would limit their use in harsh environments.
2. Compatibility and Integration: it can be difficult to incorporate low-carbon materials into the current energy systems. Current technologies may have compatibility problems that call for major redesigns or modifications. For instance, adding new sustainable materials to photovoltaic cells might need adjustment of the system architecture and manufacturing procedures.

5.3.3.2 Economic Challenges

a. Greater Initial Costs: the cost of sustainable sourcing, cutting-edge manufacturing techniques, and constrained economies of scale can make the production of low-carbon materials costlier. For instance, producing bio-based materials may initially cost more than producing conventional materials.
b. Market Acceptance: it can be challenging to get the market to accept new materials. New materials may be reluctantly adopted by industries and consumers because of resistance to change, ignorance, or perceived risks. The use of low-carbon materials in energy-harvesting applications may be slowed down as a result.

5.3.3.3 Supply Chain Issues

1. Resource Availability: there may not be as much raw material available for low-carbon production. For example, growing some bio-based

materials could conflict with food production or use a lot of water and land, which would be problematic for sustainability.

2. Supply Chain Complexity: it can be difficult to develop a low-carbon supply chain that is sustainable. Strong tracking and certification systems are needed to make sure that all phases of the supply chain – from the extraction of raw materials to the disposal of end-of-life materials – adhere to sustainability principles.[14,15]

5.3.3.4 *Environmental and Social Impacts*

1. Land Use and Biodiversity: the production of bio-based materials may have unanticipated effects on biodiversity and land use, among other environmental issues. For instance, extensive crop cultivation for bio-based products may result in habitat loss and deforestation.
2. Social Impacts: the displacement of communities and changing job patterns are just two examples of the social effects of the switch to low-carbon materials that need to be taken into account. Ethical application of these materials requires community involvement and fair labour practices.

5.3.3.5 *Technological Barriers*

1. Research and Development: to get around the technical constraints of low-carbon materials, a significant amount of research and development is needed. This entails creating new materials with improved qualities, enhancing production techniques, and expanding output to satisfy market demands.
2. Life Cycle Analysis: to completely comprehend the environmental effects of low-carbon materials, a thorough life cycle analysis is required. To make sure that the overall advantages outweigh any drawbacks, this analysis must take into account every stage of the material's life cycle.

5.3.3.6 *Regulatory and Policy Challenges*

The adoption of low-carbon materials may face difficulties due to disparate regulatory standards in various locations. Global trade and adoption can only be facilitated by harmonizing standards and regulations. Encouraging the development and use of low-carbon materials requires strong policy support. This entails offering rewards for R&D, financing for environmentally friendly manufacturing, and laws requiring the use of low-carbon materials in specific applications.

Low-carbon materials are critical to the development of sustainable energy systems in the future because of their properties. Their non-toxicity, durability, renewability, recyclability, low embodied energy, and application efficiency place them in a critical position for lowering the carbon footprint of energy-harvesting technologies. Utilizing low-carbon materials has several benefits, including reduced adverse effects on the environment, increased sustainability, better energy efficiency, financial gains, compliance with regulations, and beneficial social effects.

However, adopting low-carbon materials is not without its difficulties. To fully realize the potential of these materials, obstacles related to technology, economics, supply chains, the environment, society, and technology must be removed. A coordinated effort from researchers, industry stakeholders, policymakers, and consumers is needed to overcome these obstacles. We can progress the incorporation of low-carbon materials into energy-harvesting systems and make a positive impact on a more sustainable future by encouraging innovation, improving market acceptance, and putting supportive policies into place.[16]

5.4 Types of Low-carbon Energy-harvesting Materials

Low-carbon materials are essential for reducing the environmental impact of energy-harvesting technologies in the search for sustainable energy solutions. With an emphasis on photovoltaic materials specifically, this section investigates a variety of low-carbon energy-harvesting materials. Developments in the field of photovoltaic (PV) materials, which are essential for converting solar energy into electricity, are propelling the creation of energy systems that are more sustainable and efficient.

5.4.1 Photovoltaic Materials

At the forefront of renewable energy technologies are materials called photovoltaics, which directly convert sunlight into electricity. An overview of photovoltaic energy conversion, an investigation of low-carbon photovoltaic material types, and an analysis of performance and efficiency considerations are all covered in this section.

5.4.1.1 Overview of Photovoltaic Energy Conversion

5.4.1.1.1 Principle of Operation. The photovoltaic effect, in which light photons are absorbed by a substance, excites electrons and produces electron–hole pairs, is the foundation of photovoltaic energy conversion (Figure 5.2). An internal electric field within the photovoltaic cell then divides these free carriers, producing an electric current flow.

The fundamental parts of a solar cell consist of:

- Absorber Layer: captures photons and generates electron–hole pairs.
- P–N Junction: creates an electric field that separates the electron–hole pairs.
- Electrodes: collect and transport the generated current.

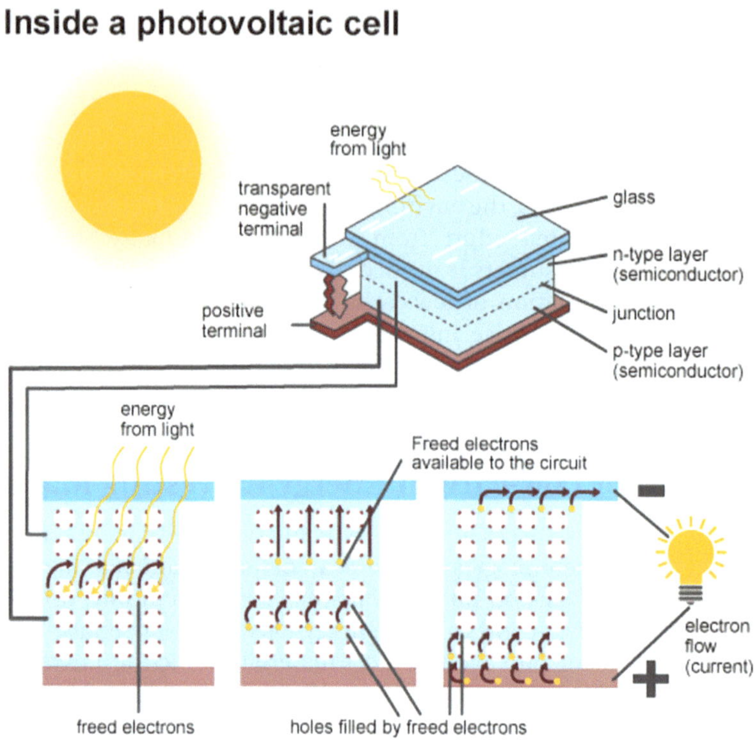

Figure 5.2 Photovoltaic energy conversion. Reproduced from ref. 17.

5.4.1.1.2 Photovoltaic Cell Structure. The typical structure of a photovoltaic cell consists of several layers:

1. Anti-reflective Coating: reduces the reflection of light to increase absorption.
2. Front Contact: transparent conductive layer that allows light to enter while conducting electricity.
3. Absorber Layer: main layer where light absorption and carrier generation occur.
4. Back Contact: conductive layer that collects the electrons.
5. Encapsulation: protects the cell from environmental factors.

5.4.1.1.3 Energy Conversion Efficiency. The efficiency of a photovoltaic cell is determined by the fraction of incident sunlight converted into electrical energy. Factors influencing efficiency include:

1. Material Properties: bandgap, absorption coefficient, and charge carrier mobility.
2. Cell Design: layer thickness, surface texturing, and junction quality.
3. External Conditions: temperature, spectrum, and light intensity.[18,19]

5.4.1.2 *Types of Low-carbon Photovoltaic Materials*

The growth of low-carbon photovoltaic materials emphasizes reducing the environmental impact while maintaining or improving efficiency. Key types include organic solar cells and perovskite solar cells.

5.4.1.2.1 Organic Solar Cells. Organic solar cells (OSCs) make use of organic molecules or polymers as the active layer for light absorption and charge transport for electricity production. Essential characteristics include:

1. Material Composition: organic semiconductors, such as conjugated polymers and small molecules, are used as active layers.
2. Flexible Substrates: OSCs can be fabricated on flexible substrates, making them suitable for lightweight and portable applications.
3. Low-Temperature Processing: manufacturing OSCs often requires lower temperatures, reducing energy consumption and carbon emissions.

5.4.1.2.1.1 Advantages of Organic Solar Cells

a. Lightweight and Flexible: ideal for applications where weight and flexibility are crucial, such as portable electronics and wearable devices.
b. Eco-friendly Manufacturing: organic materials are derived from abundantly available materials. As a result, fewer toxic chemicals are produced during the production process and there is lower energy consumption compared to traditional silicon-based cells.
c. Customizable Properties: the chemical structure of organic semiconductors can be customized to optimize light absorption, charge transport, and energy levels.

5.4.1.2.1.2 Challenges and Limitations

1. Efficiency: as compared to inorganic counterparts, OSCs typically have lower efficiencies, with current research focusing on improving power conversion efficiency.
2. Stability: the long-term performance of OSCs is still a concern because the organic materials are prone to degradation under environmental stressors like oxygen, moisture, and UV radiation.
3. Scalability: the production scaling up to industrial levels requires addressing big issues related to uniformity, reproducibility, and cost, while lab-scale production is feasible.[20]

5.4.1.2.2 Perovskite Solar Cells. Perovskite solar cells (PSCs) have emerged as a promising technology due to their high efficiency and relatively low-cost production compared to the traditional ones. Perovskites are a class of materials with a specific crystal structure, commonly represented as ABX_3, where 'A' and 'B' are cations and 'X' is an anion (Figure 5.3).

The material of the perovskite solar cell can be made of three types.

a. Hybrid Organic–Inorganic Perovskites: the most common type, with organic cation (*e.g.*, methylammonium) and inorganic components (*e.g.*, lead iodide).
b. Fully Inorganic Perovskites: use only inorganic elements, providing better stability against environmental factors.
c. Mixed Halide Perovskites: incorporate different halides (*e.g.*, bromine, iodine) to regulate the optical and electronic properties.

5.4.1.2.2.1 Advantages of Perovskite Solar Cells

1. High Efficiency: PSCs have achieved rapid improvements in efficiency, with laboratory-scale devices from 3% to 25% in the last decade.
2. Low Production Cost: solution-based processing techniques, such as inkjet printing and spin-coating, enable scalable manufacturing and cost-effectiveness.
3. Versatility: perovskites can be deposited on various substrates, including flexible and lightweight materials.

5.4.1.2.2.2 Challenges and Limitations

1. Stability: perovskite materials are sensitive to oxygen, moisture, and heat, which can lead to degradation and reduce the lifespan.

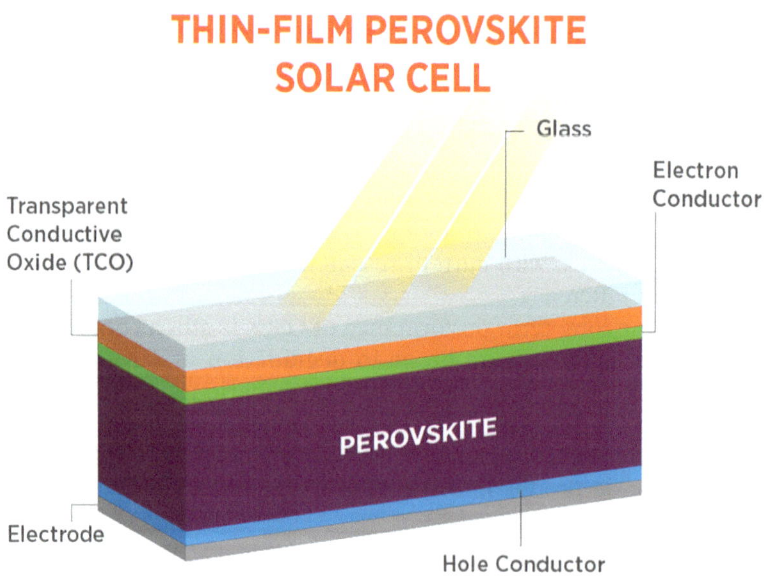

Figure 5.3 Dissecting a perovskite solar cell: exploring its layers for efficient sunlight conversion. Reproduced from ref. 21.

2. Toxicity: the utilization of lead in most perovskite compositions leads to environmental and health concerns, encouraging research into lead-free alternatives.
3. Scalability: as lab-scale efficiencies are high, attaining the similar performance in large-area devices requires addressing issues related to defect control and uniformity.[22,23]

5.4.1.3 Performance and Efficiency Considerations

The performance and efficiency of low-carbon photovoltaic materials are influenced by several factors, including device architecture, material properties, and environmental conditions.

5.4.1.3.1 Material Properties

a. Bandgap: the optimal bandgap for a single-junction solar cell is around 1.3–1.4 eV, balancing light absorption and voltage output. Materials with adjustable bandgaps, like perovskites, allow for tuning to achieve higher efficiencies.
b. Absorption Coefficient: organic materials and perovskites typically exhibit high absorption coefficients, which enables thin-film devices, reducing material usage and costs.
c. Carrier Mobility: high carrier mobility reduces recombination losses, improving charge transport and overall efficiency. While inorganic materials generally have higher mobilities, advances in organic and perovskite materials are closing the gap.

5.4.1.3.2 Device Architecture

a. Single-Junction *vs.* Multi-Junction: multi-junction devices, which stack multiple layers with different bandgaps, can achieve higher efficiencies by capturing a broader spectrum of sunlight. However, they are more complex and costly to produce.
b. Tandem Cells: different photovoltaic materials combined in a tandem configuration to improve efficiency. For example, pairing a perovskite top cell with a silicon bottom cell leverages the strengths of both materials.
c. Surface Engineering: techniques like anti-reflective coatings and surface texturing reduce reflection losses and enhance light absorption.

5.4.1.3.3 Environmental Conditions. High temperatures lead to reduced efficiency of photovoltaic cells due to increased recombination rates and reduced voltage output. Materials with better thermal stability, like inorganic perovskites, perform better in hot climates. Additionally, the outcome of photovoltaic cells varies with light intensity and spectrum. Low light conditions, such as indoor lighting or cloudy days, require materials with good low-light performance.

5.4.1.3.4 Research and Development. Ongoing research aims to address the challenges and enhance the performance of low-carbon photovoltaic materials. Key areas of focus include:

1. Stability Improvements: developing materials and encapsulation techniques to enhance the stability of organic and perovskite solar cells.
2. Efficiency Enhancements: exploring new material compositions, device architectures, and processing techniques to improve efficiency.
3. Scalability and Manufacturing: advancing scalable manufacturing methods and ensuring consistency in large-area production.

5.4.1.3.5 Environmental and Health Safety. It is crucial to find non-toxic alternatives to hazardous materials, such as lead in perovskites, to ensure safety and compliance with environmental regulations. Henceforth, the exploration of low-carbon photovoltaic materials reveals significant potential for advancing sustainable energy technologies. Organic solar cells and perovskite solar cells offer promising pathways to high efficiency and low environmental impact. Despite facing challenges, including stability, scalability, and toxicity concerns, on-going research and development is paving the way for more efficient, durable, and environmentally friendly photovoltaic solutions. The integration of these low-carbon materials into energy-harvesting systems is a critical step toward reducing the global carbon footprint and transitioning to a sustainable energy future. By addressing the technical and economic challenges, enhancing material properties, and leveraging innovative device architectures, the potential of low-carbon photovoltaic materials can be fully realized, contributing to a cleaner and more sustainable world.[24,25]

5.4.2 Thermoelectric Materials

In the broader context of sustainable energy solutions, low-carbon materials are pivotal for minimizing environmental impacts while enhancing energy conversion efficiencies. This section focuses on thermoelectric materials as a significant category of low-carbon energy-harvesting materials. Thermoelectric materials convert thermal energy directly into electrical energy, making them an essential component of various renewable energy and waste heat recovery applications. An in-depth exploration of thermoelectric energy conversion principles, different types of low-carbon thermoelectric materials, and the optimization of thermal conductivity and power factor is given as below.

5.4.2.1 Principles of Thermoelectric Energy Conversion

Thermoelectric materials offer a unique solution for energy conversion by exploiting temperature gradients to generate electrical power.

5.4.2.1.1 Fundamental Concepts. Thermoelectric energy conversion is based on the Seebeck effect, Peltier effect, and Thomson effect, which collectively describe how temperature differences can be converted into electric voltage and *vice versa*.

a. Seebeck Effect: when a temperature gradient is applied across a thermoelectric material, it generates a voltage difference. This phenomenon is quantified by the Seebeck coefficient (S), which represents the voltage produced per unit temperature difference. The process of a thermoelectric generator, using the Seebeck effect is represented in Figure 5.4.
b. Peltier Effect: when an electric current flows through a thermoelectric material, it absorbs or releases heat at the junctions, depending on the direction of the current. The Peltier coefficient (Π) quantifies this heat transfer.
c. Thomson Effect: when an electric current flows through a conductor with a temperature gradient, it generates or absorbs heat along the length of the conductor. This effect is characterized by the Thomson coefficient (μ).[27]

5.4.2.1.2 Thermoelectric Figure of Merit (*ZT*). The efficiency of thermoelectric materials is often evaluated using the dimensionless figure of merit, *ZT*, defined as:

$$ZT = \frac{S^2 \sigma T}{K}$$

where S is the Seebeck coefficient, σ is the electrical conductivity, T is the absolute temperature, κ is the thermal conductivity.

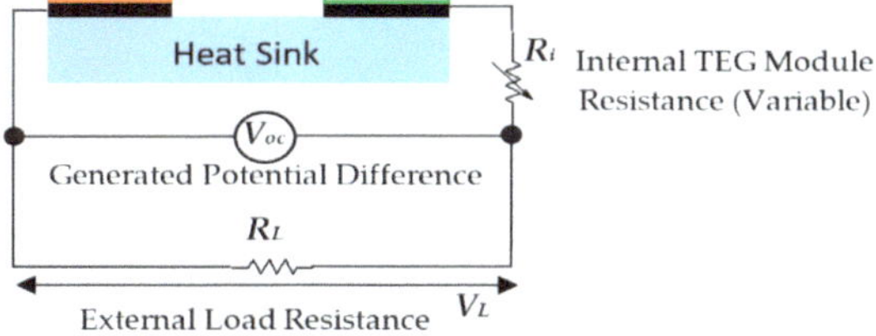

Figure 5.4 Diagram of a thermoelectric generator: converting heat into electricity using the Seebeck effect. Reproduced from ref. 26, https://doi.org/10.3390/en11123315, under the terms of the CC BY 4.0 license, https://creativecommons.org/licenses/by/4.0/.

A high *ZT* value indicates better thermoelectric performance, achieved by maximizing the Seebeck coefficient and electrical conductivity while minimizing thermal conductivity.[28]

5.4.2.1.3 Thermoelectric Modules. A thermoelectric module typically consists of an array of thermocouples connected electrically in series and thermally in parallel. These modules can be used for power generation (Seebeck effect) or cooling (Peltier effect) as represented in Figure 5.5.

Key components of a thermoelectric module include:

a. N-type and P-type Semiconductors: materials that have the majority of electrons (n-type) or holes (p-type) as charge carriers.
b. Electrical Interconnects: these are the conductive materials which connect the semiconductor elements to form the thermocouples.
c. Substrates and Insulation: these are the materials that provide structural support and electrical insulation.[30]

5.4.2.2 *Low-carbon Thermoelectric Materials*

The development of low-carbon thermoelectric materials maintaining high performance focuses on minimizing the environmental impact. Key categories include organic thermoelectric materials and nanocomposites.

5.4.2.2.1 Organic Thermoelectric Materials. Organic thermoelectric materials, composed of conductive polymers and organic molecules, offer several advantages due to their flexibility, low cost, and potential for low-temperature processing.

a. Conductive Polymers: polymers like poly(3,4-ethylenedioxythiophene) (PEDOT), poly(3-hexylthiophene) (P3HT), and polyaniline (PANI) are

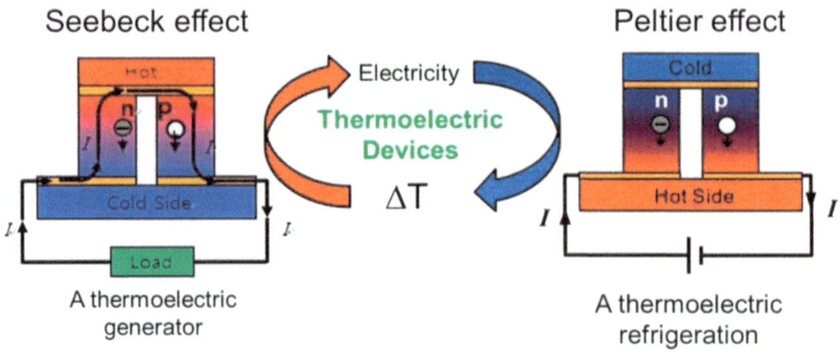

Figure 5.5 Thermoelectric module. Reproduced from ref. 29, https://doi.org/10.3389/fchem.2021.762896, under the terms of the CC BY 4.0 license, https://creativecommons.org/licenses/by/4.0/.

widely studied for their thermoelectric properties. These materials are flexible, lightweight, and can be processed using solution-based techniques.

b. Molecular Design: the thermoelectric performance of organic materials can be customized by modifying the molecular structure, such as introducing dopant side chains or enhancing electrical conductivity and Seebeck coefficient.
c. Composite Materials: combining organic polymers with inorganic nanomaterials (*e.g.*, carbon nanotubes, graphene) can improve thermoelectric performance by enhancing electrical conductivity and optimizing the Seebeck coefficient.[3]

5.4.2.2.2 Advantages of Organic Thermoelectric Materials

1. Low Environmental Impact: organic materials are derived from abundantly present elements and can be synthesized using environmentally friendly processes, reducing the carbon footprint.
2. Flexibility and Processability: these materials can be fabricated into flexible devices, making them suitable for wearable electronics and other applications requiring conformability.
3. Cost-effective Production: the solution-based processing techniques, such as printing and coating, enable low-cost manufacturing of large-area thermoelectric devices.

5.4.2.2.3 Challenges and Limitations

1. Efficiency: as compared to inorganic materials, organic materials generally exhibit lower thermoelectric efficiency (ZT values), which limits their applicability in high-performance devices.
2. Stability: the long-term performance of organic thermoelectric materials is affected because organic materials degrade under environmental stressors like heat, moisture, and UV radiation.
3. Scalability: as lab-scale efficiencies are high, attaining similar performance in large-area devices requires addressing issues related to defect control and uniformity.

5.4.2.2.4 Nanocomposites. Nanocomposites, which combine nanoscale materials with bulk materials, offer a promising approach to enhancing thermoelectric performance. These materials leverage the unique properties of nanostructures to achieve high ZT values.

a. Quantum Dots and Nanowires: incorporating nanowires or quantum dots into a matrix material can enhance the Seebeck coefficient and reduce thermal conductivity due to quantum confinement effects and phonon scattering.
b. Layered Structures: designing materials with layered structures, such as superlattices and heterostructures, can improve the thermoelectric performance by optimizing charge carrier mobility and reducing thermal conductivity.

c. Hybrid Nanocomposites: combining organic and inorganic nanomaterials can create hybrid composites that benefit from the advantages of both material types, achieving high electrical conductivity and low thermal conductivity.[31]

5.4.2.2.5 Advantages of Nanocomposites

a. High Performance: nanocomposites can achieve high *ZT* values due to the synergistic effects of nanostructuring, which enhances electrical properties and reduces thermal conductivity.
b. Versatility: these materials can be engineered for specific applications by adjusting the composition, size, and distribution of nanostructures.
c. Scalability: advancements in nanofabrication techniques, such as sol–gel processes and chemical vapour deposition, enable the scalable production of nanocomposites.

5.4.2.2.6 Challenges and Limitations

1. Complexity: the fabrication of nanocomposites involves complex processes and precise control over nanostructure formation, increasing technical challenges and production costs.
2. Stability: nanostructures can degrade over time. Hence, it is important to ensure the long-term stability of nanocomposites under operational conditions.
3. Integration: integrating nanocomposites into practical devices requires addressing issues related to compatibility with existing technologies and manufacturing processes.[32]

5.4.2.3 Thermal Conductivity and Power Factor Optimization

Optimizing thermal conductivity and power factor is critical for enhancing the efficiency of thermoelectric materials. This section discusses strategies for achieving these optimizations.

5.4.2.3.1 Thermal Conductivity. Reducing thermal conductivity while maintaining high electrical conductivity is essential for improving the *ZT* value. Strategies include:

a. Phonon Scattering: introducing defects, alloying, and creating nanostructures can scatter phonons, reducing thermal conductivity. Nanocomposites and layered materials are particularly effective in this regard.
b. Superlattices and Heterostructures: designing materials with superlattice or heterostructure architectures can create interfaces that scatter phonons and reduce thermal conductivity without significantly affecting electrical conductivity.
c. Alloying and Doping: incorporating alloying dopants or elements can disrupt the lattice structure, increasing phonon scattering and reducing thermal conductivity.[33]

5.4.2.3.2 Power Factor. The power factor ($S^2\sigma$) combines the Seebeck coefficient and electrical conductivity, and optimizing both parameters is crucial for enhancing thermoelectric performance. Strategies include:

a. Carrier Concentration Optimization: adjusting the carrier concentration through doping or alloying can optimize the Seebeck coefficient and electrical conductivity. High carrier concentration improves electrical conductivity but can reduce the Seebeck coefficient, so a balance must be struck.
b. Band Structure Engineering: modifying the electronic band structure of materials can enhance the Seebeck coefficient and electrical conductivity. Techniques such as bandgap engineering, using quantum wells, and creating resonance levels are employed to optimize the band structure.
c. Composite Design: combining materials with complementary properties can create composites with optimized power factors. For example, hybrid composites of organic and inorganic materials can achieve high Seebeck coefficients and electrical conductivity.[34]

5.4.2.3.3 Applications

1. Organic Thermoelectric Devices: applications include wearable electronics, flexible sensors, and low-power energy harvesters. Research focuses on improving the stability and efficiency of organic materials for practical use.
2. Nanocomposite Thermoelectric Generators: used in waste heat recovery systems, industrial processes, and automotive applications. Advances in nanocomposite design aim to enhance efficiency and durability under operational conditions.
3. High-performance Thermoelectric Modules: developing high-*ZT* materials for power generation and cooling applications. Innovations in material design, such as superlattices and heterostructures, drive improvements in performance.

5.4.2.3.4 Future Directions. The field of low-carbon thermoelectric materials continues to evolve, with ongoing research aimed at overcoming current limitations and enhancing performance. Future directions include:

1. Advanced Material Design: exploring new material compositions, nanostructures, and fabrication techniques to achieve higher *ZT* values and better stability.
2. Scalable Manufacturing: developing cost-effective and scalable manufacturing methods for large-area production of thermoelectric devices.
3. Environmental and Health Safety: ensuring the sustainability and safety of thermoelectric materials by reducing the use of toxic elements and optimizing recycling processes.

4. Integrated Systems: integrating thermoelectric materials into broader energy systems, such as smart grids and hybrid energy harvesters, to maximize their impact on energy efficiency and sustainability.

Therefore, thermoelectric materials represent a promising avenue for low-carbon energy harvesting, offering the potential to convert waste heat into usable electricity and contribute to sustainable energy solutions. Organic thermoelectric materials and nanocomposites provide versatile and environmentally friendly options for various applications. However, challenges related to efficiency, stability, and scalability must be addressed to fully realize their potential. By optimizing thermal conductivity and power factor, leveraging advanced material design, and integrating these materials into practical devices, the field of thermoelectric materials can make significant strides toward a sustainable energy future. Ongoing research and innovation will be crucial in overcoming current limitations and unlocking new opportunities for low-carbon energy-harvesting technologies.[32,35]

5.4.3 Piezoelectric Materials

Piezoelectric materials have gained significant attention in the realm of energy harvesting due to their ability to convert mechanical energy into electrical energy. This section explores the fundamental principles of piezoelectricity, highlights low-carbon piezoelectric materials, and discusses their applications in energy harvesting from mechanical vibrations.

5.4.3.1 Introduction to Piezoelectricity

Piezoelectricity is the ability of certain materials to generate an electric charge in response to applied mechanical stress. This phenomenon is intrinsic to materials with non-centrosymmetric crystal structures, where the positive and negative charge centres do not coincide. The mechanism of both effects is represented in Figure 5.6.

a. Direct Piezoelectric Effect: when mechanical stress is applied to a piezoelectric material, it results in the generation of an electric charge. This effect is the basis for energy-harvesting applications.
b. Inverse Piezoelectric Effect: when an electric field is applied to a piezoelectric material, it induces mechanical deformation. This property is utilized in actuators and sensors.[36,37]

5.4.3.1.1 Piezoelectric Coefficients. The efficiency of the piezoelectric effect is quantified by piezoelectric coefficients, which describe the relationship between mechanical stress and electric displacement (direct effect) or electric field and strain (inverse effect). These coefficients vary depending on the material and the direction of applied stress.

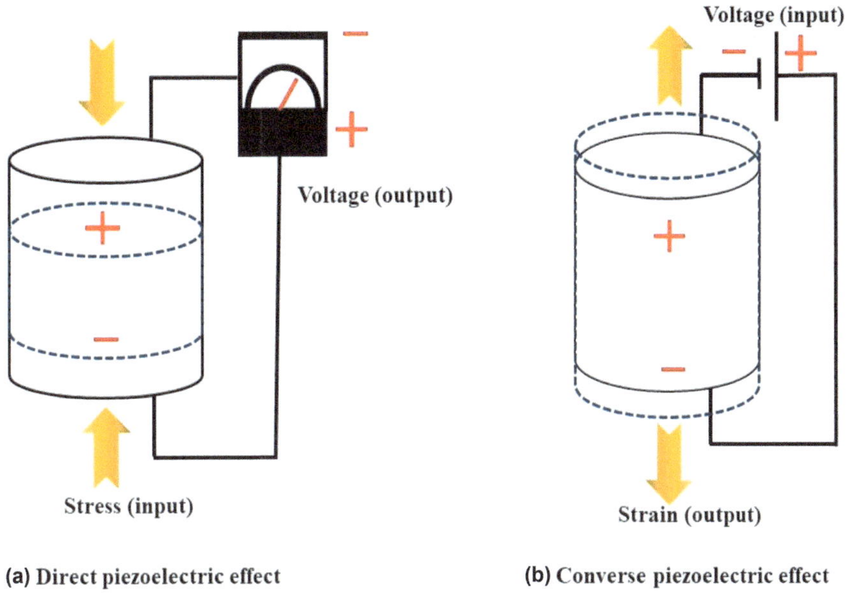

Figure 5.6 Diagrammatic representation of the direct and inverse piezoelectric effects. Reproduced from ref. 38 with permission of Elsevier, Copyright 2024.

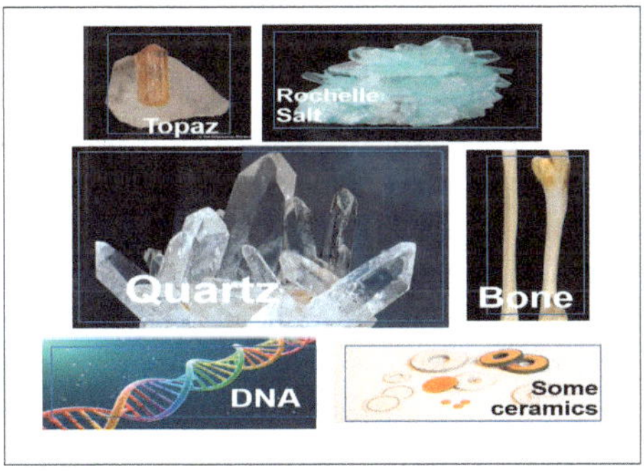

Figure 5.7 Examples of piezoelectric materials.

5.4.3.1.2 Piezoelectric Materials. Common piezoelectric materials are explained below and also represented in Figure 5.7:

1. Crystals: quartz, Rochelle salt, and tourmaline are natural piezoelectric crystals. Synthetic crystals like lithium niobate and lithium tantalate are also widely used.
2. Ceramics: lead zirconate titanate (PZT) and barium titanate are the most used piezoelectric ceramics due to their high piezoelectric coefficients.

3. Polymers: polyvinylidene fluoride (PVDF) and its copolymers are notable piezoelectric polymers, valued for their flexibility and ease of processing.[39,40]

5.4.3.2 Low-carbon Piezoelectric Materials

The development of low-carbon piezoelectric materials focuses on reducing the environmental impact while maintaining effective energy conversion capabilities. This includes exploring biodegradable polymers, natural materials, and innovative composites.

5.4.3.2.1 Biodegradable Polymers. Biodegradable polymers are increasingly being studied for their piezoelectric properties. These materials decompose naturally in the environment, reducing long-term waste and environmental impact. Examples of such polymers include polylactic acid (PLA) and polyhydroxyalkanoates (PHAs). PLA is a biodegradable polymer derived from renewable resources like corn starch. It exhibits piezoelectric properties and is used in various sustainable energy-harvesting applications. PHAs are fully biodegradable because they are produced by microorganisms. They show promising piezoelectric properties, making them suitable for eco-friendly energy-harvesting devices.

5.4.3.2.2 Natural Materials. Natural materials such as bamboo and cellulose-based composites offer sustainable alternatives to traditional piezoelectric materials. Some of the examples are bamboo and cellulose-based composites. With intrinsic piezoelectric properties, bamboo is a renewable resource. Its natural abundance and rapid growth make it an attractive material for low-carbon piezoelectric applications. Cellulose, derived from plant cell walls, is another renewable resource with piezoelectric potential. Composites incorporating cellulose nanofibers show enhanced piezoelectric performance and sustainability.[41]

5.4.3.2.3 Innovative Composites. Combining natural materials with traditional piezoelectric substances can create composites that leverage the best properties of both.

1. Hybrid Composites: the performance can be enhanced while maintaining a lower environmental footprint by mixing biodegradable polymers with inorganic piezoelectric materials, such as PZT or barium titanate.
2. Nanostructured Composites: incorporating piezoelectric nanoparticles into biodegradable or natural matrices can improve energy conversion efficiency and mechanical properties.[37,42]

5.4.3.3 Applications in Energy Harvesting from Mechanical Vibrations

Piezoelectric materials are well-suited for harvesting energy from mechanical vibrations in various environments. Key applications include:

5.4.3.3.1 Wearable Electronics. Piezoelectric materials can be integrated into wearable devices to harvest energy from human motion.

a. Smart Textiles: embedding piezoelectric films or fibres into clothing can generate power from movements like walking or running.
b. Portable Health Monitors: piezoelectric materials in wearable health monitors can harvest energy from bodily movements, reducing the need for battery replacements.

5.4.3.3.2 Structural Health Monitoring. Piezoelectric materials can be used in structural health-monitoring systems to convert mechanical vibrations from bridges, buildings, or other infrastructure into electrical signals for data collection and analysis.

a. Sensors: piezoelectric sensors embedded in structures can detect stress, strain, and vibrations, providing real-time data for maintenance and safety.
b. Energy Harvesters: these materials can also serve as self-powered sensors, converting ambient vibrations into electrical energy to power monitoring devices.

5.4.3.3.3 Industrial Applications. In industrial settings, from machinery vibrations to power sensors and small devices, piezoelectric materials can capture energy.

a. Machine Health Monitoring: piezoelectric sensors can monitor the health of industrial machines by detecting vibrations and converting them into electrical signals.
b. Waste Energy Harvesting: capturing vibrational energy from industrial processes can help reduce energy waste and improve overall efficiency.

5.4.3.3.4 Automotive Applications. Piezoelectric materials can be used in vehicles to harvest energy from engine vibrations and road-induced movements.

a. Tire Pressure-Monitoring Systems (TPMS): integrating piezoelectric sensors in TPMS can generate power from tire vibrations, eliminating the need for battery replacements.
b. Energy Harvesting from Road Vibrations: embedding piezoelectric materials in car suspensions or road surfaces can capture energy from vehicle movements, potentially powering auxiliary systems.[43–45]

5.4.4 Triboelectric Materials

Triboelectric materials generate electrical charge through contact and separation with another material, a phenomenon known as triboelectricity. This section covers the basics of triboelectricity, highlights low-carbon triboelectric materials, and discusses strategies for improving triboelectric performance.[46]

5.4.4.1 Basics of Triboelectricity

Triboelectricity arises from the transfer of electrons between two materials with differing electron affinities when they come into contact and then separate.

a. Triboelectric Series: materials are ranked in the triboelectric series based on their tendency to gain or lose electrons. Materials at opposite ends of the series generate the highest charges when brought into contact.
b. Contact Electrification: when two materials come in contact, electrons transfer from one material to the other, which leads to charge accumulation. A voltage difference is created upon separation, due to the electrostatic charges.
c. Triboelectric Nanogenerators (TENGs): devices that convert mechanical energy into electrical energy using the triboelectric effect are known as triboelectric nanogenerators. They consist of two dissimilar materials that repeatedly contact and separate, generating electrical power.[47–49]

5.4.4.2 Low-carbon Triboelectric Materials

Developing low-carbon triboelectric materials focuses on using sustainable, biodegradable, and natural materials to minimize environmental impact.

5.4.4.2.1 Cellulose-based Materials. Cellulose, a natural polymer found in plant cell walls, is a promising material for triboelectric applications due to its abundance, biodegradability, and suitable triboelectric properties.

a. Nanocellulose: nanocellulose, derived from cellulose, can be used to create high-performance triboelectric materials and exhibits excellent mechanical properties. It can be processed into films, fibres, and composites.
b. Cellulose Derivatives: derivatives like cellulose acetate and carboxymethyl cellulose can enhance triboelectric performance through chemical modification, making them suitable for various applications.

5.4.4.2.2 Natural Fibres. Natural fibres derived from plants and animals offer eco-friendly alternatives for triboelectric materials. Cotton and silk have intrinsic triboelectric properties and can be used in TENGs to harvest energy from human motion or environmental vibrations. Wool and hemp fibres also exhibit good triboelectric performance and are sustainable options for creating environmentally friendly energy-harvesting devices.[50]

5.4.4.2.3 Biodegradable Polymers. Biodegradable polymers, like those used in low-carbon piezoelectric materials, can also serve as triboelectric materials. Some examples are polylactic acid (PLA) and polycaprolactone (PCL). PLA can generate triboelectric charges when paired with materials of differing electron affinities. It is used in various applications due to its biodegradability and ease of processing. PCL is another biodegradable polymer with suitable triboelectric properties. It can be processed into films and composites for energy-harvesting applications.[51–53]

5.4.4.3 Strategies for Improving Triboelectric Performance

Optimizing the performance of triboelectric materials involves enhancing their charge generation and retention capabilities. Strategies include:

5.4.4.3.1 Surface Modification. Their charge transfer efficiency can be enhanced by modifying the surface properties of triboelectric materials.

a. Surface Roughness: the contact area and charge transfer during contact and separation can be enhanced by increasing surface roughness through physical or chemical treatments.
b. Chemical Functionalization: the charge generation can be improved by introducing functional groups or coatings that have high electron affinity differences. For example, functionalizing cellulose with amine or carboxyl groups can enhance its triboelectric performance.

5.4.4.3.2 Material Composition. Combining materials with complementary properties can create composites with enhanced triboelectric performance.

a. Hybrid Composites: the charge transfer and mechanical properties can be improved by mixing natural fibres with conductive materials, such as graphene or carbon nanotubes.
b. Multilayer Structures: the charge generation and retention can be maximized by designing multilayer TENGs with alternating layers of high and low electron affinity materials.[54–56]

5.4.4.3.3 Device Design. Optimizing the design of triboelectric devices can improve their energy-harvesting efficiency.

a. Structural Engineering: creating micro- or nanostructures on the surfaces of triboelectric materials can increase the contact area and enhance charge transfer.
b. Electrode Configuration: using conductive electrodes with optimized shapes and placements can improve the efficiency of charge collection and transfer in TENGs.

5.4.4.3.4 Environmental Adaptation. Adapting triboelectric materials to specific environmental conditions can enhance their performance in real-world applications.

1. Humidity Control: the stability and efficiency of triboelectric materials can be improved by managing humidity levels, as excessive moisture can reduce charge transfer efficiency.
2. Temperature Adaptation: developing materials and devices that can operate efficiently across a range of temperatures ensures reliable performance in various environments.[57–59]

5.5 Applications of Low-carbon Energy-harvesting Materials

Low-carbon energy-harvesting materials have diverse applications across various fields. This section explores their use in wearable electronics, Internet of Things (IoT) devices, green building technologies, and sustainable energy solutions for developing regions.

5.5.1 Wearable Electronics

Wearable electronics benefit greatly from low-carbon energy-harvesting materials, as they provide a sustainable power source for various devices.

5.5.1.1 Smart Textiles

Smart clothing that generates power from human motion can be created in textiles by integrating piezoelectric, triboelectric, or thermoelectric materials.

a. Health Monitoring: wearable devices with embedded sensors can monitor vital signs, track physical activity, and detect health anomalies, all powered by energy harvested from movements.
b. Personal Electronics: devices like fitness trackers, smartwatches, and wireless earphones can benefit from self-powered systems, reducing the need for frequent recharging.

5.5.1.2 Flexible and Stretchable Electronics

Low-carbon energy-harvesting materials enable the development of flexible and stretchable electronics that conform to the body and generate power from mechanical deformation.

a. Skin Patches: flexible piezoelectric or triboelectric patches can adhere to the skin and harvest energy from movements, providing power for medical sensors and transdermal drug delivery systems.
b. Wearable Sensors: the environmental conditions, posture, and body movements, generating power from everyday activities, can be monitored by embedding stretchable sensors in clothing.[60]

5.5.2 Internet of Things (IoT) Devices

IoT devices, which rely on constant connectivity and data transmission, can benefit from sustainable energy-harvesting technologies to power sensors and communication modules.

5.5.2.1 Environmental Monitoring

IoT devices equipped with energy-harvesting materials can monitor environmental parameters like temperature, humidity, air quality, and noise levels.

a. Remote Sensors: in remote or hard-to-reach locations, deploying self-powered sensors can provide continuous monitoring without the need for battery replacements.
b. Smart Agriculture: IoT devices in agriculture can monitor crop health, soil moisture, and weather conditions, powered by energy harvested from environmental vibrations or solar energy.

5.5.2.2 Smart Homes and Cities

Energy-harvesting materials can power smart home and city applications, enhancing efficiency and reducing energy consumption.

a. Smart Lighting: in streetlights and indoor lighting, integrating energy-harvesting systems can reduce dependence on grid power by capturing ambient energy from vibrations and light.
b. Building Automation: self-powered sensors can monitor building conditions, such as temperature, occupancy, and energy usage, optimizing heating, cooling, and lighting systems.[61]

5.5.3 Green Building Technologies

Green building technologies aim to reduce the environmental impact of construction and operation. Energy-harvesting materials play a crucial role in achieving this goal.

5.5.3.1 Building-integrated Photovoltaics (BIPV)

Low-carbon photovoltaic materials, such as organic solar cells and perovskite solar cells, can be integrated into building facades, windows, and roofs to generate renewable energy.

a. Solar Facades: BIPV systems on building exteriors can generate electricity while serving as building envelopes, reducing energy consumption and enhancing aesthetics.
b. Transparent Solar Windows: transparent photovoltaic materials can be used in windows to generate power from sunlight while allowing natural light to enter buildings.

5.5.3.2 Energy-harvesting Floors

Piezoelectric and triboelectric materials embedded in flooring systems can capture energy from foot traffic and mechanical vibrations.

a. Smart Floors: floors in high-traffic areas like shopping malls, airports, and office buildings can generate power from the movements of people, contributing to the building's energy needs.
b. Vibration Harvesters: floors in industrial settings can capture energy from machinery vibrations, reducing waste energy and providing additional power for facility operations.

5.5.3.3 Thermoelectric Building Systems

Thermoelectric materials can be integrated into building systems to capture and convert waste heat into electrical energy.

a. HVAC Systems: thermoelectric generators can be incorporated into heating, ventilation, and air conditioning (HVAC) systems to capture waste heat and improve overall energy efficiency.
b. Building Envelope: thermoelectric materials integrated into walls and roofs can convert temperature gradients between the interior and exterior into usable power.[62,63]

5.5.4 Sustainable Energy Solutions for Developing Regions

In developing regions, access to reliable and affordable energy is often limited. Low-carbon energy-harvesting materials offer sustainable solutions to address these challenges.

5.5.4.1 Off-grid Power Systems

In remote and rural areas, energy-harvesting materials can provide off-grid power solutions, enhancing energy access and supporting development.

a. Solar-powered Systems: to provide electricity for lighting, communication, and small appliances in off-grid communities, low-carbon photovoltaic materials can be used in solar panels.
b. Thermoelectric Generators: thermoelectric systems can convert waste heat from cooking stoves or industrial processes into electricity, providing a sustainable power source.[64]

5.5.4.2 Portable Energy Harvesters

Portable energy-harvesting devices can offer reliable power for mobile phones, medical devices, and other essential electronics in regions with limited infrastructure.

a. Wearable Chargers: wearable devices with integrated energy harvesters can provide power on the go, reducing reliance on grid electricity and enhancing mobility.
b. Portable Solar Chargers: in remote locations, lightweight, flexible solar panels using low-carbon photovoltaic materials can be easily transported and deployed to provide power.

5.5.4.3 Sustainable Agriculture

Energy-harvesting materials can support sustainable agriculture by providing power for irrigation systems, sensors, and other agricultural technologies.

a. Solar-powered Irrigation: photovoltaic systems can power irrigation pumps, reducing the need for fossil fuels and improving water management in farming.
b. Self-powered Sensors: sensors with integrated energy harvesters can monitor soil conditions, crop health, and weather, optimizing agricultural practices and improving yields.[65]

5.6 Challenges and Future Directions

Despite the potential of low-carbon energy-harvesting materials, several challenges need to be addressed to achieve their widespread adoption and commercialization. This section explores scalability and commercialization challenges, integration with existing energy infrastructure, and emerging trends in the field.

5.6.1 Scalability and Commercialization Challenges

Scaling up the production and commercialization of low-carbon energy-harvesting materials involves overcoming technical, economic, and regulatory barriers.

5.6.1.1 Technical Challenges

Enhancing the performance and durability of low-carbon materials to meet industrial standards is crucial for their adoption in practical applications. For commercialization, the development of scalable and cost-effective manufacturing processes for producing high-quality materials is essential.

5.6.1.2 Economic Barriers

1. Cost Competitiveness: to gain market acceptance, low-carbon materials must be cost-competitive with conventional materials. Reducing production costs through economies of scale and technological advancements is necessary.

2. Market Demand: creating market demand for products incorporating low-carbon energy-harvesting materials requires raising awareness about their benefits and potential applications.[66]

5.6.1.3 Regulatory and Standardization Issues

1. Regulatory Approval: particularly in highly regulated industries like healthcare and construction, obtaining regulatory approvals for new materials and technologies can be time-consuming and complex.
2. Standardization: for ensuring quality, reliability, and interoperability, establishing industry standards for low-carbon energy-harvesting materials and devices is essential.

5.6.2 Integration with Existing Energy Infrastructure

Integrating low-carbon energy-harvesting materials with existing energy infrastructure requires addressing compatibility, reliability, and grid integration issues.

5.6.2.1 Compatibility with Existing Systems

a. Retrofit Solutions: developing retrofit solutions that allow existing infrastructure to incorporate energy-harvesting materials without significant modifications can facilitate adoption.
b. Hybrid Systems: to enhance reliability and ensure a smooth transition to sustainable energy solutions, hybrid systems that combine conventional and renewable energy need to be developed.

5.6.2.2 Reliability and Maintenance

a. Long-term Performance: ensuring the long-term performance and durability of energy-harvesting materials is critical for their integration into infrastructure with long lifespans.
b. Maintenance Requirements: minimizing maintenance requirements through robust material design and self-healing technologies can reduce operational costs and improve reliability.

5.6.2.3 Grid Integration

1. Energy Storage: integrating energy-harvesting systems with energy storage solutions, such as batteries and supercapacitors, can ensure a stable and continuous power supply.
2. Smart Grids: leveraging smart grid technologies to manage and distribute power generated from energy-harvesting systems can enhance efficiency and reliability.[63,67]

5.6.3 Emerging Trends and Future Prospects

The field of low-carbon energy-harvesting materials is rapidly evolving, with emerging trends and future prospects pointing toward continued advancements and innovations.

5.6.3.1 Advanced Material Design

a. Nanotechnology: utilizing nanotechnology to create materials with enhanced properties and performance, such as nanostructured composites and quantum dots, can drive improvements in energy-harvesting efficiency.
b. Bio-inspired Materials: drawing inspiration from nature to design materials with unique properties, such as self-cleaning surfaces and hierarchical structures, can lead to innovative energy-harvesting solutions.

5.6.3.2 Multi-Functional Materials

a. Hybrid Energy Harvesters: developing multi-functional materials that can harvest energy from multiple sources, such as light, heat, and vibrations, can maximize energy capture and versatility.
b. Smart Materials: creating materials with integrated sensing, self-healing, and adaptive properties can enhance the functionality and resilience of energy-harvesting systems.

5.6.3.3 Sustainability and Circular Economy

To reduce the environmental impact and support circular economy principles, the use of eco-friendly and biodegradable materials in energy-harvesting applications needs to be prioritized. Developing recycling and reuse strategies for energy-harvesting materials can extend their life cycle and reduce waste.[12,68,69]

5.7 Conclusion

This chapter has explored the diverse types of low-carbon energy-harvesting materials, including piezoelectric, triboelectric, photovoltaic, and thermoelectric materials. The chapter highlighted the principles, innovative low-carbon options, and their applications in various fields such as wearable electronics, IoT devices, green building technologies, and sustainable solutions for developing regions.

5.7.1 Importance of Continued Research and Development in the Field

Continued research and development in low-carbon energy-harvesting materials are crucial for overcoming existing challenges, improving performance, and expanding their applications. Advancements in materials science, manufacturing processes, and device integration will be key to unlocking the full potential of these materials and contributing to global sustainability efforts.

5.7.2 Final Remarks on the Potential Impact of Low-carbon Energy-harvesting Materials on Global Sustainability

Low-carbon energy-harvesting materials have the potential to significantly impact global sustainability by providing renewable, eco-friendly, and decentralized energy solutions. Future perspectives can focus on advanced material designs and integrated systems for better efficiency and stability; cost-effective manufacturing; cleaner production, *etc.* Their widespread adoption can reduce reliance on fossil fuels, minimize environmental impact, and enhance energy access in underdeveloped regions. Due to these properties of low-carbon energy-harvesting materials, their implementation and usage also aligns with the United Nation Sustainable Developmental Goals of access to affordable and clean energy; sustainable cities and communities; and climate action. As research and innovation continue to drive progress in this field, low-carbon energy-harvesting materials will play a vital role in shaping a sustainable energy future.

References

1. Y. Liang, R. Kleijn, A. Tukker and E. Van der Voet, *Renewable Sustainable Energy Rev.*, 2022, **161**, 112334.
2. V. Pecunia, S. R. P. Silva, J. D. Phillips, E. Artegiani, A. Romeo, H. Shim, J. Park, J. Kim, J. S. Yun and G. C. Welch, *J. Phys. Mater.*, 2023, **6**(4), 042501.
3. M. Wu, S. Wu, Y. Cai, R. Wang and T. Li, *Energy Storage Mater.*, 2021, **42**, 380–417.
4. J. T. Liberty, A. Anil, S. J. Ijimdiya, M. J. Kwaji and R. U. Ijimdiya, *Nano-Struct. Nano-Objects*, 2024, **38**, 101216.
5. J. M. R. Tan, F. E. Osterloh and L. Wong, *J. Mater. Chem. A*, 2024, **12**(12), 6795–6796.
6. S. Bairagi, M. Shahadat, D. M. Mulvihill and W. Ali, *Nano Energy*, 2023, **111**, 108414.
7. Y. Bai, H. Jantunen and J. Juuti, *Adv. Mater.*, 2018, **30**(34), 1707271.
8. P. Martín Sánchez, F. J. Rodríguez Sánchez and E. Santiso Gómez, *Sensors*, 2019, **20**(3), 647.
9. S. Kim, S. J. Choi, K. Zhao, H. Yang, G. Gobbi, S. Zhang and J. Li, *Nat. Commun.*, 2016, **7**(1), 1–7.
10. B. Pozo, J. I. Garate, J. Á. Araujo and S. Ferreiro, *Electronics*, 2019, **8**(5), 486.
11. Y. Liang, R. Kleijn, A. Tukker and E. Van Der Voet, *Renewable Sustainable Energy Rev.*, 2021, **161**, 112334.

12. K. Calautit, D. S. Nasir and B. R. Hughes, *Renewable Sustainable Energy Rev.*, 2021, **147**, 111230.
13. R. Kleijn, E. Van Der Voet, G. J. Kramer, L. Van Oers and C. Van Der Giesen, *Energy*, 2011, **36**(9), 5640–5648.
14. D. Lingam, A. R. Parikh, J. Huang, A. Jain and M. Minary-Jolandan, *Int. J. Smart Nano Mater.*, 2013, **4**(4), 229–245.
15. M. H. Ahmadi, M. Ghazvini, M. A. Nazari, M. A. Ahmadi, F. Pourfayaz, G. Lorenzini and T. Ming, *Int. J. Energy Res.*, 2019, **43**(4), 1387–1410.
16. H. Ryu, J. Yoon and W. Kim, *Adv. Mater.*, 2019, **31**(34), 1802898.
17. U.S. Energy Information Administration, https://www.eia.gov/energyexplained/solar/photovoltaics-and-electricity.php (last accessed August 2024).
18. A. Goetzberger, C. Hebling, and H. Schock, *Mater. Sci. Eng., R*, 2003, **40**(1), 1–46.
19. F. F. Ahmad, C. Ghenai and M. Bettayeb, *Sustainable Energy Technol. Assessments*, 2021, **47**, 101430.
20. L. Xie, W. Song, J. Ge, B. Tang, X. Zhang, T. Wu and Z. Ge, *Nano Energy*, 2021, **82**, 105770.
21. https://www.energy.gov/eere/solar/perovskite-solar-cells (last accessed August 2024).
22. A. Polman, M. Knight, E. C. Garnett, B. Ehrler and W. C. Sinke, *Science*, 2016, **352**(6283), aad4424.
23. K. Rogdakis, M. Loizos, G. Viskadouros and E. Kymakis, *Mater. Adv.*, 2022, **3**(18), 7002–7014.
24. H. Zhu, J. Wei, K. Wang and D. Wu, *Sol. Energy Mater. Sol. Cells*, 2009, **93**(9), 1461–1470.
25. A. G. Macedo, L. P. Christopholi, A. E. X. Gavim, J. F. De Deus, M. M. Teridi, A. R. B. M. Yusoff and W. J. Da Silva, *J. Mater. Sci.: Mater. Electron.*, 2019, **30**, 15803–15824.
26. S. Memon and K. N. Tahir, *Energies*, 2018, **11**(12), 3315.
27. M. D'Angelo, C. Galassi and N. Lecis, *Energies*, 2023, **16**(17), 6409.
28. R. A. Kishore and S. Priya, *Materials*, 2018, **11**(8), 1433.
29. T. Wu, J. Kim, J.-H. Lim, M.-S. Kim and N. V. Myung, *Front. Chem.*, 2021, **9**, 762896.
30. T. Deng, P. Qiu, T. Yin, Z. Li, J. Yang, T. Wei and X. Shi, *Adv. Mater.*, 2024, **36**(13), 2311278.
31. E. Ghafari, F. Severgnini, S. Ghahari, Y. Feng, E. J. Lee, C. Zhang, X. Jiang and N. Lu, *Multifunct. Nanocompos. Energy Environ. Appl*, 2018, **1**, 173–202.
32. J. R. Sootsman, D. Y. Chung and M. G. Kanatzidis, *Angew. Chem.*, 2009, **48**(46), 8616–8639.
33. B. Russ, A. Glaudell, J. J. Urban, M. L. Chabinyc and R. A. Segalman, *Nat. Rev. Mater.*, 2016, **1**(10), 1–14.
34. M. F. Sanad, A. E. Shalan, S. O. Abdellatif, E. S. A. Serea, M. S. Adly and M. A. Ahsan, *Top. Curr. Chem.*, 2020, **378**(6), 48.

35. W. Liu, J. Hu, S. Zhang, M. Deng, C. Han and Y. Liu, *Mater. Today Phys.*, 2017, **1**, 50–60.
36. A. Ali, S. Ali, H. Shaukat, E. Khalid, L. Behram, H. Rani, W. A. Altabey, S. A. Kouritem and M. Noori, *Results Eng.*, 2024, **21**, 101777.
37. C. R. Bowen, H. A. Kim, P. M. Weaver and S. Dunn, *EES*, 2014, **7**(1), 25–44.
38. B. Das and S. Mohanty, *Sens. Actuators, A*, 2024, **366**, 114939.
39. F. Ali, W. Raza, X. Li, H. Gul and K. Kim, *Nano Energy*, 2019, **57**, 879–902.
40. A. Tabak, B. Safaei, A. Memarzadeh, S. Arman and C. Kizilors, *J. Vib. Eng. Technol.*, 2023, **12**, 3155–3192.
41. J. Wang, J. Wu, J. Zhang, L. Guan, H. Feng, K. Zhu, Y. Zhang, H. Sun, Y. Cheng and L. Zhang, *ACS Appl. Electron. Mater.*, 2024, **6**(3), 1562–1573.
42. H. Liu, *et al.*, *Appl. Phys. Rev.*, 2018, **5**(4), 041306.
43. H. Li, C. Tian and Z. D. Deng, *Appl. Phys. Rev.*, 2014, **1**(4), 041301.
44. Z. Li, J. Roscow, H. Khanbareh, G. Haswell and C. Bowen, *Adv. Energy Sustainability Res.*, 2024, **5**(5), 2300235.
45. S. Pandey, J. Zalke, R. Nandanwar and A. Verma, *J. Eng. Sci. Technol. Rev.*, 2018, **11**(4), 77–83.
46. Z. Zhao, L. Zhou, S. Li, D. Liu, Y. Li, Y. Gao, Y. Liu, Y. Dai, J. Wang and Z. L. Wang, *Nat. Commun.*, 2021, **12**(1), 4686.
47. G. Khandelwal, N. P. M. J. Raj and S. Kim, *Adv. Energy Mater.*, 2021, **11**(33), 2101170.
48. Y. Li, Y. Luo, H. Deng, S. Shi, S. Tian, H. Wu, J. Tang, C. Zhang, X. Zhang, J. Zha and S. Xiao, *Adv. Mater.*, 2024, 2314380.
49. S. Priya, *J. Electroceram.*, 2007, **19**(1), 167–184.
50. C. Gao, T. Liu, B. Luo, C. Cai, W. Zhang, J. Zhao, J. Yuan, P. Fatehi, C. Qin and S. Nie, *Nano Energy*, 2023, **111**, 108426.
51. H. Shaukat, A. Ali, S. Bibi, S. Mehmood, W. A. Altabey, M. Noori and S. A. Kouritem, *Energy Rep.*, 2023, **9**, 4306–4324.
52. C. Zhang, W. Zhang, G. Du, Q. Fu, J. Mo and S. Nie, *Chem. Eng. J.*, 2023, **452**, 139259.
53. R. Zhang and H. Olin, *EcoMat*, 2020, **2**(3), 12062.
54. C. Cai, B. Luo, Y. Liu, Q. Fu, T. Liu, S. Wang and S. Nie, *Mater. Today*, 2022, **52**, 299–326.
55. N. Hao, Y. Liu, C. Cai, Y. Shao, X. Meng, G. Du, B. Luo, Z. Wei, J. Wang, S. Wang and S. Nie, *Nano Energy*, 2024, **122**, 109335.
56. Z. L. Wang, J. Chen and L. Lin, *EES*, 2015, **8**(8), 2250–2282.
57. X. Cheng, W. Tang, Y. Song, H. Chen, H. Zhang and Z. L. Wang, *Nano Energy*, 2019, **61**, 517–532.
58. C. Gayner and K. K. Kar, *Prog. Mater. Sci.*, 2016, **83**, 330–382.
59. J. Luo and Z. L. Wang, *EcoMat*, 2020, **2**(4), e12059.
60. A. J. Bandodkar, I. Jeerapan and J. Wang, *ACS Sens.*, 2016, **1**(5), 464–482.
61. P. Kamalinejad, C. Mahapatra, Z. Sheng, S. Mirabbasi, V. C. M. Leung and Y. L. Guan, *IEEE Commun. Mag.*, 2015, **53**(6), 102–108.
62. L. Guo and Q. Lu, *Renewable Sustainable Energy Rev.*, 2017, **72**, 761–773.
63. T. Mori and S. Priya, *MRS Bull.*, 2018, **43**(3), 176–180.

64. H. Ghasemi, G. Ni, A. M. Marconnet, J. Loomis, S. Yerci, N. Miljkovic and G. Chen, *Nat. Commun.*, 2014, **5**(1), 1–7.
65. B. K. Sovacool, S. H. Ali, M. Bazilian, B. Radley, B. Nemery, J. Okatz and D. Mulvaney, *Science*, 2020, **367**, 30–33.
66. S. Chu and A. Majumdar, *Nature*, 2012, **488**, 294–303.
67. O. Vidal, B. Goffé and N. Arndt, *Nat. Geosci.*, 2013, **6**(11), 894–896.
68. J. Calderon, M. Bazilian, B. Sovacool, K. Hund, S. Jowitt, T. Nguyen, A. Månberger, M. Kah, S. Greene, C. Galeazzi, K. Awuah-Offei, M. Moats, J. Tilton and S. Kukoda, *Renewable Sustainable Energy Rev.*, 2020, **124**, 109789.
69. S. Yun, Y. Zhang, Q. Xu, J. Liu and Y. Qin, *Nano Energy*, 2019, **60**, 600–619.

CHAPTER 6

Low-carbon Materials for Electrochemical Water Splitting

NGUYEN VIET KHOA, NGUYEN MINH VIET,*
NGUYEN MINH PHUONG, PHUONG THAO,
NGUYEN VAN NOI AND THANH-DONG PHAM

VNU-University of Science, Vietnam National University, Hanoi, Vietnam
*Email: nguyenminhviet@hus.edu.vn

6.1 Introduction

During the 18th century, the British scientist Henry Cavendish made the initial discovery of hydrogen. He achieved this by allowing the metal zinc to mix with hydrochloric acid, resulting in the formation of hydrogen. This element was subsequently named by the French chemist Antoine Lavoisier in recognition of Cavendish's finding. Subsequently, he submitted a proposal to the Royal Society of London, in which he ignited a gas in the presence of air, resulting in the production of water. This experiment ultimately led to his identification of the constituent elements of water, namely hydrogen and oxygen. This set the first step in the human discovery of hydrogen. Thus, many scientists have researched the application of hydrogen in daily life, such as the hydrogen balloon by Jacques Alexander Cesar Charles. Subsequently, this scientist made the groundbreaking revelation that water may be electrolyzed to generate hydrogen through the application of an electric current.[18–20] This procedure was designated as "electrolysis." In the 19th century, Swiss chemist Christian Friedrich Schoenbein discovered the fuel cell effect when hydrogen reacts with oxygen, which produces an electric current. In 1845, Sir William

Low Carbon Materials: Paving a Sustainable Path for a Greener Future
Edited by Santanu Patra, Sudheesh K. Shukla and Mika Sillanpää

Published by the Royal Society of Chemistry, www.rsc.org

Grove, an English scientist and judge, successfully illustrated Schoenbein's findings by constructing a functional "gas battery." He was awarded the title "Father of the Fuel Cell" in recognition of his accomplishment. In the 1920s, German scientist Rudolf Erren modified the internal combustion engines of vehicles, buses, and submarines to run on hydrogen or hydrogen mixtures by harnessing the reaction of hydrogen. In the same year, J. B. S. Haldane, a British scientist, introduced the idea of renewable hydrogen as an energy source in his research. He suggested that excess power generated by large power stations during windy weather could be utilized to separate water into oxygen and hydrogen through electrolysis. In 1959, Francis Thomas Bacon utilized hydrogen as a rocket fuel cell for NASA. This fuel was employed during the space shuttle missions in the late 1950s and early 1960s.

With the development of technology, the demand for energy has increased daily, while fossil fuel sources are limited. As the consumption of fossil fuel increases, the environment becomes increasingly contaminated, necessitating the development of renewable energy sources. To this day, hydrogen has been applied in various aspects of life.[20] Hydrogen possesses the highest gravimetric energy density of any known energy carrier, making it a promising, sustainable, and clean fuel.[21,22] With a notable energy conversion efficiency, this fuel produces solely water as a byproduct when utilized in a fuel cell, resulting in zero pollution.[23,24] After Germany, the Netherlands is Europe's second-largest producer of grey hydrogen (made from natural gas). Hydrogen has been used in the chemical and other process industries on a large-scale basis for decades.[25] Most of it, by far, is used as a raw material to produce urea (*via* ammonia) to make fertilizer. The petroleum industry also needs hydrogen to refine oil (to remove sulfur) to produce fuel or upgrade the heavy oil fraction.

Furthermore, hydrogen is increasingly needed to produce biofuels, such as bio-methanol, or as a reducing agent in the metal and semiconductor industries.[20] This demonstrates the indisputable significance of hydrogen as an affordable and environmentally friendly source of energy (Figure 6.1). However, the primary procedure to produce hydrogen still needs natural gas as a raw ingredient, and the byproduct CO_2 released cannot be captured in this process.[20,25] Thus, developing a green method to synthesize hydrogen that meets the economic demand and environmental criteria is a critical problem for scientists.

In recent years, an alternative way to synthesize green hydrogen utilizing only renewable energy, such as water electrolysis, has been considered a leap forward in developing eco-friendly energy.[20,23,24] However, water electrolysis technologies cannot compete with the low-cost traditional production of hydrogen *via* natural gas.[1] The drawback of the process comes from the energy-intensive behavior of this process.[1,2] Hence, the right way to produce hydrogen while making it as green as possible is to develop the procedure in a renewable method by using alternative renewable energy sources. In pursuit of this concept, extensive research has been conducted on the electrochemical water-splitting process. This process involves the hydrogen evolution reaction (HER) and the oxygen evolution reaction (OER), which

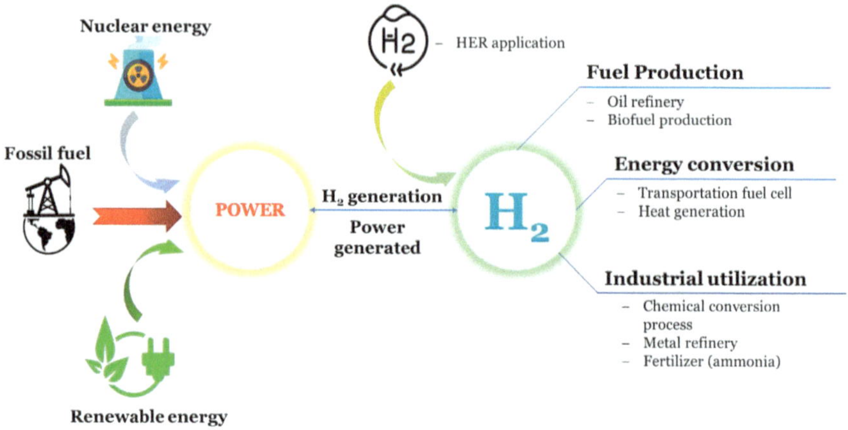

Figure 6.1 Hydrogen production and application.

enable the generation of hydrogen and oxygen, respectively. The aim is to overcome the constraints associated with the electrolysis process of water. The HER process happens in the cathode, while the OER process happens in the anode. Due to the nature of the catalyst mechanisms, OER catalysts are more favorable than HER catalysts.[3] As a result, the role of catalysis is crucial for the electrochemical water-splitting cell and fuel cell. Significant advancements in fuel and electrolysis water-splitting cells have opened new avenues for a sustainable transition to carbon-neutral operations.[26] The current standard for electrolyzers uses a cathode made of platinum (Pt) and anodes made of a mixture of ruthenium dioxide (RuO_2) and iridium dioxide (IrO_2) to enhance the HER and OER, respectively.[3] From the industrial perspective, it is a high-cost material due to the complications in the production process of the catalyst for the anode and cathode. The expense of noble metal elements also contributes to the downfall of this catalyst in industrial processes. Thus, it is greatly advantageous and a significant innovation to develop a catalyst that can split water which is globally effective, using elements derived from abundant sources or renewable sources. This addresses many challenging issues facing energy and environmental sustainability.

6.2 Electrochemical Water-splitting Process

Low-carbon or carbon-negative materials are the kinds of materials in which the overall production process emits low or no carbon in the natural carbon rotation. This catalyst production process's raw materials come from natural minerals, biomass waste, and electronic waste (Figure 6.2).

6.2.1 Low-carbon Materials for Water-splitting Process

Numerous research studies on non-noble metal and nonmetal-based nanomaterials have been conducted, and some positive catalysts for the

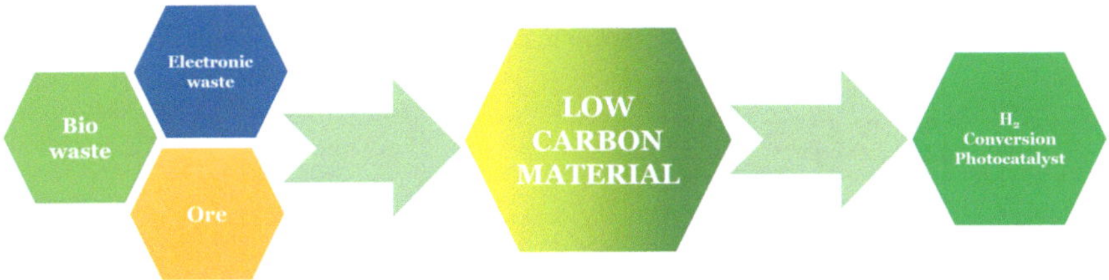

Figure 6.2 Concept of low-carbon materials.

effective electrochemical water-splitting process have been found.[27,28] Note that most catalysts mentioned are produced using chemicals or precursors that are usually expensive and acquired from chemical shops or firms. It is crucial to bear in mind that the chemicals you acquire are derived from natural resources or minerals, and the process of obtaining them entails complex procedures in metallurgy, purification, synthesis, and beneficiation.[29,30] Alternatively, using naturally occurring minerals as precursors for the synthesis of highly functional electrocatalysts has attracted increasing attention. Additionally, by lowering the high cost and detrimental environmental effects of the conventional catalyst preparation process, which entails intricate procedures, the development of affordable electrocatalysts from natural resources effectively achieves the eco-designed goal. A recent advanced study indicates that certain materials containing transition metals have shown significant electrochemical activity for OER and/or HER through the implementation of environmentally friendly techniques such as sulfidation and ball milling on naturally occurring materials after treatment (Figure 6.3).[31–33] The catalyst from this process exhibits high performance in electrocatalyst properties *via* their internal and external characteristics.

As an illustration, Jie Yin and colleagues[4] showed the utilization of transition metal chalcogenides in the process of water splitting. They documented the various compositions present in the lattice structure of the material, which can have an impact on its electrical capabilities. These materials enhance the water-splitting performance but also show poor conductivity, activity, and stability in various conditions. This restricts its utilization in industrial processes on a wide scale. Nevertheless, it is possible to modify this situation by implementing an effective and systematic approach to produce durable and reactive transition metal chalcogenides. Their objective was to alter the properties of transition metal chalcogenides by various synthesis methods, such as solid-phase chemical synthesis, wet-chemical syntheses, and the hot-injection approach. Another strategy they use is modifying the material's morphology from nanowires to nanosheets and then 3D nanostructures for favorable expression of the active site and adsorption properties. Then, they propose the structure modification *via* phase change, high index crystal face, and amorphous structure. Finally, they try to optimize the electronic structure of the transitional metal chalcogenides (defects engineering, fabrication of a nano interface, stress regulation).

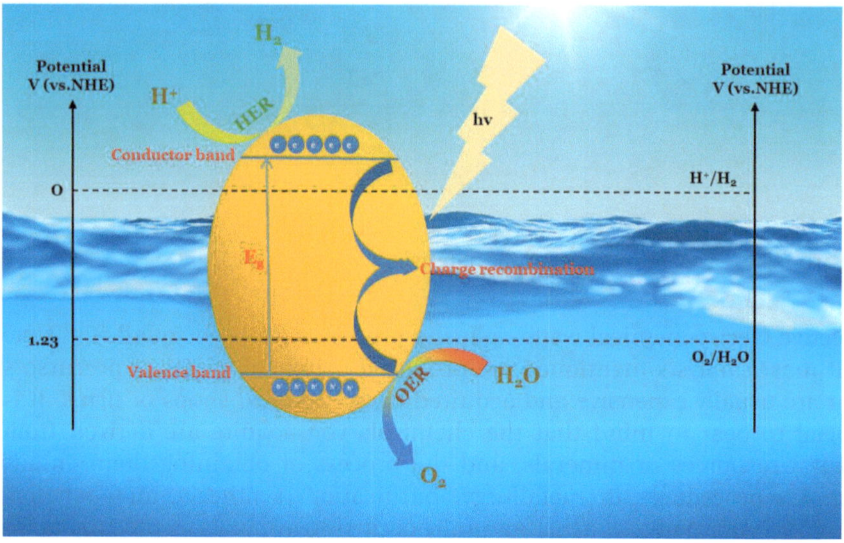

Figure 6.3 Photocatalyst's water-splitting reaction.

In conclusion, the unique qualities of metal chalcogenide nanomaterials include their large-scale manufacture, high electrical transport, readily controllable structure, and optimal atomic arrangement. However, it is still difficult to manage how those heterostructure nanocrystals are prepared so that the content of various components is finely regulated. Additional investigation is required to develop novel functioning composites using a range of substances such as metallic substances, metal oxides, chalcogenides, nitrides, and carbides. This research should focus on hybrid engineering with metal chalcogenides.

Rabia Jamil[5] and her colleagues have elucidated the function of nitrilc transition metals in the advancement of electrochemical water splitting. They stated that adding nitrogen (N) to transition-metal lattices increases their metallic conductivity, opens new catalytic sites, and increases their resilience to corrosion in both acidic and alkaline conditions. Due to the advanced characteristics of N, numerous transition metal nitrides have been extensively studied as potential contenders for electrochemical water splitting. Considerable focus has been given to the development of sophisticated heterostructures in order to enhance their inherent and external characteristics through the manipulation of composition and phase, the engineering of interfaces, and the control of facets and flaws. Metal nitrides can be considered as interstitial alloys when nitrogen atoms are introduced as dopants into the crystal lattice of a metal. Due to its lower electronegativity and energy required for nitride production compared to oxides, nitrogen can only mix with a restricted range of electropositive elements. The scarcity of stable and efficient metal nitrides is attributed to their limited occurrence, despite their strong tendency to undergo water electrolysis. The presence of metallic bonding in transition metal nitrides is accountable for the significant electron transport and remarkable resistance to corrosion. The ionicity and modification of the d-band electrons of metals are crucial factors in electrochemical processes. The presence of electronegative nitrogen atoms in transition metal nitrides influences the energy dynamics of adsorption and desorption of reaction intermediates by altering the interactions between metal atoms and the electronic configuration of catalysts. As a result, they play a crucial role in the process of water splitting. The interaction energies between the catalyst surface and the reaction intermediates also impact the material's energy splitting and H_2O adsorption characteristics.

When a metal–carbon covalent bond is formed, the interstitial atoms of carbon atoms received electrons by charge transfer progressions from metal atoms. It is hypothesized that the altered electronic configurations of the metal atoms promote a reduction in both the overpotential and the Tafel slope during the catalysis of HER. The Group VIB metals consist of chromium, molybdenum, and tungsten. The metal atoms have d-orbitals that contain an equal number of electrons and are not completely filled, resulting in an outer electron configuration of $nd^5(n^{+1})s^1$. When the five valence electrons of the d-level of the metal combine with the four outermost electrons of carbon ($2s^22p^2$), the d-orbitals of the metal become wider and resemble those of platinum ($5d^96s^1$). Therefore, the carbide materials of Cr, Mo, and W exhibit catalytic properties that are comparable to those of Pt when used in the process of HER. The utilization of transition metal carbides in the process of HER has been extensively documented in contemporary scientific research. The progress of this material is accelerating as materials science advances. The synthesis methods are limited to only the single transition metal carbides and can be expanded to include compositions with extremely effective catalytic materials, such as N,S co-doped graphene, carbon nitride, and nanoporous materials. Below, we present a detailed overview of the individual advancements made in several transition metal carbides. Because both W and Pt are elements in the sixth period of the periodic table, there has been extensive research on W carbides in the field of HER catalysis as a potential alternative to Pt since the 1930s. The prevalent variants of W carbides include WC and W_2C, both exhibiting a hexagonal crystal structure. The W carbides possess a band structure that is comparable to that of Pt and exhibit excellent resistance to poisoning by H_2S and CO. At a current density of 10 $mA\,cm^{-2}$, the overpotential was minimized to 200 mV in a 1 M H_2SO_4 solution. The Tafel slope, which was measured at 118 $mV\,dec^{-1}$, revealed that the reaction followed a Volmer–Heyrovsky mechanism. Molybdenum is a significant carbide that belongs to the category of transition metals. It exhibits different valence states, including +2, +4, and +6, which are frequently encountered. Therefore, several kinds of Mo carbides are present, including Mo_2C, MoC, and Mo_3C_2. Molybdenum carbides have been extensively utilized in diverse catalytic reactions due to innate chemical stability, including applications in industrial desulfurization, water–gas shift reactions, and hydrogenation processes.

While molybdenum carbides are suitable for HER catalysis throughout a broad pH range, their catalytic performance is inferior to that of platinum (Pt) or tungsten carbide (WC) due to the challenge of achieving tiny particle sizes. Increased particle size will decrease the specific surface area, which clearly affects the distribution of active sites in the catalyst. Therefore, there are multiple techniques available to decrease the particle size of this substance. Various methods have been employed to reduce the size of molybdenum carbide particles, including reduction *via* CH_4–H_2, sol–gel manipulation process, chemical vapor deposition (CVD), microwave-assisted thermolytic approach, and various nanoparticle substrate-supported pyrolysis methods. Chromium, the first element of Group VIB, is characterized by its polyvalency and its behavior as a chemically active metal. When Cr is exposed to air or water during the HER period, it undergoes easy oxidation and creates a dense coating of Cr_2O_3 on its surface. Chromium has excellent efficiency in HER catalysis when it takes the form of Cr carbides, including CrC, Cr_3C_2, Cr_7C_3, and $Cr_{23}C_6$. Among them, the hexagonal Cr_7C_3 demonstrates exceptional activity as a catalyst.

Metal sulfides have been intensively investigated as a potential catalyst material for OER in water-splitting reactions. This is due to their many active sites, strong electrical conductivity, and high intrinsic activity.[6] The catalysts used for OER were referred to as pre-catalysts. The *in situ* reaction will transfer them to the actual active phase, which is induced by the high oxidation potentials. All of these catalysts are primarily assessed and examined for their potential in OER. However, their catalytic functioning is currently subpar, and it is imperative to enhance the stability of these catalysts. The excellent performance of metal sulfide in OER can be attributed to its abundant composition, distinctive lattice structure, and efficient electron-transferring ability. The metal sulfide contains sulfur atoms that can form S–S bonds through a quickly reversible redox process. This leads to a flexible phase structure that enhances not only the charge capacity but also the transportation of electrons. Metal sulfides can be categorized into layered and nonlayered structures based on their crystal structures. Sulfides containing metals such as Mo, W, Ta, Nb, *etc.* always exhibit a layered structure similar to graphite. On the other hand, overall transition-metal sulfides-containing metals like Fe, Co, Ni, *etc.* primarily form nonlayered structures. A conventional layered metal sulfide has a sandwich-like arrangement, with a single layer of metal connected to two layers of sulfur. The metal sulfide unit cells are vertically stacked on top of each other through the comparatively frail van der Waals forces. As a result, exfoliation can be used to obtain a single layer of metal sulfide. The metal sulfides that are not layered have a structure similar to pyrite or marcasite, with the metal atoms forming octahedral bonds with the neighboring sulfur atoms. These distinct architectures will unavoidably result in variations in their electrical and electrochemical characteristics. Nevertheless, these immaculate metal sulfides have several flaws, including a limited number of active sites, inadequate electrical conductivity, and instability when subjected to operational conditions. Several techniques have been devised to create metal sulfides with excellent performance, such as mono-metal sulfides, bimetallic sulfides, polymetallic sulfides, doping, and heterostructures. The significant association among the nanostructures and their practical qualities motivates further advancements in their application in OER.

Transition-metal phosphides, namely those with M_xP_y composition where M represents Fe, Co, Ni, Cu, Mo, and W, show great potential as catalysts for the hydrogen evolution reaction.[7] These catalysts are particularly attractive because they do not contain noble metals. They are capable of maintaining a high level of activity, stability, and Faradaic efficiency of almost 100% for the process of hydrogen evolution in a solution with any pH value. The crucial features of a high-quality catalyst for the HER include efficient proton trapping, effective bonding of atomic hydrogen, and the ability to facilitate the desorption of H_2. [NiFe] hydrogenases have high activity in catalyzing the HER at ambient temperature. The exceptional reactivity of the bimetallic nickel-iron phosphide (NiFeP), which exhibits several redox sites resulting in electrochemical bidirectionality and excellent stability, was the main feature of this material.[8] The justification for designing heterostructures with different combinations of elements and nanocomposite morphologies was to enhance the OER activity. Phosphorus-based substances have been identified as a potential substitute for metal oxides. A recent study indicates that phosphates with the right ability to be wetted can enhance the catalytic activity of an OER catalyst. For example, the biscuit-shaped NiFePi/P catalyst demonstrates a 230 mV overpotential, enabling it to achieve a current density of 10 $mA\,cm^{-2}$. The initial phosphates undergo a transformation into various types of phosphides, leading to the formation of transition metal phosphides (TMPs). These TMPs commonly serve as the active catalyst for the OER. The shape of phosphide materials is crucial in the process of water oxidation, since differences in the distribution of nanostructures result in varying levels of O_2 generation activity. Therefore, the synthetic technique for producing these materials must be carefully adjusted to guarantee an optimal catalytic reaction.

In conclusion, these electrocatalysts all have some disadvantages in their water-splitting performance. Fortunately, the water-splitting capability can be enhanced by modifying the electrocatalyst used. However, this process takes many chemical resources, which can produce more pollutants and increase the production cost. Thus, alternative pathways to synthesize significantly high-performance electrocatalysts are crucial to be a competitive material to traditional noble metal catalysts. As the economy and environment cannot be separated from each other, the trend of producing materials from green sources has become more crucial than ever. By using renewable sources to synthesize materials, we can achieve a balance in both economic profit and maintaining the environment. Low-carbon materials or carbon-negative materials are the kind of materials in which the overall production process emits low or no carbon to the natural carbon rotation. This catalyst production process's raw materials come from natural minerals, biomass waste, and electronic waste.

6.2.2 Electrocatalysts from Natural Minerals

By directly using natural minerals as the source for the synthetization of electrocatalyst material. To reduce costs, we can purify the substance into a pure compound for modifying/synthesizing the electrocatalyst. Direct synthesis also reduces the carbon footprint of the collecting process while maintaining a high-purity product, which yields high water-splitting performance.

Chi Zhang and colleagues[9] have published a study on the efficient and rapid fabrication of cost-effective mineral-based two-dimensional electrocatalysts for the HER. The catalyst exhibits a high current density of 1000 mA cm^{-2} when an overpotential of 412 mV is applied for the hydrogen evolution process. In order to showcase the practicality of efficient production techniques, the researchers utilized molybdenite concentrate as the initial material. This concentrate primarily consists of bulk MoS_2, which is predominantly found in the form of molybdenite mineral extracted directly from an open-pit mine. A molybdenite concentrate is produced by an initial flotation process, which can be utilized for the production of industrial-grade MoS_2 as well as high-purity MoS_2. They enhance the composition of the 2D MoS_2 electrocatalyst by including Mo_2C nanoparticles on both the edge and surface. This modification leads to a highly efficient HER with a significant current density of 1000 mA cm^{-2} at 412 mV, a low Tafel slope of 60 mV dec^{-1}, and excellent stability over a period of 24 h. The reported production code is 30 times less than the commercial Pt/C electrocatalyst.

Bharathi Konkena and colleagues[10] successfully utilized the natural ore pentlandite ($Fe_{4.5}Ni_{4.5}S_8$) as an upfront 'rock' electrode catalyst for HER under acidic conditions with an overpotential of 280 mV at 10 mA cm^{-2}. Additionally, it obtains a charge as low as 190 mV after 96 h of electrolysis due to the depletion of sulfur on the surface, which may alter the electronic structure of the activated catalyst–nickel–iron centers. The 'rock' catalyst displays an unpredicted catalytic activity with comparable overpotential and Tafel slope to some well-developed metallic or nanostructured catalysts. Remarkably, the 'rock' catalyst suggests high current densities ($\leq$650 mA cm^{-2}) without any deficiency in activity within 170 h. The superior HER efficiency of pentlandites as 'rock' electrode labels this ore as a gifted electrocatalyst for the upcoming green/hydrogen-based economy.

6.2.3 Electrocatalyst from Biomass Waste

The utilization of biomass and waste plastic shows potential in supporting the worldwide objective of achieving a carbon-neutral and waste-free human society.[11] Biomass encompasses a wide variety of natural organic substances, derived from either plants or animals, which have the potential to be utilized as chemical or energy sources. By 2060, it will be possible to generate up to 240 exajoules (1 exajoule = 24 million tons of oil equivalent) of energy annually through the utilization of wood harvesting, and agricultural and municipal trash. Nevertheless, the proportion of energy obtained from biomass that is utilized in industrial consumption remains very low, such as approximately 9% in the European Union in 2017. The proportion of chemicals obtained from biomass is expected to grow and make up 22% of the global chemical business by 2025. Plastics are synthetic materials that have made a remarkable impact on the advancement of humanity. Nevertheless, the majority of plastics employ synthetic polymers produced from petroleum as their primary components, which gives rise to significant environmental concerns. Out of the total amount of plastic ever produced, which is 8600 million metric tons, 59% has been disposed of, either by being thrown into landfills or by accumulating in natural environments like oceans. Only 17% has been recycled or burned to generate energy. Through meticulous analysis of biomass and plastics, it becomes evident that they share significant similarities in terms of their composition and structure, both at a large scale and a small scale. When comparing a piece of wood to a disposable mask, both can be classified as composite materials that exhibit secondary interactions between polymer chains, such as hydrogen bonding and cross-linking. The intermolecular interactions between polymers, such as cellulose and lignin, are crucial characteristics of both raw biomass and waste plastics. The interpolymer interactions have a significant impact on the polymer's properties and frequently enhance its resistance to chemical degradation or recalcitrance. The significant structural resemblances between biomass and plastic lead to notable parallels between the two substances in terms of bond activation chemistry and catalytic conversion. Therefore, biomass and plastic have historically and are currently employed in similar methods for conversion. Therefore, by considering the catalyst conversion of biomass and waste plastics from a comprehensive perspective, encompassing various technologies in selective degradation and bond activation chemistry for both feedstocks, and utilizing biomass and plastic as starting materials, it is possible to significantly reduce the carbon footprint. In fact, it could even result in a negative carbon output within a natural carbon circular system.

Jitendra N. Tiwari[12] and colleagues have conducted research on creating catalysts from yeast biomass-derived materials. These catalysts, known as multi-heteroatoms (nitrogen, sulfur, and phosphorus) doped carbon (MHC) catalysts, are made from waste biomass. The purpose of this research is to develop catalysts that are efficient, environmentally friendly, and cost-effective in order to enhance the sustainability of hydrogen production. They offer an efficient hydrogen production technology that may be implemented in the industrial sector. The MHC catalyst, which contains Fe_3O_4, exhibits exceptional activity in the oxygen evolution reaction. To achieve complete water splitting, a solar voltage of 1.74 V is necessary to provide a current of approximately 30 mA. Additionally, the system must demonstrate exceptional long-term stability for at least 12 hours in the presence of external sunshine and for 58 hours in its absence. This approach holds great promise for the advancement of sustainable energy.

Plastic is commonly utilized as the feedstock for hydrogen production rather than as a catalyst for the water-splitting process, owing to its exceptional durability and resistance. Hua Zhou[13] and colleagues have successfully utilized polyethylene terephthalate (PET) plastic to produce valuable commodity chemicals, namely potassium diformate and terephthalic acid, as well as H2 fuel through the process of electrocatalytic upcycling. An initial techno-economic study indicated that this process is profitable when the ethylene glycol (EG) component of PET is selectively electro-oxidized to formate with a selectivity of over 80% at a high current density of over 100 mA cm^{-2}. A cobalt phosphide ($CoNi_{0.25}P$) electrocatalyst, modified with nickel, is created to attain a current density of 500 mA cm^{-2} at 1.8 V in a reactor utilizing a membrane-electrode assembly. This electrocatalyst demonstrates an efficiency of over 80% in terms of Faradaic efficiency and selectivity toward the formation of formate. This study reveals a viable method for utilizing waste PET to create high-value items through upcycling.

6.2.4 Electrocatalysts from Electronic Waste

As technology advances, the increasing demand for energy has become a significant issue, not to mention the problem of electronic waste. Electronic garbage (e-waste) is a significant contributor to environmental degradation, exacerbating the problem by introducing heavy metals into landfills and contaminating water systems.[14] Over the past two decades, the continuous introduction of electronic items with intelligent designs, features, and technologies has resulted in the quick obsolescence of numerous electric and electronic appliances. Furthermore, the lifespan of electronic devices has been significantly reduced, leading to a substantial increase in waste production annually. This reduction in lifespan can be attributed to factors such as consumer satisfaction, marketing strategies, and compatibility concerns. E-wastes are a major contributor to the presence of heavy metals and persistent organic pollutants in the environment. If not carefully handled, they can lead to significant damage to ecosystems over time. Hence, the collection, retrieval, and reutilization of electronic trash for the advancement of environmentally friendly device design is crucial for the effective management of electronic waste. In this report, we aimed to use copper extracted from discarded electric cables as a cost-effective and technically viable current collector for applications involving OER and HER.

Therefore, they require catalysts that can be reused from electronic waste in order to achieve a water-splitting process that is economically advantageous. A recent study demonstrated the recycling of electronic trash to produce a catalyst for the HER with a significant overpotential of 178 mV at a current density of 10 mA cm^{-2}, as well as an overpotential of 220 mV for the OER that remained stable for 36 hours.[15] They revealed components that are widely utilized as primary materials in the production of electronic devices. Indium, silver, zinc, gallium, tellurium, arsenic, and germanium are projected to expand over the next century and are categorized as a significant menace. Co and Cr are facing an increasing menace as a result of their widespread utilization, whereas Cu, Mg, Ni, Li, and Mn are categorized as potential supply vulnerabilities due to their limited accessibility. These officially recognized endangered elements exhibit exceptional electrocatalytic water-splitting capabilities. In their study, Yi *et al.*[14] demonstrated the practical and cost-effective use of Cu extracted from waste electric cables as a potential current collector for the overall water-splitting reaction. They successfully obtained excellent electrochemical performance for both HER and OER using a simple electrodeposition method. This involved creating a highly efficient nickel–cobalt–phosphide (NiCoP) electrocatalyst on a copper wire. Hence, the incorporation of the non-noble catalyst onto the SCW resulted in an improvement of both the OER and HER performances in alkaline conditions at a current density of 10 mA cm^{-2}, achieving values of 220 mV and 178 mV, respectively. These values surpass the activity of the most advanced electrocatalysts for OER (IrO_2/RuO_2) and HER (Pt). In addition, the electrocatalytic activity of the produced electrodes was monitored to stimulate an alkaline electrolyzer. This was done at a current density of 10 mA cm^{-2} and a low voltage of 1.59 V for a duration of one day.

Therefore, it is crucial to recover nanomaterials from any use of nanotechnology in order to achieve sustainable waste management and provide a constant supply of raw materials. Nanomaterials are essential for energy storage applications like supercapacitors and rechargeable batteries. The materials utilized in the supercapacitive application are costly. As an example, the price of cobalt oxide rose from $20 per kilogram in 1998 to $60 per kilogram in 2017.[16] Metal oxides are utilized in electrical storage systems due to their high capacitance value, rendering them excellent choices. The first focus of research was on ruthenium oxide, primarily because of its exceptional surface area and conductivity.[17] The capacitance exhibited is 1340 F g^{-1}, which is in close proximity to its optimal theoretical capacitance of 1400 F g^{-1}.

6.3 Conclusions

Because of their detrimental environmental effects, fossil fuels have become recognized as significant air pollutant sources in recent years. Hence, hydrogen energy produced from water electrolysis technologies is the most promising energy source as a renewable fuel for electricity or heat production. Due to the overall production process emitting low or no carbon to the natural carbon rotation, low-carbon materials show great promise for electrochemical water splitting, including hydrogen production. This also contributes to both the United Nations Sustainable Development Goals related to business, innovation, and infrastructure, as well as the goal of promoting climate action. In conclusion, this chapter reviews recent work on applying eco-friendly low-carbon materials for electrochemical water splitting to produce renewable energy. One of the most intriguing research avenues in this field is using natural minerals, biomass waste, and electronic waste, which greatly promise to improve human quality of life and the environment.

References

1. I. Slobodkin, E. Davydova, M. Sananis, A. Breytus and A. Rothschild, *Nat. Mater.*, 2024, **23**, 398–405.
2. S. Shiva Kumar and H. Lim, *Energy Rep.*, 2022, **8**, 13793–13813.
3. B. H. R. Suryanto, Y. Wang, R. K. Hocking, W. Adamson and C. Zhao, *Nat. Commun.*, 2019, **10**, 5599.
4. J. Yin, J. Jin, H. Lin, Z. Yin, J. Li, M. Lu, L. Guo, P. Xi, Y. Tang and C.-H. Yan, *Adv. Sci.*, 2020, **7**, 1903070.
5. R. Jamil, R. Ali, S. Loomba, J. Xian, M. Yousaf, K. Khan, B. Shabbir, C. F. McConville, A. Mahmood and N. Mahmood, *Chem Catal.*, 2021, **1**, 802–854.
6. R. He, X. Huang and L. Feng, *Energy Fuels*, 2022, **36**, 6675–6694.
7. A. Li, Y. Sun, T. Yao and H. Han, *Chem. – Eur. J.*, 2018, **24**, 18334–18355.
8. T. Aziz, M. A. Haque, S. Saha, B. Mondal, S. Jain and A. Dutta, *Energy Fuels*, 2023, **37**, 18291–18309.
9. C. Zhang, Y. Luo, J. Tan, Q. Yu, F. Yang, Z. Zhang, L. Yang, H.-M. Cheng and B. Liu, *Nat. Commun.*, 2020, **11**, 3724.
10. B. Konkena, K. junge Puring, I. Sinev, S. Piontek, O. Khavryuchenko, J. P. Dürholt, R. Schmid, H. Tüysüz, M. Muhler, W. Schuhmann and U.-P. Apfel, *Nat. Commun.*, 2016, **7**, 12269.
11. K. Lee, Y. Jing, Y. Wang and N. Yan, *Nat. Rev. Chem.*, 2022, **6**, 635–652.
12. J. N. Tiwari, N. K. Dang, S. Sultan, P. Thangavel, H. Y. Jeong and K. S. Kim, *Nat. Sustainability*, 2020, **3**, 556–563.
13. H. Zhou, Y. Ren, Z. Li, M. Xu, Y. Wang, R. Ge, X. Kong, L. Zheng and H. Duan, *Nat. Commun.*, 2021, **12**, 4679.
14. V. R. Jothi, R. Bose, H. Rajan, C. Jung and S. C. Yi, *Adv. Energy Mater.*, 2018, **8**, 1802615.
15. R. Seif, F. Z. Salem and N. K. Allam, *Environ., Dev. Sustainability*, 2024, **26**, 5473–5508.
16. M. B. J. G. Freitas and E. M. Garcia, *J. Power Sources*, 2007, **171**, 953–959.
17. E. A. A. Aboelazm, N. Mohamed, G. A. M. Ali, A. S. H. Makhlouf and K. F. Chong, in *Waste Recycling Technologies for Nanomaterials Manufacturing*, ed. A. S. H. Makhlouf and G. A. M. Ali, Springer International Publishing, Cham, 2021, pp. 91-123
18. S. Dai, P. Shen, W. Deng and Q. Yu, *Electronics*, 2024, **13**, 3370.
19. A. Ajanović, *Oxford Open Energy*, 2023, **2**, oiad013.
20. A. Risco-Bravo, C. Varela, J. Bartels and E. Zondervan, *Renewable Sustainable Energy Rev.*, 2023, **189**, 113930.
21. C. Tarhan and M. A. Çil, *J. Energy Storage*, 2021, **40**, 102676.
22. S. Van Renssen, *Nat. Clim. Change*, 2020, **10**, 799–801.
23. I. Staffell, D. Scamman, A. V. Abad, P. Balcombe, P. E. Dodds, P. Ekins, N. Shah and K. R. Ward, *Energy Environ. Sci.*, 2018, **12**, 463–491.

24. N. L. Garland, D. C. Papageorgopoulos and J. M. Stanford, *Energy Procedia*, 2012, **28**, 2–11.
25. J. Hwang, K. Maharjan and H. Cho, *Int. J. Hydrogen Energy*, 2023, **48**, 28629–28648.
26. T. Koivunen, A. Khosravi and S. Syri, *Energy*, 2023, **284**, 128624.
27. S. Paygozar, A. S. R. Aghdam, E. Hassanizadeh, R. Andaveh and G. B. Darband, *Int. J. Hydrogen Energy*, 2022, **48**, 7219–7259.
28. S. Rhatigan, M.-C. Michel and M. Nolan, *J. Phys. Energy*, 2020, **2**, 042002.
29. U. B. Demirci and P. Miele, *J. Cleaner Prod.*, 2013, **52**, 1–10.
30. I. Dincer, *Int. J. Hydrogen Energy*, 2011, **37**, 1954–1971.
31. Y. Zheng, Y. Jiao, M. Jaroniec and S. Z. Qiao, *Angew. Chem., Int. Ed.*, 2014, **54**, 52–65.
32. Z. Chen, X. Duan, W. Wei, S. Wang and B.-J. Ni, *J. Mater. Chem. A*, 2019, **7**, 14971–15005.
33. F. Sun, Q. Tang and D.-E. Jiang, *ACS Catal.*, 2022, **12**, 8404–8433.

CHAPTER 7

Low-carbon Materials for Healthcare Monitoring

M. A. ÖZBEK,[a] A. A. TOPÇU,[b] F. YILMAZ[c] AND A. DENIZLI*[a]

[a] Department of Chemistry, Hacettepe University, Ankara, Turkey; [b] Medical Laboratory Program, Vocational School of Health Services, Aksaray University, Aksaray, Turkey; [c] Department of Chemistry Technology, Bolu Abant Izzet Baysal University, Bolu, Turkey
*Email: denizli@hacettepe.edu.tr

7.1 Introduction

The 17 Sustainable Development Goals (SDGs) of the United Nations are at the heart of the 2030 Agenda and were adopted by the 193 member states of the United Nations General Assembly in 2015 (https://www.un.org/sustainabledevelopment/development-agenda/). The SDGs were proposed to establish the foundations of sustainable development, aiming to advance global health and well-being, eliminate all forms of inequality and poverty, and support climate action, promote quality education, gender equality, peace, and social justice. The SDGs' objectives are closely linked to all research disciplines, including the humanities, engineering, medical, statistics, and economics, and should be the focus of future science and technology. The complexity and multifaceted nature of the SDGs' targets frequently require contributions from various disciplines and a multidisciplinary, transdisciplinary approach.[1]

Low-carbon materials play a crucial role in promoting sustainability by addressing environmental, social, and economic dimensions. The United Nations' SDGs are primarily centered around several key points: affordable and clean energy; low-carbon materials are essential for advancing renewable energy technologies. For example, the materials used in solar panels,

Low Carbon Materials: Paving a Sustainable Path for a Greener Future
Edited by Santanu Patra, Sudheesh K. Shukla and Mika Sillanpää

Published by the Royal Society of Chemistry, www.rsc.org

wind turbines, and energy-storage systems can greatly reduce the carbon footprint of energy production. This facilitates the transition to cleaner energy sources, enhances energy efficiency, and reduces reliance on fossil fuels. Industry, innovation, and infrastructure; incorporating low-carbon materials into infrastructure projects, such as construction and transportation, promotes sustainable industrialization. These materials help reduce emissions from these sectors, which are typically high due to the use of conventional materials like concrete and steel. Innovations in low-carbon materials can result in more sustainable and resilient infrastructure. Sustainable cities and communities; urbanization poses a significant challenge, but utilizing low-carbon materials in building and construction can help create more sustainable cities. These materials help reduce the urban carbon footprint, improve air quality, and promote more sustainable urban development practices. Responsible consumption and production on climate action and environment; low-carbon materials promote responsible consumption and production by decreasing the environmental impact of manufacturing processes. They encourage the use of recycled and renewable resources, which helps to minimize waste and reduce the overall carbon footprint of products. One of the most immediate effects of low-carbon materials is their impact on climate action. By lowering greenhouse gas emissions, these materials help mitigate climate change and are crucial for meeting targets established by international climate agreements, such as the Paris Agreement. Also, the production and use of low-carbon materials can decrease pollution and environmental degradation. This helps safeguard ecosystems both on land and in water, preserving biodiversity and maintaining ecological balance, and human populations as well. Economic growth: the development and adoption of low-carbon materials can stimulate economic growth by generating new markets and job opportunities. This encompasses areas such as research and development, manufacturing, and the implementation of sustainable practices across various industries. Health; low-carbon materials enhance health outcomes by decreasing pollution and improving indoor air quality. This is especially important in urban areas, where conventional construction materials can worsen air quality and contribute to related health problems. In conclusion, low-carbon materials are essential for achieving the UN's Sustainable Development Goals. They advance sustainable development by cutting carbon emissions, supporting cleaner energy, encouraging sustainable industrial practices, and improving overall quality of life. By fostering innovation and responsible resource use, low-carbon materials can make a significant contribution to a more sustainable and resilient future.[2]

Lately, useful and comprehensive health tracking has become important. Traditional public health systems face challenges of incomplete, delayed, and inefficient medical services. Interest in flexible and wearable devices is increasing due to their potential applications in wearable human health-monitoring and care systems. Leveraging advanced carbon materials can enhance the development of wearable and flexible devices that offer multiple functions and outstanding performance, enabling comprehensive, real-time monitoring of human health and disease prevention.[3–5] The increasing demand for environmentally friendly and sustainable technology has made the development of low-carbon materials for biomedical sensors a critical issue in recent years. Because of their special features like excellent electrical conductivity, biocompatibility, mechanical strength, ease of production, and functionalization, carbon-based materials have attracted interest and their characteristics render them perfect options for application in biomedical sensors that require precise and reliable measurements.

Biosensors are instruments that measure the chemical and biological reactions with the generation of measurable signals depending on analyte concentrations and are widely used in environmental monitoring, medical research, and healthcare applications. One of the most significant difficulties in the development of biological sensors is the need for materials that are not only sensitive but also environmentally friendly; so, low-carbon materials, on the other hand, suggest a potential solution to provide a sustainable alternative to traditional sensor components derived from non-renewable resources.[6,7] Overall, there is a lot of potential to reduce the environmental impact of healthcare technology without sacrificing their usefulness by investigating low-carbon materials for biomedical sensors. Accurate disease monitoring and a greener future in healthcare can be achieved through the integration of sustainable material choices into sensing devices. Furthermore, this integration can improve patient care.

In this regard, researchers have been investigating a range of low-carbon materials for possible use in biomedical sensors, including graphene, carbon nanotubes, and conductive polymers. These materials have great sensing capabilities due to their large surface area-to-volume ratio and capacity to interact with biological molecules at the molecular level. Moreover, low-carbon materials can be easily modified with particular receptors or biomolecules to improve their selectivity toward target analytes or biomarkers. This adaptability makes it possible to customize sensor platforms to meet particular requirements for medical diagnosis.[8,9]

In this chapter, special attention is given to explain the properties of carbon-based materials and their use in biomedical applications, to sample literature studies about these uses, and to gain a different perspective for new studies. The purpose of this chapter is to discuss the fundamental logic, types, and biomedical sensor applications of carbon-based materials as key concepts.

7.2 Nanocarbon-based Materials

There has been an increasing focus on low-carbon materials, with the goal of decreasing the carbon footprint from their manufacturing and consumption. Low-carbon materials are characterized by their reduced environmental impact, particularly in terms of CO_2 emissions, in comparison to traditional materials.[10] Due to the rapid advancement of nanotechnology, there has been a wide variety of nanocarbon-based materials available, such as fullerenes, carbon nanotubes, carbon nanofibers, graphene, and more.

7.2.1 Graphite

Graphite, one of the natural forms of carbon, with sp^2 hybridization has carbon layers formed by the covalent and metallic bonding of each layer and its carbon layers that are stacked forms in AB sequence linked *via* a van der Waals interaction thanks to the delocalized π orbital.[11] Graphite consists of alpha and beta forms that have hexagonal and rhombohedral geometries, respectively; although the two forms have the same physical properties, their layer stacks are quite different from each other.[12]

As depicted in Figure 7.1, which corresponds to the crystal form of graphite, the atoms of graphite are formed with a hexagonal structure located in each layer and the unit cells consist of four atoms, two of them (A and B) located on one layer plane and the others (A′ and B′) are in the same location plane like A and B atoms shifted by half the crystallographic *c*-axis. Also, A and A′ are the neighbor atoms in the adjacent layer planes, while B and B′ don't have the same positions as the A and A′ atoms.[11]

Because of the anisotropic feature of graphite dependent on its bonding mode this brings graphite good features such as electric conductivity and

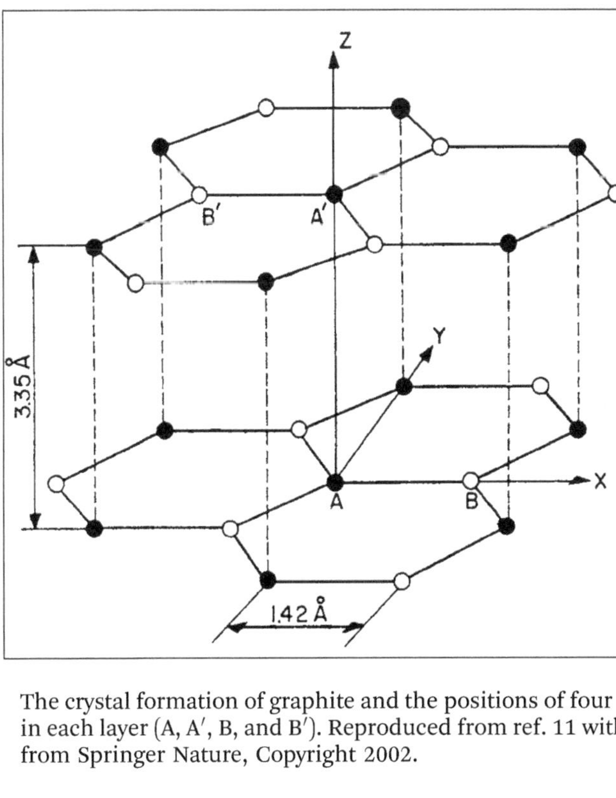

Figure 7.1 The crystal formation of graphite and the positions of four atoms found in each layer (A, A′, B, and B′). Reproduced from ref. 11 with permission from Springer Nature, Copyright 2002.

thermal properties, and makes graphite involved in some reactions known as intercalation.[11]

In other words and in detail, the thermal property of graphite is based on its anisotropic feature, especially the lattice vibration, and this property of graphite is examined *via* the Debye equation [eqn (7.1)].[13]

$$K = bCpvL \tag{7.1}$$

where b is a constant, C is the specific heat per unit volume of the crystal, v is the heat-transporting acoustic wave speed, and L is the free path for wave scattering.

However, the vibration severity in the planes and temperature are the main factors affecting the thermal conductivity of graphite.[13] In addition to this thermal conductivity, graphite is a good electric conductor material thanks to the free electrons moving; however, this key feature of graphite is related to the planes of graphite.[12,13] Graphite with an anisotropy feature can be involved in some reactions as a reactant and be able to form graphite intercalation compounds that are capable of higher electrical performances as compared to regular graphite.[11]

7.2.1.1 *Fabrication Routes*

Artificial or synthetic graphite is commonly fabricated *via* carbonization and graphitization processes,[14] while the carbonization process is known as a heating cycle, the carbon precursors such as polymers including cellulose, polyimide and aromatic hydrocarbons including petroleum fractions, and naphthalene are slowly heated at a high-temperature, *i.e.*, 900 °C or 1300 °C to decompose the carbon sources.[13] On the other hand, graphitization, including direct graphitization and catalytic graphitization, is accomplished in higher temperature conditions in the range of 1000–3000 °C using graphitizable and non-graphitizable carbon sources to obtain graphite by transforming the amorphous carbon sources.[14]

7.2.2 Carbon Nanotubes

Nanotechnology has been implemented to manipulate materials with nanodimensions to design a wide range of innovative materials with different desirable functional features which is driven by fundamental research science fields. Consequently, the inclusion of nanotechnology in various technological sectors such as sensors, electronic devices, energy, and materials science has greatly facilitated the quality of human life by their successful applications.

One of these nanostructured materials, carbon nanotubes (CNTs), as a member of carbon allotropes, with nanosized diameters and tube-like structures, were discovered by Iijima in 1991, which originated from a single

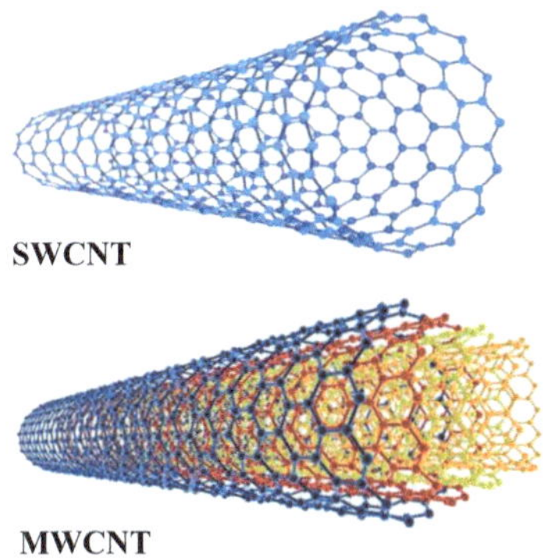

Figure 7.2 The formation of SWCNTs and MWCNTs. Reproduced from ref. 17 with permission from The Royal Society of Chemistry.

graphite which was rolled into a hollow cylinder form in a tube-like structure.[15–17] These graphite layers look like rolled-up, unending unbreakable hexagonal mesh structures with carbon atoms located at the apexes of the hexagonal structures.[15]

CNTs are classified as single-walled carbon nanotubes (SWCNTs) and multiwalled carbon nanotubes (MWCNTs),[16,17] which are depicted in Figure 7.2. SWCNTs with single graphene layers that contain solely carbon atoms are prepared by rolling of a tubular shell of graphene sheets, which are made of benzene-type hexagonal rings of carbon atoms, whereas MWCNTs with multigraphene layers have stacked morphologies and are made up by rolling up graphene sheets into concentric cylinders. Although a SWCNT has usually 10 atoms around its surface and the thickness of a SWCNT is the same as the thickness of the atom, an MWCNT with a larger and stacky form contains many single-walled tubes.[16]

Despite the nanoscale diameter of their thickness, their lengths can reach to centimeters, and furthermore, by changing the rolling positions of graphene along the lattice vectors during the preparation steps, other forms of SWCNTs, such as armchair, zigzag, and chiral CNTs (Figure 7.3) can be prepared.[16,17]

7.2.2.1 *Fabrication Routes*

The synthesis of these nanomaterials is achieved by arc discharge, laser ablation, and chemical vapor deposition methods; the arc discharge and laser ablation methods depend on the condensation of hot carbon gases after the evaporation of carbon atoms *via* high-temperature and laser sources, respectively; however, the use of catalytic agents such as nickel, iron, and cobalt is necessary for the generation of individual SWCNTs during the arc discharge method.[15]

The chemical vapor deposition method (CVD) consists of the decomposition of carbon sources, *e.g.*, methane and ethylene, *via* the

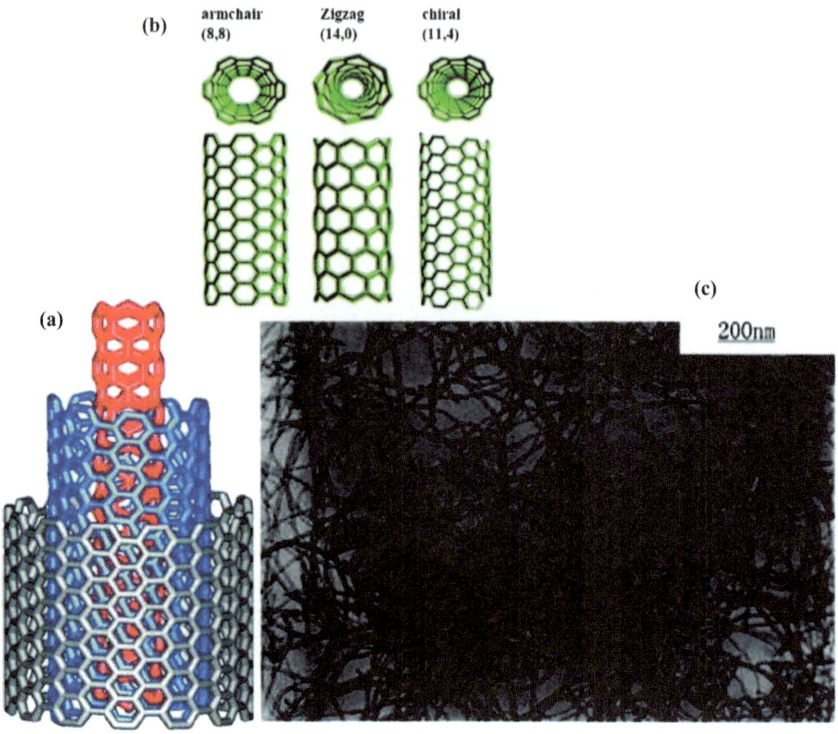

Figure 7.3 (a) MWCNTs, (b) different types of CNTs dependent on rolling up graphite, and (c) transmission electron microscope (TEM) image of purified MWCNTs with carboxylation functionalization. Reproduced from ref. 17 with permission from The Royal Society of Chemistry.

catalyzation of metallic nanoparticles that are used to promote nucleation sites where the initiation of CNT growth is enhanced. This CVD method offers the large-scale production of CNTs with a lower temperature as compared to the above-mentioned methods and during the production process of the proposed methods, the wall number, length, diameter, and fabricated amounts can be well optimized.[15,16]

7.2.2.2 Key Features for Biosensing Applications

The sp^2 covalent bonds of CNTs provide them with outstanding features including mechanical strength, and CNTs with 1 TPa of Young's modulus and up to 100 GPa of textile strength, which is stronger than steel;[18] however, CNTs with low weights are elastic materials and can protect their original shapes and formations under certain limits.[19] Furthermore, CNTs show metallic and semiconducting features along their tubular axis, and with the adjustment of some parameters, *i.e.*, diameter, alignment, aspect ratio, and chirality, it is possible to control the electric conductivity of these nano-sized materials.[19,20]

7.2.3 Graphene

Graphene (GR), a one-atom-thick, 2D material is referred to as a miraculous material which has taken the attention of scientists because of its many incredible properties such as high surface area, elasticity, thermal conductivity, optical transmittance capability, and so on, after the invention of the purest form of carbon.[21]

The GR structure looks like a hexagonal honeycomb (Figure 7.4) and during the formation of GR in a hexagonal honeycomb lattice, the carbon atoms bond together *via* sp^2 hybridization; however, the lattice includes two similar sublattices that are constituted by bonding carbon atoms to each other.[21,22]

As the structural formation of GR is examined with tight bonding band (TBM) theory, GR valance electrons (the arrangement of three of four valence electrons) form σ and π bonds (the fourth valance electron) that give GR its exceptional features such as elasticity, flexibility, robustness, and electric properties; furthermore, GR differs from other materials because of its 2D lattice and incomparable band structure.[21]

7.2.3.1 Band Structure and Properties of Graphene

GR is a zero band structure because of Fermi energy that takes place by the decoupling of three π bonds from the σ bonds due to the inversion of symmetry,[21,23,24] and the band of GR is constituted with conduction and

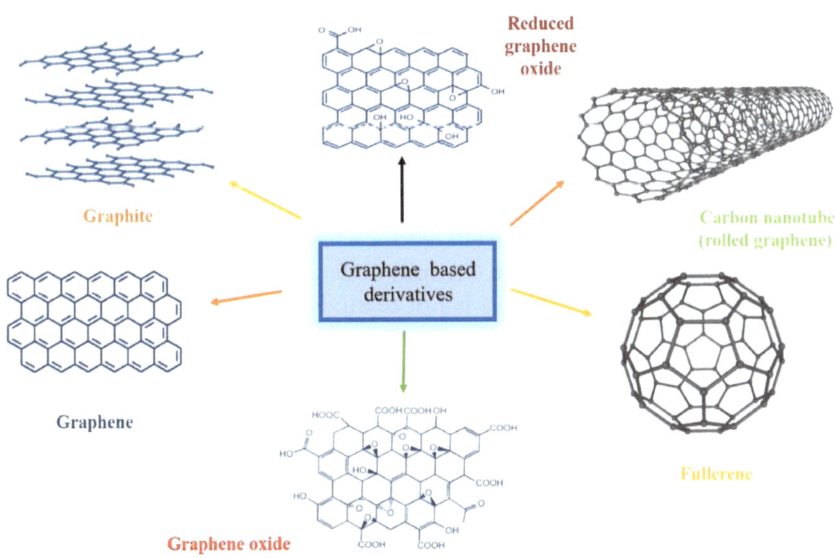

Figure 7.4 Graphene and its derivative forms. Reproduced from ref. 22, https://doi.org/10.1016/j.mtadv.2022.100208, under the terms of the CC BY 4.0 license, https://creativecommons.org/licenses/by/4.0/.

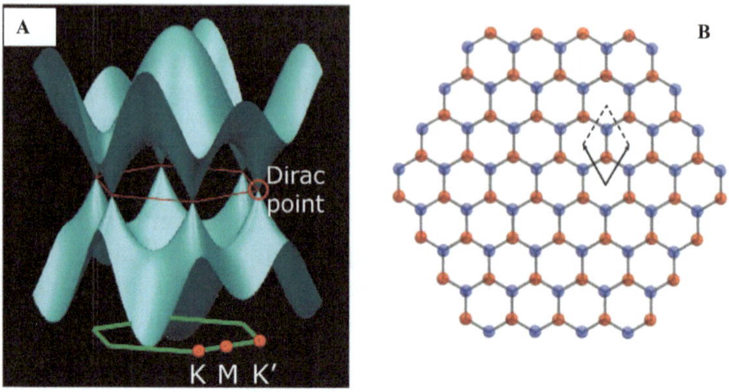

Figure 7.5 (A) Represents the Dirac, *K*, and *K*′ points of GR and (B) represents the sub-lattices (illustrated with red and blue colors) of GR. Reproduced from ref. 25 with permission from Elsevier, Copyright 2012.

valence bands. The valence bands are molded by bonding π states, whereas anti-π bonding states form the conduction band, and these orthogonal states touch at six points referred to as Dirac points (Figure 7.5A).[25] The sub-lattices of GR are presented in Figure 7.5B. The degeneration of these states occurs at six points located on the corners of the Brillouin zone defined as *K* and *K*′ points.[21]

The Brillouin zone describes the GR Fermi surface that is zero dimensional and contains six infinite points; hence, GR is a semimetal material because of the reasons mentioned above and its electric feature is independent of the changing of the *K* and *K*′ valley states; however, this is only related to the phase reversal symmetry.[21,26,27]

The electronic structure of GR is explained by the Hamiltonian equation which is illustrated in eqn (7.2).[24]

$$H = VF\sigma\hbar k \tag{7.2}$$

where *VF* is the Fermi velocity of GR, *k*, the wave factor of the electron, and σ, a spinor-like-wave function. The GR spin or wave feature depends on the two atoms found in the unit cell and is not related to spin rising; therefore, this phenomenon could be explained using a pseudo-spin, which has the same velocity direction of the group and describes the electron population found in A and B sites. Unlike the real-spin, pseudo-spin reversal is not permitted during the carrier interactions, resulting in backscattering inhibition of GR, weak electron coupling, and high optical phonon frequencies, which give GR a good transport feature.[25]

7.2.3.2 *Fabrication Process*

The top-down and bottom-up methods are the basic routes of GR fabrication; the top-down method is based on the exfoliation of graphite (a starting material) *via* chemical, mechanical, chemical/electrochemical exfoliation, and chemical/electrochemical fabrication routes, which allows for large-scale production of GR in a straightforward manner using graphite or carbon sources.[26]

Among them, the mechanical exfoliation method was the first Nobel prize method used in obtaining a monolayer GR sheet,[26–28] and during the fabrication of this method, graphite is placed on a substrate and continuously peeled with the help of an adhesive tape until the monolayer is created.[27,28] The electrochemical exfoliation method needs an anode, cathode, electrolyte, power supply components, and a carbon source, the anode is first oxidized then exfoliated by applying positive and negative voltages to the anode depending on the peeling process.[26]

On the other hand, bottom-up methods for GR fabrication consist of CVD and arc discharge approaches; among these routes, single and few layers of GR can be produced with a CVD route that utilizes Ni and Cu ions as a substrate; however, the experimental conditions can be optimized during the formation of GR due to the utilization of metal substrates.[26–29]

7.2.4 Carbon Black

As with the other forms of carbon, carbon black contains more than 90% pure carbon elements and is defined as a type of aggregate (Figure 7.6) with the formation of tiny and spherical forms of carbon atoms. The diameter of black carbon particles is in the range of 10–500 nm, and the particles with larger sizes lead to the formation of a graphitic structure.[30]

The outer surface of black carbon shows more graphitic structure as compared to its center, while carbon black with non-parallel graphitic layers is quite different from graphite.[12] The carbon atoms in the aggregates that locally exhibit a quasi-crystalline turbostratic structure with an interlayer spacing of about 0.35 nm can be visualized in overlapping graphitic layers *via* the higher magnification and surface modification of carbon black, such as introducing functional groups, can be accomplished using the graphitic edge found in each layer.[30]

7.2.4.1 Fabrication Routes and Specifications of Carbon Black

Carbon black is manufactured using different petroleum products such as coal tar and fluid catalytic cracking tar, and a small amount is produced using vegetable oils.[12] For the production of carbon black, the lampblack, oil furnace, acetylene, thermal, and impingement approaches are employed. Lampblack is the oldest method to produce carbon black based on burning aromatic oils in shallow pans with a limited air supply. Despite the very limited use of lampblack, it is possible to manufacture particles with 70–100 nm diameters. The oil furnace method is a rapid and continuous process employed in producing black carbon aggregates by converting highly aromatic feedstock oils at high temperatures. During the manufacturing steps, the

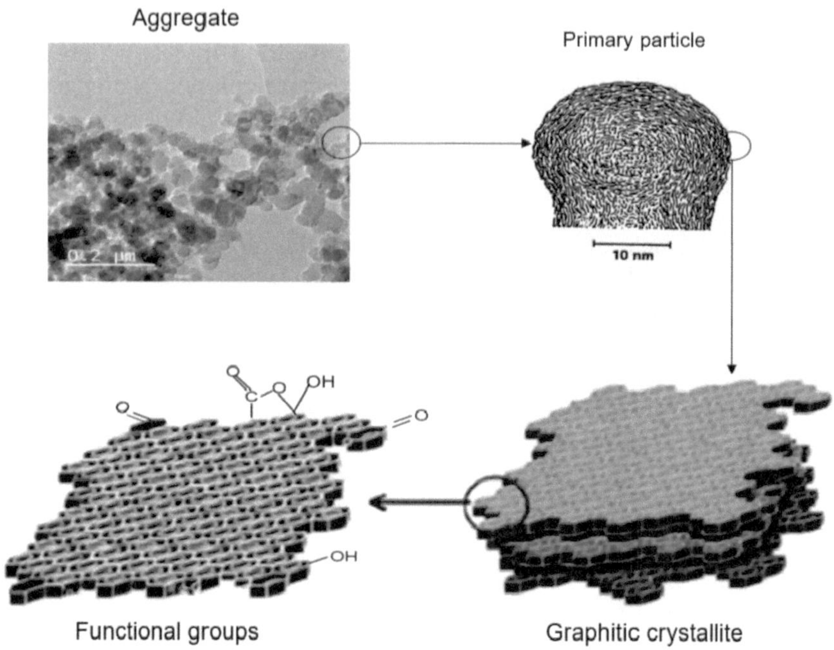

Figure 7.6 Aggerate form of carbon black, primary particle formation, graphitic crystallite, and introduction of functional groups into carbon black.[30] Reproduced from ref. 30 with permission from Elsevier, Copyright 2020.

flame in a reactor is used with the combination of natural gas or oil and an excess of air, and then the feedstock oil is injected into the flame. After a while, the cooling process is applied until the carbon black is collected. The acetylene gas process involves the manufacture of carbon black at high temperature where before the fabrication process, the reactor is preheated to approximately 800 °C, and after that, acetylene gas is passed through the reactor and the temperature is increased to above 2500 °C. The thermal process is a cyclic-type process and carbon black with the larger sizes (250–500 nm) is fabricated with the help of the decomposition of oil and natural gases in the absence of air. During the fabrication process, the generator is preheated and after the production cycle, a steam purge is applied to remove the carbon black and hydrogen gas from the reactor and then, after the cooling step, the carbon black is collected. The impingement approach is rarely used compared with the other processes and, in this approach, fan-type flames are produced from the burning of natural gas, and the flames are utilized in the production and collection of carbon black.[31]

Even though carbon black is widely used in the rubber industry, it is also useful for thermal and electric applications because of its high porosity and compatibility, and the choice of fabrication process mentioned above is of great importance for carbon black performance during usage due to several factors.

7.2.5 Carbon Fibers and Their Properties

Carbon fiber is another allotrope of carbon with a long history dating back to the 1860s and was used in early incandescent light bulbs; however, the use of tungsten in light bulb filaments led to a reduction in its popularity until the 1950s. After that, several researchers produced carbon-based materials and these studies have been chronologically reviewed by Peijs and coworkers.[32]

Carbon fiber contains more than 90% (wt) of carbon and has good mechanical features including high tensile properties, compressive strength, and low thermal expansion, and shows electrical and thermal conductivity, temperature, and chemical resistance.[33] Because of these suitable features, the use of carbon fibers has been growing in several areas, for instance the automobile, aerospace, and military sectors, construction of turbine blades, and also in sensor applications.[34,35]

Although the structure of carbon fiber is similar to that of graphite, the layer planes of carbon fiber can be a turbostratic, graphitic, or hybrid structure based on the manufacturing process and the use of precursors. For instance, the graphitic structure of carbon fibers can be fabricated *via* mesophase pitch (MS), whereas a turbostratic structure can be formed using polyacrylonitrile (PAN) precursors.[34]

7.2.5.1 Fabrication of Carbon Fibers

Today, more than 90% of PAN and a small amount of lignin and cellulose are employed in producing carbon fibers that are generally produced *via* controlled pyrolysis of stabilized precursor fibers.[34] In this approach, first, the precursor fibers are stabilized in air by an oxidation process at 200–400 °C, following that, carbonization is accomplished at around 1000 °C in an inert atmosphere to remove H, O, N, and the other non-carbon elements. Finally, the carbonized fibers can be further graphitized at around 3000 °C to produce carbon fibers; however, other processes like electrospinning can be applied to fabricate carbon nanofibers.[35]

7.2.6 Fullerene

Another form of carbon allotrope, fullerene, is a symmetrical zero-dimensional (0D) nanomaterial with sp^2 carbon hybridization that contains carbon atoms in ellipsoid, hollow sphere, or tube forms.[36–38] The carbon atoms found in the fullerene have π configuration and are arranged in hexagonal and pentagonal rings.[37,38] These 0D nanomaterials can be found in various forms such as C_{24}, C_{28}, C_{60}, C_{70}, and C_{120} depending on the number of carbon atoms; however, the first discovered form of fullerene, C_{60}, and C_{70} are the most studied and more stable forms of fullerenes.[36,38]

7.2.6.1 *Fabrication and Properties of Fullerene*

The fabrication routes of fullerene are dependent mainly on high-temperature and proper pressure to reduce the carbon sources into the smallest particles before the formation of fullerene; for instance, the laser ablation and arc discharge routes need high temperature and proper pressure control during the fabrication of fullerene. Radio frequency and direct vaporization routes are other approaches for fullerene fabrication, and during the fabrication of fullerene *via* the radio frequency route, radio frequency-powered plasma is used to evaporate the metal ions and carbon sources in the fullerene formation zone, whereas sunlight is employed in vaporization of hydrocarbon to fabricate fullerenes *via* large solar furnaces.[36]

The tensile strength, good thermal and electrical conductor properties, high stability, and reaction capability of fullerene make it a promising material for various applications from medical applications to electronic studies.[37,38]

7.2.7 Carbon Dots and Fabrication

Like fullerenes, carbon dots (CDs) with less than 20 nm diameter, are quasi-0D carbon-based materials exhibiting fluorescence properties and quantum yields.[39] These promising materials can be found in several forms, *e.g.*, graphene quantum dots (GQDs), carbon quantum dots (CQDs), and carbonized quantum dots (CQDs), based on their different forms, nanostructures, properties, and fabrication processes.[39,40]

The top-down and bottom-up routes are the main production routes for CDs, and the choice of fabrication process has several advantages and disadvantages when compared with each other. During the top-down methods, graphene, carbon black, carbon fiber, and GO are used as carbon sources for CD fabrication using several methods such as oxidative cutting, hydrothermal synthesis, microwave-assisted cutting, solvothermal synthesis, ultrasonication methods, and electrochemical exfoliation with the aid of oxidative agents like acids, H_2O_2, high temperature and pressure, organic solvents, and ultrasonication depending on the fabrication route; however, GO with plenty of oxygen-containing groups is highly preferable among the carbon sources during the fabrication of CDs.[41]

Fabrication using bottom-up methods, including the carbonization, hydrothermal, solvothermal, and microwave-assisted methods, is generally based on carbonization processes that are employed in fabricating CDs using small molecules, *i.e.*, glucose, citric acid, urea, *etc.*[41]

7.2.7.1 *Key Properties of Carbon Dots for Biosensing Applications*

The most common reason for using CDs in the various application areas, but especially in biosensor technology, is the photoluminescence (PL) and electronic properties thanks to the electron transitions across the sp^2 band gaps.[41] Furthermore, the low-toxicity, high quantum yield (QY), affordability, stability, and biocompatibility of CDs make them favorable materials as compared with the other fluorescent materials, *e.g.*, traditional quantum dots and dyes.[39] Therefore, from these key features, these promising materials have paved the way for several applications, including in biosensor platforms.

7.3 Sensor Types Used in Low-carbon Materials

In recent years, nanotechnology has experienced notable growth in sensor applications. There is a perception that integrating these advancements, alongside nano-scale materials, holds great promise for enhancing sensor performance. Nano-scale materials display a multitude of fascinating physical and chemical characteristics. In recent decades, the discovery of nanometer-sized materials has opened up new frontiers in the exploration of condensed matter physics. Changes in the electronic configurations within various nano-scaled systems give rise to diverse sensing applications and their corresponding responses. These applications span across electrochemical sensing, biosensing, amperometry, piezoelectric sensing, and more. The responses vary depending on the carbon-based nanomaterials (CNMs) such as carbon nanotubes, graphene sheets, carbon dots, carbon nanoflowers, *etc.* Recent trends in research within this domain will underscore these distinctions.[42] Carbon can form numerous chains with varying lengths and electronic configurations, as depicted in Figure 7.7. Carbon-based nanomaterials frequently appear in the forms of hollow spheres, ellipsoids, or tubes. Fullerenes are carbon nanomaterials that exist in spherical and ellipsoidal forms, while nanotubes refer to the cylindrical variants. These particles hold significant potential for applications in biomedical analyses.[13]

A sensor identifies changes in a physical stimulus and converts them into a signal that can be evaluated, measured, and/or recorded. Recently, nanotechnology has seen notable adoption in sensor applications. Utilizing these technologies alongside nanoscale materials is thought to hold promise for enhancing sensor performance. Nanosensors consist of two primary components: a recognition element made of nanomaterials such as nanometals, nanotubes, nanowires, nanoparticles, *etc.*, directly linked to a physical transducer such as voltammetric, amperometric, conductometric, spectrophotometric, *etc.* The wide range of nanomaterial systems available for use as recognition agents enables the development of specific nanosensors for a diverse array of analytes. Biosensors are classified according to their transducers, which include optical, electrochemical, electrical, thermal, and mass-change sensors.

7.3.1 Optical Biosensors

Optical biosensors, the prevailing type among biosensors, are discussed here, including a concise categorization and an explanation of their operational principles. Furthermore, various other optical biosensor systems are

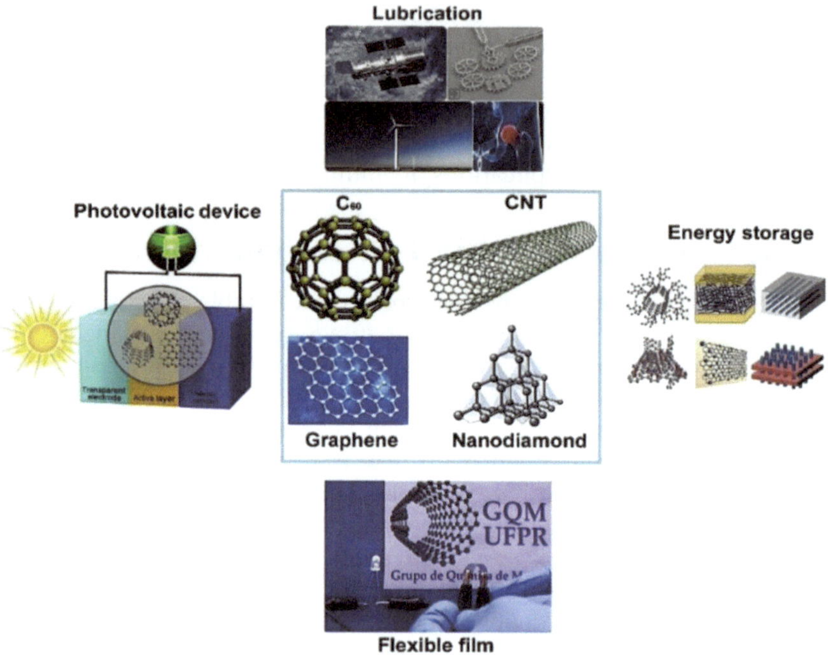

Figure 7.7 Nanomaterials with advanced technological use: fullerene C60, CNT, graphene, and nanodiamond. Reproduced from ref. 43 with permission from Elsevier, Copyright 2017.

outlined, including interferometric, ellipsometric, and reflectometric interference spectroscopy, along with surface-enhanced Raman scattering biosensors. Additionally, fluorescence and bioluminescent optical fiber biosensors are discussed.

7.3.1.1 Surface Plasmon Resonance (SPR) and Localized Surface Plasmon Resonance (LSPR)-based Biosensors

The SPR approach relies on detecting alterations in refractive index resulting from molecular interactions occurring at a metal surface or other conductive materials *via* surface plasmon waves (Figure 7.8). SPR biosensors comprise optical components (such as a light source, prism, and gold sensor) as well as fluidic elements. The laser, serving as the light source for SPR, upon interacting with the gold surface, triggers the stimulation of unbound electrons on the metal surface, thereby generating a plasmonic wave between the gold surface and the fluid layer. Until now, SPR optical-based detection has been demonstrated as a powerful technique for the real-time and label-free detection of proteins, DNA, aptamers, peptides, and cell surface markers, as well as for assessing kinetic data in interactions involving small molecules.

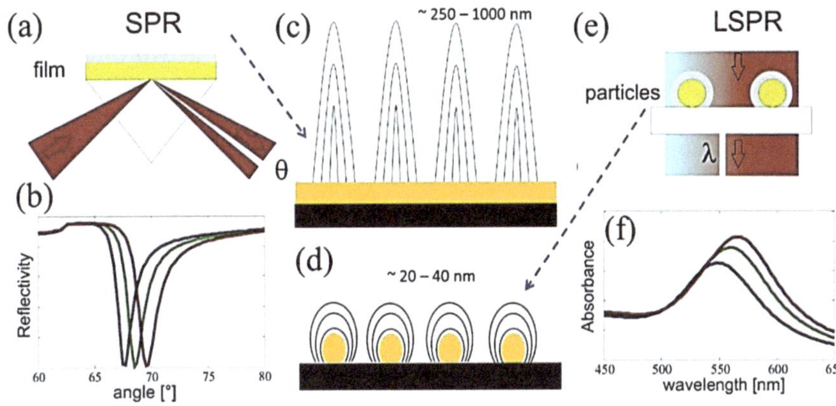

Figure 7.8 The mechanisms of SPR (a–c) and LSPR (d–f). Reproduced from ref. 46 with permission from the Royal Society of Chemistry.

In SPR-based biosensors, there is no requirement for a precise electrochemical cell or a functional carbon electrode for analysis. Collective oscillations induced by laser light on nanometric metallic thin films such as gold, silver, or copper on the SPR platform can be analyzed using polarized laser light and a prism. SPR responses correlate with the reflected light at specific incidence angles, particularly when steep slopes are observed in the reflectivity curves. Through SPR-based techniques, changes in mass and refractive index on the specific capture probe can be quantified in response units (RU), corresponding to signals of 10^{-6} refractive index units (RIU) on the transducer.[44]

Biosensors based on localized surface plasmon resonance (LSPR) rely on the local dielectric properties and alterations in the refractive index of the surrounding environment of a gold chip (Figure 7.8). These advancements have led to significant progress in diagnostic research. Over the past decades, SPR and LSPR sensors have emerged as key label-free technologies. These sensors find applications in various fields such as molecular diagnostics, clinical diagnostics, toxin and small-molecule sensing, environmental monitoring, and food safety assessment. By tracking changes in the angle or wavelength of p-polarized light absorbed by the surface plasmon, changes in the dielectric constant or refractive index near the surface of the plasmonic material are recorded.[45]

7.3.1.2 Optical Interferometry

Interferometry facilitates the detection and examination of binding events without relying on costly and intrusive labeling methods. When two or more light waves overlap, they generate an interference pattern. By examining these patterns, we can explore the characteristics of both the light waves and the material they interacted with. Interferometry has found applications in the biological sciences as a means to monitor and quantify molecular interactions. In this domain, interferometry presents a twofold benefit: it serves as a highly sensitive technique without necessitating the employment of costly molecular labels. Interferometric measurements outperformed SPR when a probe other than an antibody was utilized and provided exclusive capability for achieving the highest resolution for low-molecular-weight compounds. Interferometric optical biosensors including Mach–Zehnder interferometers, Young interferometers, and Hartman interferometers are capable of measuring binding interactions in free solution while requiring minimal sample volumes and maintaining a simple and cost-effective setup. However, these sensors are susceptible to environmental factors, particularly ambient vibrations and fluctuations in temperature. Laser interferometry is renowned for its remarkable precision in measurement, attaining displacement resolutions as minute as femtometers (fm). The advancement of micro-electro-mechanical systems (MEMS) technology has led to the emergence of numerous outstanding optical interferometric MEMS sensors. These sensors play a crucial role in various fields such as intelligent manufacturing, aerospace, and beyond. Optical interferometric MEMS accelerometers are experiencing rapid development, primarily due to their notable advantages such as high sensitivity, minimal noise, and resistance to electromagnetic interference. Numerous types of MEMS accelerometers, which rely on optical interferometric principles, are under development. Extensive research is being conducted on optical interferometric chip design, micro-/nano-fabrication processes, signal demodulation algorithms, and range extension technologies.[47]

7.3.1.3 *Surface-enhanced Raman Scattering Spectroscopy*

Raman spectroscopy has been proved to be a potent technique for the identification, characterization, and quantification of unidentified objects. This method provides fast, non-destructive, and non-invasive analysis, yielding distinctive fingerprint spectra and chemical insights for substances in almost any state. In comparison to other spectral detection methods like infrared spectroscopy, Raman spectroscopy presents notable advantages by exhibiting minimal interference from water and necessitating little sample pretreatment. Yet, its broader applicability remains constrained by relatively low sensitivity, stemming from the Raman signal's diminished intensity. Surface-enhanced Raman scattering (SERS) has significantly overcome the mentioned limitations, demonstrating high sensitivity, minimal interference from water backgrounds, shorter testing times, and decreased sample preparation requirements. Consequently, SERS is esteemed as a highly promising analytical method for the targeted detection of chemical and biological molecules across various matrices, providing ultra-high sensitivity. Recently, it has attracted significant attention as a vital tool for identifying and detecting trace targets in areas such as analytical chemistry, biomedical diagnostics, food quality evaluation, environmental monitoring, and national security. As molecules adsorb to rough noble metal surfaces, like silver (AgNPs) or gold nanoparticles (AuNPs), the local plasmon resonance effects on these surfaces can amplify the Raman signal through an electromagnetic mechanism. The efficacy of SERS predominantly depends on the plasmonic characteristics of metal nanostructures, alongside their morphology, size, aggregation state, and ambient conditions.[48] There are three SERS probe types: (i) Raman reporter molecules; (ii) plasmonic metal nanoparticles coupled with Raman reporters; and (iii) sandwiched metal nanoparticle@Raman reporter@silica nanoparticles. Surface-enhanced Raman scattering (SERS) is considered one of the most sensitive analytical techniques, resulting from the adsorption of the target of interest onto a plasmonic surface.[49]

7.3.1.4 Fluorescent Biosensors

Fluorescence is commonly employed as a detection technique for biological molecules because of its high sensitivity and user-friendly nature. Fluorescent techniques are highly significant in biomedicine because they provide direct insights into the localization of molecules within a wide range of living cells and organisms. Observation of molecular interactions through fluorescence resonance energy transfer (FRET) has led to the development of many fluorescent biosensors containing various nanomaterials. FRET depends on the energy transfer between a donor and acceptor, which is distance-dependent, highly specific, and sensitive. This makes it well-suited for assessing molecular interactions.[50] FRET, a non-radioactive process, involves the transfer of energy from a luminescent donor to an acceptor with a lower-energy excited state when they are within a spatial range of 10–100 Å. In this process, an energetically excited donor system emits photons, raising the energy state of an electron in the acceptor system to different vibrational levels of the singlet excited state. This occurs as a result of the overlap of the absorption spectrum of the acceptor system with the fluorescence emission spectrum of the donor molecule. Generally, a fluorophore or a non-fluorescent molecule is used as the acceptor molecule (Figure 7.9).[51]

Low-carbon materials are gaining recognition as a promising category of photo-luminescent nano-materials, attributed to their outstanding optical, electrical, and chemical characteristics, along with their versatility in various environmental applications. Carbon dots (CDs) exhibit distinctive and attractive properties, including exceptional stability, minimal toxicity, water solubility, and derivability. Combining CDs with FRET leads to the creation of exceptionally sensitive radiometric fluorescence sensor probes, offering promising applications in bio-imaging, metal detection, membrane dynamics, and environmental sensing.

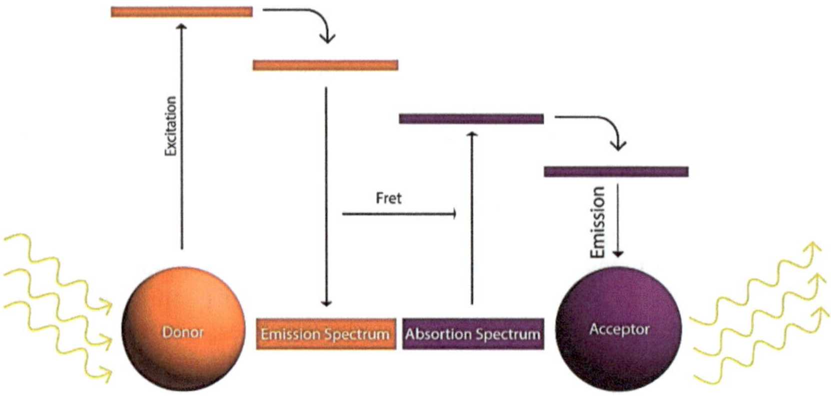

Figure 7.9 FRET mechanism.

7.3.2 Electrochemical Biosensors

Low-carbon-based materials hold promise for the development of electrochemical biosensors with their abundant surface areas and they are well-suited for immobilizing biomolecules, improving electron transfer rates, and aiding in the detection of target analytes through electrochemical signals. The exploration of the electrochemical properties of bioactive molecules is rapidly evolving as a significant scientific field. There is a growing interest within the biotechnology and biomedical engineering sectors to develop more accurate and dependable voltammetric/amperometric biosensors. An increasingly promising direction in biosensor advancement involves incorporating carbon-based materials, such as one-dimensional carbon nanotubes, two-dimensional graphene, and graphene oxide, into working electrodes.

7.3.2.1 Amperometric and Voltammetric Biosensors

Oxidation or reduction reactions occurring on the surface of the working electrode produce current (i), and continuous measurement of the produced current density is provided by amperometric biosensors. The electrode materials may consist of graphite, noble metals, and modified variants of carbon or conducting polymers. However, appropriate analytes must be added to the medium to facilitate the electrochemical reaction at the working electrode to be used for the selected analyte. Certain analytes, such as those based on proteins, may not function effectively as redox partners in such reactions. Reports also indicate that the uncompensated intrinsic resistance of DC measurements may make this technique inadequate in terms of specificity and selectivity for applications requiring high sensitivity.

The subclass of amperometric biosensing also encompasses voltammetry, wherein the current is measured while controlling variations in the potential. Voltammetric biosensors rely on electro-analytical chemistry methodologies, where the quantitative detection of analytes occurs through the modulation of potential, and the resultant current is measured as the analyte undergoes electrochemical reactions at the surface of the working electrodes. Essentially, a voltammetric biosensor functions as an analytical tool that transforms a biological reaction into a measurable current signal. The components of a standard voltammetric biosensor consist of a working electrode (various types of carbon such as platinum, gold, carbon fiber, epoxy graphite, graphene, glassy carbon, commercial screen printing electrode, *etc.*), a counter electrode, and a reference electrode. Voltammetric methods are favored for biosensing due to their exceptional sensitivity, convenience, excellent selectivity, straightforward designs, affordability, and rapid analysis capabilities. Voltammetric biosensors utilize electro-analytical chemistry methods, where the quantitative detection of analytes occurs by altering the potential and measuring the ensuing current as the analyte undergoes electrochemical reactions on the surface of the working electrode. Various techniques can be employed to vary the potential in voltammetric sensing, including linear sweep, differential staircase, normal pulse, reverse pulse, and differential pulse methodologies. Linear sweep cyclic voltammetry (CV) is the most commonly used method to determine the redox potential and electrochemical reaction rate occurring in analyte solutions. Several factors influence the overall shape of a cyclic voltammogram (CV), including the reduction potential of the analyte, sweep rate, electrolyte composition, analyte isomerization, electrode surface properties, stability of the reduced/oxidized analyte, and electrochemical reversibility of the analytc.[52]

7.3.2.2 Potentiometric and Conductometric Biosensors

Biosensors incorporating potentiometric transducers detect generated potential difference across an ion-selective membrane that separates two solutions with minimal current flow for measuring analyte activities in a range. The operational concept of potentiometric biosensors is based on the occurrence of current flow within an electrochemical reaction, facilitated by the application of a ramp voltage to the electrodes immersed in the solution. Potentiometric biosensors are well-suited to miniaturization and modern silicon fabrication techniques.[53]

Most commercially available biosensors use pH electrodes, ion-selective electrodes (ISEs), glass electrodes, and metal oxides. Researchers are increasingly directing their efforts toward developing innovative membrane compositions using carbon-based materials, polyvinyl chloride (PVC), and specialized ionophores to augment their capabilities and applications. These membranes are predominantly employed in ISEs and ion-sensitive field-effect transistors (ISFETs). ISFETs are equipped with an ion-sensitive surface. When there is any interaction between the semiconductor and ions, it leads to a modification in the surface's electrical potential, which can subsequently be quantified. In the construction of the ISFET, a selectively permeable polymer layer covering the sensor electrode is utilized. Through this layer, ions diffuse, altering the surface potential.

Conductometric transducers in biosensors can also fall under the category of potentiometric transducers. Conductometry involves measuring alterations in the conductivity of ionic species in response to the bio-recognition event. Although they have certain advantages, such as the ability to operate without the need for reference electrodes, the ability to operate with low-amplitude alternating voltages, immunity to light interference, and ease of integration and miniaturization, applications of conductometric transducing methods in wearable biosensing are limited. These can be summarized as follows: high signal-to-noise ratios (>2%), which leads to reduced sensitivity; low specificity resulting in inability to discriminate between simultaneous reactions; and polarization at the electrodes due to the double-layer capacitance formed during the reaction. Moreover, the response value of these biosensors is significantly influenced by environmental factors such as pH, ionic strength, and buffer capacity.

7.3.2.3 Impedimetric and Capacitive Biosensors

Impedimetric biosensors measure the impedance created by redox reactions at the analyte–electrode interface triggered by biological recognition. This method involves applying a minor perturbation bias to detect alterations in the oscillations of the current response. Impedimetric measurements necessitate defining a mathematical transfer function for electrochemical impedance. Furthermore, impedimetric transduction gives rise to the electrochemical impedance spectroscopy (EIS) technique, a widely utilized biosensing technology that involves scanning across perturbation frequencies to measure the resulting impedance changes at the interface between the bio-receptor and the analyte.[54]

7.3.3 Thermal Sensors

Biosensors employing thermal, calorimetric, or thermometric transducers quantify changes in energy (heat, q) occurring within a system and its surroundings. These biosensors are built by attaching bio-elements onto temperature sensors, where the energy change resulting from bio-recognition processes within the system and its surroundings (*i.e.*, heat exchange) is assessed.

After biosensors are created by immobilizing bio-elements on temperature sensors, the energy change (*i.e.* heat change) caused by biological recognition of the system and its environment is determined. The working principle of these biosensors involves the acquisition of analyte solution. During the operation of these biosensors, the analyte solution is taken from the substrate inlet and then the temperature is measured using a thermal transducer such as a microelectromechanical system, thermocouple, resonator, or thermophile. The solution passes through a packed column housing immobilized enzymes, facilitating enzyme-catalyzed reactions that result in either heat loss or generation. The subsequent temperature measurement is conducted by a distinct thermal transducer as the solution advances toward the outlet.[55]

7.3.4 Gravimetric/Piezoelectric/Mass-sensitive Biosensors

Piezoelectric biosensors are created through the integration of a biocomponent with piezoelectric transducers, typically employing a quartz crystal coated with gold electrodes. Piezoelectricity is a reversible phenomenon wherein the application of mechanical stress induces deformation or vibration in the material, leading to the accumulation of electrical charge. Piezoelectric biosensors operate by detecting frequency changes that occur on the surface of the transducer. As the target analyte binds to the material, the mass of the crystal component changes, resulting in changes in the resonance frequency. To pursue this objective, microcantilevers that detect alterations in bending or vibrational frequency caused by physical, chemical, or biological stimuli can be categorized as gravimetric or mass-sensitive biosensors. While precision and sensitivity are advantageous qualities, a common drawback of mass-sensitive transducers is their susceptibility to temperature fluctuations, which can lead to the degradation of these characteristics at extremely low or high temperatures. Figure 7.10 shows a schematic representation for mounting of a piezoelectric sensor.[56]

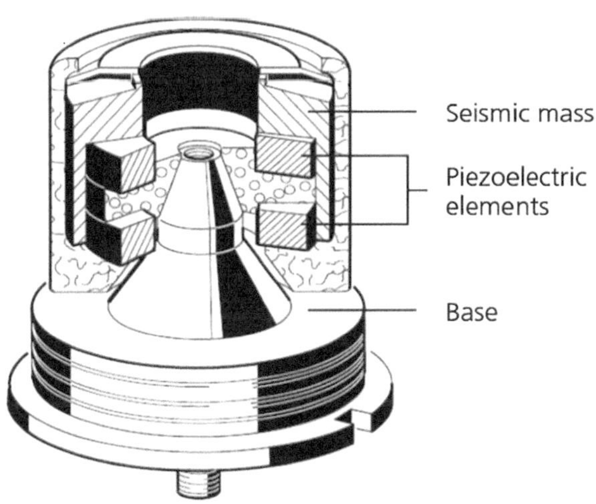

Figure 7.10 Mounting a piezoelectric sensor.[56] Reproduced from ref. 56 with permission from Springer Nature, Copyright 2024.

7.4 Applications of Nanocarbon-based Materials for Biosensing

Carbon-based materials are increasingly becoming attractive nanomaterials with a wide range of uses, such as carbon nanotubes, graphene, and carbon nanofibers. These materials have several important uses in the biomedical sciences, diagnostics, tissue engineering, biosensors, and drug-delivery systems. Recently, there has been a lot of interest in the use of nanocarbon-based materials in biomedical analyses, as well as in applications using real samples, such as human serum, blood, urine, *etc.*[57,58]

Carbon nanotubes are evaluated in different fields, especially clinical diagnosis, environmental monitoring, and food safety, due to their advantageous properties such as biocompatibility, fast electron transfer kinetics, chemical inertness, which make them a suitable material for biomedical applications. MWCNTs have a wide surface area and strong conductivity due to their cylindrical form, which makes it easier for analytes to interact and adsorb on them.[15,59]

Tabei *et al.*[60] designed an electrochemical sensor for the identification of the antioxidant drug dipyridamole in aqueous solution. Here, the electrophoretic deposition approach was used to modify a hollow-fiber membrane with silver/multiwalled carbon nanotube/polyaniline nanocomposite. After detailed characterization, optimization studies were carried out to emphasize the importance of effects such as pH and potential scan rate, the detection limit was found to be 1×10^{-3} μM.

Zn-metal–organic frameworks and multiwalled carbon nanotubes were effectively constructed using electrodeposition techniques on a gold electrode to create a non-enzymatic glucose sensor by Xue *et al.*[61] These researchers also investigated the effect of different parameters such as selectivity, stability, and reproducibility on the sensor performance. The fabricated sensor demonstrated a linear response to glucose concentrations ranging from 0.020 to 8.14 mM with a detection limit of 0.037 mM.

Considering the advantages of the molecular imprinting method, Duan *et al.*[62] developed an electrochemical sensing platform based on nitrogen-doped carbon nanotubes loaded with nickel nanoclusters to determine the level of cortisol in saliva samples. In the study using *o*-phenylenediamine as the functional monomer, they reached a detection limit of 2.37×10^{-15} M in the linear concentration range of 10^{-14}–10^{-9} M.

Utilizing an ionic liquid-based terpolymer/iron oxide/multiwalled carbon nanotube nanocomposite electrode, Resende *et al.*[62] reported a study on a new electrochemical sensor for dopamine detection. In this study, selective detection of dopamine was achieved in the presence of other interfering inorganic ions, especially ascorbic acid, thanks to the nanocomposite electrode with increased electrocatalytic activity. The detection limit for dopamine in the linear range of 1–200 $\mu mol\,L^{-1}$ was calculated as 63 $nmol\,L^{-1}$.

In the next study, Ozhikandathi *et al.*[63] constructed an LSPR sensor with the combination of gold (Au) nanoparticles and MWCNTs to sense human recombinant bovine growth hormone. According to their experimental results, the integration of MWCNTs onto the sensor system can increase the limit of sensing capability of the LSPR sensor toward the growth hormone.

Graphene exhibits extraordinary physical properties, including high surface area, excellent thermal and electrical conductivity, high mechanical strength, and enhanced chemical stability. These attributes make graphene well-suited for a range of biomedical applications in bio-sensing and drug delivery.[57,64]

The novel sensing platform, FRET on fiber concept, was proposed by Yao *et al.*[65] to sense metal ions, single-stranded DNA (ss-DNA), and dopamine. For this purpose, the deposition of partially reduced graphene oxide film was integrated on an a-fiber optic modal interferometer and the sensing performance of constructed sensor was examined.

Pothipor *et al.*[66] published an experimental study to enhance the electrochemical signal and conductivity to detect prostate-specific antigen. To achieve this, they combined approaches of graphene-poly(3-aminobenzoic acid) for electrode modification and metal-ion-loaded porous-hollow-silver-gold nanoparticles to improve the electrochemical signal. Ultimately, the detection limit was obtained as 0.13 $pg\,mL^{-1}$.

To perform lysozyme analysis in serum, Vasilescu *et al.*[67] prepared graphene oxide-coated SPR interfaces modified with *Micrococcus lysodeikticus*. Here, lysozyme's bioactivity was exploited for sensing, in contrast to previously described sensors that rely on the definition of lysozyme *via* bioaffinity interactions. Based on the principle that this situation causes a distinctive decrease in the SPR signal, the detection limit was found to be 0.05 $\mu g\,mL^{-1}$ as a result of the experiments carried out.

Ali *et al.*[68] demonstrated how they prepared a sensor system based on the controlled aggregation of Au nanoparticles mediated by PEGylated nanosized graphene and their use as SERS nanotags for SERS-based biosensing. After a wide range of characterization studies, IgG experiments were carried out to demonstrate the effectiveness of the developed SERS nanotags. They stated that considering the relationship between IgG concentration and SERS intensity, the detection limit was lower than 31.0 fM.

In addition to their remarkably low cost and large surface area, carbon-black-based sensor systems have been reported to provide high surface area, fast charge transfer kinetics, and excellent functionality.[69]

Shanmugam *et al.*[70] synthesized porous iron oxide/carbon black composite for the determination of chlorpromazine hydrochloride, which is used as a psychotropic drug as a treatment for diseases such as psychosis, bipolar disorder, and schizophrenia. In the repeatability experiments, a negligible decrease occurred after seven measurements and the relative standard deviation was calculated as 4.67%. The detection limit with the developed electrochemical sensor was found to be 0.001 μM.

Ahmed *et al.*[71] proposed an electrochemical sensor for the effective determination of catechol, which is considered one of the environmental pollutants. For this purpose, the modification of the glassy carbon electrode used was carried out with polypyrrole/carbon black doped α-Fe_2O_3 and the characterization process was successfully completed with a wide variety of techniques such as FESEM, EDS, XRD, TEM, and FTIR. In this study, the detection limit was calculated as 52.8 ± 0.1 nM.

Abbasi *et al.*[72] developed SERS-active materials for use in the detection of anthropogenic water pollutant groups. In brief, the gold–carbon nanoparticles produced in the study were obtained by first coating carbon black with the cationic polyelectrolyte poly(L-lysine) and then exposing it to a tetrachloroauric acid solution. Thanks to the developed sensor system, it became possible to detect both anionic and cationic aromatic analytes and the detection limit was estimated as 15 μM.

Because of their large aspect ratio, strong electrical and thermal conductivity, large specific surface area, and other characteristics, carbon nanofibers have a wide range of potential applications in the fields of sensing, electrochemistry, and adsorbent materials.[52,73]

Jahromi *et al.*[74] described the development of a sensor to detect tramadol, a painkiller used to treat moderate to severe pain. After characterization studies of the carbon nanofibers used in the study, the electrochemical behavior of the modified and unmodified electrodes was evaluated by examining the effects of various parameters. The detection limit was obtained as 0.016 nM with the electrochemical sensor using a screen-printed electrode modified with electrospun carbon nanofibers.

Afzali *et al.*[75] presented a novel approach involving composite nanofibers to detect pemetrexed, which acts as an effective anticancer drug. In this study, the carbon paste electrode was produced by incorporating palladium, carbon nanofibers, ionic liquid, and Nafion. After the composite nanofibers were synthesized by the electrospinning method and characterization procedures were performed, detailed experimental studies were carried out to determine the optimum conditions for the developed sensor system. As a result, the sensor indicated an improved electrochemical response with a detection limit of 0.33 nM.

Similarly, in another drug-related study conducted with carbon nanofibers, Banan *et al.*[76] prepared molecularly imprinted polymers for the determination of amiodarone, an antiarrhythmic drug. Here, first, the synthesis of magnetic nanoparticles coated with silica and then molecularly imprinted polymers was carried out. In the study, the selectivity, stability, and sensitivity of the electrochemical sensor developed by taking advantage of the molecular imprinting technique and the superior properties of carbon nanofibers were emphasized. The detection limit value of amiodarone was calculated as 0.4 $\mu g\,mL^{-1}$ in plasma.

Since their discovery, fullerenes have been discussed for their tremendous potential for use in medicine and biology. Fullerene molecules are particularly active because they may generate a huge number of free bonds with diverse radicals, including biologically active ones.[76]

Alzheimer's disease is a severe neurological condition characterized by progressive cognitive decline, leading to memory loss, impaired speech function, and difficulty in responding to the environment. In this context,

Soylemez *et al.*[77] developed a sensitive and cost-effective enzyme-free fullerene-based sensor system for the detection of acetylcholine. A highly electroactive surface was developed with an electrochemical sensor system modified using fullerene, histidine, serine, and nickel. The detection limit of this electrochemical assay was 8.01 μM.

Table 7.1 Biomedical applications of nanocarbon-based materials and various biosensing platforms.[a]

Carbon material	Biosensor type	Analyte	Sample type	Detection limit ($mol\,L^{-1}$)	Ref.
CB	Electrochemical	Glucose	Artificial tear	7.8×10^{-6}	80
CB	Electrochemical	Interleukin-1α	Serum/saliva	3.3 fg mL^{-1}	81
CB	Electrochemical	Glutathione	Blood Blood	1.5×10^{-8} 2.6×10^{-8}	82
CB	Electrochemical	mRNA-125	Serum	1.0×10^{-11}	83
Graphene	Electrochemical	DNA	Serum	7.6 fM	84
Graphene	Electrochemical	PSA	Serum	2 pg mL^{-1} 60 pg mL^{-1}	85
Graphene	SPR	miRNA-141 Adenosine	Cancer cell extract Serum	0.1 fM 0.1 pM	86
Graphene	SPR	miRNA-21	Urine	0.3 fM	87
Graphene	QCM	Anti-PLA2R	Serum	<100 ng mL^{-1}	88
CNT	Electrochemical	Thrombin	Serum	0.1 fM	89
CNT	Electrochemical	Adenosine	Serum	1.0 nM	90
CNT	SPR	EPO and GM-CSF	Aqueous	<0.1 ng mL^{-1}	91
CNF	Electrochemical	Xanthine	Urine	0.5 μM	92
CNF	Electrochemical	Glucose	Serum	1 μM	93
CDs	SPR	Dopamine	Biological samples	0.23 μM	94
CDs	Fluorescent	Cancer cell imaging	No data	No data	95
Graphite	Electrochemical	DNA bases	Animal and plant DNA samples	0.5–1.0 μg mL^{-1}	96
Graphite	Electrochemical	Nitric oxide	Murine neuroblastoma cells	3.6 nmol L^{-1}	97
Graphite NP	Electrochemical	Glucose	Real urine sample	50 nM	98
Graphite	Electrochemical	Dopamine	Human serum and urine samples	45×10^{-4} mM	99
Graphite	Electrochemical	Paracetamol and dopamine	Artificial sweat	1.4 μmol L^{-1} (paracetamol) 78 nmol L^{-1} (dopamine)	100

[a] Abbreviations: CB, carbon black; CNT, carbon nanotube; CNF, carbon nanofiber; SPR, surface plasmon resonance; EPO, human erythropoietin; GM-CSF, human granulocyte macrophage colony-stimulating factor; CDs, carbon dots; QCM, quartz crystal microbalance; anti-PLA2R, antibody against the phospholipase A2 receptor; NP, nanoparticle.

A comprehensive study on the sensing of catechol and hydroquinone, which exhibit high toxicity to human and environmental health even at very low concentrations, was reported by Zhu *et al.*[78] The nanocomposite material created with platinum nanoparticles and fullerene contributed to increasing the catalytic activity of the proposed electrochemical sensor and thus improving its sensitivity. After determining the optimum conditions of the modified electrode, detection limits for hydroquinone and catechol were obtained as 2.19 μM and 2.97 μM, respectively, over a wide linear concentration range.

An interesting study was reported by Kalytchuk *et al.*,[79] and the researchers fabricated a thermal sensor using water-soluble nitrogen and sulfur-doped CDs to image NIH/3T3 and HeLa cell lines according to the PL property of the CDs. The proposed nanothermal sensor exhibited a good performance in the adjustment of different experimental conditions, *i.e.*, change of pH, temperature, and ionic strength.

Especially in clinical and biomedical applications, the need for carbon-based materials is increasing daily. In this context, some other current studies in the literature are summarized in Table 7.1.

7.5 Conclusions

The 2030 Agenda for Sustainable Development seeks to achieve the Sustainable Development Goals (SDGs) by harmonizing three essential aspects: economic growth, environmental protection, and social inclusion.[1] There has been a rising demand for convenient and comprehensive health monitoring recently. In the past few decades, traditional healthcare systems have faced many challenges in addressing various health-related issues. Limited resources do not allow hospitals to provide universal care and perform comprehensive health diagnostic tests on people. In these instances, the disease often advances to a severe stage before it is diagnosed. The demand for eco-friendly and sustainable products is increasing, driven by both regulatory bodies and heightened consumer awareness. The increasing focus on sustainability, environmental impact, and the demand for advanced healthcare solutions has highlighted the importance of low-carbon materials in biomedical sensing. Natural polymers like cellulose are being explored for their biocompatibility, biodegradability, and low environmental impact. Also, carbon-based materials offer excellent electrical, thermal, and mechanical properties, making them ideal for sensitive and accurate biosensors. Environmentally friendly synthesis methods, such as using plant extracts, are used to produce these carbon-based materials which are used for sensors, minimizing harmful byproducts. Low-carbon materials are used in the development of flexible and stretchable sensors for continuous health monitoring. These materials enable the creation of portable and efficient diagnostic tools that can be used in resource-limited settings such as point-of-care testing methods. Recently, outstanding new materials and manufacturing techniques have been employed to develop a wide range of biomedical sensors with exceptional performance, high sensitivity, and rapid response times. These sensors have many potential applications, including health monitoring, and have a wide range of applications in biochemical and physiological monitoring of human anatomy. Although biomedical sensors have made remarkable advancements, there remain numerous opportunities and challenges related to materials, sensing accuracy, and real-world applications. Many biomarkers, such as luteinizing hormones and other sex hormones with one-month fluctuation cycles, have durations that are too lengthy for current biomedical sensors to accurately track. Researchers frequently fail to recognize that sweat sensors take more than a minute to update the sweat composition, resulting in redundant data and higher power consumption. Therefore, before developing a biomedical sensor, a microfluidic model associated with biological markers should be created. Furthermore, the influence of individual differences and long-term biological safety on sensor performance is frequently overlooked.[101] Consequently, there is still progress to be made before achieving long-term continuous monitoring of human health indices, because monitoring of biomarkers or measurements of their levels hold great promise for the diagnosis of several diseases. Sensing of biomarkers such as proteins, glucose, creatinine, pathogens, *etc.*, plays a key role in the diagnosis of some serious diseases and monitoring the progress of the disease stage.

Biosensors that measure the biological or chemical reaction *via* measurable signals depending on analyte concentrations are the center of several sectors including medical studies due to the high selectivities, reliabilities, and ease of operations. One of the possible routes during the construction of a biosensor is to boost the detection limit and sensitivity with the aid of nanoparticles, and so carbon-based materials hold great promise. Among them, low-carbon materials with their biocompatibilities, large surface areas, excellent mechanical strengths, electric conductivities, ease of fabrication methods and production hold great promise during the design of biosensing platforms. Furthermore, these materials are eco-friendly and have less detrimental effects on the ecosystem.

In this context, herein, we have exemplified several sensor platforms which encompass optical, fluorescent, electrochemical, thermal, and mass-sensitive sensors with the combination of low-carbon materials due to the key features mentioned above to monitor value biomarkers such as proteins, and some clinical analytes including glucose, interleukin, specific cancer antigens, and cancer cells, as well.

In terms of future perspectives, the functional properties of low-carbon materials can be improved through chemical modifications, increasing their sensitivity, selectivity, and stability for biomedical applications. In order to ensure that these materials can be used safely in long-term medical applications without causing adverse effects, research should be increased to boost their biocompatibility. Scientific research should be carried out to increase the biocompatibility of these materials in order to ensure that they can be used safely in long-term medical applications without causing adverse effects. For that, encouraging collaborations between materials scientists, biomedical engineers, and healthcare professionals is essential to drive innovation and effectively address the challenges in this field. This paves the way for the safe use and biocompatibility of these materials in long-term medical applications without causing adverse effects. Finally, clear regulatory guidelines and standards should be established to ensure safety, effectiveness and widespread acceptance for the use of low-carbon materials.

In conclusion, integrating low-carbon materials into biomedical sensing offers great potential for developing sustainable, efficient, and advanced healthcare solutions. Ongoing research, innovation, and interdisciplinary collaboration will be crucial for addressing current challenges and unlocking the full potential of these materials to enhance patient outcomes and support environmental sustainability. We believe that biosensor platforms will open up new avenues and fill the gaps in healthcare sectors with the assistance of nanotechnology in the near future.

References

1. E. Bontempi, G. P. Sorrentino, A. Zanoletti, I. Alessandri, L. E. Depero and A. Caneschi, *Molecules*, 2021, **26**, 1407.
2. D. Veeman, M. V. Shree, P. Sureshkumar, T. Jagadeesha, L. Natrayan, M. Ravichandran and P. Paramasivam, *J. Nanomater.*, 2021, **2021**, 21.
3. S. Shrivastava, T. Q. Trung and N. E. Lee, *Chem. Soc. Rev.*, 2020, **49**, 1812–1866.
4. H. Wang, S. Li, H. Lu, M. Zhu, H. Liang, X. Wu and Y. Zhang, *Small Methods*, 2023, **7**, 1–21.
5. A. Sen, N. Jette, M. Husain and J. W. Sander, *Lancet*, 2020, **395**, 735–748.
6. G. Maduraiveeran, M. Sasidharan and V. Ganesan, *Biosens. Bioelectron.*, 2018, **103**, 113–129.
7. L. F. Cabeza, C. Barreneche, L. Miró, J. M. Morera, E. Bartolí and A. Inés Fernández, *Renew. Sustain. Energy Rev.*, 2013, **23**, 536–542.
8. X. Zhou and W. Cao, *Nanomaterials*, 2023, **13**, 316.
9. S. Singh and R. Melnik, *Chemosensors*, 2022, **10**, 1–43.
10. T. D. Moshood, G. Nawanir, F. Mahmud, S. Sorooshian and A. Q. Adeleke, *Clean. Eng. Technol.*, 2021, **4**, 100144.
11. D. D. L. Chung, *J. Mater. Sci.*, 2002, **37**, 1475–1489.
12. Z. U. Khan, A. Kausar and H. Ullah, *Polym. - Plast. Technol. Eng.*, 2016, **55**, 559–581.
13. H. O. Pierson, in *Handbook of Carbon, Graphite, Diamonds and Fullerenes*, 1993, pp. 43–69.
14. C. Liang, Y. Chen, M. Wu, K. Wang, W. Zhang, Y. Gan, H. Huang, J. Chen, Y. Xia, J. Zhang, S. Zheng and H. Pan, *Nat. Commun.*, 2021, **12**, 1–9.
15. N. Anzar, R. Hasan, M. Tyagi, N. Yadav and J. Narang, *Sensors Int*, 2020, **1**, 100003.

16. N. Saifuddin, A. Z. Raziah and A. R. Junizah, *J. Chem.*, 2013, 1–18.
17. M. N. Norizan, M. H. Moklis, S. Z. Ngah Demon, N. A. Halim, A. Samsuri, I. S. Mohamad, V. F. Knight and N. Abdullah, *RSC Adv*, 2020, **10**, 43704–43732.
18. Y. Fan, Y. Hou, M. Wang, J. Zheng and X. Hou, *Mater. Adv.*, 2022, **3**, 3070–3088.
19. A. Eatemadi, H. Daraee, H. Karimkhanloo, M. Kouhi, N. Zarghami, A. Akbarzadeh, M. Abasi, Y. Hanifehpour and S. W. Joo, *Nanoscale Res. Lett.*, 2014, **9**, 1–13.
20. B. Pei, W. Wang, N. Dunne and X. Li, *Nanomaterials*, 2019, **9**, 1501.
21. S. K. Tiwari, V. Kumar, A. Huczko, R. Oraon, A. D. e Adhikari and G. C. Nayak, *Crit. Rev. Solid State Mater. Sci.*, 2016, **41**, 257–317.
22. A. Hashmi, V. Nayak, K. R. Singh, B. Jain, M. Baid, F. Alexis and A. K. Singh, *Mater. Today Adv.*, 2022, **13**, 100208.
23. K. K. Gomes, W. Mar, W. Ko, F. Guinea and H. C. Manoharan, *Nature*, 2012, **483**, 306–310.
24. M. Klintenberg, S. Lebègue, M. I. Katsnelson and O. Eriksson, *Phys. Rev. B: Condens. Matter Mater. Phys.*, 2010, **81**, 1–5.
25. P. Avouris and C. Dimitrakopoulos, *Mater. Today*, 2012, **15**, 86–97.
26. K. A. Madurani, S. Suprapto, N. I. Machrita, S. L. Bahar, W. Illiya and F. Kurniawan, *ECS J. Solid State Sci. Technol.*, 2020, **9**, 093013.
27. B. L. Dasari, J. M. Nouri, D. Brabazon and S. Naher, *Energy*, 2017, **140**, 766–778.
28. K. S. Novoselov, A. K. Geim, S. V. Morozov, D. Jiang, Y. Zhang, S. V. Dubonos, I. V. Grigorieva and A. A. Firsov, *Science*, 2004, **306**, 666–669.
29. M. Saeed, Y. Alshammari, S. A. Majeed and E. Al-Nasrallah, *Molecules*, 2020, **25**, 3856.
30. Y. Fan, G. D. Fowler and M. Zhao, *J. Clean. Prod.*, 2020, **247**, 119115.
31. B. Jin, in *Carbon Black: Production, Properties and Uses*, 2011, pp. 263–278.
32. T. Peijs, R. Kirschbaum and P. J. Lemstra, *Adv. Ind. Eng. Polym. Res.*, 2022, **5**, 90–106.
33. Y. Liu and S. Kumar, *Polym. Rev.*, 2012, **52**, 234–258.
34. X. Huang, *Materials*, 2009, **2**, 2369–2403.
35. K. Li, X. Ni, Q. Wu, C. Yuan, C. Li, D. Li, H. Chen, Y. Lv and A. Ju, *Carbon-Based Fibers: Fabrication, Characterization and Application*, Springer Singapore, 2022, vol. 4.
36. M. Paukov, C. Kramberger, I. Begichev, M. Kharlamova and M. Burdanova, *Materials*, 2023, **16**, 1–21.
37. P. Bhakta and B. Barthunia, *J. Indian Acad. Oral Med. Radiol.*, 2020, **32**, 159–163.
38. A. Kausar, I. Ahmad, M. Maaza and M. H. Eisa, *Membranes*, 2023, **13**, 1–22.
39. J. Liu, R. Li and B. Yang, *ACS Cent. Sci.*, 2020, **6**, 2179–2195.
40. S. Zhu, Y. Song, J. Shao, X. Zhao and B. Yang, *Angew. Chem., Int. Ed.*, 2015, **54**, 14626–14637.

41. D. Ozyurt, M. A. l Kobaisi, R. K. Hocking and B. Fox, *Carbon Trends*, 2023, **12**, 100276.
42. S. Kurbanoglu and S. A. Ozkan, *J. Pharm. Biomed. Anal.*, 2018, **147**, 439–457.
43. W. Zhai, N. Srikanth, L. B. Kong and K. Zhou, *Carbon*, 2017, **119**, 150–171.
44. K. Nejati-Koshki, F. Fathi, A. A. Arabzadeh and A. Mohammadzadeh, *Anal. Methods*, 2023, **15**, 5441–5458.
45. J. F. Masson, *Analyst*, 2020, **145**, 3776–3800.
46. J. Liu, M. Jalali, S. Mahshid and S. Wachsmann-Hogiu, *Analyst*, 2020, **145**, 364–384.
47. M. Zhao, Y. Qi, H. Wang, Z. Xie, B. Li, H. Wang and X. Wei, *Laser Photonics Rev*, 2024, **18**, 1–34.
48. Y. Cao, Y. Sun, R. J. Yu and Y. T. Long, *Microchim. Acta*, 2024, **191**, 1–22.
49. H. Tang, C. Zhu, G. Meng and N. Wu, *J. Electrochem. Soc.*, 2018, **165**, B3098–B3118.
50. S. K. Krishnan, E. Singh, P. Singh, M. Meyyappan and H. S. Nalwa, *RSC Adv*, 2019, **9**, 8778–8781.
51. A. R. Gopal, F. Joy, V. Dutta, J. Devasia, R. Dateer and A. Nizam, *Part. Part. Syst. Charact.*, 2024, **41**, 1–24.
52. J. N. Tiwari, V. Vij, K. C. Kemp and K. S. Kim, *ACS Nano*, 2016, **10**, 46–80.
53. J. Ding and W. Qin, *TrAC Trends Anal. Chem.*, 2020, **124**, 115803.
54. E. O. Polat, M. M. Cetin, A. F. Tabak, E. B. Güven, B. Ö. Uysal, T. Arsan, A. Kabbani, H. Hamed and S. B. Gül, *Biosensors*, 2022, **12**, 385.
55. V. Naresh and N. Lee, *Sensors*, 2021, **21**, 1–35.
56. Z. Ghemari, S. Belkhiri and S. Saad, *J. Comput. Electron.*, 2024, **23**, 448–453.
57. M. A. Elkodous, S. O. Olojede, S. Sahoo and R. Kumar, *Chem. Biol. Interact.*, 2023, **379**, 110517.
58. D. Maiti, X. Tong, X. Mou and K. Yang, *Front. Pharmacol.*, 2019, **9**, 1401.
59. Y. Zhang, J. Guo, Z. Tang, C. Tang, Y. Li, X. Tao, B. Zhou, W. Chen, L. Guo, K. Tang and T. Liang, *Biosens. Bioelectron. X*, 2024, **17**, 100424.
60. P. Tabei and A. Yari, *Mater. Chem. Phys.*, 2024, **311**, 128534.
61. Y. T. Xue, Z. Chen, X. Chen, G. C. Han, X. Z. Feng and H. B. Kraatz, *Electrochim. Acta*, 2024, **483**, 144009.
62. D. Duan, H. Lu, L. Li, Y. Ding and G. Ma, *Microchem. J.*, 2022, **175**, 107231.
63. J. Ozhikandathil, S. Badilescu and M. Packirisamy, *Sci. Rep.*, 2015, **5**, 1–11.
64. R. Goodrum, H. Weldekidan, H. Li, A. K. Mohanty and M. Misra, *Adv. Ind. Eng. Polym. Res.*, 2024, **7**, 37–53.
65. B. C. Yao, Y. Wu, C. B. Yu, J. R. He, Y. J. Rao, Y. Gong, F. Fu, Y. F. Chen and Y. R. Li, *Sci. Rep.*, 2016, **6**, 1–9.
66. C. Pothipor, N. Wiriyakun, T. Putnin, A. Ngamaroonchote, J. Jakmunee, K. Ounnunkad, R. Laocharoensuk and N. Aroonyadet, *Sens. Actuators, B*, 2019, **296**, 126657.

67. A. Vasilescu, S. Gáspár, M. Gheorghiu, S. David, V. Dinca, S. Peteu, Q. Wang, M. Li, R. Boukherroub and S. Szunerits, *Biosens. Bioelectron.*, 2017, **89**, 525–531.
68. A. Ali, E. Y. Hwang, J. Choo and D. W. Lim, *Analyst*, 2018, **143**, 2604–2615.
69. L. M. C. Ferreira, P. S. Silva, K. K. L. Augusto, P. C. Gomes-Júnior, S. O. D. Farra, T. A. Silva, O. Fatibello-Filho and F. C. Vicentini, *J. Pharm. Biomed. Anal.*, 2022, **221**, 115032.
70. R. Shanmugam, J. Ganesamurthi, T. W. Chen, S. M. Chen, M. Balamurugan, M. A. Ali, A. M. Al-Mohaimeed, W. A. Al-onazi and K. Alagumalai, *Mater. Today Chem.*, 2022, **25**, 100982.
71. J. Ahmed, M. Faisal, M. Jalalah, M. Alsaiari, S. A. Alsareii and F. A. Harraz, *Colloids Surf., A*, 2021, **619**, 126469.
72. A. Abbasi, T. T. S. De Oliveira, G. D. Bothun and A. Bose, *ACS Appl. Nano Mater.*, 2020, **3**, 2605–2613.
73. Z. Wang, S. Wu, J. Wang, A. Yu and G. Wei, *Nanomaterials*, 2019, **9**, 1045.
74. Z. Jahromi, E. Mirzaei, A. Savardashtaki, M. Afzali and Z. Afzali, *Microchem. J.*, 2020, **157**, 104942.
75. M. Afzali, A. Mostafavi, R. Nekooie and Z. Jahromi, *J. Mol. Liq.*, 2019, **282**, 456–465.
76. K. Banan, S. Niknam, M. Ahmadi, S. Tabasi, M. Ghalkhani, M. Adabi and F. Ghorbani-Bidkoreph, *Microchem. J.*, 2024, **200**, 110365.
77. S. Soylemez, V. Dolgun and S. Özçubukçu, *Microchem. J.*, 2023, **193**, 109099.
78. Y. Zhu, S. Huai, J. Jiao, Q. Xu, H. Wu and H. Zhang, *J. Electroanal. Chem.*, 2020, **878**, 114726.
79. S. Kalytchuk, K. Poláková, Y. Wang, J. P. Froning, K. Cepe, A. L. Rogach and R. Zbořil, *ACS Nano*, 2017, **11**, 1432–1442.
80. F. Calegari, L. P. de Souza, M. M. Barsan, C. M. A. Brett, L. H. Marcolino-Junior and M. F. Bergamini, *Sens. Actuators, B*, 2017, **253**, 10–18.
81. E. B. Aydın and M. K. Sezgintürk, *Anal. Chim. Acta*, 2019, **1077**, 129–139.
82. F. Della Pelle, L. Blandón-Naranjo, M. Alzate, M. Del Carlo and D. Compagnone, *Talanta*, 2020, **207**, 120349.
83. G. Yammouri, J. Mandli, H. Mohammadi and A. Amine, *J. Electroanal. Chem.*, 2017, **806**, 75–81.
84. N. Gao, F. Gao, S. He, Q. Zhu, J. Huang, H. Tanaka and Q. Wang, *Anal. Chim. Acta*, 2017, **951**, 58–67.
85. P. Assari, A. A. Rafati, A. Feizollahi and R. Asadpour Joghani, *Microchim. Acta*, 2019, **186**, 484.
86. Q. Li, Q. Wang, X. Yang, K. Wang, H. Zhang and W. Nie, *Talanta*, 2017, **174**, 521–526.
87. M. L. Mujica, Y. Zhang, F. Bédioui, F. Gutiérrez and G. Rivas, *Anal. Bioanal. Chem.*, 2020, **412**, 3539–3546.
88. P. Hampitak, T. A. Jowitt, D. Melendrez, M. Fresquet, P. Hamilton, M. Iliut, K. Nie, B. Spencer, R. Lennon and A. Vijayaraghavan, *ACS Sens.*, 2020, **5**, 3520–3532.

89. M. Konari, E. Heydari-Bafrooei and M. Dinari, *Int. J. Biol. Macromol.*, 2021, **166**, 54–60.
90. S. Cao, H. Zhao, K. Chen, F. Zhou and M. Lan, *Anal. Chim. Acta*, 2023, **1260**, 341212.
91. E. G. Lee, K. M. Park, J. Y. Jeong, S. H. Lee, J. E. Baek, H. W. Lee, J. K. Jung and B. H. Chung, *Anal. Biochem.*, 2011, **408**, 206–211.
92. X. Tang, Y. Liu, H. Hou and T. You, *Talanta*, 2011, **83**, 1410–1414.
93. M. Li, L. Liu, Y. Xiong, X. Liu, A. Nsabimana, X. Bo and L. Guo, *Sens. Actuators, B*, 2015, **207**, 614–622.
94. M. Amiri, S. Dadfarnia, A. M. Haji Shabani and S. Sadjadi, *J. Pharm. Biomed. Anal.*, 2019, **172**, 223–229.
95. S. K. Bhunia, A. R. Maity, S. Nandi, D. Stepensky and R. Jelinek, *ChemBioChem*, 2016, **17**, 614–619.
96. K. L. Ng and S. M. Khor, *Anal. Chem.*, 2017, **89**, 10004–10012.
97. Y. Wu, C. Liu, X. Liu, C. Zhu, X. Dang, S. Hu and D. Zheng, *Anal. Lett.*, 2021, **54**, 2537–2550.
98. Y. Piao, D. J. Han and T. S. Seo, *Sens. Actuators, B*, 2014, **194**, 454–459.
99. C. Rajkumar, B. Thirumalraj, S. M. Chen and H. A. Chen, *J. Colloid Interface Sci.*, 2017, **487**, 149–155.
100. R. C. de Freitas, W. T. Fonseca, D. C. Azzi, P. A. Raymundo-Pereira, O. N. Oliveira and B. C. Janegitz, *Microchem. J.*, 2023, **191**, 108823.
101. M. B. Kulkarni, S. Rajagopal, B. Prieto-Simón and B. W. Pogue, *Talanta*, 2024, **272**, 125817.

CHAPTER 8

Low-carbon Materials in Biosensors Applications

XIANGYU YAO,[a] LANQING YANG,[a] BAOQING BAI,[a] XIAOMIN WANG,[b] JINHUA ZHANG,[a] LIGANG YU,[a] TAO BO,[a] YING GU*[c] AND YUKUN YANG*[a]

[a] School of Life Science, Shanxi University, Taiyuan 030006, China; [b] Institute of Pharmaceutical and Food Engineering, Shanxi University of Chinese Medicine, Yuci 030619, China; [c] Faculty of Food Science and Engineering, Kunming University of Science and Technology, Kunming 650500, China
*Emails: guying@kust.edu.cn; yangyukun@sxu.edu.cn

8.1 Introduction

Amid the increasing global concern for environmental sustainability and ecological balance, research and applications of low-carbon materials have become a prominent focus across various disciplines and fields. Against this backdrop, biosensors, as vital analytical tools, have garnered growing attention and are playing significant roles in medical diagnostics, environmental monitoring, food safety, and other fields. Traditional biosensor construction typically relies on costly and environmentally hazardous materials such as metal electrode materials like silver, gold, platinum, *etc.* The production of these metals consumes significant resources and energy, and their improper disposal can lead to environmental contamination. Additionally, these sensing elements are often embedded in plastic matrices to provide mechanical support and protection. However, commonly used plastic materials such as polyvinyl chloride (PVC) and polypropylene (PP) release harmful substances during production and are difficult to degrade, resulting in adverse environmental impacts.

Low Carbon Materials: Paving a Sustainable Path for a Greener Future
Edited by Santanu Patra, Sudheesh K. Shukla and Mika Sillanpää

Published by the Royal Society of Chemistry, www.rsc.org

With the increasing demand for environmentally friendly materials, the emergence of low-carbon materials provides a new possibility for the development of biosensors, especially in the biomedical field where safety and biocompatibility are crucial. Incorporating bio-based, carbon-based, and silk-based materials into sensor components not only improves sensor performance but also ensures environmental sustainability and reduces the ecological footprint of these devices. These low-carbon materials, derived from renewable resources, produce minimal greenhouse gas emissions during production, and often exhibit excellent biodegradability, making them ideal for sustainable sensor development. Specifically in biomedical sensors, the use of low-carbon materials offers the added benefits of reduced toxicity and enhanced biocompatibility, which are essential for safe and effective medical diagnostics and treatments.

Moreover, the integration of low-carbon materials into biosensors supports several United Nations Sustainable Development Goals (SDGs). For instance, SDG 3 (Good Health and Well-being) is promoted by the development of safer and more effective biomedical sensors. SDG 9 (Industry, Innovation, and Infrastructure) is addressed through the innovation and use of sustainable materials in sensor technology. Lastly, SDG 12 (Responsible Consumption and Production) is advanced by minimizing the environmental impact associated with the production and disposal of biosensors.

In this chapter, we will first introduce the development history, principles, construction, and advantages of biosensors, with a special emphasis on the application of low-carbon materials in biosensors, such as bio-based biosensors, paper-based biosensors, silk-based biosensors, and carbon-based biosensors. Finally, we summarize the challenges and future trends of low-carbon materials in biosensor applications. Through the exposition in this chapter, we aim to provide scientists, engineers, and policymakers with a comprehensive understanding of the latest research progress and directions in the application of low-carbon materials in biosensors. It is intended to foster the widespread application and promotion of environmentally friendly technologies in the field of biosensors, highlighting their significance in advancing both technological performance and environmental sustainability.

8.2 An Overview of Biosensors

The story begins in 1956 with the invention of a glucose biosensor based on electrochemical detection of oxygen, credited to Clark's oxygen electrode.[1] Since then, the Clark electrode has been widely employed in biosensing systems across various fields including clinical, food, and environmental applications. The Clark oxygen electrode is regarded as a milestone in the development of novel biosensing concepts, with the term "biosensor" first coined by Cammann. According to the definition provided by the International Union of Pure and Applied Chemistry (IUPAC), a biosensor refers to an analytical device that combines elements of physicochemical sensors with components capable of identifying biologically derived molecules.[2]

8.2.1 Elements Contained Within a Biosensor

The manufacture of biosensors involves multiple interdisciplinary fields, including chemistry, biology, physics, electronics, materials science, engineering, *etc.*[3] Typical components of a biosensor include (see Figure 8.1):

a. Analyte: The target substance to be detected or measured, which can be a molecule, chemical substance, or environmental condition within a living organism.
b. Bioreceptor: A biological recognition element used to interact specifically with the target molecule or organism, such as enzymes, antibodies, nucleic acid aptamers, *etc.*
c. Transducer: A key component of the biosensor, which converts the signal generated by the interaction between the biological recognition element and the target molecule into a measurable electrical or light signal. Common transducers include electrochemical sensors, optical sensors, *etc.*
d. Electronics: Electronic components used for amplifying, processing, and transmitting signals generated by the sensor, including amplifiers, converters, microprocessors, *etc.*
e. Display: A device used to display the data or results measured by the sensor, typically a digital display or computer interface.[4,5]

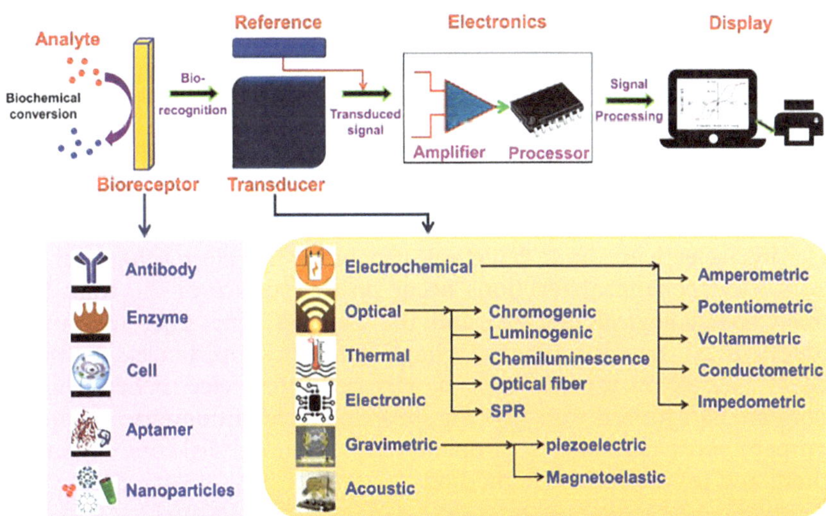

Figure 8.1 A schematic diagram of a typical biosensor, consisting of a bioreceptor, transducer, electronic system (amplifier and processor), and display (PC or printer), is shown. Various types of bioreceptors and transducers used in biosensors are also illustrated.[5] Reproduced from ref. 6, https://doi.org/10.3390/s21041109, under the terms of the CC BY 4.0 license, https://creativecommons.org/licenses/by/4.0/.

These components together constitute the biosensor system, enabling efficient detection and measurement of the target substance.

8.2.2 Characteristics of Biosensors

In order to develop efficient and powerful biosensor systems, a series of static and dynamic requirements must be met. An ideal biosensor should possess advantages such as selectivity, sensitivity, linearity, response time, and repeatability. Selectivity refers to the sensor's ability to selectively recognize target molecules or organisms, while sensitivity measures its detection capability for target substances, with higher sensitivity being preferable. A good linear response enables the sensor to generate predictable signal responses within a certain range. Response time refers to the time required to obtain results, while repeatability refers to the consistency of results when repeatedly measuring the same sample. Additionally, stability and portability are also key characteristics of biosensors. Stability involves resistance to environmental interference, while portability requires sensors to be cost-effective, miniaturized, and easy to operate.[6–8]

8.2.3 Different Types of Biosensors

Based on the transducer or the type of quantitative signal, biosensors can be classified into several types: (a) electrochemical; (b) optical; (c) gravimetric; and (d) thermal.

8.2.3.1 Electrochemical Biosensors

Electrochemical biosensors are one of the most widely researched and applied biosensors, with their principle based on the electrochemical properties between the analyte and the sensor.[9,10] These sensors exhibit high sensitivity, selectivity, and detection capability. In electrochemical biosensors, electrochemical reactions occur on the surface of the transducer between the biological receptor and the analyte, generating measurable electrochemical signals such as voltage, current, impedance, or capacitance.[11,12] Based on the different signals generated, electrochemical biosensors are categorized into (see Figure 8.2): (a) potentiometric method;[13] (b) amperometric method;[14] (c) impedimetric method;[15] (d) conductometric method; and (e) voltammetric method.[16]

8.2.3.2 Optical Biosensors

Optical biosensors integrate biological recognition elements into analytical tools, operating on the principle of generating signals proportional to the concentration of the analyte.[18,19] These recognition elements include enzymes, antibodies, aptamers, whole cells, and tissues, among others.

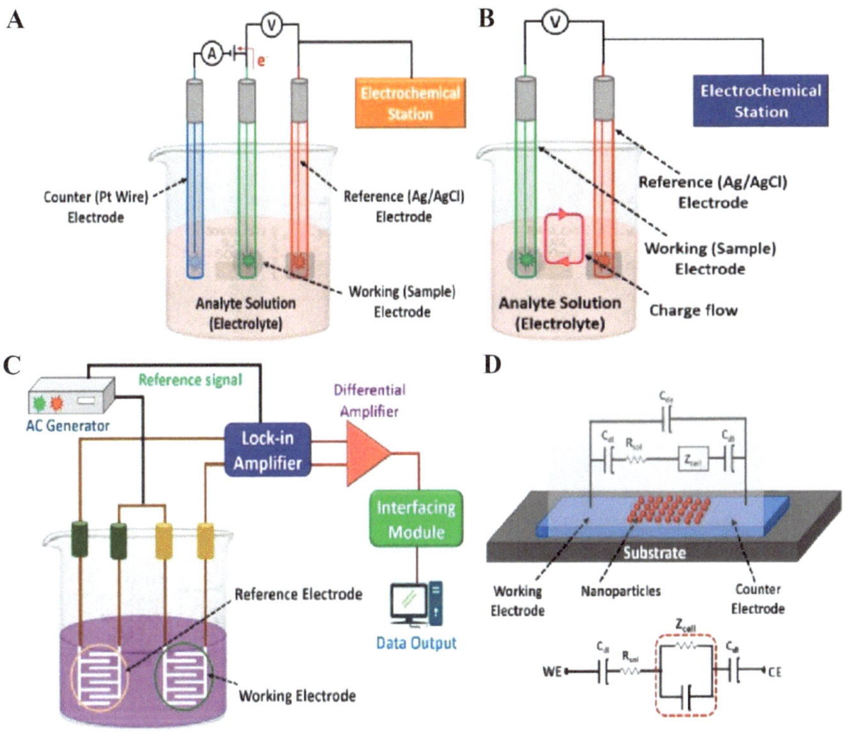

Figure 8.2 (A) Amperometric/voltammetric (B) potential, (C) conductive biosensors, and (D) impedance biosensors equivalent circuit diagrams.[17] Reproduced from ref. 18 https://doi.org/10.3390/bios12070543, under the terms of the CC BY 4.0 license, https://creativecommons.org/licenses/by/4.0/.

In optical biosensors, the physical or chemical changes induced by these biological recognition elements result in changes in optical properties such as absorption, transmission, reflection, refraction, phase, amplitude, frequency, and/or polarization.[20] Based on the principle, optical biosensors can be divided into two types: label-free and label-based. In label-free sensing, the detected signal is generated through the interaction between the analyte and the sensor, whereas in label-based sensing, light signals are generated through methods such as calorimetry, fluorescence, or luminescence.[21] Optical biosensors can be designed based on different optical principles, including surface plasmon resonance (SPR), evanescent wave (EW), fluorescence, optical waveguide interference measurement, chemiluminescence, refractive index, and surface-enhanced Raman scattering, among others. Common optical biosensors include those based on fluorescence, chemiluminescence, SPR, and optical fibers.[20] Figure 8.3 illustrates optical biosensors based on chemiluminescence, SPR, and evanescent wave (see Figure 8.3).

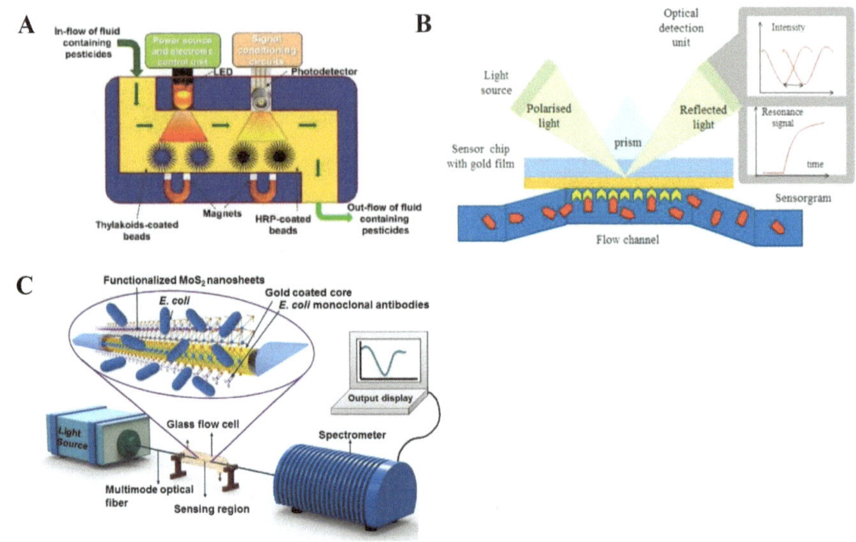

Figure 8.3 (A) Chemical luminescence biosensor.[22] Reproduced from ref. 6 https://doi.org/10.3390/s21041109, under the terms of the CC BY 4.0 license, https://creativecommons.org/licenses/by/4.0/. (B) Surface plasmon resonance (SPR) biosensor.[23] Reproduced from ref. 22 with permission from Elsevier, Copyright 2017. (C) Evanescent wave-based fiber optic biosensor schematic diagrams.[24] Reproduced from ref. 23 with permission from Elsevier, Copyright 2019.

8.2.3.3 Gravimetric Biosensors

Mass-based biosensors belong to a type of biosensor that operates based on mass, where the principle involves responding to tiny changes in the mass of the bound material (such as proteins or antibodies on the surface), resulting in measurable signals (see Figure 8.4). These sensors utilize thin piezoelectric quartz crystals, where an electric current is applied at a specific frequency, and the frequency changes caused by the mass variation are detected.[25,26]

8.2.3.4 Thermal Biosensors

Thermal biosensors utilize the fundamental characteristics of biological reactions (exothermic or endothermic), namely the measurement of the heat absorbed or released during the reaction process. The total heat absorbed/released or the temperature change (ΔT) measured by the thermal biosensor is directly proportional to the enthalpy (ΔH) generated in the biochemical reaction and the total number of product molecules (n_p), and inversely proportional to the heat capacity (C_p) of the reaction. This can be expressed as: $\Delta T = -(n_p \Delta H)/C_p$.[29–33]

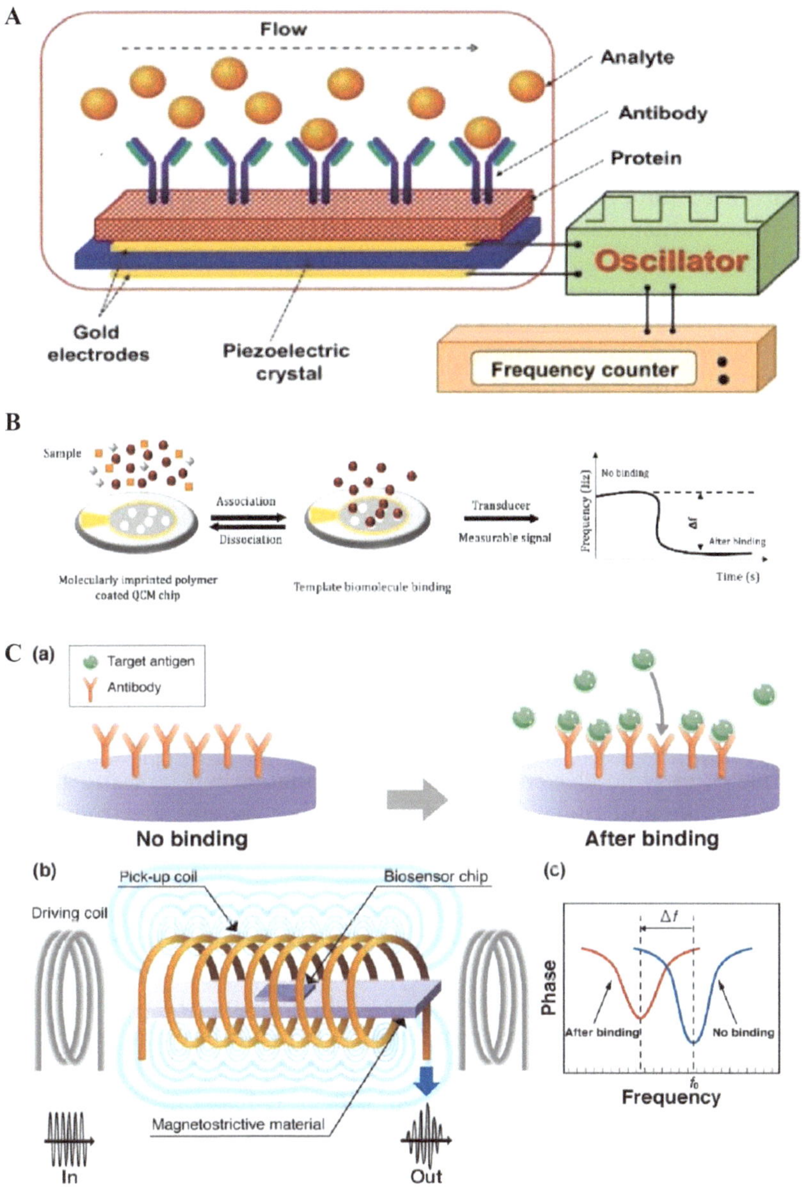

Figure 8.4 (A) Piezoelectric biosensor.[22] Reproduced from ref. 6 https://doi.org/10.3390/s21041109, under the terms of the CC BY 4.0 license, https://creativecommons.org/licenses/by/4.0/. (B) Quartz crystal microbalance biosensor.[27] Reproduced from ref. 26 https://doi.org/10.3390/chemosensors10030106, under the terms of the CC BY 4.0 license, https://creativecommons.org/licenses/by/4.0/. (C) Magnetoelastic biosensor[28] schematic diagrams. Reproduced from ref. 27 https://doi.org/10.1002/adma.202005448, under the terms of the CC BY 4.0 license, https://creativecommons.org/licenses/by/4.0/.

8.3 Bio-based Biosensors

Bio-based biosensors utilize biological materials as recognition elements for detecting and quantifying specific biomolecules, chemicals, or environmental conditions within or surrounding biological organisms. These sensors leverage the biological response mechanisms within organisms to interact with target substances, generating specific signals that are converted into measurable outputs through detection techniques. The use of biological materials as the basis for sensors embodies characteristics of being low-carbon, environmentally friendly, and sustainable. These materials are typically renewable or derived from renewable resources, resulting in lower carbon emissions during their preparation. Additionally, biological materials can degrade over their lifecycle, aligning with natural environmental processes and reducing the environmental impact. Moreover, biological materials exhibit good biocompatibility, being compatible with the biological environments within organisms, thereby minimizing adverse effects on human health. Bio-based biosensors hold broad application prospects in fields such as medical diagnostics, environmental monitoring, and food safety.

8.3.1 Enzyme-based Biosensors

Enzymes are essential proteins in biological activities, acting as biocatalysts that play vital roles in regulating chemical reactions within living cells. The principle of enzyme electrodes was first elucidated by Clark and Lyons in 1962.[34] Updike and Hicks further described the "enzyme electrode" as the pioneering current-based biosensor in 1967, which quantified glucose by combining glucose oxidase with an oxygen electrode.[35] Compared to conventional chemical catalysts, enzymes exhibit remarkable characteristics of high substrate specificity and sensitivity. These features are advantageous for the detection of specific target molecules, while their sensitivity makes them a favorable tool for detecting trace components. Enzyme-based biosensors can detect analytes in complex systems with minimal interference, making the utilization of enzyme-related properties as sensor components highly beneficial. Here, we focus on the classification, immobilization methods, and applications of enzymes in biosensors.

8.3.1.1 Methods of Enzyme Immobilization

Immobilizing enzymes can prevent their dissolution or deactivation during sensor use, while also aiding in increasing the enzyme's contact area with the analyte, enhancing sensor sensitivity and selectivity. Additionally, immobilized enzymes can streamline the sensor's operational procedures, thereby enhancing its stability and long-term reliability. Enzymes are closely associated with transducers in biosensors, and various enzyme immobilization methods have been attempted to enhance the detection performance of biosensors.[36–39] These methods mainly include physical adsorption and chemical binding. Physical adsorption immobilizes enzymes on carriers or substrates through weak interactions such as electrostatic attraction. Methods for enzyme immobilization on carriers include the use of amino, phenol, and thiol groups. Chemical binding, on the other hand, is achieved through the formation of strong chemical bonds. The stability of immobilized enzymes is influenced by various factors, including interactions between the support material and enzymes, the immobilization position and quantity, and the localized environment of the enzyme.

Enzyme immobilization methods can be divided into:

- Physical methods
  1. Adsorption
  2. Entrapment
- Chemical methods
  1. Covalent attachment
  2. Cross-linking

8.3.1.2 Applications of Enzyme-based Biosensors in Various Fields

Enzyme-based biosensors are applied in a wide range of domains, including:

1. Food quality monitoring
2. Soil quality monitoring
3. Environmental monitoring
4. Disease detection
5. Toxins of defense interest
6. Water quality management
7. Prosthetic devices
8. Drug discovery

Scientists have been continuously striving to enhance the stability, activity, efficiency, and repeatability of immobilized enzymes to broaden their applications in everyday life. Enzyme-based biosensors have shown great potential in biomedical, environmental monitoring, bioprocessing, food safety, and other fields. Rapid, reliable, and cost-effective methods are urgently needed for sample analysis in various fields, and the development of biosensor technology has significantly advanced application techniques in diverse scenarios.

In the healthcare sector, enzyme-based biosensors play a crucial role in early disease detection.[40] Various applications have been developed, including diagnostic kits for detecting human chorionic gonadotropin (hCG) in urine to diagnose pregnancy. In the commercial sector, glucose-based biosensors have dominated a significant portion of the global market.

Brands such as Accu-Chek, One Touch, Glucocard, and Freestyle have established commercial biosensor systems for monitoring blood glucose levels in diabetic patients.[41]

In environmental pollution monitoring, such as pesticide detection, enzyme-based biosensors also play a significant role. For example, optical multi-analyte biosensors are employed for pesticide detection, while DNA enzyme-based optical biosensors are utilized for detecting heavy metal pollutants (such as lead) in river water.[42]

During the monitoring of fermentation processes, enzyme-based biosensors are capable of detecting the presence of products, biomass, antibodies, by-products, enzymes, and metabolites. These sensors, known for their high repeatability, simplicity of equipment, good selectivity, and ease of operation, have become effective tools for controlling industrial processes. For instance, electrochemical biosensors based on alcohol oxidase are used to determine low concentrations of alcohol, facilitating the control of fermentation processes and storage of juices, beers, wines, and alcoholic beverages.[43]

The food and beverage industry urgently requires a rapid and cost-effective analytical method to ensure the quality of products and processes. Enzyme-based biosensors hold great promise in this field, offering accurate, rapid, and real-time analysis capabilities. One effective application is the determination of sweeteners, as they may have adverse effects on the population, such as dental issues, cardiovascular diseases, obesity, and type 2 diabetes. By analyzing the electrophysiological activity of taste epithelium, multi-channel biosensors can detect these sweeteners. Furthermore, enzyme biosensors are also utilized to monitor the aging process in beer brewing, employing cobalt phthalocyanine during beer aging. These biosensors can monitor the aging status of beer during storage.[44] Additionally, malic acid in apples, potatoes, and tomatoes can be sensed by malic dehydrogenase.[45]

8.3.1.3 Challenges in the Development of Enzyme-based Biosensors

Enzyme-based biosensors, characterized by their high specificity and simple scalability, are suitable for rapid, precise, and continuous monitoring and analysis of analytes. These macromolecules exhibit affinity for specific substrates, ensuring selective attachment to particular molecules or analytes, thereby enabling selective detection of analytes. Despite this high specificity, which allows for the detection of analytes even at low concentrations, it also presents some challenges. Different enzymes operate under different conditions, such as substrate concentration, temperature, and pH, while the presence of inhibitors can also affect their catalytic properties. Therefore, designing biosensors requires an understanding of the complexity of both the target analytes and the matrix. Moreover, contaminants and chemical interferences from sample matrices may reduce signal intensity, adding to the complexity of sensor design. Enzyme-based biosensors also face issues of deactivation, as they may undergo denaturation or aggregation due to changes in pH and temperature. Therefore, to achieve detection of biomolecules under different conditions, greater attention is needed on the stability and efficiency of enzymes.

8.3.2 Antibody-based Biosensors

The 1980s are considered the "enlightenment" decade of molecular immunology, as in the late 19th century, Paul Ehrlich first envisioned antibodies as "magic bullets" for delivering cytotoxic agents to tumor cells.[46] Antibodies have since become dynamic tools in diagnostics, therapeutics, and research. They are critical components of the immune response, used to counteract foreign substances or antigens. Due to their dynamic nature, antibodies are widely employed in biosensors as the most common diagnostic biomolecular recognition elements.[47] Antibodies play a crucial role in the defense mechanisms of eukaryotes against pathogens or foreign antigens. Their potential as analytical and diagnostic tools has been developed for over a century. They form immune complexes with specific antigens, the basis of immunoassays, aiding in the development of effective biosensors. Antibody-based sensors can rapidly and accurately detect various analytes. Although classical antibodies have long been used as biological receptors in biosensor fabrication due to their fragility, in recent years they have been engineered into more stable fragments, with increased exposure of their antigen-binding sites. In biosensing, formats constructed through antibody engineering can enhance signals, as the resistance provided by traditional antibodies far exceeds that of these fragments. Thus, signal amplification can be observed when using antibody fragments rather than full-length antibodies as biological receptors. Here, we will discuss the structure, types, and detection technologies of antibody-based biosensors.

8.3.2.1 The Structure of Antibodies

Antibodies, produced by B lymphocytes, primarily function to recognize and attach to harmful substances within an organism, thereby marking them for destruction by other components of the immune system. Antibodies adopt a Y-shaped structure, typically comprising four polypeptide chains: two light chains (25 kDa) and two heavy chains (50 kDa), connected *via* disulfide bonds and non-covalent interactions.[48] The heavy and light chains contain constant (C) domains and variable (V) domains. The unique specificity of antibodies is formed through the combination of variable heavy (VH) and variable light (VL) chains. Six complementary determining regions (CDRs), including CDR1, CDR2, and CDR3 (hypervariable regions), are found on each VH and VL domain. These chains exist in two types: lambda (λ) or kappa (κ).[49]

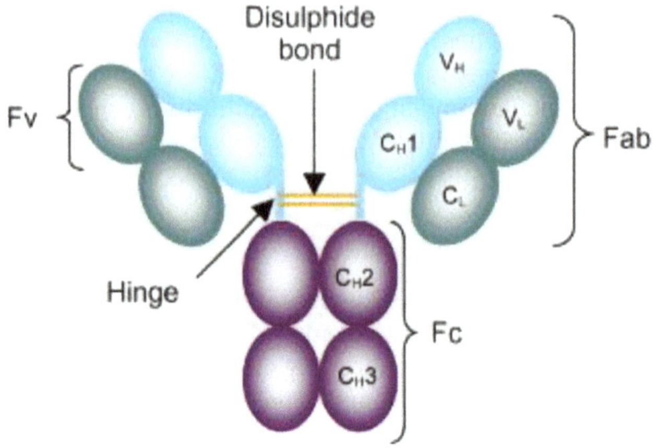

Figure 8.5 The general structure of a classical antibody typically shows that both the light chain and the heavy chain are divided into two fragments: the antigen-binding fragment (Fab) and the constant fragment (Fc).[53] Reproduced from ref. 52 with permission from Elsevier, Copyright 2006.

Therefore, the variable regions of the heavy and light chains play a crucial role in determining the specificity of the antibody to a given antigen, while the constant domains of the heavy chain determine all effector functions of the immunoglobulin. Based on the constant region of the heavy chain, antibodies are classified into five different classes: IgD, IgA, IgM, IgE, and IgG,[50] each with distinct structures and physiological functions.[51] Antibody molecules consist of three functional components: one constant fragment (Fc) and two Fab fragments. Fab fragments possess identical antigen-binding sites, while the Fc fragment provides effector functions by binding to various receptor molecules.[52] The general structure of an antibody is illustrated in Figure 8.5.

8.3.2.2 Types of Antibodies

Polyclonal antibodies (pAb) are typically produced in rabbits, goats, or sheep,[54] and they are widely used in pathogen detection because they can recognize different epitopes. However, in cases where high specificity is required, such as in pathogen detection, monoclonal or recombinant antibodies may be more suitable.

Monoclonal antibodies are produced using hybridoma technology,[55,56] typically by immunizing a mouse host. The resulting hybridomas secrete full-length antibodies targeting a single epitope, which are then analyzed to identify suitable candidates and clone single cells that produce specific antibodies for use in assay development.

Through phage display technology and antibody libraries, recombinant antibodies have been generated for the detection of various antigen structures, including proteins,[57] haptens,[58] and carbohydrate moieties.[59]

The sources of antibody libraries include synthetic, natural, and immune libraries. Synthetic libraries are generated by PCR randomization of the complementarity-determining regions (CDRs) of heavy and/or light chains, whereas natural antibody libraries are typically assembled from B cells extracted from non-immunized human donors. Immune libraries, on the other hand, are constructed from RNA isolated from spleen cells or bone marrow of hosts (such as birds, rodents, rabbits, *etc.*) and immunized with the desired antigen.

8.3.2.3 Technologies of Antibody-based Biosensors

A search of the literature reveals limited publications on biosensors related to recombinant antibodies (see Table 8.1). However, we anticipate a significant increase in such publications in the near future. Initially, many studies on biosensor transduction elements utilize commercially available antibodies (typically monoclonal or polyclonal) to showcase the technology and develop practical diagnostic systems. By integrating antibody generation/characterization with transduction, surface and engineering sciences, nanotechnology, and microfluidics, the field of recombinant antibody-based sensors is poised for rapid advancement. Key considerations include enhancing specificity, sensitivity, stability, and orientation/immobilization to optimize antibody performance. The ultimate challenge lies in employing highly specific and sensitive antibodies capable of deployment on nanodevices, enabling the determination of picomolar concentrations in nanoliter volumes for cancer or cardiac disease-related marker detection.

8.3.3 Nucleic Acid Aptamer-based Biosensors

Aptamer originates from the Latin word "aptus" meaning "fit" and the Greek word "meros" meaning "part."[67] It typically refers to DNA, RNA, or peptides that can bind to targets with high affinity and specificity, resembling

Table 8.1 Applications of antibody-based biosensors.

Analyte	Antibody form	Transducer	Ref.
Disease			
HIV-1 virion infectivity factor	scFv (VH and VHD)	Piezoelectric	60
SARS virus	scFv	Imaging ellipsometry	61
Biowarfare			
Venezuelan equine encephalitis virus	scFv	Potentiometric	62
B. anthracis S-layer protein	scFv	Resonant mirror	63
Contaminants			
Aflatoxin B1	scFv (mono- and dimeric)	SPR	64
Parathion (insecticide)	scFv	Piezoelectric	65
Atrazine	scab	Amperometric	66

antigen–antibody interactions. Compared to antibodies, aptamers offer several advantages, including low immunogenicity, ease of large-scale synthesis, consistency, physical stability, and convenient chemical modification.[68] Aptamers are typically identified from vast combinatorial libraries containing 10^{13}–10^{15} different sequences through a process called Systematic Evolution of Ligands by Exponential Enrichment (SELEX).[69] Aptamers are usually identified from large libraries through multiple rounds of cycling, retaining those with high affinity after each round.[70] These characteristics have led to significant interest in aptamers for applications such as biosensors. This chapter will introduce the preparation process of aptamers, types of aptamer biosensors, and their application research.

8.3.3.1 *The Preparation Process of Aptamers*

Aptamers are relatively short nucleic acid sequences (20–100 nucleotides long), obtained through SELEX schemes and other related *in vitro* selection strategies (see Figure 8.6).[67,71–73] In brief, large oligonucleotide libraries (typically around 10^{14} molecules) are created through primer extension reactions or PCR. The design and quality of the initial library are crucial components for successful SELEX experiments, as libraries with an equal distribution of all four nucleotides' bases maximize the sequence space to be explored.[74]

Selection experiments typically utilize random regions composed of 20–60 nucleotides, although aptamer sequences as short as eight nucleotides and as long as 228 nucleotides have been isolated.[73] The random library is either incubated with the target immobilized on a solid support or directly incubated with cells in the case of cell-SELEX.[75] Species capable of binding to the target are then retained, PCR-amplified, and the resulting enriched pool is injected

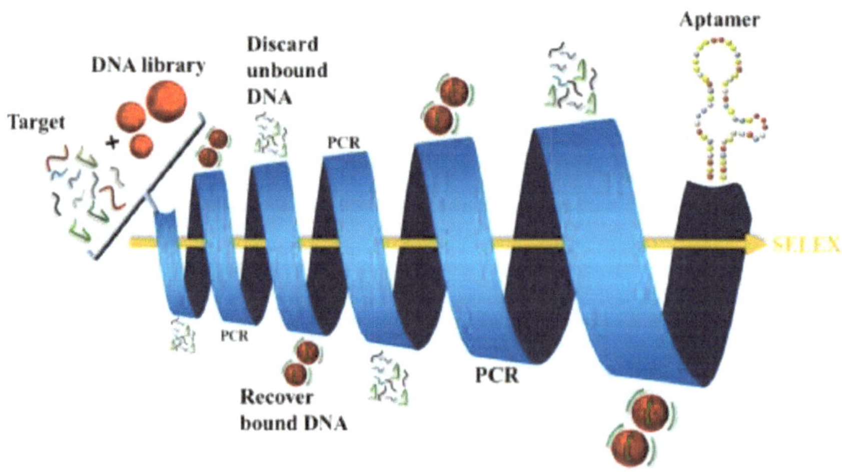

Figure 8.6 A schematic diagram of the SELEX method.[73] Reproduced from ref. 72 with permission from Elsevier, Copyright 2018.

into subsequent rounds of selection. Selective pressure or stringency can be increased through iterative cycles by gradually adjusting physicochemical parameters or by adding negative selection steps.[76] Aptamers have been employed in numerous practical applications, including therapeutics, diagnostic applications, drug-delivery systems, and are even beginning to penetrate the fields of medical imaging and DNA nanotechnology.[73]

The initial DNA library (typically containing 10^{14} molecules) is incubated together with the target immobilized on a solid-phase support. Unbound DNA molecules are discarded, while active species are recovered, amplified *via* PCR, and injected into subsequent rounds of selection. The stringency of the selection protocol can be adjusted by altering physicochemical parameters such as concentration, pH, temperature, or buffer composition. At the conclusion of the protocol, the enriched population is sequenced, and the ability of each aptamer sequence to bind to the target is assessed.[73]

8.3.3.2 Types of Aptamer Biosensors

Aptamer sensors consist of three main types: electrochemical, optical, and mass-sensitive aptamer sensors.

Electrochemical aptamer sensors immobilize biosensing aptamers on the surface of electrodes to detect analyte binding by monitoring changes in current. The detection signal can be enhanced using bio-catalytic labels to improve sensitivity.[77,78] Depending on the measured parameters such as current, conductivity, potential, and impedance, electrochemical sensors can be classified into different methods, such as voltammetry and conductometric methods. For instance, various voltammetric techniques including square wave voltammetry (SWV), differential pulse voltammetry (DPV), cyclic voltammetry (CV), and alternating current voltammetry (ACV) can be applied in voltammetric biosensors. Current biosensors can be regarded as a subclass of voltammetric biosensors as they operate under a constant applied potential (voltage), where this potential is retained at the working electrode with respect to the reference electrode and varies with changes in controlled current over time.[68]

Optical aptamer sensors utilize aptamers labeled with fluorophores, luminescent groups, enzymes, or nanoparticles, or label-free detection systems such as optical resonance and SPR. Most methods are fundamentally based on labeled approaches. Many targets can be optically revealed not only through colorimetric methods, chemiluminescence, or state-of-the-art fluorescence patterns, but also through more current unconventional optical techniques, including surface plasmon-coupled directional emission.[79,80]

Mass-sensitive aptamer sensors utilize minute changes in mass for conduction and can employ piezoelectric crystals and ferromagnetic amorphous wires or strips for mass analysis. These sensors do not require additional labeling and exhibit high detection sensitivity. Mass-based aptamer sensors have several advantages, including not requiring extra labeling and having high detection sensitivity.[81,82]

8.3.3.3 Applications of Aptamer Biosensors in Various Fields

Aptamer biosensors bring broad prospects for diagnostic analysis, primarily applied in biomarker detection, early cancer diagnosis, antibiotic, heavy metal, and toxin contamination detection, among others.

Accumulation of antibiotics in the body can lead to serious acute consequences such as aplastic anemia, cardiovascular collapse, and "gray syndrome." Hence, there is a necessity to develop a sensitive *in situ* analysis method for antibiotic detection.[83,84] Wang *et al.* investigated the analysis of tetracycline using surface plasmon resonance aptamer sensors (see Figure 8.7). They found that compared to single-stranded aptamers, tetracycline aptamers can enhance the sensitivity of the sensor and demonstrate good screening effectiveness for tetracycline in honey samples. Subsequently, the efficacy of aptamer sensors in screening tetracycline in several honey samples was determined in practical applications, showing acceptable recovery rates ranging from 80.20% to 114.3%.[85]

Early diagnosis of cancer is crucial for reducing mortality and suffering. One effective method for the initial diagnosis of cancer is the detection of cancer biomarkers. Parra *et al.* reported a novel technique based on aptamer sensors for detecting prostate-specific antigen (PSA) and free prostate-specific antigen (fPSA) emitted by cancer cells (Figure 8.8). The aptamer sensor recorded a linear response for PSA and fPSA ranging from 1 to 100 ng mL^{-1}, with a detection limit of 1.1 ng mL^{-1}.[86]

Bacterial infections and foodborne illnesses are significant public health concerns, particularly in developing countries where medical resources are scarce and public health conditions are poor. Therefore, rapid identification and detection of such bacterial pathogens are essential for implementing treatment regimens and appropriate control measures.[87] Wu *et al.* developed a visual aptamer sensor for detecting *Vibrio parahaemolyticus* in traditional seafood (see Figure 8.9). In the presence of the target, a "sandwich-type" complex of a-aptamer/target/c-aptamer/magnetic bead formation was observed through specific assays of aptamers and their corresponding targets. Subsequently, this complex was employed as a model for chain displacement

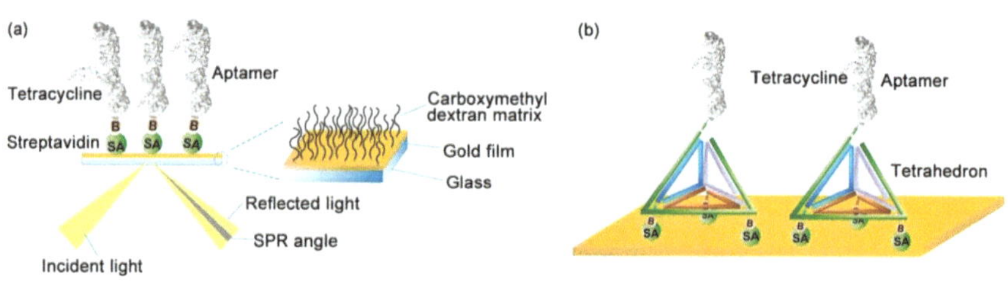

Figure 8.7 Scheme of SPR aptasensor for direct capture of tetracycline: (a) ss-aptamer-based aptasensor for direct capture of tetracycline; (b) oriented immobilization of aptamer (Apt76) assisted by DNA tetrahedron nanostructure.[85] Reproduced from ref. 84 with permission from Elsevier, Copyright 2018.

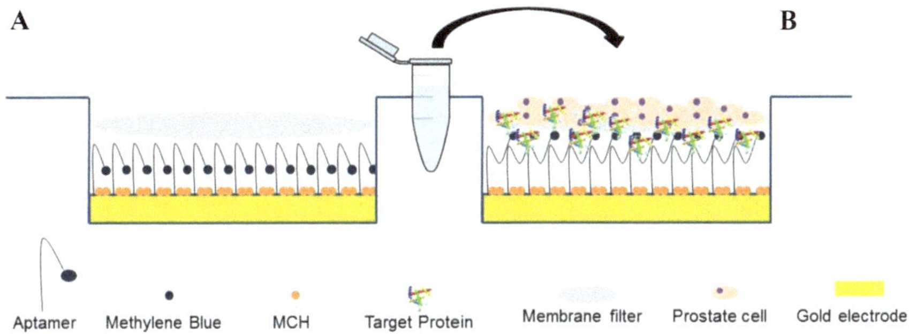

Figure 8.8 Scheme of fPSA and PSA aptamer-based electrochemical sensors. (A) Self-assembled hairpin structure, (B) hairpin changes conformation when binding to PSA or fPSA.[86] Reproduced from ref. 85 with permission from John Wiley & sons © 2018 Wiley-VCH Verlag GmbH & Co. KGaA, Weinheim.

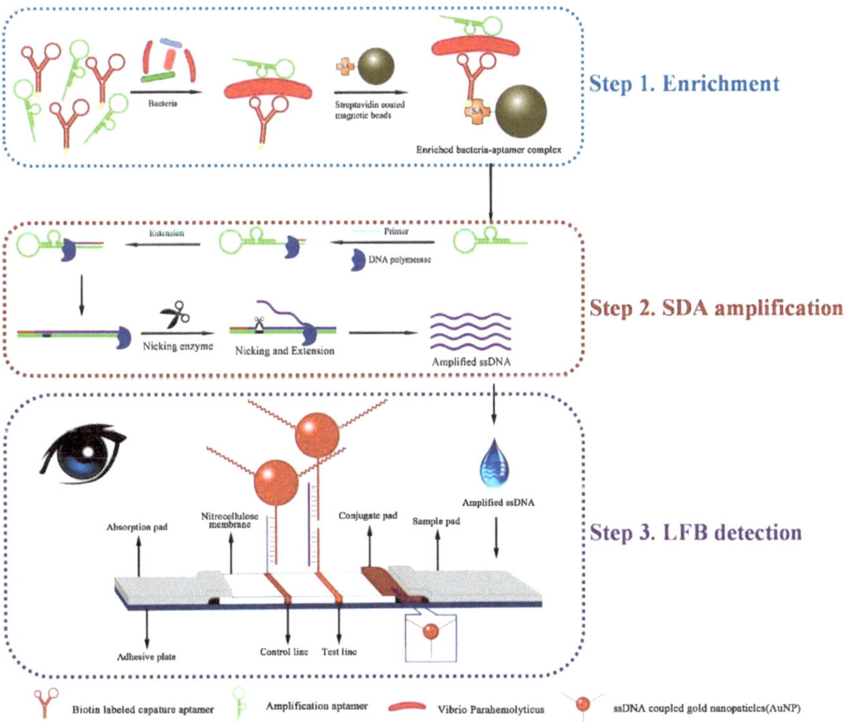

Figure 8.9 Schematic representation of the aptamer-based SDA/LFB assay for *V. parahaemolyticus* detection.[88] Reproduced from ref. 87 with permission from Elsevier, Copyright 2018.

amplification, which, combined with a lateral flow biosensor, was used for detecting *Vibrio parahaemolyticus* in pure cultures and spiked oyster homogenates.[88]

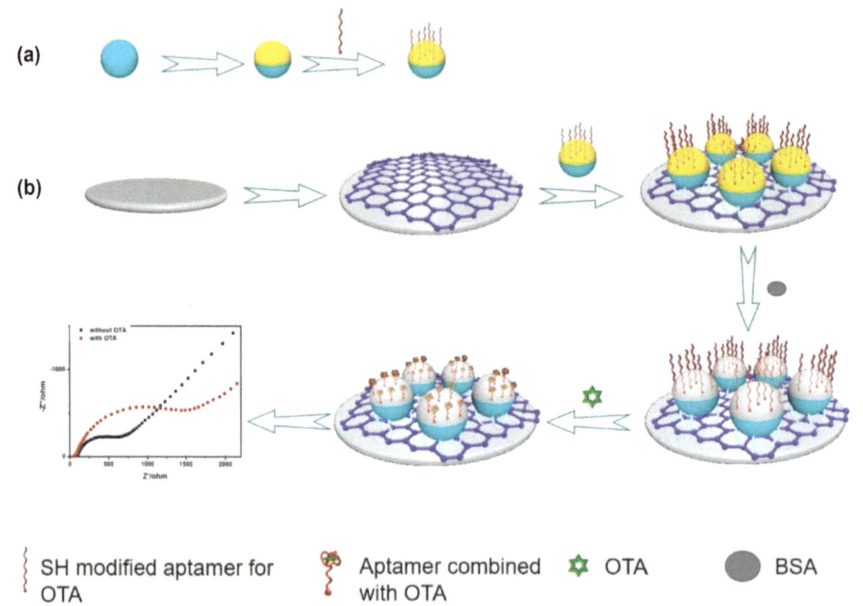

Figure 8.10 Synthesis strategy of the Janus particle (a) and a schematic diagram of the sensing principle of the ochratoxin A immunosensor (b) aptasensor.[90] Reproduced from ref. 89 with permission from Elsevier, Copyright 2019.

One of the key challenges in ensuring food safety is toxin contamination. With the increasing public awareness of healthy eating, developing an effective and convenient method for toxin contamination detection has become crucial in the fields of food safety and environmental protection.[89] The Yang team developed a Janus-functionalized glassy carbon electrode based on aptamers and immune-reactive amino groups (see Figure 8.10). This selective strategy significantly enhances the ability to exclude other interfering substances in the test solution. Their method was successfully used to detect the content of ochratoxin A in real wine samples, demonstrating high recovery rates of 95.7–100.18%.[90]

Simultaneously, the issue of environmental pollution resulting from urbanization and industrial development is becoming increasingly prominent, with heavy metal pollution being particularly notable. Once these crucial metals for industries enter the natural environment, they may affect human health through the food chain. Therefore, the study of heavy metal residues has become a focus of scientific research.[91] Zeng *et al.* designed a fluorescence biosensing system based on continuous signal amplification to trace tiny amounts of arsenic(III) by specifically binding with aptamers, eliciting a signal amplification reaction (see Figure 8.11). This aptamer sensor has been applied to the determination of arsenic in water samples and has shown convincing accuracy.[92]

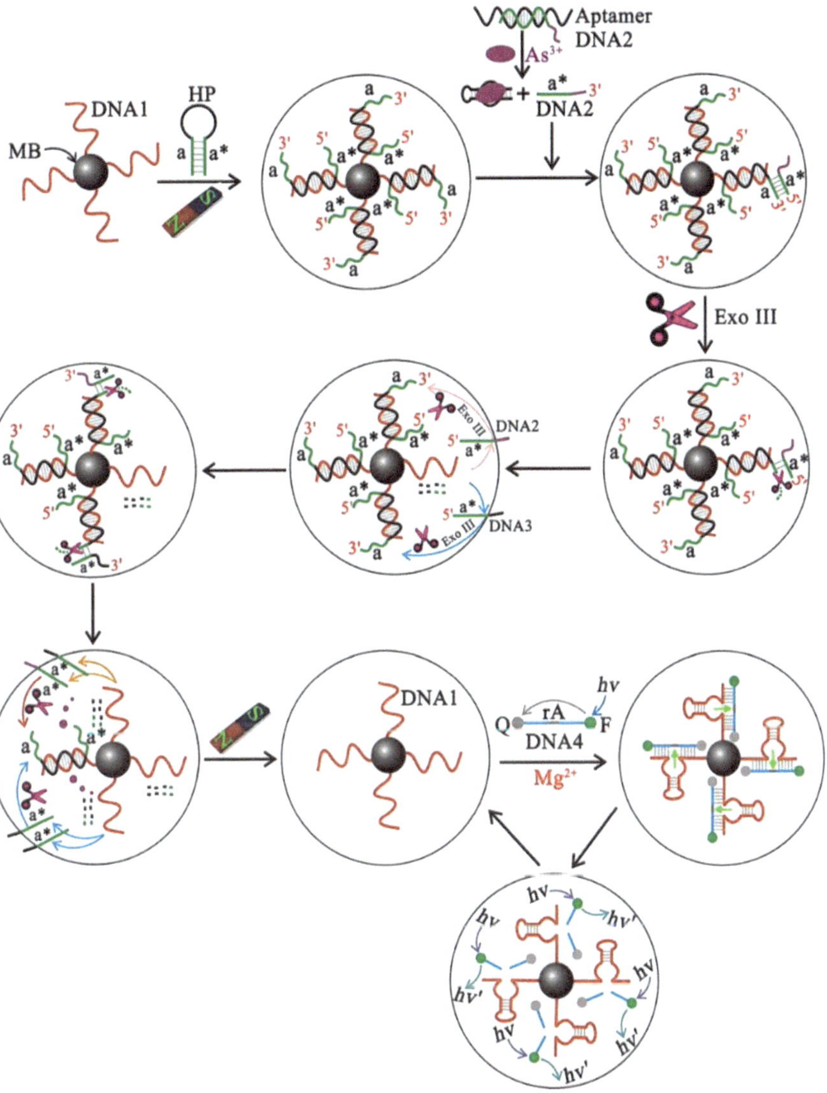

Figure 8.11 Schematic illustration of the assay principle for As^{3+} detection based on a target-triggered successive signal amplification strategy.[92] Reproduced from ref. 91 with permission from American Chemical Society, Copyright 2019.

In the field of biosensing, these devices are precisely designed to rapidly and sensitively detect microbial contamination, food toxins, harmful chemicals, and critical clinical biomarkers. The advantage of aptamers as biomolecular recognition elements lies in their high affinity for target analytes. With the advancement of research, novel structural-switching aptamers are expected to be used for detecting various clinical biomarkers and chemical threats, greatly advancing the application prospects of biosensors.

8.3.4 Cell-based Biosensors

Living cells, these complex units of life, do not only consist of various subcellular organelles, macromolecules, and supramolecular complexes, but also exhibit highly coordinated functions under strict genetic regulation, thereby executing a series of life-sustaining metabolic and defense activities. Viewing living cells as an array of numerous molecular sensors is not an exaggeration, as they can monitor and identify various stimuli in the surrounding environment and initiate a series of responses through energy conversion.[93] Since the early 1970s, cell-based components have been widely used in biosensors, serving as biomolecular recognition elements. Cells as the biomolecular recognition components of biosensors have their unique advantages: they allow flexible customization of detection strategies, have lower costs compared to traditional purified enzymes and antibodies, and simplify the production process, thereby improving cost-effectiveness. With the rapid development of synthetic biology, microfluidics, and lithography techniques, the past five years have witnessed several exciting breakthroughs in the design of cell-based sensors. Three-dimensional cell culture systems integrated with electrodes provide new perspectives for studying disease mechanisms and physiology; the gradually standardized microelectrode array (MEA) technology integrated with cardiac cells is used to assess drug-induced cardiotoxicity. However, despite significant progress, cell-based biosensors still need to overcome some challenges to realize their full potential, including cell regeneration capability, storage lifespan, heterogeneity of cell populations, the influence of environmental interference, and the potential cost increase associated with relevant auxiliary equipment. This chapter will delve into the immobilization methods of recognition element cells in cell-based biosensors, and explore their current applications and future development prospects in various fields.

8.3.4.1 Methods of Cell Immobilization in Cell-based Biosensors

Research has revealed that immobilization techniques significantly enhance the performance of biosensors, especially in terms of response time, target detection accuracy, and long-term stability.[94] Specifically, direct fixation of cells onto sensors, rather than in traditional culture vessels, enables precise quantification of cellular functions to be achieved[95] (see Figure 8.12).

Common passive fixation methods, such as physical adsorption and hydrogel encapsulation, leverage the natural adhesion ability of cells, maintaining close contact between cells and the sensing surface, thus enhancing sensor sensitivity. For instance, microbial fuel cells rely on formed

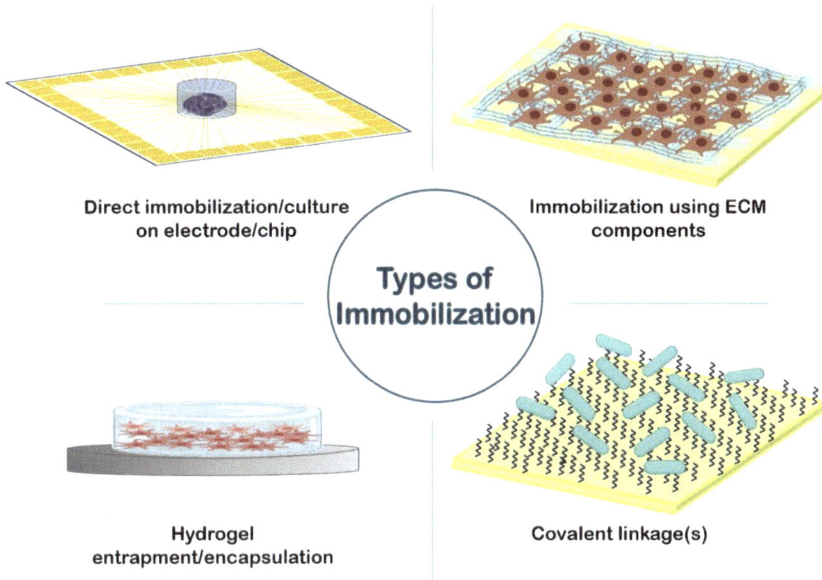

Figure 8.12 Some of the prevalent methods used for immobilization of cells.[100] Reproduced from ref. 99 with permission from Elsevier, Copyright 2019.

biofilms for monitoring water quality toxicity and biochemical oxygen demand.[96,97]

Moreover, living cells can be actively immobilized onto specific substrates through chemical methods, such as using glutaraldehyde cross-linking, to further enhance the binding strength with the sensing surface, thereby increasing detection sensitivity. Recently, Mishra *et al.* demonstrated that the use of covalent fixation techniques improves the detection stability and sensitivity of biosensors to specific chemical substances.[98] Meanwhile, Leonard *et al.* successfully immobilized *Escherichia coli* cells onto a silicon micropillar array using the interaction between biomolecules and cell surface receptors, providing a new platform for antibiotic sensitivity testing.[99]

8.3.4.2 Applications of Cell-based Biosensors in Various Fields

Biocellular sensors based on cancer cells are rapidly emerging as powerful tools for studying the characteristics of cancer cells. They enable non-destructive analysis of various cell properties, including growth, migration, cell surface interactions, cell–cell interactions, and cell–drug interactions.[101] Recently, it has been reported that three-dimensional (3D) graphene interfaces can distinguish different types of breast cancer cells at the single-cell level, thereby enhancing capture efficiency and sensing sensitivity (see Figure 8.13A).[102]

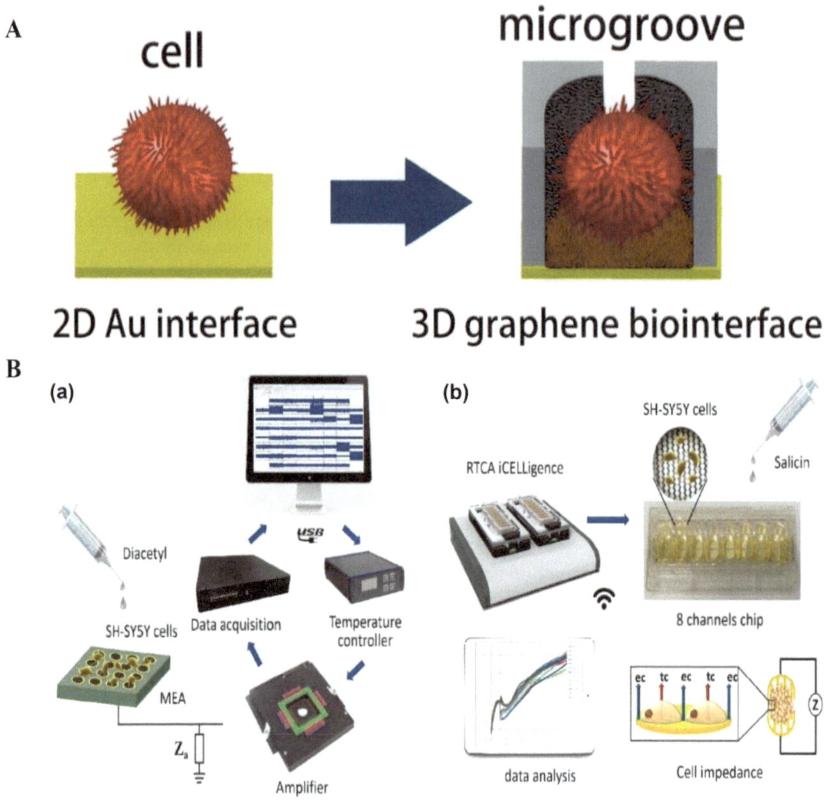

Figure 8.13 (A) Three-dimensional graphene biointerface with extremely high sensitivity to single cancer cell monitoring.[102] Reproduced from ref. 101 with permission from Elsevier Copyright, 2018. (B) Schematic of the biomimetic chemical sensors: (a) MEAs sensor system, (b) cell-impedance sensor system.[105] Reproduced from ref. 104 with permission from Elsevier, Copyright 2019.

These sensors are also used to monitor the response of normal cells to microenvironmental changes (including the presence of drugs and other bioactive molecules) (see Figure 8.13B). Determination of electrophysiological and biomechanical properties is crucial for understanding the pathology of the heart and nervous systems.[103] The technology developed by the Chowdhury team integrates optical mapping and multielectrode arrays (MEAs) for simultaneous measurement of contact potential and action potentials (APs) in *ex vivo* human cardiac slices.[104]

Taste and odor sensors, commonly referred to as "bioelectronic noses" or "tongues," play an important role in food quality and safety assessment.[106,107] Recently, Gao *et al.* reported an integrated biomimetic olfactory, gustatory, and toxicity biosensor based on human neuroblastoma SH-SY5Y cells, which utilize human cells to mimic the functions of these senses.[105]

Furthermore, cell-based biosensors have also been widely studied in drug screening, toxicity assessment, and environmental pollutant detection.[100]

8.3.5 Bacteriophage-based Biosensors

In the early 20th century, prior to antibiotics becoming the mainstream treatment for bacterial infections, Europe had already begun exploring the medical applications of bacteriophages. With scientific advancements, the utility of bacteriophages has expanded into multiple fields, including using phage typing for disease diagnosis, developing phage-based vaccines for prevention, employing phage therapy for treatment, and detecting the presence of pathogens in environmental monitoring.[108] Bacteriophages possess several advantages as biological sensing probes,[109] including: (1) specificity towards target host bacteria. (2) Efficient replication within host bacteria, generating thousands of progeny from a single infection event. (3) Tolerance to extreme conditions such as high temperatures, organic solvents, and pH variations without losing activity compared to antibodies. They can also distinguish between live and dead bacteria, which is crucial for accurate detection.[110,111] (4) Simple and inexpensive large-scale production. These characteristics make bacteriophages ideal candidates for developing efficient biosensors and bioassays. Modern bacteriophage detection schemes often involve the bacteriophage itself or its components, capable of producing direct signals such as fluorescence, colorimetric reactions, electrical signals, or luminescence by capturing and infecting target cells. Thus, bacteriophages demonstrate potential advantages in rapid and specific pathogen detection, showing distinct uniqueness and effectiveness compared to other identification tools.[112]

8.3.6 Antimicrobial Peptide-based Biosensors

Antimicrobial peptides (AMPs) are short peptide fragments found in various organisms, ranging from prokaryotes to humans. They are an indispensable part of the innate immune system, functioning to protect the host from pathogen invasion.[113] In biosensing technology, AMPs are utilized as biomolecular recognition elements for selective capture of specific bacteria.[114] AMP-based biosensors immobilize these molecules on the surface of substrates, leveraging the electrostatic properties of AMPs to attract bacterial outer membranes. Subsequent interactions with the cytoplasmic membrane, such as insertion into the membrane through amphipathic conformations, disrupt the bacterial lipid bilayer.[115]

In biosensing applications, the high affinity of AMPs for bacterial surfaces is emphasized over their antimicrobial properties. Through electrostatic and hydrophobic interactions, they can stably adhere to the surface of biosensors to achieve effective bacterial capture. Another advantage of AMPs is their wide natural source and easy accessibility, allowing selection of suitable AMP molecules from known databases. With advancements in synthesis techniques, specific sequences of AMPs can now be synthesized at reasonable costs. Meanwhile, their low molecular weight increases stability and flexibility during biosensor manufacturing and storage.[116,117] AMPs demonstrate affinity for bacteria even under harsh conditions, such as high temperatures or chemical denaturants, significantly enhancing the stability and repeatability of biosensors. These characteristics have made AMP-based biosensors a hot topic in research, with their potential applications generating widespread scientific interest.

8.3.7 Summary

Bio-based biosensors utilize biological materials as recognition elements to detect and quantify specific biomolecules, chemicals, or environmental conditions within or surrounding biological organisms. These sensors leverage the biological response mechanisms within organisms to interact with target substances, generating specific signals that are converted into measurable outputs through detection techniques. The use of biological materials as the basis for sensors embodies characteristics of being low-carbon, environmentally friendly, and sustainable. These materials are typically renewable or derived from renewable resources, resulting in lower carbon emissions during their preparation. Additionally, biological materials can degrade over their lifecycle, aligning with natural environmental processes and reducing the environmental impact. Moreover, biological materials exhibit good biocompatibility, being compatible with the biological environments within organisms, thereby minimizing adverse effects on human health. Bio-based biosensors hold broad application prospects in fields such as medical diagnostics, environmental monitoring, and food safety.

8.4 Paper-based Biosensors

From the ancient art of papermaking to modern paper-based biosensing technology, paper has played a significant role throughout human history.[118] The ancient papermaking techniques in China had profound cultural, economic, and lifestyle impacts. Paper-based biosensors, as an environmentally friendly innovative technology, have shown tremendous potential in the field of biosensing. Their uniqueness lies in the use of biodegradable paper substrates, which not only reduce energy consumption and environmental pollution during production but also significantly decrease resource requirements during use. Compared to traditional sensors based on silicon or plastic, paper-based sensors have lower carbon footprints and higher biocompatibility, demonstrating vast prospects for sustainable development. Their biodegradable nature also means they are easily disposed of after use, further alleviating environmental burdens (see Table 8.2).

Table 8.2 Nature of paper.[119] Reproduced from ref. 118 with permission from The Royal Society of Chemistry.

Nature	Result
Flexibility	The three-dimensional structures can be formed by folding.
Thickness	The small thickness allows relatively few reagents to be used in device fabrication.
Adsorption	A volume of reagent can be stored and transported in the paper matrix.
Large specific surface area	The number of immobilized enzyme molecules and colorimetric probes is increased.
Capillary action	Capillary forces can be used to transport liquids in all directions without the need for a pump.
Biodegradability	It can be biodegraded by microorganisms or incineration to prevent biological hazards.

8.4.1 The Advantages of Paper as a Substrate Material for Biosensors

Paper-based sensing technologies are particularly well-suited for low-income areas as they offer a low-cost solution that adapts to resource-limited environments. The unique physical properties of paper make it an ideal substrate material for low-cost point-of-care diagnostic devices. Paper-based detection platforms have been deployed globally, even playing vital roles in resource-constrained remote areas and emergency situations. The rapid development of paper-based flexible electronics benefits from advancements in microfluidic technology, mechanical sensing, humidity sensing, and self-powered technology, providing a solid foundation for the manufacturing and application of paper-based devices. This paves the way for new applications in medical diagnostics, health monitoring, and other fields.[119]

8.4.2 Applications of Paper-based Biosensors

Paper-based biosensors share the same advantages as paper, such as environmental friendliness, affordability, flexibility, ease of manufacturing, and biocompatibility. Therefore, many scientists are exploring their applications in various fields, including medicine, food safety, environmental monitoring, human behavior, and physiological signal collection.

8.4.2.1 Disease Diagnosis

Paper-based sensors for disease diagnosis have attracted widespread attention, and various methods have been proposed to collect results, such as colorimetric detection, fluorescence detection, chemiluminescence detection, and electrochemical detection.[120]

Colorimetric detection is particularly popular in resource-limited environments because it is simple, rapid, and does not require complex laboratory equipment.[121] Although this method typically relies on visual observation, which may be influenced by individual perception and environmental factors, it provides a reliable means for semi-quantitative measurement. Additionally, fluorescence detection, a method for qualitative or quantitative analysis based on observing the excited state of fluorescent substances, has been widely applied in paper-based sensing devices. These devices enhance detection sensitivity and accuracy through the adsorption of fluorescent labels.[122,123] For example, Tsai *et al.*[124] developed a paper-based colorimetric detection device for tuberculosis diagnosis, utilizing surface plasmon resonance effects and single-stranded DNA probe molecules hybridized with target double-stranded tuberculosis DNA to monitor the color change of gold nanoparticles (see Figure 8.14). Chen *et al.*[125] designed a rapid colorimetric sensing system for detecting commonly abused drugs such as ketamine on microfluidic paper.

In recent years, fluorescence-based paper-based biosensors have received widespread attention due to their high sensitivity and simplicity. These sensors rely on observing the excited state of fluorescent substances under UV light irradiation for qualitative or quantitative analysis. For instance, a paper-based microfluidic device with quantum dot fluorescent probes developed by Liang *et al.* is capable of detecting and distinguishing different types of tumor cells, enabling simultaneous detection of tumor cell types.[126] Furthermore, Guo *et al.* designed a fluorescence immunoassay device utilizing zinc oxide nanowires to enhance the sensitivity of fluorescence detection, thereby enabling more precise and reliable detection of cardiac biomarkers for acute myocardial infarction (see Figure 8.15).[127]

In addition to fluorescence methods, chemiluminescence has also been applied in paper-based medical diagnostic sensors. For example, Zhao *et al.*[128] developed a paper-based immunoassay device for plasma treatment, employing novel protein immobilization techniques and chemiluminescent

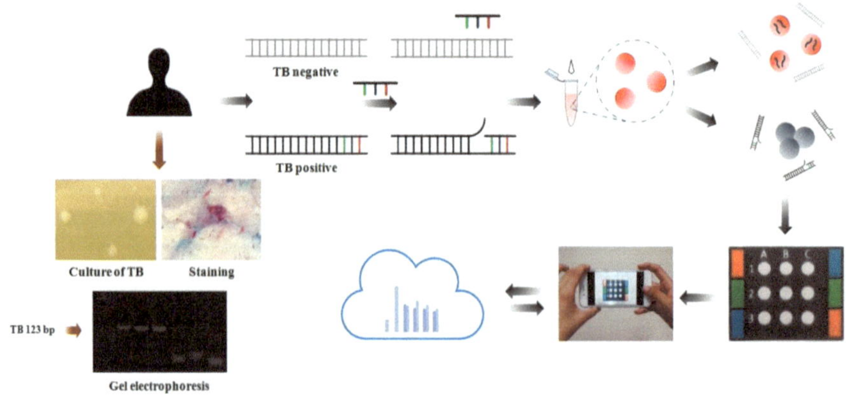

Figure 8.14 Schematic illustration of the proposed TB diagnostic method.[124] Reproduced from ref. 123 with permission from American Chemical Society, Copyright 2017.

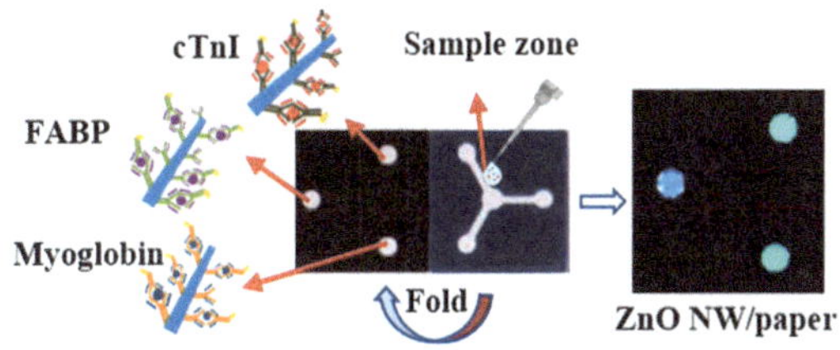

Figure 8.15 Signal-enhanced detection of multiplexed cardiac biomarkers by a paper-based fluorogenic immunodevice integrated with zinc oxide nanowires.[127] Reproduced from ref. 126 with permission from American Chemical Society, Copyright 2019.

immunoassays, successfully detecting carcinoembryonic antigen in human serum with a detection limit as low as 0.03 ng mL^{-1}, far below clinical levels.

Electrochemical analysis methods have similarly found application in paper-based biosensors. Cinti *et al.*[129] developed a paper-based biosensor containing Prussian blue nanoparticles capable of effectively detecting glucose in blood. This biosensor exhibits a broad dynamic response range, up to 25 mM (450 mg dL^{-1}), with accuracy comparable to commercial blood glucose test strips. Such devices offer a convenient self-monitoring solution, particularly suitable for diabetes patients.

While many researchers continue to strive for the development of paper sensors with broader detection ranges and higher sensitivity, advancements in technology have enabled paper-based sensors to interface more seamlessly with portable devices, reducing reliance on complex medical equipment. Nie *et al.*[130] developed an electrochemical micro-paper analytical device for rapid quantitative analysis of key biomarkers in blood or urine, such as glucose and cholesterol, streamlining the health-monitoring process.

8.4.2.2 Paper-based Wearable Sensors for Monitoring Human Physiological and Behavioral Signals

Paper-based wearable sensors hold tremendous potential in human behavior perception, physiological information monitoring, and human–computer interaction fields. Since the outbreak of the COVID-19 pandemic in 2019, the significance of these devices has become increasingly prominent. With the advancement of technology, paper-based sensing technologies based on human behavior and physiological signals have been extensively researched.

For instance, researchers such as Gao *et al.*[131] developed a fully paper-based resistive pressure sensor (see Figure 8.16A). This sensor utilizes paper coated with silver nanowires as the sensing layer, with nano-cellulose paper serving as the bottom layer substrate and top encapsulation layer for printed electrodes. This sensor not only provides accurate signal feedback

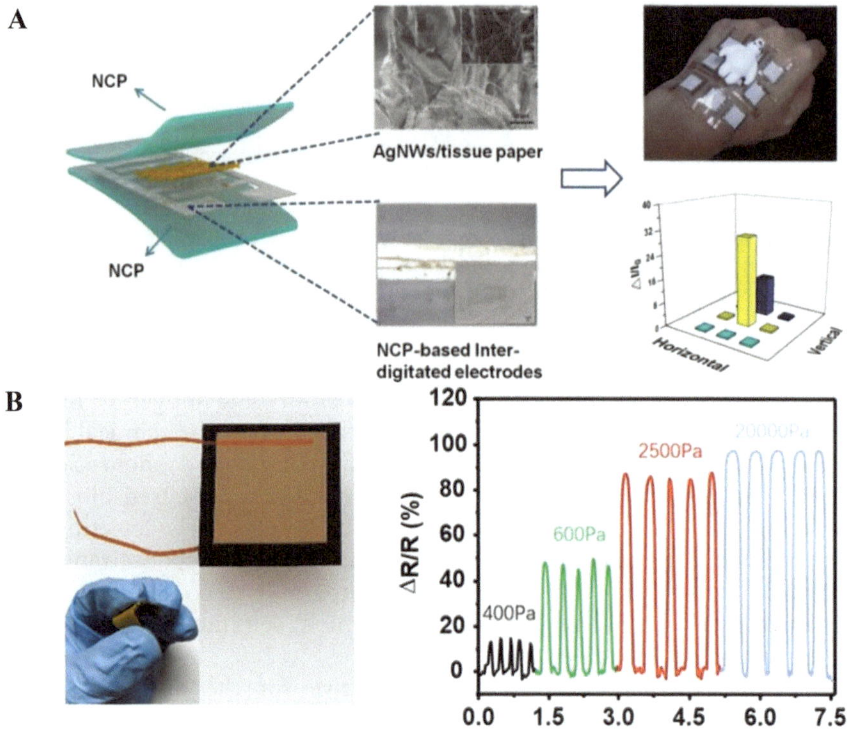

Figure 8.16 (A) All-paper-based flexible and wearable piezoresistive pressure sensor.[131] Reproduced from ref. 130 with permission from American Chemical Society, Copyright 2019. (B) Graphene-paper pressure sensor for detecting human motions.[131] Reproduced from ref. 131 with permission from American Chemical Society, Copyright 2017.

during bending and pressing but can also be attached to the skin to monitor physiological signals such as arterial heartbeats and throat movements.

Additionally, Tao *et al.*[132] combined multiple layers of graphene oxide solution with paper materials using a thermal reduction method to produce multi-layer graphene paper, thereby creating a high-performance pressure sensor (see Figure 8.16B). This sensor fully exploits the air channels between graphene paper layers and the microporous structure of the paper, resulting in a significant resistance change under pressure. By selecting appropriate types of graphene paper and the number of paper layers, this device can greatly enhance pressure sensitivity and accurately monitor pulses, respiration, and various movement states.

8.4.2.3 *Food Safety Testing*

Food safety directly impacts health and well-being, making rigorous testing at every step of food production crucial. Paper-based sensors, due to their

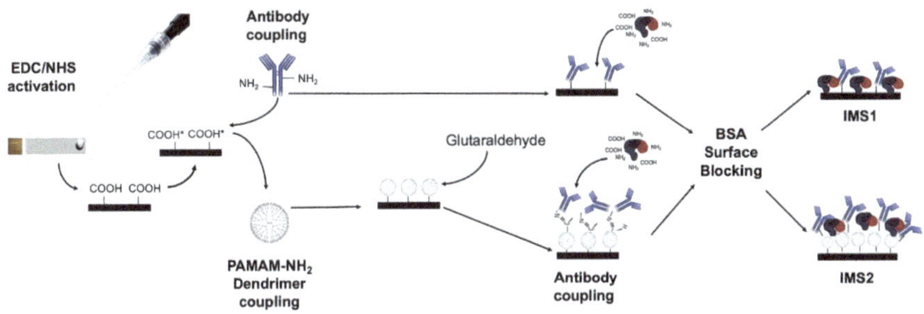

Figure 8.17 Schematic representation of the two immunosensing interfaces (IMS) implemented (IMS1 and IMS2), showing the different modification steps performed to antibody coupling and electrode surface blocking.[136] Reproduced from ref. 135 with permission from Elsevier, Copyright 2019.

low cost and ease of operation, particularly in immobilizing or freeze-drying various biomolecules, have become a research hotspot in food safety analysis. Although most paper-based sensors are still in the laboratory development stage, their practical potential is undeniable.[133,134]

Mahato *et al.*[135] developed a paper-based immunosensor for detecting alkaline phosphatase in dairy products. The device allows for semi-quantitative analysis visible to the naked eye and quantitative estimation *via* a smartphone, with a detection range of 10–1000 U mL^{-1} and a detection limit of 0.87 U mL^{-1}. Silva *et al.*[136] reported a novel paper-based sensing platform for detecting *Salmonella typhimurium* (see Figure 8.17). The logarithmic correlation between the displacement of potential caused by the ion flux blockade effect induced by antigen–antibody binding and the concentration of *Salmonella typhimurium* in the sample was demonstrated, showing excellent sensitivity with a detection limit of 5 cells per mL, proving its effectiveness in practical applications such as apple juice detection.

8.4.2.4 *Environmental Monitoring*

Environmental pollution is an inevitable byproduct of industrialization, including factory emissions, domestic wastewater, vehicle exhaust, pesticide use, *etc.*, all of which severely impact the quality of water, air, and soil. Therefore, rapid detection of environmental pollutants is particularly important, and paper-based sensing devices, with their simplicity, cost-effectiveness, and speed, provide new tools for environmental monitoring.

Taking metal ions as an example, excessive metal ions can have adverse effects. Especially heavy metal ions, even in trace amounts, are highly toxic. Their contamination poses a significant threat to the environment and public health worldwide, potentially disrupting ecosystems and adversely affecting global sustainability. The paper-based microfluidic electrochemical sensing device developed by Nie *et al.* utilizes photolithography to manufacture microchannels and employs carbon and Ag/AgCl conductive inks to

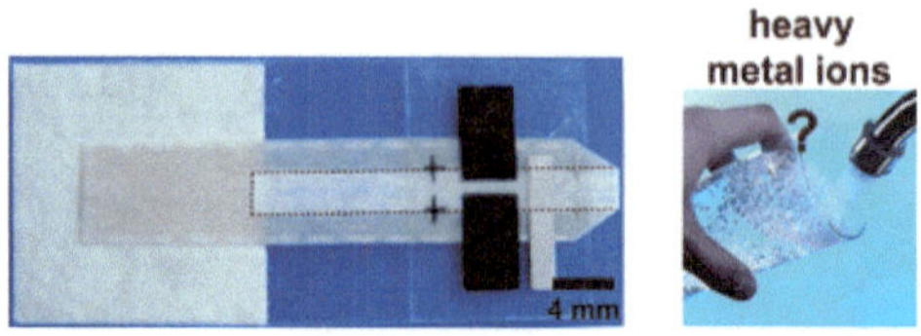

Figure 8.18 Microfluidic paper-based electrochemical devices (μPEDs). The μPEDs are capable of quantifying the concentrations of various analytes in aqueous solutions, including biological fluids such as urine, serum, and blood.[137] Reproduced from ref. 136 with permission from The Royal Society of Chemistry.

print electrodes, enabling precise determination of heavy metal ions and other analytes in water, such as glucose (see Figure 8.18). This low-cost device is particularly suitable for resource-limited areas.[137] Additionally, Feng *et al.* constructed a paper-based device based on fluorescence methods capable of semi-quantitatively detecting multiple heavy metal ions, intuitively displaying ion concentrations through color changes, providing a convenient on-site detection solution.[138]

8.4.3 Outlook and Conclusion

The most significant advantage of paper-based sensors lies in the inherent properties of paper itself, such as low cost, high biocompatibility, and biodegradability. These characteristics make paper-based sensors particularly important in areas such as health monitoring, food safety, and environmental protection, especially when considering the practical needs of remote and underdeveloped regions. In developing these sensors, a key factor is to ensure the simplicity of the manufacturing process, ensuring that even untrained individuals can easily use and interpret the results. This is crucial for the successful market penetration of paper-based sensors. However, to achieve commercialization and large-scale production of these sensors, a series of challenges need to be overcome:

i. Reproducibility: Sensors from different production batches must provide reproducible detection performance within acceptable error ranges.
ii. Environmental adaptability: Most current paper-based sensors are primarily used in laboratory environments. For marketization, it is necessary to ensure that these devices can adapt to environmental changes, guaranteeing the accuracy and reliability of test results.
iii. Targeted development: The primary development goal of low-cost paper-based detection devices is to meet the demand for real-time disease detection in underdeveloped regions. Therefore, development must fully consider the actual needs of these regions, ensuring the reliability, convenience, and applicability of the devices.

By addressing these issues, paper-based sensors will be able to serve a wider global audience, especially in resource-limited environments, providing effective and economical health solutions and further promoting progress in global public health and safety.

8.5 Silk-based Biosensors

With the increasing demand for sustainable development and environmentally friendly technologies, natural biomaterials have garnered attention in the field of biosensors due to their low carbon emissions, wide availability, low cost, and excellent biocompatibility. Among these materials, silk shows tremendous potential in the development of biosensors due to its unique properties.

Silk is a continuous long fiber formed by the solidification of silk fluid secreted by silkworms during the cocooning process. It is an ancient natural biomaterial and one of the earliest animal fibers utilized by humans. The biocompatibility of silk has led to its widespread application in targeted drug delivery and tissue engineering.[139] Its biodegradable nature, where it degrades into amino acids and peptides under the action of proteases, gives silk special value in medical materials. Furthermore, by adjusting the crystallinity of silk fibroin molecules, the degradation rate of the material can be precisely controlled.[140] In terms of mechanical properties, the highly crystalline protein structure of silk makes it one of the strongest natural fibers.[141] Researchers such as Professor Kaplan from Tufts University have made significant progress in processing silk into various forms such as films, gels, microspheres, sponges, *etc.*,[142–144] providing important material foundations for the development of multifunctional implantable biosensor substrates.

8.5.1 Traditional Biosensors Based on Silk Substrates

8.5.1.1 Electrochemical Biosensors Based on Silk Substrates

Due to their high sensitivity, selectivity, rapid response, and excellent compatibility with the substances being tested, electrochemical biosensors are widely used in the manufacturing of biosensors.[145] Among these sensors, the material of the working electrode directly influences its performance. However, traditional electrode materials such as noble metals, metal oxides, carbon black, and graphene are often expensive, lack flexibility, and exhibit poor wettability.[146]

To overcome these limitations, researchers have begun exploring the use of natural silk. Xu *et al.* utilized silk gel conductive ink printed with conductive polymers and vascular endothelial growth factor (VEGF) antibody on a silk fibroin protein substrate to successfully fabricate impedance sensors (see Figure 8.19). This novel sensor demonstrates high selectivity in

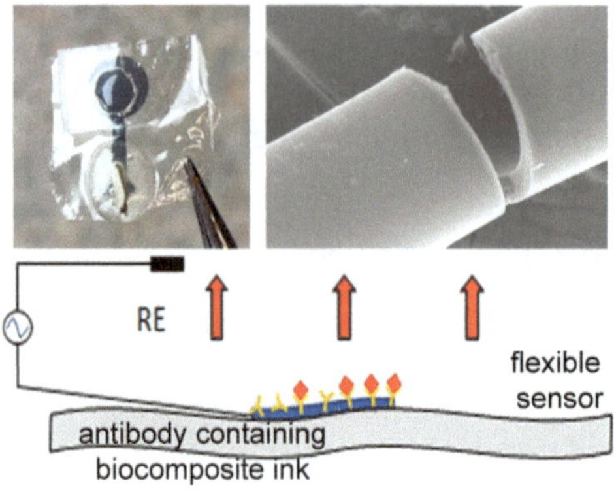

Figure 8.19 Flexible biosensors for the impedimetric detection of protein targets using silk-conductive polymer biocomposites.[147] Reproduced from ref. 146 with permission from American Chemical Society, Copyright 2019.

detecting VEGF, highlighting the potential of silk-based materials in the high-precision biosensing field.[147]

8.5.1.2 Colorimetric Biosensors Based on Silk Substrates

Traditional sensor substrates are typically composed of synthetic polymers such as polydimethylsiloxane (PDMS), polyethylene oxide (PEO), and polyethylene terephthalate (PET), but these materials often lack biocompatibility.[148,149] Recent studies have modified regenerated silk fibroin protein by adding glycerol and 1-(3-dimethylaminopropyl)-3-ethylcarbodiimide hydrochloride (EDC), followed by blending with polypyrrole and drying to produce a novel pH-responsive silk fibroin protein film colorimetric sensor (see Figure 8.20). This sensor is suitable for monitoring the pH of human sweat to assess human health conditions, and its excellent biocompatibility and stable mechanical properties ensure comfort and environmental friendliness during prolonged wear.[150]

8.5.1.3 Fluorescent Biosensors Based on Silk Substrates

Fluorescent proteins are commonly used in biosensors due to their unique luminescent properties. However, traditional fluorescent proteins require UV/infrared excitation, and their potential carcinogenicity limits their widespread use in medical applications.[151] In this regard, the team led by Choi at the National Institute of Agricultural Sciences in Korea has developed photosensitive transgenic silk by combining far-red fluorescent protein (mKate2) with silk fibroin (see Figure 8.21). This further processed composite flexible film (mKate2 silk film) can be activated by visible light, generating reactive oxygen species through photocatalysis, used for cell imaging and cancer treatment. This technology exploits the optical properties of natural silk (strong light scattering) to enhance the interaction between light and

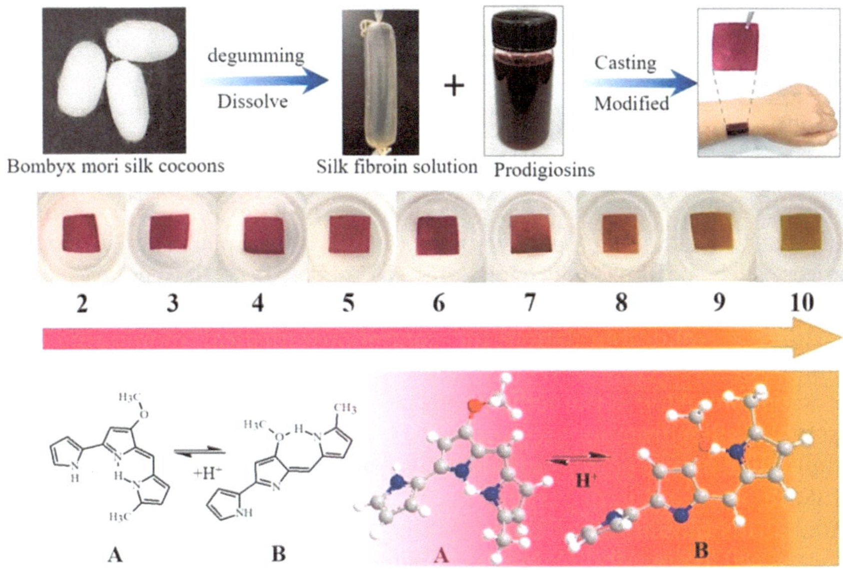

Figure 8.20 pH-Responsive discoloration silk fibroin films based on prodigiosin from microbial fermentation.[150] Reproduced from ref. 149 with permission from Elsevier, Copyright 2022.

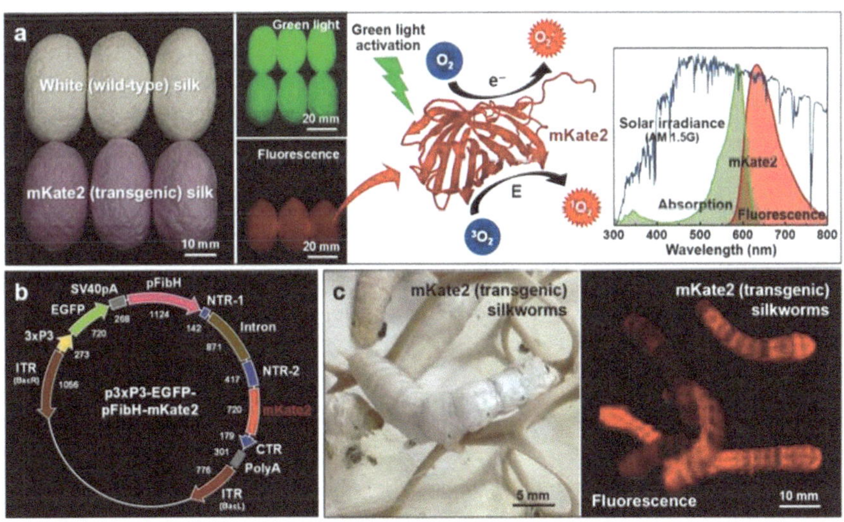

Figure 8.21 Schematic illustration of reactive oxygen species (ROS)-generating mKate2 (transgenic) silk under green light activation. Superoxide ($O_2^{\bullet-}$) and singlet oxygen ($^1O_2$) are generated by mechanisms of electron (e^-) transfer and energy (E) transfer, respectively. Photographs of white (wild-type) and mKate2 (transgenic) silk cocoons and fluorescent image of mKate2 silk cocoons. Green light belongs to the peak wavelength range of the solar spectrum.[152] Reproduced from ref. 151 https://doi.org/10.1002/advs.201700863, under the terms of the CC BY 4.0 license, https://creativecommons.org/licenses/by/4.0/.

fluorescent proteins within transgenic silk fibers, demonstrating the potential of genetic engineering in manufacturing functional silk materials.[152]

8.5.2 Implantable Silk-based Biosensors

Leveraging the excellent properties of silk, such as good biocompatibility, biodegradability, outstanding mechanical performance, and processability, researchers have developed various forms of implantable electronic devices, such as films, hydrogels, and scaffolds. These devices, with their unique properties, become the ideal choice for implantable biosensors, providing a solid foundation for continuous medical monitoring. For example, Cui *et al.* combined natural silk fibroin (SF) with the oxygen-sensitive chromophore Pd(II) tetramethylphenylporphyrin (PdBMAP) to prepare an insoluble silk/chromophore composite film for oxygen sensing. The amphiphilic nature of the composite film stabilizes the oxygen sensitivity of PdBMAP, enabling it to respond to changes in dissolved oxygen concentration. This biosensor not only exhibits good *in vivo* biocompatibility and *in vitro* cell compatibility but also demonstrates mechanical properties suitable for subcutaneous implantation in rats (see Figure 8.22).[153]

On the other hand, the 3D scaffold based on silk fibroin developed by Wang *et al.* demonstrates its potential in guiding neural regeneration. Using electrospinning technology, they mixed *Bombyx mori* silk fibroin, poly(ε-caprolactone-*co*-lactide) (PLCL), and graphene oxide (GO) to prepare a composite nanofiber 3D scaffold with multiple channels and layered structure. This biomimetic nerve conduit has been shown in cell experiments and *in vitro* animal studies to promote Schwann cell migration, proliferation, and myelin formation, effectively facilitating *in vivo* nerve regeneration.[154]

8.5.3 Conclusion and Future Perspectives

In the material selection for biosensors, silk has attracted considerable attention as one of the promising natural polymer materials in the field of

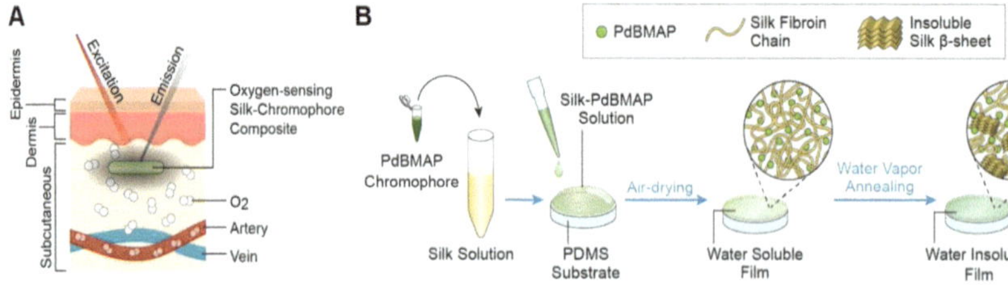

Figure 8.22 (A) Diagram of the continuous oxygen-monitoring platform by the phosphorescent silk–chromophore composite film. (B) Schematic outlining the silk–chromophore film production.[153] Reproduced from ref.152 with permission from John Wiley & Sons, © 2022 Wiley-VCH GmbH.

biosensors due to its outstanding mechanical properties, good biocompatibility, biodegradability, and tunable morphology. Its low carbon emissions, environmental friendliness, and sustainability make it particularly noteworthy in the application of biosensors. In the medical sensing field, researchers have achieved a series of remarkable results in using silk as the substrate for implantable sensors, providing a new path for environmental protection and sustainable development. Although the potential of 3D silk fibroin materials in biosensor development has not been fully explored, their environmental friendliness and degradability open up new possibilities. By genetically modifying the primary structure of silk fibroin macromolecules, functionalized silk can be obtained, providing new insights for the green development of biosensors. Despite the many advances in this field and a wealth of research for reference, the complexity and interdisciplinary nature of biosensor material design remain the primary challenges. In the future, the search for new materials and material design strategies will be crucial directions in biosensor research, which will help achieve a wider and more sustainable application of biosensors in the real world. By integrating the unique properties of biomaterials with technological innovation, strides can be made toward more environmentally friendly and efficient biosensing solutions.

8.6 Carbon-based Biosensors

As the demand for sustainable development and eco-friendly technologies continues to grow, the application of low-carbon materials in the field of biosensors has garnered widespread attention. Carbon-based materials, in particular, are highly regarded due to their low carbon emissions, environmental friendliness, and sustainability, as well as their enormous potential for applications. Over the past few years, carbon nanomaterials such as graphene, carbon nanotubes (CNTs), graphene quantum dots (GQDs), carbon dots (CDs), and fullerenes have demonstrated outstanding performance in biosensors. The advantages of surface functionalization, rapid electron transfer capability, biocompatibility, and electrochemical stability of these materials enable them to be compatible with the biological environment while reducing adverse effects on human health and the environment. Therefore, the application of these low-carbon materials in the manufacturing of biosensors not only contributes to reducing carbon emissions and resource consumption but also enhances the performance and sustainability of sensors, making a positive contribution to environmental protection and sustainable development.

8.6.1 Types of Carbon Nanomaterials and Their Applications

In the field of biosensors, carbon-based materials have become a hot topic of research. Carbon nanomaterials, including graphene, carbon nanotubes (CNTs), graphene quantum dots (GQDs), carbon dots (CDs),[155] and fullerenes, have shown enormous potential due to their unique structures and properties.

8.6.1.1 Graphene

Graphene is a two-dimensional honeycomb structure composed of single-layer carbon atoms hybridized in sp^2 orbitals, easily extractable from graphite. Its high surface area to volume ratio, excellent surface functionalization capability, high mechanical strength, as well as outstanding thermal and electrical conductivity, make it outstanding in biosensors.[156] The optical properties of graphene, such as polarization-related effects and tunable broadband absorption, have been applied in space optical sensors, surface plasmon resonance (SPR) sensors, and optical fiber SPR sensors for detecting anticancer drugs, cell lines, and single cells.[157]

Graphene can be functionalized by various materials such as metal oxides, polymer films, metal nanoparticles, quantum dots, and ionic liquids. For instance, composite materials of MnO_2 and graphene exhibit improved dispersibility and superior conductivity of graphene nanoparticles due to the structural characteristics of MnO_2.[158] In the research conducted by Gao *et al.*, a poly-L-lysine (PLL)-functionalized graphene field-effect transistor biosensor was prepared by functionalizing PLL on the channel surface of the biosensor and immobilizing DNA probes through electrostatic force (see Figure 8.23). They found that using minute amounts of human serum samples, these biosensors could achieve a detection range of five orders of magnitude for miRNA, providing an effective method for rapid diagnosis and screening of breast cancer.[159]

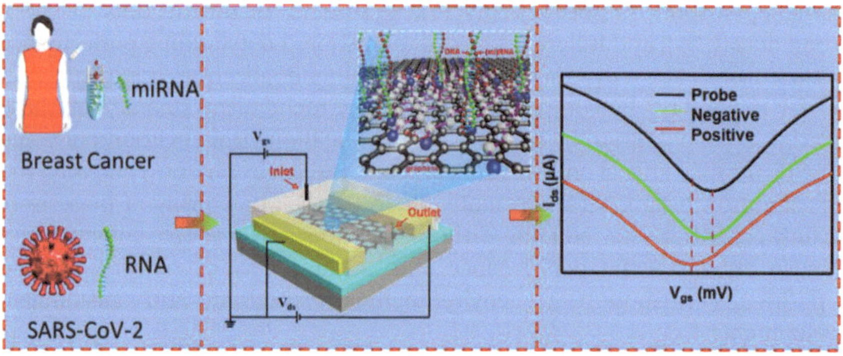

Figure 8.23 Poly-L-lysine-modified graphene field-effect transistor biosensors for ultrasensitive breast cancer miRNAs and SARS-CoV-2 RNA detection poly-L-lysine-modified graphene field-effect transistor biosensors for ultrasensitive breast cancer miRNAs and SARS-CoV-2 RNA detection.[159] Reproduced from ref. 158 with permission from American Chemical Society, Copyright 2022.

8.6.1.2 Graphene Oxide, Reduced Graphene Oxide, and Their Derivatives

Graphene is not only a material but also a diverse material group, with the most important members including graphene oxide (GO) and reduced graphene oxide (rGO).[160]

GO is widely utilized in optical biosensors due to its unique ability to quench photoluminescence. This property improves with increasing carbon/oxygen ratio.[161] Furthermore, rGO nanocomposites combined with metal nanoparticles such as Au, Ag, Ce, and Ti open up new possibilities for applications in biosensors. These composite materials exhibit synergistic effects of various material properties, resulting in excellent performance.[162] In particular, cerium oxide (CeO_2) has emerged as an ideal choice for manufacturing point-of-care biosensing devices due to its biocompatibility, conductivity, large surface area, and high oxygen transferability.[163]

8.6.1.3 Carbon Nanotubes

Carbon nanotubes (CNTs) are one-dimensional carbon allotropes composed of curled graphene sheets, with diameters around 1 nm and lengths ranging from 1 to 100 nm. As one of the most prominent carbon allotropes following graphene, CNTs are renowned for their high surface area to volume ratio, excellent mechanical properties, extremely high electrical conductivity (10^4–10^7 S m^{-1}), and superior optical properties. These characteristics enable CNT-based biosensors to detect ultra-low concentrations of biomolecules with fast response times and long lifespans. There exist two types of carbon nanotubes: multiwalled carbon nanotubes (MWCNTs) and single-walled carbon nanotubes (SWCNTs). SWCNTs exhibit fluorescence quenching and photoluminescence properties, making them suitable for optical biosensors.[164] For instance, an optical sensor developed by Williams and his team utilizes the optical properties of SWCNTs to detect urokinase plasminogen activator (uPA) through gap modulation in antibody binding, which is highly effective in early detection of prostate cancer.[165]

8.6.1.4 Carbon Dots

Carbon dots (CDs) are zero-dimensional (0D) carbon materials with average sizes smaller than 20 nanometers, known for their high stability, good biocompatibility, high quantum yield, and thermal stability, and have been extensively studied. The performance of CDs highly depends on the synthesis methods of surface functional groups. CDs have been widely applied in various fields such as energy applications, bioimaging, phototherapy, drug delivery, and sensors, demonstrating their multifunctionality and high efficiency.[166,167]

Zarei-Ghobadi and colleagues designed an optical biosensor to detect the genome of human T-lymphotropic virus type 1 (HTLV-1), a retrovirus causing human cancers. They exploited the photoluminescence properties of CDs in the presence of Fe_3O_4-capped Au (Fe_3O_4@Au). The detected CDs adsorbed on the surface of Fe_3O_4@Au quenched the fluorescence emission of CDs. Upon adding the target virus to the system, CDs were desorbed, and fluorescence emission was restored, termed as the detection signal. One drawback of this sensor is the presence of other viruses such as HIV and hepatitis B, and its sensitivity is not significant.[168]

8.6.1.5 Fullerenes

Fullerenes, with their distinctive symmetrical cage structures, exist in various sizes such as C60, C80, and C20. Among them, C60 is the most common form, consisting of 60 carbon atoms with a diameter of approximately 7.0×10^{-10} meters. This zero-dimensional material exhibits high thermal stability (up to 1000 °C), long-lived triplet states, and significant nucleophilic and electrophilic properties.[169] In the construction of biosensors, individual C60 typically does not directly participate but serves as a mediator to facilitate electron transfer in the electrode, achieved through conjugation with biomolecules. However, this conjugation or functionalization often needs to be performed externally or internally on the carbon cage, depending on the nature of the analyte and the type of sensor.[170]

Functionalization of fullerenes is a crucial step to enable their application in biosensors. Functionalization can be achieved through various methods, such as modification of C60 nanoparticles (FC60 NPs) with 10-phenanthrene tetracarboxylic dianhydride and L-cysteine to increase their water solubility and active sites. Subsequently, further enhancement of current response and electrocatalytic activity can be achieved by decorating FC60 NPs with gold nanoparticle-carrying Prussian blue. This composite material can be utilized for constructing efficient electrochemical aptasensors.

The study by Han *et al.* serves as an example (see Figure 8.24), where they designed an electrochemical aptasensor for detecting cancer-related proteins. By immobilizing aptamers on a glassy carbon electrode in combination with gold nanoparticles and onion-like mesoporous graphene sheets (O-GS), they were able to efficiently capture and detect specific cancer biomarkers. Additionally, they developed a tracking label, achieving detection of platelet-derived growth factor B-chain with extremely low detection limits (0.6 pM) and a wide detection range (0.002–40 nM) using functionalized C60 nanoparticles (FC60 NPs) as tracers.[171] The development of these fullerene-based aptasensors not only demonstrates the immense application potential of fullerenes in the field of biosensors but also showcases the significant role of nanotechnology in modern medical diagnostics.

8.6.2 Summary

Carbon-based biosensors are currently undergoing rapid development, leveraging their environmentally friendly and low-carbon-emission

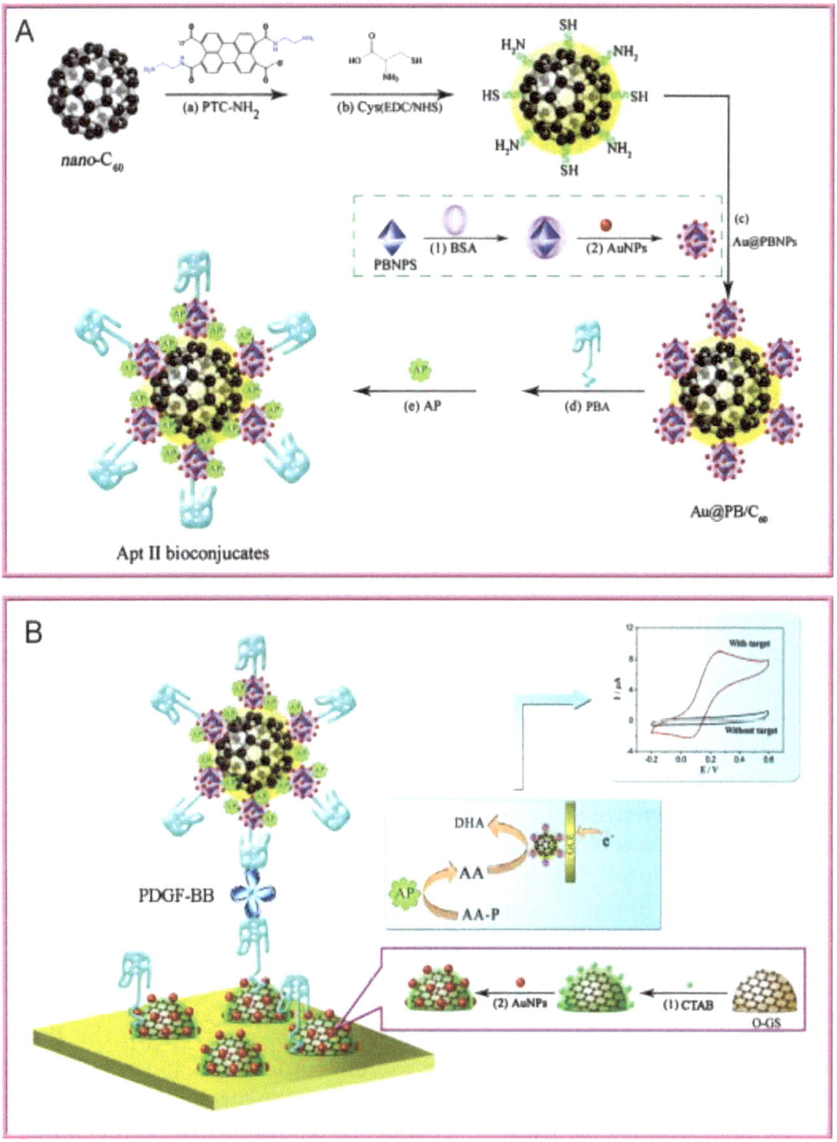

Figure 8.24 Schematic illustration of the stepwise aptasensor fabrication process and the dual-signal amplification mechanism.[171] Reproduced from ref. 170 with permission from Elsevier, Copyright 2013.

advantages, showcasing tremendous application potential in medical diagnostics and environmental monitoring. These sensors primarily utilize carbon materials such as graphene, carbon nanotubes, graphene oxide, reduced graphene oxide, their derivatives, and fullerenes. These materials possess excellent electrochemical stability, rapid electron transfer capabilities, and outstanding biocompatibility, enabling biosensors based on them to demonstrate outstanding performance in sensitivity, accuracy, and environmental friendliness. Through continuous material innovation and engineering techniques, carbon-based biosensors are driving significant technological prospects in advancing healthcare diagnostics and environmental protection.

8.7 Challenges and Future Trends

In the current era of pursuing sustainable development and environmentally friendly technologies, low-carbon materials in the field of biosensors have garnered widespread attention. Biosensors, as key technologies, have extensive applications in medical diagnostics, environmental monitoring, and food safety, playing an important role in achieving sustainable development goals. Therefore, leveraging the excellent characteristics of low-carbon materials, such as environmental friendliness, renewability, and degradability, will bring new opportunities and challenges to the development of biosensor technology.

Low-carbon materials hold immense potential in the field of biosensors, yet their practical application still faces challenges. Firstly, enhancing sensor performance, including sensitivity, selectivity, and stability, requires new material design and process improvements. Secondly, the lack of standardization in the field of biosensors poses challenges in production and application processes, demanding the establishment of a comprehensive standardization system. Additionally, despite the apparent cost-effectiveness of low-carbon materials, the high manufacturing and application costs of biosensors remain a challenge. Lastly, biosensors need to overcome sample complexity and interference issues in practical applications, necessitating the development of more robust and adaptable systems. By addressing these challenges, we can further promote the development of biosensor technology and expand its applications in medicine, the environment, and food industries.

Despite facing challenges, the potential for development in the field of biosensors remains enormous. With the continuous advancement and innovation of science and technology, future biosensors will exhibit several development trends, providing broader possibilities for applications in various fields. Firstly, sensors will evolve toward multifunctionality, capable of simultaneously detecting multiple biomolecules or environmental conditions, thus expanding their application scope in medicine, environmental monitoring, and beyond. Additionally, with the ongoing development of the Internet of Things (IoT) and artificial intelligence (AI) technologies, biosensors will become smarter and more portable, enabling real-time monitoring and remote data transmission, thereby enhancing their practicality and operability. The application of nanotechnology will further drive the development of biosensors, achieving higher sensitivity, selectivity, and stability, thereby providing greater room for improvement in biosensor performance. Lastly, with the popularity of wearable technology, biosensors will pay more attention to comfort and portability on the body surface, providing more convenient and effective solutions for personal health monitoring.

These trends will collectively propel the development of the biosensor field, offering broader prospects for the future of biosensor technology. Particularly, with the increasing global demand for health monitoring and environmental protection, the high integration and intelligence of biosensors will play a crucial role in personal health management, disease prevention, environmental protection, and industrial process control. For example, highly sensitive sensors can detect pathological changes in diseases early, while intelligent data processing can provide real-time feedback and predictions, thereby achieving more precise health status monitoring and disease management. Meanwhile, in the field of environmental monitoring, biosensors can provide real-time monitoring of air quality, water quality, and soil conditions, providing scientific basis for environmental policy formulation and natural resource management. Additionally, with the advancement of biosensor technology, they will also demonstrate their unique value in agriculture, food safety, and bio-defense, contributing to the continuous innovation and progress of relevant industries through higher efficiency and response speed.

References

1. L. C. Clark Jr., R. Wolf, D. Granger and Z. Taylor, *J. Appl. Physiol.*, 1953, **6**, 189–193.
2. D. R. Thevenot, K. Tóth, R. A. Durst and G. S. Wilson, *Biosens. Bioelectron.*, 1999, **71**, 2333–2348.
3. A. Khosla and B. L. Gray, *Mater. Lett.*, 2012, **45**, 1203–1206.
4. N. Bhalla, P. Jolly, N. Formisano and P. Estrela, *Essays Biochem.*, 2016, **60**, 1–8.
5. V. Naresh and N. Lee, *Sensors*, 2021, **21**, 1109.
6. D. R. Thévenot, K. Toth, R. A. Durst and G. S. Wilson, *Biosens. Bioelectron.*, 2001, **16**, 121–131.
7. C. Dincer, R. Bruch, E. Costa-Rama, M. T. Fernández-Abedul, A. Merkoçi, A. Manz, G. A. Urban and F. Güder, *Adv. Mater.*, 2019, **31**, e1806739.
8. A. P. Turner, *Chem. Soc. Rev.*, 2013, **42**, 3184–3196.
9. D. Grieshaber, R. MacKenzie, J. Vörös and E. Reimhult, *Sensors*, 2008, **8**, 1400–1458.
10. Y. Gu, Y. Li, D. Ren, L. Sun, Y. Zhuang, L. Yi and S. Wang, *Food Front.*, 2022, **3**, 453–479.
11. A. Shanker, K. Lee and J. Kim, *Nat. Mater.*, 2015, **14**, 295–300.
12. Y. Li, X. Chen, Y. Lin, Y. Yang, L. Zhang, P. Zhao, C. Wang, J. Fei and Y. Xie, *eFood*, 2023, **4**, e64.
13. E. Zdrachek and E. Bakker, *Anal. Chem.*, 2019, **91**, 2–26.

14. D. G. Rackus, M. H. Shamsi and A. R. Wheeler, *Chem. Soc. Rev.*, 2015, **44**, 5320–5340.
15. H. S. Magar, R. Y. A. Hassan and A. Mulchandani, *Sensors*, 2021, **21**, 6578.
16. N. Elgrishi, K. J. Rountree, B. D. McCarthy, E. S. Rountree, T. T. Eisenhart and J. L. Dempsey, *J. Chem. Educ.*, 2018, **95**, 197–206.
17. M. B. Kulkarni, N. H. Ayachit and T. M. Aminabhavi, *Biosensors*, 2022, **12**, 543.
18. P. Damborský, J. Švitel and J. Katrlík, *Essays Biochem.*, 2016, **60**, 91–100.
19. N. Khansili, G. Rattu and P. M. Krishna, *Sens. Actuators, B*, 2018, **265**, 35–49.
20. C. Chen and J. Wang, *Analyst*, 2020, **145**, 1605–1628.
21. T. H. Bui, B. Thangavel, M. Sharipov, K. Chen and J. H. Shin, *Chemosensors*, 2023, **11**(9), 468.
22. V. Naresh and N. Lee, *Sensors*, 2021, **21**, 1109.
23. T. Chen, G. Cheng, S. Ahmed, Y. Wang, X. Wang, H. Hao and Z. Yuan, *Talanta*, 2017, **175**, 435–442.
24. S. Kaushik, U. K. Tiwari, S. S. Pal and R. K. Sinha, *Biosens. Bioelectron.*, 2019, **126**, 501–509.
25. K. Cali, E. Tuccori and K. C. Persaud, *Methods Enzymol.*, 2020, **642**, 435–468.
26. P. W. Walton, M. R. O'Flaherty, M. E. Butler and P. Compton, *Biosens. Bioelectron.*, 1993, **8**, 401–407.
27. S. Akgönüllü, E. Özgür and A. Denizli, *Biosens. Bioelectron.: X*, 2022, **10**, 106.
28. F. Narita, Z. Wang, H. Kurita, Z. Li, Y. Shi, Y. Jia and C. Soutis, *Adv. Mater.*, 2021, **33**, 2005448.
29. S. C. Gopinath and T. Lakshmipriya, *Nanobiosensors for biomolecular targeting*, Elsevier, 2018.
30. M. Yakovleva, S. Bhand and B. Danielsson, *Anal. Chim. Acta*, 2013, **766**, 1–12.
31. K. Ramanathan and B. Danielsson, *Biosens. Bioelectron.*, 2001, **16**, 417–423.
32. R. Marks, C. Lowe, D. C. Cullen, H. H. Weetall and I. Karube, *Handbook of biosensors and biochips*, Wiley, 2007.
33. S. Vasuki, V. Varsha, R. Mithra, R. Dharshni and S. Abinaya, *Am. Int. J. Res. Sci., Technol., Eng. Math.*, 2019, **1**, 262–264.
34. L. C. Clark Jr. and C. Lyons, *Ann. N. Y. Acad. Sci.*, 1962, **102**, 29–45.
35. S. J. Updike and G. P. Hicks, *Nature*, 1967, **214**, 986–988.
36. G. Rocchitta, A. Spanu, S. Babudieri, G. Latte, G. Madeddu, G. Galleri, S. Nuvoli, P. Bagella, M. I. Demartis and V. Fiore, *Sensors*, 2016, **16**, 780.
37. P. Walter, B. Podsiadły, B. Wałpuski and M. Jakubowska, Proceedings SPIE 10808, Photonics Applications in Astronomy, Communications, Industry, and High-Energy Physics Experiments, 2018, p. 108083R.
38. N. R. Mohamad, N. H. C. Marzuki and N. A. Buang, *Biotechnol. Biotechnol. Equip.*, 2015, **29**, 205–220.
39. H. H. Nguyen and M. Kim, *Appl. Sci. Convergence Technol.*, 2017, **26**, 157–163.

40. R. Srivastava and A. B. Joshi, Patentdocs stay tuned to the technology, 2012.
41. A. Heller and B. Feldman, *Chem. Rev.*, 2008, **108**, 2482–2505.
42. S. Rodriguez-Mozaz, M.-P. Marco, M. L. De Alda and D. Barceló, *Appl. Chem.*, 2004, **76**, 723–752.
43. V. Alferov, M. Zaitsev, O. Ponomareva, T. Kuznetsova and T. Rogova, *J. Anal. Chem.*, 2011, **66**, 1205–1211.
44. M. Ghasemi-Varnamkhasti, M. L. Rodríguez-Méndez, S. S. Mohtasebi, C. Apetrei, J. Lozano, H. Ahmadi, S. H. Razavi and J. Antonio de Saja, *Food Control*, 2012, **25**, 216–224.
45. L. A. Terry, S. F. White and L. J. Tigwell, *J. Agric. Food Chem.*, 2005, **53**, 1309–1316.
46. L. Loo, M. K. Robinson and G. P. Adams, *Cancer J.*, 2008, **14**, 149–153.
47. S. arah A. Goodchild, Tracey Elizabeth Love, N. eal A. E. Hopkins and Carl Nicholas Mayers, *Adv. Appl. Microbiol.*, 2005, **58**, 185–226.
48. N. A. E. Hopkins, Antibody Engineering for Biosensor Applications, in *Recognition Receptors in Biosensors*, ed. M. Zourob, Springer, New York, 2010.
49. C. J. Padoa, N. J. Crowther, J. W. Thomas, T. R. Hall, L. M. Bekris, C. Torn, M. Landin-Olsson, E. Ortqvist, J. P. Palmer, A. Lernmark and C. S. Hampe, *Clin. Exp. Immunol.*, 2006, **74**, S51–S62.
50. P. J. Conroy, S. Hearty, P. Leonard and R. J. O'Kennedy, *Semin. Cell Dev. Biol.*, 2009, **20**, 10–26.
51. H. A. Parray, S. Shukla, S. Samal, T. Shrivastava, S. Ahmed, C. Sharma and R. Kumar, *Int. Immunopharmacol.*, 2020, **85**, 106639.
52. M. L. Chiu, D. R. Goulet, A. Teplyakov and G. L. Gilliland, *Antibodies*, 2019, **8**, 55.
53. C. J. Padoa and N. J. Crowther, *Diabetes Res. Clin. Pract.*, 2006, **74**, S51–S62.
54. M. Leenaars and C. F. Hendriksen, *ILAR J.*, 2005, **46**, 269–279.
55. G. Köhler and C. Milstein, *Nature*, 1975, **256**, 495–497.
56. P. N. Nelson, G. M. Reynolds, E. E. Waldron, E. Ward, K. Giannopoulos and P. G. Murray, *Mol. Pathol.*, 2000, **53**, 111.
57. H. R. Hoogenboom, *Nat. Biotechnol.*, 2005, **23**, 1105–1116.
58. S. Townsend, W. J. Finlay, S. Hearty and R. O'Kennedy, *Biosens. Bioelectron.*, 2006, **22**, 268–274.
59. K. Sakai, Y. Shimizu, T. Chiba, A. Matsumoto-Takasaki, Y. Kusada, W. Zhang, M. Nakata, N. Kojima, K. Toma and A. Takayanagi, *Biochemistry*, 2007, **46**, 253–262.
60. J. M. Encarnação, L. Rosa, R. Rodrigues, L. Pedro, F. A. da Silva, J. Gonçalves and G. N. Ferreira, *J. Biotechnol.*, 2007, **132**, 142–148.
61. C. Qi, J.-Z. Duan, Z.-H. Wang, Y.-Y. Chen, P.-H. Zhang, L. Zhan, X.-Y. Yan, W.-C. Cao and G. Jin, *Biomed. Microdevices*, 2006, **8**, 247–253.
62. W.-G. Hu, H. G. Thompson, A. Z. Alvi, L. P. Nagata and M. R. Suresh, *J. Immunol. Methods*, 2004, **289**, 27–35.

63. T. E. Love, C. Redmond and C. N. Mayers, *J. Immunol. Methods*, 2008, **334**, 1–10.
64. L. Dunne, S. Daly, A. Baxter, S. Haughey and R. O'Kennedy, *Spectrosc. Lett.*, 2005, **38**, 229–245.
65. J. Horáček, S. D. Garrett and P. Skládal, *Food Agric. Immunol.*, 1998, **10**, 363–374.
66. K. Grennan, G. Strachan and A. J. Porter, *Anal. Chim. Acta*, 2003, **500**, 287–298.
67. A. D. Ellington and J. W. Szostak, *Nature*, 1990, **346**, 818–822.
68. S.-R. Yan, M. M. Foroughi, M. Safaei, S. Jahani, N. Ebrahimpour, F. Borhani, N. Rezaei Zade Baravati, Z. Aramesh-Boroujeni and L. K. Foong, *Int. J. Biol. Macromol.*, 2020, **155**, 184–207.
69. L. Gold, N. Janjic, T. Jarvis, D. Schneider, J. J. Walker, S. K. Wilcox and D. Zichi, *Cold Spring Harbor Perspect. Biol.*, 2012, **4**, a003582.
70. H. Ulrich and C. Wrenger, *Methods Mol. Biol.*, 2013, **986**, 17–39.
71. C. Tuerk and L. Gold, *Science*, 1990, **249**, 505–510.
72. D. L. Robertson and G. F. Joyce, *Nature*, 1990, **344**, 467–468.
73. P. Röthlisberger and M. Hollenstein, *Adv. Drug Delivery Rev.*, 2018, **134**, 3–21.
74. M. Blind and M. Blank, *Mol. Ther.–Nucleic Acids*, 2015, **4**, e223.
75. K. Sefah, D. Shangguan, X. Xiong, M. B. O'Donoghue and W. Tan, *Nat. Protoc.*, 2010, **5**, 1169–1185.
76. S. K. Silverman, *Chem. Commun.*, 2008, **30**, 3467–3485.
77. N. K. Singh, S. K. Arya, P. Estrela and P. Goswami, *Biosens. Bioelectron.*, 2018, **117**, 246–252.
78. Y. N. Khonsari and S. Sun, *Mater. Sci. Eng., C*, 2019, **96**, 146–152.
79. S. M. Taghdisi, N. M. Danesh, M. Ramezani, A. S. Emrani and K. Abnous, *ACS Appl. Mater. Interfaces*, 2018, **10**, 12504–12509.
80. H. Liu, Y. Zhang, Y. Dong and X. J. T. Chu, *Talanta*, 2019, **199**, 507–512.
81. P.-J. J. Huang and J. Liu, *J. Anal. Test.*, 2022, **6**, 20–27.
82. X. Yan, H. Chen, G. Du, Q. Guo, Y. Yuan and T. Yue, *Food Front.*, 2022, **3**, 428–452.
83. F. Li, Y. Guo, X. Wang and X. Sun, *Biosens. Bioelectron.*, 2018, **115**, 7–13.
84. J.-B. Jiao, Q. Kang, T. Du, J.-L. Cao, Z.-H. Xiao, L.-F. Wei, C.-M. Zhao, L. He, X.-J. Du and S. Wang, *Food Front.*, 2023, **4**, 1472–1481.
85. S. Wang, Y. Dong and X. Liang, *Biosens. Bioelectron.*, 2018, **109**, 1–7.
86. J. P. R. L. L. Parra, B. P. Crulhas, C. R. Basso, F. K. Delella, G. R. Castro and V. A. Pedrosa, *Electroanalysis*, 2018, **30**, 2869–2877.
87. S. Muniandy, S. J. Teh, J. N. Appaturi, K. L. Thong, C. W. Lai, F. Ibrahim and B. F. J. B. Leo, *Bioelectrochemistry*, 2019, **127**, 136–144.
88. W. Wu, M. Zhou, H. He, C. Liu, P. Li, M. Wang, Y. Liu, X. Hao and Z. Fang, *Sens. Actuators, B*, 2018, **272**, 550–558.
89. M. Baghayeri, R. Ansari and M. Nodehi, *Microchim. Acta*, 2018, **185**, 1–9.
90. Y. J. Yang, Y. Zhou, Y. Xing, G. M. Zhang, Y. Zhang, C. H. Zhang, P. Lei, C. Dong, X. Deng, Y. He and S. M. Shuang, *Talanta*, 2019, **199**, 310–316.

91. J. Wen, Z. Li, N. Luo, M. Huang, R. Yang and G. Zeng, *Ecotoxicol. Environ. Saf.*, 2018, **162**, 184–191.
92. L. Zeng, D. Zhou, J. Gong, C. Liu and J. Chen, *Anal. Chem.*, 2019, **91**, 1724–1727.
93. B. Wang, M. Barahona and M. Buck, *Biosens. Bioelectron.*, 2013, **40**, 368–376.
94. S. Asif, A. Chaudhari, P. Gireesh-Babu, P. Roy Chaudhuri and R. Sen, *Mater. Today: Proc.*, 2016, **3**, 3492–3497.
95. T. Yoetz-Kopelman, Y. Dror, Y. Shacham-Diamand and A. Freeman, *Sens. Actuators, B*, 2016, **232**, 758–764.
96. G. Gao, J. Qian, D. Fang, Y. Yu and J. Zhi, *Anal. Chim. Acta*, 2016, **924**, 21–28.
97. T. Catal, S. Yavaser, V. Enisoglu-Atalay, H. Bermek and S. Ozilhan, *Bioresour. Technol.*, 2018, **268**, 116–120.
98. A. Mishra, J. Kumar and J. S. Melo, *Biosens. Bioelectron.*, 2017, **87**, 332–338.
99. H. Leonard, S. Halachmi, N. Ben-Dov, O. Nativ and E. Segal, *ACS Nano*, 2017, **11**, 6167–6177.
100. N. Gupta, V. Renugopalakrishnan, D. Liepmann, R. Paulmurugan and B. D. Malhotra, *Biosens. Bioelectron.*, 2019, **141**, 111435.
101. Y. Y. Pan, N. Hu, X. Wei, L. Gong, B. Zhang, H. Wan and P. Wang, *Biosens. Bioelectron.*, 2019, **130**, 344–351.
102. X. Wang, A. Liu, Y. Xing, H. Duan, W. Xu, Q. Zhou, H. Wu, C. Chen and B. Chen, *Biosens. Bioelectron.*, 2018, **105**, 22–28.
103. C. Denning, V. Borgdorff, J. Crutchley, K. S. Firth, V. George, S. Kalra, A. Kondrashov, M. D. Hoang, D. Mosqueira and A. Patel, *Biochim. Biophys. Acta, Mol. Cell Res.*, 2016, **1863**, 1728–1748.
104. R. A. Chowdhury, K. N. Tzortzis, E. Dupont, S. Selvadurai, F. Perbellini, C. D. Cantwell, F. S. Ng, A. R. Simon, C. M. Terracciano and N. S. Peters, *Sci. Rep.*, 2018, **8**, 6947.
105. K. Gao, F. Gao, L. Du, C. He, H. Wan and P. Wang, *Bioelectrochemistry*, 2019, **128**, 1–8.
106. D. Ha, Q. Sun, K. Su, H. Wan, H. Li, N. Xu, F. Sun, L. Zhuang, N. Hu and P. Wang, *Sens. Actuators, B*, 2015, **207**, 1136–1146.
107. M. Son and T. H. Park, *Biotechnol. Adv.*, 2018, **36**, 371–379.
108. I. U. Haq, W. N. Chaudhry, M. N. Akhtar, S. Andleeb and I. Qadri, *Virol. J.*, 2012, **9**, 1–8.
109. S. D. Alcaine, K. Law, S. Ho, A. J. Kinchla, D. A. Sela and S. R. Nugen, *Biosens. Bioelectron.*, 2016, **82**, 14–19.
110. P. Bárdy, R. Pantůček, M. Benešík and J. Doškař, *J. Appl. Microbiol.*, 2016, **121**, 618–633.
111. Z. Altintas, J. Pocock, K.-A. Thompson and I. E. Tothill, *Biosens. Bioelectron.*, 2015, **74**, 996–1004.
112. E. A. Mack, D. Tong and K. Credit, *Int. J. Health Geographics*, 2017, **16**, 37.
113. J. Zhan, L. Wang, Y. Zhu, H. Gao, Y. Chen and J. Chen, *et al.*, *ACS Appl. Mater. Interfaces*, 2018, **10**, 35830–35837.

114. M. Hoyos-Nogués, F. J. Gil and C. Mas-Moruno, *Molecules*, 2018, **23**, 1683.
115. S. A. Onaizi and S. S. Leong, *Biotechnol. Adv.*, 2011, **29**, 67–74.
116. C. A. S. Andrade, J. M. Nascimento and I. S. Oliveira, *Colloids Surf., B*, 2015, **135**, 833–839.
117. Z. Qiao, Y. Fu, C. Lei and Y. Li, *Food Control*, 2020, **112**, 107116.
118. J. Hu, S. Wang, L. Wang, F. Li, B. Pingguan-Murphy, T. J. Lu and F. Xu, *Biosens. Bioelectron.*, 2014, **54**, 585–597.
119. Z. Yao, P. Coatsworth, X. Shi, J. Zhi, L. Hu, R. Yan, F. Güder and H.-D. Yu, *Sens. Diagn.*, 2022, **1**, 312–342.
120. Y. Hou, C.-C. Lv, Y.-L. Guo, X.-H. Ma, W. Liu, Y. Jin, B.-X. Li, M. Yang and S.-Y. Yao, *J. Anal. Test.*, 2022, **6**, 247–273.
121. S. Kim and H. D. Sikes, *ACS Appl. Mater. Interfaces*, 2019, **11**, 28469–28477.
122. C. Wang, X. Gao and S. Wang, *Anal. Bioanal. Chem.*, 2020, **412**, 611–620.
123. G. G. Morbioli, T. Mazzu-Nascimento, A. M. Stockton and E. Carrilho, *J. Anal. Chim. Acta*, 2017, **970**, 1–22.
124. T. T. Tsai, C. Y. Huang, C. A. Chen, S. W. Shen, M. C. Wang, C. M. Cheng and C. F. Chen, *ACS Sens.*, 2017, **2**, 1345–1354.
125. S. Kummari, L. R. Panicker, J. Rao Bommi, S. Karingula, V. Sunil Kumar, K. Mahato and K. Y. Goud, *Biosensors*, 2023, **13**.
126. L. Liang, M. Su, L. Li, F. Lan, G. Yang, S. Ge, J. Yu and X. Song, *Sens. Actuators, B*, 2016, **229**, 347–354.
127. X. Guo, L. Zong, Y. Jiao, Y. Han, X. Zhang, J. Xu, L. Li, C.-w Zhang, Z. Liu and Q. Ju, **91***Anal. Chem.*, 2019, 9300–9307.
128. M. Zhao, H. Li, W. Liu, Y. Guo and W. Chu, *Biosens. Bioelectron.*, 2016, **79**, 581–588.
129. W. Dungchai, O. Chailapakul and C. S. Henry, *Anal. Chem.*, 2009, **81**, 5821–5826.
130. Z. Nie, F. Deiss, X. Liu, O. Akbulut and G. M. Whitesides, *Lab Chip*, 2010, **10**, 3163–3169.
131. L. Gao, C. Zhu, L. Li, C. Zhang, J. Liu and H.-D. Yu, *ACS Appl. Mater. Interfaces*, 2019, **11**, 25034–25042.
132. L.-Q. Tao, K.-N. Zhang, H. Tian, Y. Liu, D.-Y. Wang, Y.-Q. Chen and Y. Yang, *ACS Nano*, 2017, **11**, 8790–8795.
133. T. Ozer, C. McMahon and C. S. Henry, *Annu. Rev. Anal. Chem.*, 2020, **13**, 85–109.
134. Y.-X. Li, H.-Y. Qin, C. Hu, M.-M. Sun, P.-Y. Li, H. Liu, J.-C. Li, Z.-B. Li, L.-D. Wu and J. Zhu, *J. Anal. Test.*, 2022, **6**, 431–440.
135. K. Mahato and P. Chandra, *Biosens. Bioelectron.*, 2019, **128**, 9–16.
136. N. F. Silva, C. M. Almeida, J. M. Magalhães, M. P. Gonçalves, C. Freire and C. Delerue-Matos, *Biosens. Bioelectron.*, 2019, **141**, 111317.
137. Z. Nie, C. A. Nijhuis, J. Gong, X. Chen, A. Kumachev, A. W. Martinez, M. Narovlyansky and G. M. Whitesides, *Lab Chip*, 2010, **10**, 477–483.
138. L. Feng, H. Li, L.-Y. Niu, Y.-S. Guan, C.-F. Duan, Y.-F. Guan, C.-H. Tung and Q.-Z. Yang, *Talanta*, 2013, **108**, 103–108.

139. Y. Cao and B. Wang, *Int. J. Mol. Sci.*, 2009, **10**, 1514–1524.
140. C. Guo, C. Li and D. L. Kaplan, *Biomacromolecules*, 2020, **21**, 1678–1686.
141. F. Chen, D. Porter and F. Vollrath, *Acta Biomater.*, 2012, **8**, 2620–2627.
142. X. Zhang, L. Xiao, Z. Ding, Q. Lu and D. L. Kaplan, *ACS Biomater. Sci.*, 2021, **7**, 2337–2345.
143. X. Wang and D. L. Kaplan, *Macromol. Biosci.*, 2011, **11**, 100–110.
144. S. Sundelacruz, M. Levin and D. L. Kaplan, *Stem Cell Rev. Rep.*, 2009, **3**, 231–246.
145. K. Mahato and J. Wang, *Sens. Actuators, B*, 2021, **344**, 130178.
146. S. Emaminejad, W. Gao, E. Wu, Z. A. Davies, H. Yin Yin Nyein, S. Challa, S. P. Ryan, H. M. Fahad, K. Chen, Z. Shahpar, S. Talebi, C. Milla, A. Javey and R. W. Davis, *Proc. Natl. Acad. Sci. U. S. A.*, 2017, **114**, 4625–4630.
147. M. Xu and V. K. Yadavalli, *ACS Sens.*, 2019, **4**, 1040–1047.
148. Y. C. Lai, B. W. Ye, C. F. Lu, C. T. Chen, M. H. Jao, W. F. Su, W. Y. Hung, T. Y. Lin and Y.-F. Chen, *Adv. Funct. Mater.*, 2016, **26**, 1286–1295.
149. M. Y. Lee, J. Hong, E. K. Lee, H. Yu, H. Kim, J. U. Lee, W. Lee and J. H. Oh, *Adv. Funct. Mater.*, 2016, **26**, 1445–1453.
150. J. Liu, M. Yang, J. Tan, Y. Yin, Y. Yang and C. Wang, *Dyes Pigm.*, 2022, **198**, 109994.
151. T. Laviv, B. B. Kim, J. Chu, A. J. Lam, M. Z. Lin and R. Yasuda, *Nat. Methods*, 2016, **13**, 989–992.
152. J. W. Leem, J. Park, S.-W. Kim, S.-R. Kim, S. H. Choi, K.-H. Choi and Y. L. Kim, *Adv. Sci.*, 2018, **5**, 1700863.
153. T. Falcucci, K. F. Presley, J. Choi, V. Fizpatrick, J. Barry, J. KishoreSahoo, J. T. Ly, T. A. Grusenmeyer, M. J. Dalton and D. L. Kaplan, *Adv. Funct. Mater.*, 2022, **32**, 2202020.
154. J. Wang, Y. Cheng, H. Wang, Y. Wang, K. Zhang, C. Fan and H. Wang, *Acta Biomater.*, 2020, **117**, 180–191.
155. Y. Sun, T. Yue and Y. Yuan, *eFood*, 2023, **4**(1), e65.
156. A. P. Johnson, C. Sabu, N. K. Swamy and A. Anto, *Biosens. Bioelectron.*, 2021, **184**, 113245.
157. Z. Li, W. Zhang and F. Xing, *Int. J. Mol. Sci.*, 2019, **20**, 2461.
158. P. Butmee, G. Tumcharern, G. Thouand, K. Kalcher and A. Samphao, *Bioelectrochemistry*, 2020, **132**, 107452.
159. J. Gao, C. Wang, C. Wang, Y. Chu, S. Wang, M. Y. Sun, H. Ji, Y. Gao and Y. Wang, *Anal. Chem.*, 2022, **94**, 1626–1636.
160. E. Morales-Narváez, L. Baptista-Pires, A. Zamora-Gálvez and A. Merkoçi, *Adv. Mater.*, 2017, **29**, 1604905.
161. A. Merkoçi and E. Morales-Narváez, *Adv. Mater.*, 2019, **31**, 0935–9648.
162. Y. Zhu, Q. Tian, X. Li, L. Wu, A. Yu, G. Lai, L. Fu, Q. Wei, D. Dai and N. Jiang, *Biosensors*, 2021, **11**, 462.
163. F. Nemati, M. Rezaie, H. Tabesh, K. Eid, G. Xu, M. R. Ganjali, M. Hosseini, C. Karaman, N. Erk and P.-L. Show, *Environ. Res.*, 2022, **208**, 112685.

164. J. Ackermann, J. T. Metternich, S. Herbertz and S. Kruss, *Angew. Chem., Int. Ed.*, 2022, **61**, e202112372.
165. R. M. Williams, C. Lee and D. A. Heller, *ACS Sens.*, 2018, **3**, 1838–1845.
166. J. Liu, R. Li and B. Yang, *ACS Cent. Sci.*, 2020, **6**, 2179–2195.
167. C. Lin, X. Song, W. Ye, T. Liu, M. Rong and L. Niu, *J. Anal. Test.*, 2024, **8**, 95–113.
168. M. Zarei-Ghobadi, S.-H. Mozhgani, F. Dashtestani, A. Yadegari, F. Hakimian, M. Norouzi and H. Ghourchian, *Sci. Rep.*, 2018, **8**, 15593.
169. S. Goodarzi, T. Da Ros, J. Conde, F. Sefat and M. Mozafari, *Mater. Today*, 2017, **20**, 460–480.
170. S. Afreen, K. Muthoosamy, S. Manickam and U. Hashim, *Biosens. Bioelectron.*, 2015, **63**, 354–364.
171. J. Han, Y. Zhuo, Y. Chai, R. Yuan, Y. Xiang, Q. Zhu and N. Liao, *Biosens. Bioelectron.*, 2013, **46**, 74–79.

CHAPTER 9

Low-carbon Materials for Electrochemical Biomedical Sensors

ÇIĞDEM KANBEŞ DINDAR,[a] SELENAY SADAK,[a,b] HÜLYA SILAH[c] AND BENGI USLU*[a]

[a] Ankara University, Faculty of Pharmacy, Department of Analytical Chemistry, Ankara, Turkey; [b] Ankara University, The Graduate School of Health Sciences, 06110, Ankara, Türkiye; [c] Bilecik Şeyh Edebali University, Faculty of Science, Department of Chemistry, Bilecik, Turkey
*Email: buslu@pharmacy.ankara.edu.tr

9.1 Introduction

Electrochemical biomedical sensors are instruments utilized in the electrochemical biomedical domain to detect, measure, or assess signals from biological or non-biological sources.[1] These sensors have a wide range of applications, including health-monitoring systems and medical diagnostic instruments.[2] These sensors are utilized for patient monitoring, diagnosis, drug delivery and therapy, rehabilitation, and numerous research projects.[3] Therefore, in addition to improving the technological progress of the healthcare system, they also enhance patient comfort and facilitate the creation of efficient and dependable healthcare solutions.[4] Also, electrochemical biomedical sensors should possess certain fundamental characteristics to be feasible in the health sector. These include the capacity to accurately detect target analytes at low concentrations, biocompatibility for direct use on the body, and miniaturization to enable integration into wearable devices.[1,5,6]

Low Carbon Materials: Paving a Sustainable Path for a Greener Future
Edited by Santanu Patra, Sudheesh K. Shukla and Mika Sillanpää

Published by the Royal Society of Chemistry, www.rsc.org

The significance of biological sensors in the healthcare sector and their potential to enhance different facets of healthcare is evident from these considerations. Continuous advancement and enhancement of these sensors are crucial to enhance the standard of healthcare and elevate the quality of life for patients.[7,8] Due to these factors, a variety of materials are being highlighted for the purpose of developing sensors that are more sensitive and precise.

Electrochemical biomedical sensors could monitor or measure various biological factors, depending on their specific role and design. Thus, these sensors play an important role in diagnosing, monitoring, and treating diseases in healthcare.

Carbon-based materials are prominent in this area. Carbon-based materials offer greater catalytic qualities to sensors when used as surface modification agents.[9–11] These materials also have many other superior features, such as a wide working range, chemical resistance, high biological compatibility, and reusability. Nevertheless, their most significant drawback is their contribution to carbon emissions.[12,13] Researchers have prioritized the study of ecologically benign materials that mimic carbon's behavior without adding to the carbon footprint, due to carbon-based materials' essential features.

Electrochemical biomedical sensors should be made from recyclable or renewable materials, engineered to save energy, and produced with little greenhouse gas emissions. In this context, low-carbon materials shine.[14] Low-carbon materials refer to carbon materials that have a high efficiency level and do little damage to the environment and human health. In the future, smart materials are anticipated to rely on low-carbon materials (Figure 9.1).[15] "Low-carbon" materials can reduce carbon emissions in electrochemical biomedical sensor production and use. Low-carbon approaches can also optimize energy consumption by integrating sensors with smart systems for optimal operation.[16] Thus, the low-carbon idea in sensor technologies reduces environmental implications during material selection, manufacturing, and use.[17]

Electrochemical biomedical sensors use optical and electrochemical approaches as measuring tools. Electrochemical sensors are prominent due to their compact dimensions and user-friendly nature. Furthermore, over the last 20 years, a wide range of nanomaterials, including metals, conducting

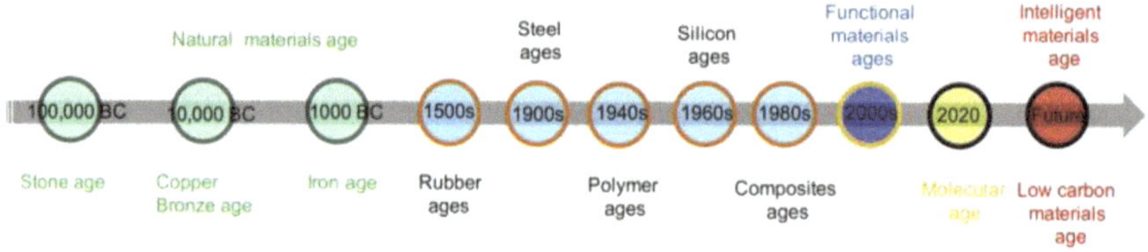

Figure 9.1 The schedule for significant materials. Reproduced from ref. 15 with permission from Elsevier, Copyright 2020.

polymers, metal oxides, and metal-organic and carbon-based nanomaterial frameworks, have been used in electrochemical investigations to improve analytical performance. Carbon-based research is particularly noteworthy in this perspective. Employing non-toxic materials and minimizing carbon use are crucial in developing sensor techniques that can be brought to market. Low-carbon materials are an effective answer in this context.

As a consequence, the focus of this chapter is on the most recent developments and trends in low-carbon-based electrochemical biomedical sensors. As a result, current advances and trends in electrochemical biomedical sensors, and the significance of using low-carbon materials sensor investigations for electrochemical biomedical analysis are the subject of this chapter. The composition and structure of low-carbon materials utilized in the study of electrochemical biomedical applications are described and discussed in detail. Finally, the advantages, limitations, and prospects of low-carbon materials in electrochemical biomedical sensors are presented to explain the benefits, challenges, and prospects of modification in sensor studies for both *in vitro* and *in vivo* detection.

9.2 Properties and Importance of Low-carbon Materials

Owing to environmental problems induced by the consumption of fossil resources and their depletion, the idea of sustainable improvements utilizing environmentally harmless materials is being accepted.[18] Recently, efforts have been made to improve adaptable and low-cost methodologies for addressing environmental apprehensions utilizing both multifunctional and sustainable materials.[19]

Carbon is one of the most crucial elements for humanity. It is important in people's lives as well as for industrial procedures as a raw material. Because of their various electronic features, carbon matters have a broad range of molecular structures and features in accordance with their C–C bonding.[20]

Carbon is one of the most versatile elements in the periodic table, acknowledged for its tetravalent shell. Its inimitable electronic structure ensures it is difficult to entirely lose or gain electrons for creating fully unoccupied/occupied orbitals in chemical reactions. By combining varied hybridizations of sp^3, sp^2, and sp^1 or accommodating other foreign atoms in carbon hybrids, wanted features can be modified, and investigators studying carbon materials have been carrying this out for a remarkable amount of time.[21]

Presumably, carbon-based materials are the predominant group of materials existing in native and industrial activities, with a high influence on contemporary society because of their convenient employments and suitable characteristics for countless utilizations for a carbon-neutral condition.[22,23]

Also, materials commonly used today, such as ceramics and polymers, piezoelectrics, thermoelectrics, magnetics, and semiconducting substances, are no longer new to us. Numerous state-of-the-art apparatuses and implementations based on these materials are in utilization. Nevertheless, these materials and their apparatuses have some disadvantages, such as deficiencies in many functionalities and adaptability, such as self-control, sustainability, and intelligence, which are essential for the progressively complicated modern ways of life.[24]

In recent years, low-carbon materials have gained tremendous attention, thanks to the presence of plenty of native sources, various excellent properties, and their several important advantages.[25]

In order to ensure the sustainable development of humanity and to mitigate climate change, it is necessary to develop low-carbon materials and to give due importance to this issue. The term 'carbon' here refers not to elemental carbon, but to the potential for carbon dioxide emissions. The low-carbon materials are defined as matters with low embodied and operational carbon. Specifically, these materials are less detrimental to the Earth's biosphere and in turn to human conditions. It is clear that the forthcoming materials will be both low-carbon and intelligent.[24]

To decrease the environmental negative effects of natural resources and toxicity, both carbon-based nanomaterials and carbonaceous carbon are being recommended as alternates, together with the usage of renewable carbon sources.[12] Investigations and studies have demonstrated that native and green sources can be used for synthesizing low-carbon materials, to create cost-effective, environmentally friendly, and resource-saving operations for large-scale manufacture.[19]

Sensor systems enhance the capacity of physical and digital environments around us to detect changes and to transmit information related to these changes. These systems are designed to significantly enhance the accessibility and quality of human life across nearly all domains.[26] The sensor is a type of detection/determination apparatus that can transform the acquired knowledge into an electrical signal or other signal output. The requirements for sensor selectivity, sensitivity, and implementation range are high. Hence, the electrode, as the sensitive element of the sensor, plays a decisive role in determining the sensor's selectivity, sensitivity, accuracy, and detection limits.[27]

Recently, thanks to lower-cost, single-use electrodes, various analyses can be performed easily. However, the manufacture and utilization of these electrodes is environmentally suspect because of some toxic chemicals and materials employed and discarded after only a single application. Therefore, many scientists have carried out research toward improving the sustainability of these biosensors.[28]

In available researches, many scientists are raising awareness about carbonaceous materials as electrode materials for different implementations because of the enormous developments in their various properties and benefits.[29] Carbon-based electrodes have various advantages for the design of sensors that can be categorized as: (i) nominal background current and broad potential range boundaries can be obtained from their redox processes on the carbon surface. This situation indicates that carbon has a wide boundary for sensing. (ii) They have high producibility and stability properties. Producibility can be determined through the carbon material preparation procedure. At the same time, stability can be achieved when they are employed in various background currents, electrode areas, electron transfer rates, and analytical replies. (iii) When the situation is examined in terms of electron transfer kinetics, carbon-based materials show a wide range of different electron transfer properties depending on the type of raw material from which they are produced.[30,31]

The history of employing carbon as an electrode material dates back to the beginning of the 19th century, when Humphrey Davy utilized a graphite electrode for the electrolytic extraction of alkali metals. The material in the form of carbon paste was used for the first time in electroanalysis by R. N. Adams in 1958.[11,32,33] Carbon-based materials symbolize a class of significant raw materials for diverse industry applications. Because of their high electrical conductivity and high chemical and thermal stability, as well as their biocompatibility, many kinds of carbon-based materials, such as graphene (GR) or reduced graphene oxides (rGO), carbon fibers (CFs), carbon dots (CDs), active carbons (ACs), and biomass-derived carbons, have been intentionally prepared for various implementation objectives.[34,35]

Novel materials are broadly investigated for implementation in the design of electrochemical sensors. Especially nanomaterials used for modifying surfaces or electrodes are generally electrochemically inactive, meaning they do not participate in electrical reactions, and have high electrical conductivity over a wide voltage range.[36]

In recent years, there has been a growing awareness of the use of low-carbon materials in sensor systems. Especially, renewable biomass has gained much more attention as a key pioneer in the production of sustainable carbon materials due to its abundant availability and economic importance.[37,38] Biomass-derived carbon materials have been prepared for various uses in diverse dimensions and sizes, which, together with some other properties such as surface area, conductivity, porosity, and stability, control the sensor efficiency. The most common categorization depends on their dimensions.[39]

i. The zero-dimensional nanomaterial (named as 0D) has all its dimensions in nanoscale, sizes of which are an average size of less than 10 nm, and include 0D compass cubic, nanorod, spherical, polygon, hollow sphere, metal, and core–shell nanomaterials, as well as quantum dots (QDs). These nanomaterials are formed of a carbon-based core with various functional groups linked to the surface such as carboxyl, hydroxyl, amino, *etc.*
ii. One-dimensional nanomaterials (named as 1D), these materials involve only one non-nanoscale dimension, for example, fibrous, tubular, nanowire, nanofiber, polymeric, metallic, *etc.*
iii. Two-dimensional nanomaterials (named as 2D), these include just one nanoscaled length and can be single-layered or multi-layered

types. These carbon-based materials have sp^2 hybridization, and high active surface edges, so they are appropriate for energy storage and conversion.

iv. Three-dimensional (named as 3D) materials have three dimensions larger than 100 nm, and these materials facilitate a continuous electron current for efficient electrical activity.[39–41]

Biosensors have sometimes faced challenges in developing new methods to meet the goals of green chemistry and sustainability. The use of these devices is a clear example of enabling immediate and direct analysis to prevent contamination, which is one of the fundamental principles of green chemistry. Furthermore, the requirement to have a minimum negative ecological impact has driven the usage of eco-friendly chemicals such as solvents, materials, reagents, and processes as regular applications in this area for the design and use of biosensors.[42]

Residuals, in particular biowaste, can be a worthy resource of new carbon materials. Renewable carbon materials, such as biomass-derived carbons, have obtained notable attention recently as possible electrode materials for numerous electrochemical apparatuses, including supercapacitors and batteries. The significance of these renewable carbon materials as electrodes can be ascribed to their sustainability, low price, high surface area and purity, and tailored features.[43]

As is well known, biomass is a compound of carbon elements with myriad other elements (O, H, N, S, P, *etc.*) that generate living biological organisms in nature. These carbon materials with a permeable structure and produced from biomass have excellent potential for many applications.[10]

Native materials have been utilized in designing electrodes for sensors and provide various benefits, such as a combination of mechanical robustness and flexibility. Especially, the arrangement of plant-based biomass carbon has decreased the dependency on fossil fuels and encouraged environmental sustainability. The most common materials are reproduced from biological resources and include wool, cotton, silk, linen, hemp, and chitin. Being biological in nature, these are biocompatible and sustainable.[44,45]

These materials, especially those with biopolymer structure, create materials for developing environmentally sustainable electrodes. These materials have some special properties, such as self-cleaning and restoring features, that prohibit interference from environmental influences, for example, heat, moisture, etc., so they consequently have increased accuracy. Additionally, greater cost-effectiveness, lower molecular weight, and suitable biological features allow for large-scale fabricating of sensor systems.[7]

The surface area, pore size distribution, and performance of the electrode material are determined by the type of biomass used, the activation method applied, and the functional groups present on the surface of the carbon material.[46,47]

The synthesis of these carbon materials involves a multi-step process that requires the use of expensive raw materials needing pretreatment. The most commonly used methods are biomass pyrolysis, carbonization, physical activation, chemical activation, and hydrothermal carbonization.[36,38]

Among the different processes for nanostructured carbon synthesis, pyrolysis is a straightforward and general thermochemical procedure that disintegrates the carbon resources into smaller components without the presence of oxygen. Conventionally, this process is utilized for the fractionation of petroleum products, but recently, pyrolysis has gained attention because of its modifiable and renovated procedure for the transformation of agricultural waste into economically and eco-friendly beneficial products.[48]

The carbonization method is a main procedure to synthesize porous carbon materials. In particular, mesoporous carbon materials have gained extensive importance on account of the great qualities of high specific surface area, good conductivity, tunable permeable structure, perfect stability, and diverse applications in electrochemistry, catalysis, separation, removal, and adsorption of various pollutants, environmental remediation, energy storage and conversion, and biomedicine.[49] The hydrothermal carbonization method is a kind of hydrothermal processing of materials. Similar to other hydrothermal methods, this process usually takes place in an autoclave. High temperatures, generally in the range of 150–350 °C, are applied and it behaves as both a catalyst and a solvent for the breaking down of a diversity of starting materials, such as lignin, carbohydrates, and biomass.[50,51]

Various types of biomass, such as shells, chitosan, seeds, husks, rice, sawdust, *etc.*, yield better carbon precursors as they are very cheap and can be easily processed to prepare carbon materials. One of the most important factors in preparing carbon materials is the process of activation. This stage influences the physicochemical features and performance of the materials. Improvement of the surface features such as porosity, surface area, pore diameter, and volume, as well as the modification of the carbon surface with functional groups can be succeeded by the procedure of activation.[52] The activation of carbon is carried out through two methods: physical and chemical activation. The physical activation method is a properly operated two-stage procedure to activate carbon precursors, which includes the pyrolysis carbonization process of raw matters (at below 1000 K) followed by a controlled gasification procedure at relatively high temperatures (at >1150 K) in the presence of oxidizing gases like carbon dioxide, air, steam, or mixtures of them.[53] H_2SO_4, H_3PO_4, $ZnCl_2$, NaOH, $FeCl_3$, KOH, and K_2CO_3 are used as chemical activating reagents. However, H_3PO_4 is the most commonly used chemical activating agent during the synthesis of carbon from biomass due to its convenient recovery, low environmental negative effects, and higher carbon yield.[54]

Green synthesis methods utilizing bioorganisms, domestic and agricultural wastes, and sewage sludge follow the principles of eco-friendly chemistry and are based on the usage of non-toxic and safer reagent/compounds.[55]

Compared to the other carbon resources and synthesis methods, biomass-derived carbon procedures are straightforward and economically workable. Through physical/chemical activation and carbonization utilizing chemical and physical treatments, biomass becomes a superb activated carbon matter with the ability to have various chemical structures.[47,56,57]

9.3 Utilizing Low-carbon Material for the Development of an Electrochemical Biomedical Sensor

9.3.1 Chitosan and Its Utilization in Electrochemical Biomedical Sensors

Chitosan, an organic compound derived from polysaccharides, exhibits hydrophilic properties, possesses a minimal carbon footprint, and finds extensive application in the field of biological investigation.[58,59] It is typically obtained from the skeletal remains of marine organisms and is a biological molecule containing nitrogen.[60] Currently, chitosan, a polymer that can be broken down naturally, is more favored in comparison to polymers derived from synthetic sources. These polymers provide characteristics that make them easy to utilize, environmentally sustainable, and highly affordable.[61] Chitosan, known for its non-toxic nature, possesses distinctive electrochemical and physical characteristics.[62] These properties encompass the capacity to undergo functionalization, as well as possessing mechanical resistance and chemical and thermal stability. As a result, it can be utilized in a wide variety of areas, including fuel cells, biomaterials, and drug-delivery systems. Despite this, chitosan's low proton conductivity is a drawback when using these materials.[63] Nevertheless, this issue can be resolved by functionalizing it with other groups and it can also serve as a robust ionic conductor.[64] In recent times, chitosan has been referenced in a great deal of research.

Magnetic nanoparticles are used in conjunction with other nanoparticles to enhance the electrochemical conductivity of chitosan. Of these nanoparticles, Fe_3O_4 is the most well-known and widely used. The nanoparticles demonstrate excellent electrical and magnetic characteristics as a result of electron transfer occurring between iron(II) and iron(III).[65] Some applications for appropriately coated magnetic nanoparticles include medication delivery and clinical diagnostics.[66] The extremely high chemical activity of these nanoparticles makes them extremely oxidation-prone, with the fact that they have a large surface area.[67] In addition to their various applications, these advantages have led to using them in electromagnetic absorption, cell separation, and resistors. Graphene is another material preferred to be used with chitosan. Graphene is a substance composed of a solitary layer of carbon atoms arranged in a honeycomb-shaped crystal lattice.[68] Graphene has the capability to form many-dimensional structures, including nanotubes, fullerenes, and nanosheets. It is not the porous structure of graphene that is responsible for its high surface area; rather, it is the interconnecting channels that exist between the graphene layers that are responsible for contributing to this surface area.[69] Graphene exhibits a diverse array of applications in the realm of electrochemistry. The utilization of its sophisticated electrical, mechanical, electrochemical, optical, and thermal characteristics is obvious. In 2024, Dinu *et al.* fabricated a gold electrode by incorporating chitosan, Fe_3O_4, and graphene as shown in Figure 9.2.[70] Their goal in creating this electrode was to detect tannic acid, a pollutant, in the environment. One material that is particularly harmful to both humans and wildlife is tannic acid, which also tends to accumulate. The surface morphology of the sensor was analyzed using scanning electron microscopy (SEM), energy-dispersive X-ray spectroscopy (EDX), and X-ray diffraction (XRD) techniques. Chitosan was employed as a stabilizer in the investigation to enhance the functioning of the electrode. Graphene enhanced electron transport on the electrode surface, resulting in improved sensitivity. The concentration of tannic acid was measured using the differential pulse voltammetry (DPV) technique. The detection range was determined to be between 0.01 and 1000 μM, while the detection limit was found to be

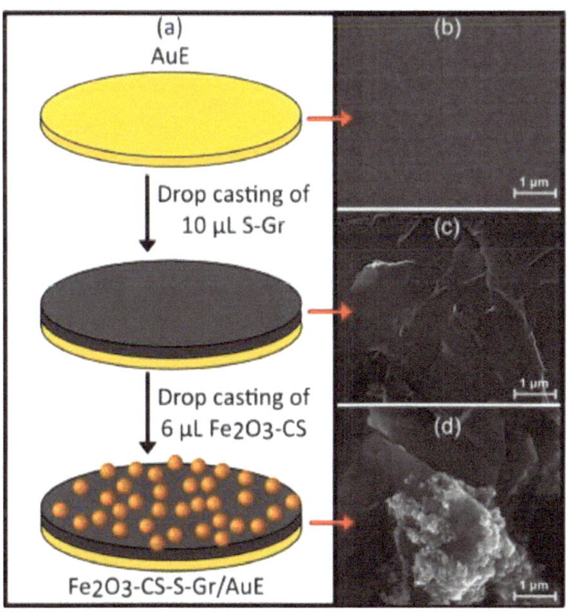

Figure 9.2 (a) The preparation steps of Fe_2O_3-CS-S-Gr/AuE and SEM images of (b) the gold electrode (AuE), (c) S-Gr, and (d) Fe_2O_3-CS.[70] Retrieved with permission. Reproduced from ref. 70, https://doi.org/10.1016/j.apsadv.2024.100602, under the terms of the CC BY 4.0 license, https://creativecommons.org/licenses/by/4.0/.

3.6×10^{-3} μM. The implemented technique enabled the identification of the substance in surface water samples with a recovery rate ranging from 97.7% to 99.5%.

Molecularly imprinted polymers (MIPs) are gaining significant interest and are extensively utilized. MIPs are typically created through the polymerization of the desired analyte and a compatible monomer on a surface.[71] To conduct the analysis, cavities that are appropriate for the target analyte are formed on the polymerized surface. The dimensions of these empty spaces are unique to the substance being analyzed. Consequently, techniques carried out employing MIPs exhibit a great degree of selectivity and sensitivity. The development of sensors appropriate for highly accurate and precise analysis is accomplished expediently and specifically through the utilization of MIPs.[72] The methods that have been developed are also characterized by the stability, repeatability, and specificity to the analyte. By combining MIPs with different nanoparticles, it is feasible to achieve enhanced sensitivity and faster response times. AuNPs are one of those nanoparticles. Gold nanoparticles are commonly favored in electrochemical investigations because of their optical and electrochemical characteristics, which enhance the sensor's sensitivity, selectivity, and adaptability. By employing the most straightforward electrochemical deposition technique, gold nanoparticles are deposited onto the electrode surface, enabling the attainment of the intended thickness of the gold layer on the surface.[73] Thirabowonkitphithan *et al.* modified an SPCE to detect procyanin using MIP, chitosan, and AuNPs as shown in Figure 9.3.[74]

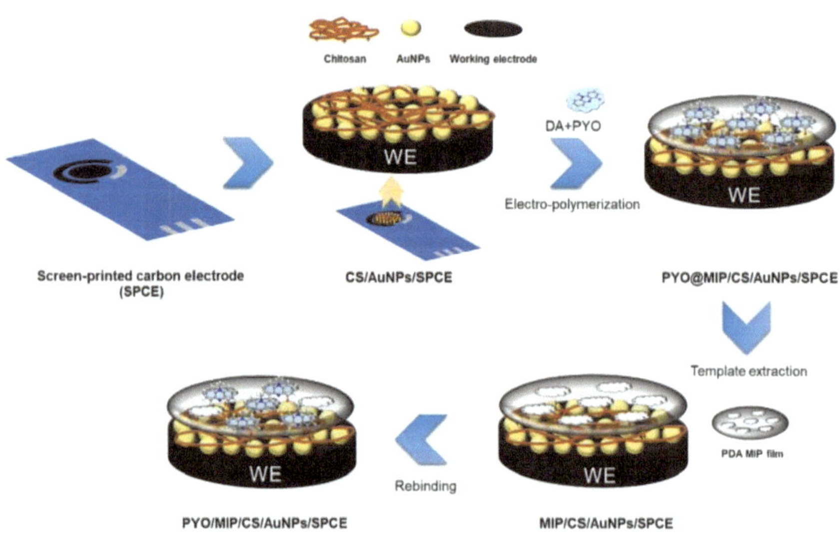

Figure 9.3 The process of preparation of the MIP-coated working electrode (MIP/CS/AuNPs/SPCE) and rebinding of pyocyanin to the MIP-coated electrode.[74] Retrieved with permission. Reproduced from ref. 74, https://doi.org/10.1016/j.eurpolymj.2024.112892, under the terms of the CC BY 4.0 license, https://creativecommons.org/licenses/by/4.0/.

Procyanin is a secondary metabolic product produced by *Pseudomonas aeruginosa*, which is a prevalent cause of health issues. Researchers found that increasing the electrocatalytic activity of SPCEs by electrochemically depositing AuNPs on their surfaces enhanced their surface area and, in turn, their electrocatalytic activity. Chitosan facilitated strong adhesion of the nanoparticles to the surface. The MIP layer was created using electrochemical polymerization, resulting in enhanced sensitivity and selectivity of the sensor. Electrochemical and analytical techniques such as EDX, Raman spectroscopy, and XPS were employed to accomplish surface characterization. The SWV method was used to determine the analyte. The linear determination range was determined to be 1–100 μM, and the detection limit was found to be 0.74 μM. The sensor was effectively utilized for bacterial culture samples, with a recovery rate ranging from 93% to 103%. Electrochemical biomedical application examples of chitosan-based electrochemical sensor studies conducted in recent years are given in Table 9.1.

9.3.2 Cellulose and Its Utilization in Electrochemical Biomedical Sensors

Cellulose is the predominant natural polymer found on Earth, offering a sustainable, compatible with living organisms, and economically efficient green material. Cellulose is the primary constituent of plant cell walls, playing a crucial role in preserving the plant's rigidity and robustness. This substance is a tasteless and scentless, white fibrous substance that usually does not dissolve in organic solvents or water. Cellulose is widely recognized as the predominant carbohydrate present in nature, occurring in various forms in plants, animals, algae, fungi, and minerals. The cellulose parent gel is often classified into three fundamental categories: cellulose renewal, functional cellulose derivatives, and nanocellulose. The phrase "cellulose-based froth and aerogels in order" will be used to designate the categories of cellulose precursor gel. Cellulose-based foam generally has a very small surface area, characterized by macropores that have pore diameters of up to 50 nm. Cellulose-based aerogel is produced through convective drying or freeze–drying methods, using precursors of cellulose. It possesses a relatively large specific surface area, high porosity (up to 90%), and tiny pores with pore sizes that vary from 2 nm to 50 nm. The aerogel is created by employing supercritical drying on cellulose precursors. Cellulose's unique properties such as its abundance in nature, crystalline structure, and insolubility in water make it suitable for nanotechnology. Cellulose, especially nanocellulose, is notable for its ability to create separators with a mesoporous structure.[103]

Cellulose is the dominant compound found in abundance worldwide and is a polymer that may be renewed and broken down by natural processes. Fibers made from cellulose possess a strong hydrogen-bonding network consisting of microfibers and fibrils, which creates an internal polymer that

Table 9.1 Chitosan-based electrochemical sensor studies in electrochemical biomedical applications between 2023–2024.[a]

Analyte	Surface layer content	Method	Linear range	Limit of detection	Real sample	Ref.
Tannic acid	Fe_2O_3–CS–S–Gr/AuE	DPV	0.01–1000 μM	3.6×10^{-3} μM	Surface water samples	70
Procyanin	MIP/CS/AuNPs/SPCE	SWV	1–100 μM	0.74 μM	Bacterial culture samples	74
Naproxen	$SrTi_{0.7}Fe_{0.3}O_3$ perovskite/chitosan/GCE	CV	1.99–130.84 μM	0.50 μM	Human plasma samples	75
Triclosan	Ni–Ag_2S-Chit/LabSPE	DPV	0.05 nM–90 μM	0.02 nM	Mouthwash and eczema cream	76
Alpha-fetoprotein	Chit-Fc-SAu/ Co/Fe MOF-TB/PDA-Ab2/GCE	SWV	0.001–200 ng mL^{-1}	0.34 pg mL^{-1}	Human serum samples	77
Carbohydrate antigen 19-9	Au–MoO_3-Chi/PG/SPE	DPV	0.0025–0.1 U mL^{-1} and 0.1–1.0 U mL^{-1}	1.0 mU mL^{-1}	Human serum samples	78
Carcinoembryonic antigen			0.001–0.01 ng mL^{-1} and 0.01–1.0 ng mL^{-1}	0.5 pg mL^{-1}		
Matrix metalloproteinase 9	Fc-CMCS/DEX-ALD hydrogel	DPV	50–1000 ng mL^{-1}	73.08 ng mL^{-1}	Cell secretion samples	79
Glucose	AuNPs/CS/Gr/GOx/IrOx/PGE	CV	50–1000 μM	30 μM	Urine	80
Imatinib	Chitosan/rGO/GCE	DPV	1–300 μM	7.3 nM	Serum and urine samples	81
Glucose	SBO/FCS-GCE	DPV	0.008–4.848 mM	0.057 μM	Human serum samples	82
Metadoxine	CuO/CH/GCE	SWV	1.99–29.56 μg L^{-1}	0.64 μg L^{-1}	Pharmaceutical dosage form and human blood serum	83
Acetaminophen	ZnO@Chitosan/GCE	CV	10–50 μM	0.71 μM	Pharmaceutical dosage form	84
Chloramphenicol	MIPs-Apt/AuNPs/CS-MWNTs/GCE	DPV	10^{-8}–10^{-2} g L^{-1}	3.3×10^{-9} g L^{-1}	Sewage, milk, and honey	85
Glucose	PGOx@MXene/CS/GCE	DPV	0.03–16.5 mM	3.1 μM	Human blood serum samples	86
α-Lipoic acid	PV-CS/f-MWCNTs/GCE	DPV	0–3000 μM	0.012 μM	Biological fluids and pharmaceutical specimens	87

Methadone	CS/CNT/GCE	DPV	40–1000 μM	0.019 μM	Pears and blood samples	88
Dopamine	CP/Fe_3O_4–GO/chi	SWV	3–32 μM	90 nM	Dermis-mimicked gel	89
N-Acetyl-4-aminophenol	AuNPs@rGO/CTS/GCE	DPV	0.50–39.31 μM	0.06 μM	Pharmaceutical formulation samples	90
BPA	CMWNT-Fc-PMo10V2/Chit/GCE	DPV	0.06–100 μM	0.035 μM	Milk samples	91
Glucose	PUC-PANI-ZnO@GOx/PPE	CA	0.01–9.48 mM	7.31 μM	Human serum samples	92
Dopamine	CF_3SO_3H@EA-JUC-1000/CS/SCE	CV	0.1–500 μM	50 nM	Bovine serum samples	93
Glucose	Nif/Cs/GO@Cu	DPV	5 μM–4 mM	83 nM	Human serum samples	94
Glucose	PB/GOx-AuNPs-CS/AuE	Amp	0.025–2.00 mM	1.62 μM and 7.16 μM	Human blood samples	95
Rutin	[$C_n$mim][AA]/MWCNTs/CS/GCE	DPV	0.5–5 μM and 5–100 μM	0.05 μM	Black tartary buckwheat tea	96
Staphylococcus aureus	SA/GO/CS/β-CD/GCE	DPV	30–300 CFU per mL	12 CFU per mL	—	97
Glucose	Cs/Cu-MOF/SPCE	Amp	2–1700 μM	2 μM	Saliva samples	98
Glucose	CuO/AC/CTS/GCE	Amp	0.2–2400 μM	0.1 μM	—	99
Tyrosine	SPCE/SBA-15/Fe_3O_4/polyaniline/chitosan	DPV	0.1–250 μM	0.05 μM	Milk samples	100
Diclofenac sodium	CS/TiO_2/CPE	DPV	0.2–10 μM and 10–100 μM	0.013 μM	Commercial tablets and synthetic urine samples	101
Serotonin	e-CS/MWCNTs/GCE	DPV	9–1000 μM	10 μM	Human saliva samples	102

[a] Abbreviations: AC, activated carbon; Ag_2S, silver sulfide; Amp, amperometry; Apt, aptamer; Au, gold nanoparticles; AuE, gold electrode; AuNPs, gold nanoparticles; BPA, bisphenol A; CA, chronoamperometry; CF_3SO_3H, trifluoromethanesulfonic acid; Chi, chitosan; Chit, chitosan; CMCS, carboxymethyl chitosan; Co, cobalt; CP, carbon paste electrode; CS, chitosan; Cu, copper; CuO, copper oxide; CV, cyclic voltammetry; DEX-ALD, dextran-aldehyde; DPV, differential pulse voltammetry; EA-JUC-1000, amino-functionalized cuprum-based MOF; e-CS, electrodeposited chitosan; Fc, ferrocenecarboxaldehyde; FCS, functionalized chitosan; Fe, iron; Fe_2O_3, iron(III) oxide; Fe_3O_4, iron(II) oxide; GCE, glassy carbon electrode; GO, graphene oxide; Gox, glucose oxidase; Gr, graphene; IrOx, iridium oxide; LabSPE, laboratory-fabricated screen printed electrode; MIP, molecularly imprinted polymers; MOF, metal–organic framework; MoO_3, molybdenum trioxide; MWNTs, multi-walled carbon nanotubes; Ni, nickel; PANI, polyaniline;, PB, Prussian blue; PG, porous graphene; PGE, paper-based gold electrode; PGOx, GOx polynanogel; $PMo_{10}V_2$, ferrocene-$H_5PMo_{10}V_2O_{40}$; PPE, platinum plate electrode; PUC, polyurethane-chitosan; PV-CS, poly(vanillin-*co*-chitosan); rGO, reduced graphene oxide; SA, *Staphylococcus aureus*; Sau, spindle gold; SBA-15, Santa Barbara amorphous 15; SBO, 1,3,5-triazine-bis(4-methylbenzenesulfonyl)-hydrazone; SCE, saturated calomel electrode; S-Gr, sulfur-doped graphene; SPCE, screenprinted carbon electrode; SPE, screenprinted electrode; TiO_2, titanium dioxide; ZnO, zinc oxide; β-CD, β-cyclodextrin.

is insoluble in both water and typical organic solvents. Substituting the hydroxyl groups results in the production of soluble derivatives, such as the following: ethyl cellulose (EC), cellulose acetate (CA), sodium carboxymethyl cellulose (Na-CMC), and more. These compounds may be utilized as fillers in LIBs. In addition, as contrasted with traditional separators, polymers based on cellulose exhibit extremely excellent ion diffusion properties and electrolyte humidity when they are utilized as separators.[104]

The objective of the reclamation and reuse of cellulose waste has the aim of mitigating global environmental concerns and reducing the deterioration of natural resources. Because of its incredible potential for uses in electrochemical power plants, the utilization of cellulose trash was therefore regarded as a critical part of the circular economy idea. Bio-wastes contain high levels of carbon, making them excellent materials for producing chemicals and nanomaterials. This is why the use of carbon-based substances from biological waste for sustainable energy development has become important in materials research. Pyrolysis and hydrothermal processing are among the methods used to convert biological waste into carbon compounds for diverse applications.[105]

Akhtera *et al.* used a silver–gold bimetallic alloy nanoparticle electrochemical detection platform that integrated nanocellulose and 3D polypyrrole for the detection of ciprofloxacin (CIF) using electrochemical methods.[10] Nanocellulose (NC) is an organic polymer that is widely used in the construction of electrochemical sensors because of its abundant supply, compatibility, eco-friendliness, and ability to create films.

The hydrogen-binding capacity of NC improves the binding and detection of target analytes in biosensing applications. Three-dimensional polypyrrole (3DPPY) is a conductive polymer that is primarily manufactured by chemical polymerization. Current research is focused on incorporating polypyrrole into NC to create hybrid materials that include enhanced chemical characteristics, environmental stability, and electrical conductivity. This is a prominent area of study within the field of conductive polymers. The hydrogen bonding between the OH– and NH groups of NC and 3DPPY creates a strong structure with a large surface area and high porosity. In addition, the incorporation of 3DPPY into NC fibers enhances the formation of conductive channels inside the nanocomposite, hence facilitating the rapid movement of charges and ions and leading to a highly effective electrochemical reaction (Scheme 9.1).[106]

Bimetallic gold–silver nanoparticles (Ag–Au) provide a cost-effective substitute for monometallic gold nanoparticle (Au) catalysts, with usually superior catalytic activity compared to single metals. The Ag–Au alloy's bimetallic structure has high-energy surface areas that enable fast and concurrent identification of analytes. The incorporation of the Ag–Au alloy inside the NC-3DPPY polymer matrix guarantees the enduring stability of the nanoparticles, preventing them from clumping together over time. Furthermore, the polymer matrix provides supplementary functionality to hybrid metal nanoparticles, hence enhancing many biological and catalytic

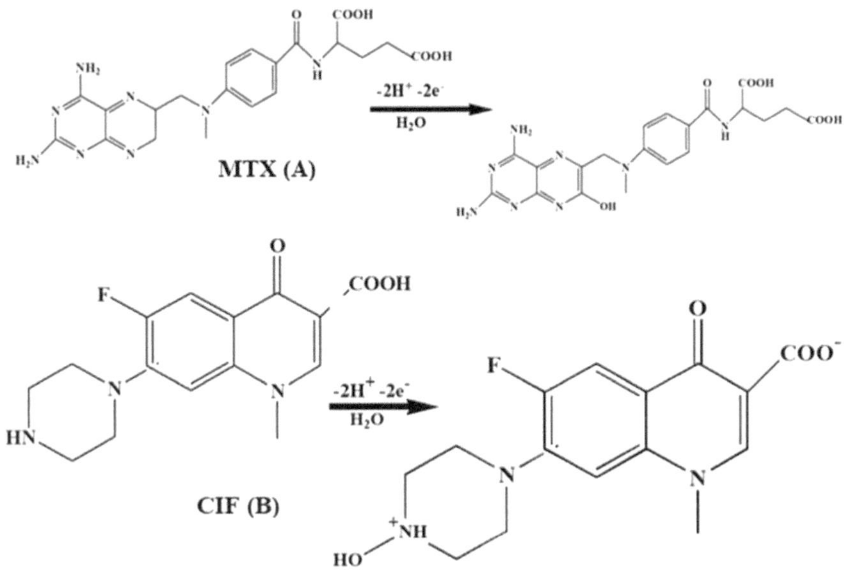

Scheme 9.1 Probable oxidation mechanism of MTX (A) and CIF (B).

sensing applications. Furthermore, the NC-3DPPY polymeric matrix has a significant impact on the distribution of the bimetallic Ag–Au alloy, enhancing the effectiveness of load mobilization by creating a continuous 3D pathway for electrons and serving as a matrix for immobilizing biomolecules or target analytes.[107]

NC-3DPPY composites are created by acid hydrolysis, whereas Ag–Au bimetallic alloys are formed through chemical reduction. The NC-3DPPY@Ag–Au nanocomposite structure was created by ultrasonication, and the electrochemical properties of the nanocomposites were studied using cyclic voltammetry (CV) and square wave voltammetry (SWV). The nanocomposites were analyzed using field emission scanning electron microscopy (FESEM) and high-resolution transmission electron microscopy (HR-TEM). The results revealed a porous structure and interconnected nanoparticles in the produced samples.

Under optimum settings of scan speed, pH, and supporting electrolyte conditions, the produced sensor demonstrated a low limit of detection (LOD) of 0.017 μM and a broad linear detection range of 0.5–50 μM. The electrochemical sensor that was created has exceptional stability, remarkable repeatability, and a high level of selectivity in detecting CIF even when other species are present. The research found that when using simulated blood serum, the analysis of CIF in tablets and water samples showed outstanding performance and reasonable recovery rates for trace detection.

Electrochemical biomedical application examples of cellulose-based electrochemical sensor studies conducted in recent years are given in Table 9.2.

Table 9.2 Cellulose-based electrochemical studies in electrochemical biomedical applications between 2019–2024.[a]

Analyte	Surface layer content	Method	Linear range	Limit of detection	Real sample	Ref.
H_2O_2	PANI/D-CMC/ZnO	DPV	0.50–5 µM	0.45 µM	Milk	108
Glucose	Pt/RGO/CNC/GCE	CV	0.005–8.5 mM	2.1 µM	Human serum	109
Dopamine	GO@CMC/MgO	DPV	7.4–0.95 µM	0.2 µM	Human serum and urine	110
Glucose	Cu/LIG/EC	CV	0–2.5 mM	0.34 µM	Cola, Sprite, black tea	111
Amsacrine	MIP/BC-AgNPs/GCE	DPV	30–110 µM	1.51 µM	Urine	112
Ochratoxin A	Apt/Cel NFM/SPCE	DPV	0.002–2 nM	0.81 pM	Cold brew coffee samples	113
Lactate	PB/Nafion/LOx/BC/SPE	CV	1–24 mM	1.331×10^{-3} M	Blood sample	114
4,4′-Methylene diphenyl diamine	AuNPs/CNCs/SPE	CV	0.12–100 µM	0.057 µM	—	115
Carcinoembryonic antigen	CNC-PPy gel/GCE	CV	1 fM–200 nM	0.06 fM	Human serum	116
Ciprofloxacin	NC-3DPPY@Ag–Au/SPE	SWV	0.5–50 µM	0.017 µM	Blood serum, tablet and water sample	117

[a] Abbreviations: 3DPPY, three-dimensional polypyrrole; AgNPs, silver nanoparticles; Apt, aptasensor; AuNPs, gold nanoparticles; BC, bacterial cellulose; Cel NFMs, Cellulose acetate; CMC, carboxymethylcellulose; CNC, cellulose nanocrystal; CV, cyclic voltammetry; D-CMC, dialdylated carboxymethylcellulose; DPV, differential pulse voltammetry; EC, ethyl-cellulose; GCE, glass carbon electrode; GO, graphene oxide; H_2O_2, hydrogen peroxide; LIG, laser-induced graphene;, LOx, lactate oxidase; MgO, magnesium oxide; MIP, molecularly imprinted polymer nanofibrous membranes; NC, nanocellulose; PANI, polyaniline; PB, Prussian blue; PPy, polypyrrole; Pt NPs, platinum nanoparticles; SPCE, screen-printed carbon electrode; SWV, square wave voltammetry.

9.3.3 Silk and Its Applications in Electrochemical Biomedical Sensors

Silk, specifically referred to as silk proteins like silk fibroin and silk fibroin hydrogels in research, exhibits exceptional biological compatibility, mechanical characteristics, and biodegradability.[118] Compared to synthetic hydrogels, they possess the benefit of biological activity.[118] Nevertheless, these materials lack electrochemical conductivity and hence require the addition of conductive chemicals to enhance their conductivity. These conductive materials encompass substances like graphene and PEDOT. Electrochemical biomedical applications involve the healing of skin and the release of drugs. Silk proteins exhibit advantageous characteristics, including high thermomechanical strength and resilience under pressure.[119] Furthermore, these proteins have the ability to dissolve in water-based solvents, enabling the creation of diverse structures such as membranes, films, and sponges. Also, the enzymatic degradation of these proteins under normal physiological settings is advantageous as it aids in the development of environmentally friendly electronics and the reduction of waste generation.[120]

In 2022, Chen *et al.* created a screen-printing plate specifically designed for the purpose of measuring glucose levels.[121] The researchers altered the surface of the plate by including silk nanofibrils, decreased graphene oxide, and glucose oxidase. Silk nanofibrils are naturally occurring compounds characterized by their low density and extraordinary durability.[122] These materials, in conjunction with other materials they are paired with, draw attention due to their exceptional mechanical qualities, well-structured composition, compatibility, and applications in medicine and biology.[123] Reduced graphene oxide sheets (rGOs) are notable for their significant active surface area, excellent electrical conductivity, and remarkable durability.[124] rGOs are insoluble in water and tend to form precipitates. Therefore, it is crucial to identify the suitable resolution for dispersing the material.[124] The work involved conducting surface characterizations using AFM (atomic force microscopy) and TEM (transmission electron microscopy). The resulting composite exhibited exceptional stability. The cause of this phenomenon was attributed to the interaction between reduced graphene oxide (rGO) and the hydrophobic areas present in silk fibroin. The detection of glucose was conducted using the CV method. The linear detection range was determined to be from 0 to 100 μM, while the detection limit was measured to be 300 nM. The procedure was successfully utilized on human sweat samples.

Furthermore, electrochemical biomedical application examples of silk-based electrochemical sensor studies conducted in recent years are given in Table 9.3.

9.4 Conclusion and Future Perspectives

Electrochemical biomedical sensors are among the most commonly used sensors, as they are popular owing to their user-friendly nature and quick

Table 9.3 Silk-based electrochemical sensor studies in electrochemical biomedical applications between 2019–2024.[a]

Analyte	Surface layer content	Method	Linear range	Limit of	detection	
Real sample	Ref.					
Glucose	MWCNT/SF/Pt NPs/GOx	Amperometry	0.1–3.75 mM	0.1 µM	—	125
Chloramphenicol	Fe-Silk PNC/PGE	DPV	1–200 µM	0.57 µM	Shrimps, carp, chicken samples	126
2-Nitroaniline	Sn–Bi_2O_3/SF	DPV	0.009–785.7 µM	3.5 nM	River samples	127
Glucose	GOD/PB/RGO/SF	CV	65.57 µM–10 mM	65.57 µM	Serum, salvia, sweat and urine samples	128
Glucose	SNF/RGO/GOx	CV	0–100 µM	300 nM	Human sweat samples	129
Dopamine	GO-SF/CC	DPV	100–900 µM	0.41 µM	—	130

[a] Abbreviations: CC, carbon cloth; GCE, glassy carbon electrode; GOD, glucose oxidase; GOx, glucose oxidase; MWCNT, multi-walled carbon nanotube; PB, Prussian blue; PGE, pyrolytic graphite electrodes; PNC, porous carbon nanosheets; Pt NPs, platinum nanoparticles; RGO, reduced graphene oxide; SF, silk fibroin; SNF, silk nanofibril.

results. Thus, these sensors play an important role in diagnosing, monitoring, and treating diseases in healthcare. Electrochemical biomedical sensors can monitor or measure various biological factors, depending on their specific role and design. It is crucial to explore the applications of low-carbon materials in this industry and the research conducted in this area. Recently, there has been a significant increase in the adoption of low-carbon materials as a strategy to mitigate global environmental damage. Low-carbon materials are crucial for achieving objectives such as sustainable development, mitigating climate change, and minimizing the utilization of environmentally harmful substances. When selecting materials for sensor applications, it is advisable to prioritize low-carbon materials that enable rapid, cost-effective, and consistent analyses, while also being highly portable and user-friendly. This chapter discusses the roles of low-carbon materials in sensor research and their applications in electrochemical biomedical sensor studies.

Electrochemical sensors have great potential as a result of their ongoing advancements and the incorporation of novel materials. Nevertheless, there is still room for improvement in electrochemical sensors. Researchers in this field are largely concentrating on the use of rapid, sensitive, and cost-effective nanomaterials. Nevertheless, in the context of a biological sensor, the toxicity and antibacterial characteristics of the material have equal significance to its sensitivity. Nevertheless, the range of low-carbon materials accessible in this particular situation is restricted. Researchers must expand the range of materials suitable for low-carbon electrochemical biomedical sensors.

References

1. T. Yokota, K. Fukuda and T. Someya, Recent Progress of Flexible Image Sensors for Biomedical Applications, *Adv. Mater.*, 2021, **33**, 1–26.
2. D. Murzin, D. J. Mapps, K. Levada, V. Belyaev, A. Omelyanchik, L. Panina and V. Rodionova, Ultrasensitive magnetic field sensors for biomedical applications, *Sensors*, 2020, **20**(6), 1569.
3. G. L. Cote, R. M. Lec and M. Pishko, Emerging biomedical sensing technologies and their applications, *IEEE Sens. J.*, 2003, **3**(3), 251–266.
4. M. Pirzada and Z. Altintas, Recent Progress in Optical Sensors for Biomedical Diagnostics, *Micromachines*, 2020, **11**(4), 356.
5. M. Seçkin, A. Ç. Seçkin and Ç. Gençer, Biomedical Sensors and Applications of Wearable Technologies on Arm and Hand, *Biomed. Mater. & Devices*, 2023, **1**, 443–455.
6. P. Mohankumar, J. Ajayan, T. Mohanraj and R. Yasodharan, Recent developments in biosensors for healthcare and biomedical applications: A review, *Measurement*, 2021, **167**, 108293.
7. N. Nath, S. Chakroborty, D. P. Vishwakarma, G. Goga, A. S. Yadav and R. Mohan, Recent advances in sustainable nature-based functional

materials for biomedical sensor technologies, *Environ. Sci. Pollut. Res.*, 2024, **31**(46), 57289–57313.
8. A. M. Upadhyaya, M. K. Hasan, S. Abdel-Khalek, R. Hassan, M. C. Srivastava, P. Sharan, S. Islam, A. M. E. Saad and N. Vo, A Comprehensive Review on the Optical Micro-Electromechanical Sensors for the Biomedical Application, *Front. Public Health*, 2021, **9**, 759032.
9. H. Jirimali, J. Singh, R. Boddula, J.-K. Lee and V. Singh, Nano-Structured Carbon: Its Synthesis from Renewable Agricultural Sources and Important Applications, *Materials*, 2022, **15**, 3969.
10. T. D. Nguyen, M. T. N. Nguyen and J. S. Lee, Carbon-Based Materials and Their Applications in Sensing by Electrochemical Voltammetry, *Inorganics*, 2023, **11**, 81.
11. S. Michalkiewicz, A. Skorupa and M. Jakubczyk, Carbon Materials in Electroanalysis of Preservatives: A Review, *Materials*, 2021, **14**, 7630.
12. M. Freitas, L. P. da Silva, P. M. S. M. Rodrigues and J. Esteves da Silva, Sustainable Technological Applications of Green Carbon Materials, *Sustainable Chem.*, 2024, **5**, 81–97.
13. Z. Liu, Z. Deng, S. Davis and P. Ciais, Monitoring global carbon emissions in 2022, *Nat. Rev. Earth Environ.*, 2023, **4**, 205–206.
14. O. Das, Á. Restás, V. Shanmugam, G. Sas, M. Försth, Q. Xu, L. Jiang, M. S. Hedenqvist and S. Ramakrishna, Demystifying Low-Carbon Materials, *Mater. Circ. Econ.*, 2021, **3**, 26.
15. K. Liu, M. Tebyetekerwa, D. Ji and S. Ramakrishna, Intelligent materials, *Matter Opin.*, 2020, **3**, 590–593.
16. Y. Guo, L. Luo, T. Liu, L. Hao, Y. Li, P. Liu and T. Zhu, A review of low-carbon technologies and projects for the global cement industry, *J. Environ. Sci.*, 2024, **136**, 682–697.
17. Z. Zhang, G. Hu, X. Mu and L. Kong, From low carbon to carbon neutrality: A bibliometric analysis of the status, evolution and development trend, *J. Environ. Manage.*, 2022, **322**, 116087.
18. F. Foroughi, E. Rezvani Ghomi, F. Morshedi Dehaghi, R. Borayek and S. Ramakrishna, A Review on the Life Cycle Assessment of Cellulose: From Properties to the Potential of Making It a Low Carbon Material, *Materials*, 2021, **14**, 714.
19. S. Chawla, P. Rai, T. Garain, S. Uday and C. M. Hussain, Green Carbon Materials for the Analysis of Environmental Pollutants, *Trends Environ. Anal. Chem.*, 2022, **33**, e00156.
20. G. S. dos Reis, S. H. Larsson, H. P. de Oliveira, M. Thyrel and E. Claudio Lima, Sustainable Biomass Activated Carbons as Electrodes for Battery and Supercapacitors—A Mini-Review, *Nanomaterials*, 2020, **10**, 1398.
21. J. Li, D. Yin and Y. Qin, Carbon materials: structures, properties, synthesis and applications, *Manuf. Rev.*, 2023, **10**, 13.
22. N. López-Salas and M. Antonietti, Carbonaceous Materials: The Beauty of Simplicity, *Bull. Chem. Soc. Jpn.*, 2021, **94**, 2822–2828.

23. S. Vaz Jr and D. T. de Souza, Sustainable carbonaceous raw materials: A contribution to reduce the negative environmental impacts of chemical industry, *Sustainable Chem. Environ.*, 2024, **5**, 100058.
24. K. Liu, M. Tebyetekerwa, D. Ji and S. Ramakrishna, Intelligent Materials, *Matter*, 2020, **3**, 590–593.
25. M. Hatimuria, P. Phukan, S. Bag, J. Ghosh, K. Gavvala, A. Pabbathi and J. Das, Green Carbon Dots: Applications in Development of Electrochemical Sensors, Assessment of Toxicity as Well as Anticancer Properties, *Catalysts*, 2023, **13**(3), 537.
26. M. Javaid, A. Haleem, S. Rab, R. Pratap Singh and R. Suman, Sensors for daily life: A review, *Sens. Int.*, 2021, **2**, 100121.
27. J. Liu, S. Bao and X. Wang, Applications of Graphene-Based Materials in Sensors: A Review, *Micromachines*, 2022, **13**, 184.
28. R. D. Crapnell and C. E. Banks, Electroanalysis overview: Addressing the green credentials in the use of electroanalytical sensors, *Green Carbon*, 2023, **1**, 85–93.
29. G. Zhou, F. Li and H.-M. Cheng, Progress in flexible lithium batteries and future prospects, *Energy Environ. Sci.*, 2014, **7**, 1307–1338.
30. B. Uslu and S. Ozkan, Electroanalytical Application of Carbon Based Electrodes to the Pharmaceuticals, *Anal. Lett.*, 2007, **40**, 817–853.
31. P. K. Jiwanti, B. Y. Wardhana, L. G. Sutanto and M. F. Chanif, A Review on Carbon-based Electrodes for Electrochemical Sensor of Quinolone Antibiotics, *ChemistrySelect*, 2022, **7**(15), e202103997.
32. M. Zhou, Y. Zhai and S. Dong, Electrochemical Sensing and Biosensing Platform Based on Chemically Reduced Graphene Oxide, *Anal. Chem.*, 2009, **81**, 5603–5613.
33. R. L. McCreery, ChemInform Abstract: Advanced Carbon Electrode Materials for Molecular Electrochemistry, *Chem. Rev.*, 2008, **108**(7), 2646–2687.
34. C. Ma, M. Zhang, Y. Ding, Y. Xue, H. Wang, P. Li and D. Wu, Green Production of Biomass-Derived Carbon Materials for High-Performance Lithium–Sulfur Batteries, *Nanomaterials*, 2023, **13**, 1768.
35. B. Uslu, *Electroanalytical Applications of Quantum Dot-Based Biosensors*, Elsevier, 2021.
36. V. S. Bhat, S. Supriya and G. Hegde, Review—Biomass Derived Carbon Materials for Electrochemical Sensors, *J. Electrochem. Soc.*, 2020, **167**, 037526.
37. M. Sillanpää and C. Ncibi, in *A Sustainable Bioeconomy*, Springer International Publishing, Cham, 2017, pp. 185–231.
38. H. Krishnamoorthy, R. Ramyea, A. Maruthu, K. Kandasamy, M. Michalska and S. K. Kandasamy, Synthesis methods of carbonaceous materials from different bio-wastes as electrodes for supercapacitor and its electrochemistry - A review, *Bioresour. Technol. Rep.*, 2022, **19**, 101187.
39. C. Onfray and A. Thiam, Biomass-Derived Carbon-Based Electrodes for Electrochemical Sensing: A Review, *Micromachines*, 2023, **14**, 1688.

40. K. Zúñiga, G. Rebollar, M. Avelar, J. Campos-Terán and E. Torres, Nanomaterial-Based Sensors for the Detection of Glyphosate, *Water*, 2022, **14**, 2436.
41. T. A. Saleh, Nanomaterials: Classification, properties, and environmental toxicities, *Environ. Technol. Innovation*, 2020, **20**, 101067.
42. P. Yáñez-Sedeño, S. Campuzano and J. M. Pingarrón, Electrochemical (bio)sensors: promising tools for green analytical chemistry, *Curr. Opin. Green Sustainable Chem.*, 2019, **19**, 1–7.
43. A. T. S. C. Brandão, S. State, R. Costa, P. Potorac, J. A. Vázquez, J. Valcarcel, A. F. Silva, L. Anicai, M. Enachescu and C. M. Pereira, Renewable Carbon Materials as Electrodes for High-Performance Supercapacitors: From Marine Biowaste to High Specific Surface Area Porous Biocarbons, *ACS Omega*, 2023, **8**, 18782–18798.
44. P. Rajkumar, V. Thirumal, G. Radhika, R. M. Gnanamuthu, R. Subadevi, M. Sivakumar, K. Yoo and J. Kim, Eco-friendly production of carbon electrode from biomass for high performance Lithium and Zinc ion capacitors with hybrid energy storage characteristics, *Mater. Lett.*, 2024, **354**, 135320.
45. H. C. Ates, P. Q. Nguyen, L. Gonzalez-Macia, E. Morales-Narváez, F. Güder, J. J. Collins and C. Dincer, End-to-end design of wearable sensors, *Nat. Rev. Mater.*, 2022, **7**, 887–907.
46. S. Saini, P. Chand and A. Joshi, Biomass derived carbon for supercapacitor applications: Review, *J. Energy Storage*, 2021, **39**, 102646.
47. L. Manjakkal, A. Jain, S. Nandy, S. Goswami, J. Tiago Carvalho, L. Pereira, C. H. See, S. C. Pillai and R. A. Hogg, Sustainable electrochemical energy storage devices using natural bast fibres, *Chem. Eng. J.*, 2023, **465**, 142845.
48. H. Jirimali, J. Singh, R. Boddula, J.-K. Lee and V. Singh, Nano-Structured Carbon: Its Synthesis from Renewable Agricultural Sources and Important Applications, *Materials*, 2022, **15**, 3969.
49. Y. Feng, Y. Chen, Z. Wang and J. Wei, Synthesis of mesoporous carbon materials from renewable plant polyphenols for environmental and energy applications, *New Carbon Mater.*, 2022, **37**, 196–222.
50. L. Zhang, A. A. S. Gonçalves and M. Jaroniec, Synthesis of nanoporous carbonaceous materials at lower temperatures, *Front. Chem.*, 2023, **11**, 1277826.
51. A. Jain, R. Balasubramanian and M. P. Srinivasan, Hydrothermal conversion of biomass waste to activated carbon with high porosity: A review, *Chem. Eng. J.*, 2016, **283**, 789–805.
52. K. Malini, D. Selvakumar and N. S. Kumar, Activated carbon from biomass: Preparation, factors improving basicity and surface properties for enhanced CO_2 capture capacity – A review, *J. CO_2 Util.*, 2023, **67**, 102318.
53. Z. Gao, Y. Zhang, N. Song and X. Li, Biomass-derived renewable carbon materials for electrochemical energy storage, *Mater. Res. Lett.*, 2017, **5**, 69–88.

54. I. Neme, G. Gonfa and C. Masi, Activated carbon from biomass precursors using phosphoric acid: A review, *Heliyon*, 2022, **8**, e11940.
55. S. Bilge, L. Karadurmus, A. Sınağ and S. A. Ozkan, Green synthesis and characterization of carbon-based materials for sensitive detection of heavy metal ions, *TrAC, Trends Anal. Chem.*, 2021, **145**, 116473.
56. M. Yang, D. S. Kim, S. B. Hong, J.-W. Sim, J. Kim, S.-S. Kim and B. G. Choi, MnO_2 Nanowire/Biomass-Derived Carbon from Hemp Stem for High-Performance Supercapacitors, *Langmuir*, 2017, **33**, 5140–5147.
57. H. Wang, Z. Xu, A. Kohandehghan, Z. Li, K. Cui, X. Tan, T. J. Stephenson, C. K. King'ondu, C. M. B. Holt, B. C. Olsen, J. K. Tak, D. Harfield, A. O. Anyia and D. Mitlin, Interconnected Carbon Nanosheets Derived from Hemp for Ultrafast Supercapacitors with High Energy, *ACS Nano*, 2013, **7**, 5131–5141.
58. Z. Wu, W. Feng, Y. Feng, Q. Liu, X. Xu, T. Sekino, A. Fujii and M. Ozaki, Preparation and characterization of chitosan-grafted multiwalled carbon nanotubes and their electrochemical properties, *Carbon*, 2007, **45**, 1212–1218.
59. H. Yadav and R. Malviya, Chitosan in biomedicine: A comprehensive review of recent developments, *Carbohydr. Polym. Technol. Appl.*, 2024, **8**, 100551.
60. L. Pu, J. Zhang, N. Kuate, L. Jiresse, Y. Gao, H. Zhou, N. Naik and P. Gao, N - doped MXene derived from chitosan for the highly effective electrochemical properties as supercapacitor, *Adv. Compos. Hybrid Mater.*, 2022, **5**, 356–369.
61. A. S. Bhatt, D. K. Bhat and M. S. Santosh, Electrochemical properties of chitosan – Co_3O_4 nanocomposite films, *J. Electroanal. Chem.*, 2011, **657**, 135–143.
62. İ. Kaya, B. Ayten and A. Ömer, Synthesis and electrochemical properties of chitosan-polyphenol composites, *React. Funct. Polym.*, 2020, **154**, 104667.
63. V. Vijayalekshmi and D. Khastgir, Fabrication and comprehensive investigation of physicochemical and electrochemical properties of chitosan-silica supported silicotungstic acid nanocomposite membranes for fuel cell applications, *Energy*, 2018, **142**, 313–330.
64. S. B. Aziz, A. S. F. M. Asnawi, M. Fakhrul, Z. Kadir, S. M. Alshehri, T. Ahamad, Y. M. Yusof and J. M. Hadi, Structural , Electrical and Electrochemical Properties of Glycerolized Biopolymers Based on Chitosan (CS): Methylcellulose (MC) for Energy Storage Application, *Polymers*, 2021, **13**(8), 1183.
65. Z. Jie, X. Ying, B. Nian and C. Wu, Hydrothermal synthesis and self-assembly of magnetite (Fe_3O_4) nanoparticles with the magnetic and electrochemical properties, *J. Cryst. Grow.*, 2008, **310**, 5453–5457.
66. L. Fu, V. P. Dravid and D. L. Johnson, Self-assembled (SA) bilayer molecular coating on magnetic nanoparticles, *Appl. Surf. Sci.*, 2001, **181**, 173–178.

67. H. Zhang, X. Zhong, J. Xu and H. Chen, Fe_3O_4/Polypyrrole/Au Nanocomposites with Core/Shell/Shell Structure: Synthesis, Characterization, and Their Electrochemical Properties, *Langmuir*, 2008, 13748–13752.
68. S. Alwarappan, A. Erdem, C. Liu and C. Li, Probing the Electrochemical Properties of Graphene Nanosheets for Biosensing Applications, *J. Phys. Chem. C*, 2009, 8853–8857.
69. J. Yan, T. Wei, B. Shao, F. Ma, Z. Fan, C. Zheng, Y. Shang, W. Qian and F. Wei, Electrochemical properties of graphene nanosheet/carbon black composites as electrodes for supercapacitors, *Carbon*, 2010, **48**(6), 1731–1737.
70. L. A. Dinu, A. M. Baracu, E. I. Geana, C. Parvulescu, M. C. Stoian, O. Brincoveanu, C. Pachiu and S. Kurbanoglu, Integrated nanozyme electrochemical sensor for the detection of tannic acid: An advanced approach to rapid and efficient environmental monitoring, *Appl. Surf. Sci. Adv.*, 2024, **21**, 100602.
71. C. Malitesta, E. Mazzotta, R. A. Picca, A. Poma, I. Chianella and S. A. Piletsky, MIP sensors – the electrochemical approach, *Anal. Bioanal. Chem.*, 2012, **402**, 1827–1846.
72. L. Wang, M. Pagett and W. Zhang, Molecularly imprinted polymer (MIP) based electrochemical sensors and their recent advances in health applications, *Sens. Actuators Rep.*, 2023, **5**, 100153.
73. M. Libansky, J. Zima, J. Barek, A. Reznickova, V. Svorcik and H. Dejmkova, Basic electrochemical properties of sputtered gold film electrodes, *Electrochim. Acta*, 2017, **251**, 452–460.
74. P. Thirabowonkitphithan, S. Hajizadeh, W. Laiwattanapaisal and L. Ye, Detection of Pseudomonas aeruginosa infection using a sustainable and selective polydopamine-based molecularly imprinted electrochemical sensor, *Eur. Polym. J.*, 2024, **209**, 112892.
75. R. Mahmoudi-maleki, M. R. Majidi, H. Sohrabi, E. Mahmoudi, H. Fooladvand, A. Coruh and A. Niaei, Exploring the potential of $SrTi_{0.7}Fe_{0.3}O_3$ perovskite/Chitosan nanosheets for the development of a label-free electrochemical sensing assay for determination of naproxen in human plasma samples, *Anal. Biochem.*, 2024, **690**, 115513.
76. R. Cetiner, A. Sarilmaz, F. Ozel, S. Z. Bas and M. Ozmen, A new Bluetooth-assisted sensor based on lab-made screen-printed electrode modified with Ni–Ag_2 S for electrochemical detection of triclosan, *Electrochim. Acta*, 2024, **489**, 144223.
77. H. Sun, M. Wang, H. Tan, H. Liu, M. Liu and S. Li, Ratiometric electrochemical immunoassay based on 2D Co/Fe MOF decorated with toluidine blue and Fc-labeled Schiff base for accurate assay of alpha-fetoprotein in clinical serum, *Talanta*, 2024, **273**, 125876.
78. S. Cotchim, S. Kongkaew and P. Thavarungkul, A dual-electrode label-free immunosensor based on in situ prepared Au–MoO_3-Chi/porous graphene nanoparticles for point-of-care detection of cholangiocarcinoma, *Talanta*, 2024, **272**, 125755.

79. J. Wang, H. Zhang, H. Hu, S. Hu and L. Ma, An enzyme-responsive hydrogel of ferrocene-grafted carboxymethyl chitosan as a soft electrochemical sensor for MMP-9 detection, *Int. J. Biol. Macromol.*, 2024, **268**, 131582.
80. Y. Li, Y. Kong, Y. Hu, Y. Li, R. Asrosa, W. Zhang, B. D. Boruah, A. K. Yetisen, A. Davenport, T. C. Lee and B. Li, A paper-based dual functional biosensor for safe and user-friendly point-of-care urine analysis, *Lab Chip*, 2024, **24**, 2454–2467.
81. Y. Chen, Z. Liu and D. Bai, Determination of imatinib as anticancer drug in serum and urine samples by electrochemical technique using a chitosan/graphene oxide modified electrode, *Alexandria Eng. J.*, 2024, **93**, 80–89.
82. M. Ramesh, S. Umamatheswari, P. M. Vivek, C. Sankar and R. Jayavel, Synthesis of silver - bismuth oxide encapsulated hydrazone functionalized chitosan ($AgBi_2O_3$/FCS) nanocomposite for electrochemical sensing of glucose, H_2O_2 and Escherichia coli O157: H7, *Int. J. Biol. Macromol.*, 2024, **264**, 130533.
83. M. S. Sengar, P. Kumari, N. Sengar, S. P. Satsangee and R. Jain, Non-enzymatic electrochemical sensing platform based on metal oxide-loaded biopolymer for voltammetric measurement of hepatoprotective metadoxine drug in pharmaceutical formulation and human blood serum, *Biosens. Bioelectron.: X*, 2024, **17**, 100453.
84. S. S. Medany, M. A. Hefnawy, S. A. Fadlallah and R. M. El, Zinc oxide – chitosan matrix for efficient electrochemical sensing of acetaminophen, *Chem. Pap.*, 2024, 3049–3061.
85. L. Geng, J. Sun, M. Liu, J. Huang, J. Dong, Z. Guo, Y. Guo and X. Sun, Molecularly imprinted polymers-aptamer electrochemical sensor based on dual recognition strategy for high sensitivity detection of chloramphenicol, *Food Chem.*, 2024, **437**(2), 137933.
86. X. Tong, L. Jiang, Q. Ao, X. Lv, Y. Song and J. Tang, Highly stable glucose oxidase polynanogel@MXene/chitosan electrochemical biosensor based on a multi-stable interface structure for glucose detection, *Biosens. Bioelectron.*, 2024, **248**, 115942.
87. S. Lu, K. Zhang, Y. Liu, X. Zhan and R. Savari, Polymeric nanocomposite electrode for enhanced electrochemical detection of α-lipoic acid: Application in neuroinflammation prevention and clinical analysis, *Environ. Res.*, 2024, **245**, 117369.
88. G. Ma and G. Tian, Electrochemical determination of Methadone by use of chitosan and carbon nanostructured electrode: application to doping monitoring between athletes in biological and food samples, *J. Food Meas. Charact.*, 2024, **18**, 2030–2039.
89. M. R. Keerthanaa, L. R. Panicker, R. Narayan and Y. G. Kotagiri, Biopolymer-protected graphene-Fe_3O_4 nanocomposite based wearable microneedle sensor: toward real-time continuous monitoring of dopamine, *RSC Adv.*, 2024, **14**(10), 7131–7141.
90. M. H. Rahman, A. Rashed, N. I. Nayem, M. A. Rahaman, J. Ahmed, M. Faisal, M. Jalalah and F. A. Harraz, Nanogold-decorated reduced

graphene oxide/chitosan composite for electrochemical sensing of *N*-acetyl-4-aminophenol, *Mater. Chem. Phys.*, 2024, **314**, 128915.
91. X. Zhai, Q. Li, X. Zhao, N. Yao, L. He, D. Zhao, J. Sha, Z. Miao and S. Liu, Electrochemical sensing of bisphenol A based on carboxylated multi-walled carbon nanotubes-ferrocene-$H_5PMo_{10}V_2O_{40}$ with chitosan for signal amplification, *Diamond Relat. Mater.*, 2024, 142.
92. D. Shrestha, T. Nayaju, B. Kumar, B. Maharjan, K. Kang, P. M. Bacirhonde, C. H. Park and C. S. Kim, Fabrication of flexible glucose sensor based on heterostructure ZnO nanosheets decorated PU/Chitosan-PANI hybrid nanofiber, *Microchem. J.*, 2024, **197**, 109915.
93. M. Zhang, W. Tan, X. Wu, C. Wan, C. Wen, L. Feng, F. Zhang and F. Qu, A dual – functional cuprum coordination framework for high proton conduction and electrochemical dopamine detection, *Microchim. Acta*, 2024, **191**, 67.
94. F. Shi, S. Hu, J. Li, F. Wang and N. Chen, Glucose and UA sensing based on Cu nanoparticle decorated Nif/GO flexible electrode, *Microchim. Acta*, 2024, **191**, 7.
95. Z. Peng, Z. Li, J. Liao, Z. Zhang, Y. Song, C. Xia, Y. Xia and Z. Wang, Study of electrochemical biosensor for determination of glucose based on Prussian blue/gold nanoparticles-chitosan nanocomposite film sensing interface, *Meas. Sci. Technol.*, 2023, **35**, 015125.
96. J. Tang, L. Peng, A. Ali, S. Zhao, Z. Zeng, K. Yuan and S. Yao, Electrochemical detection of rutin in black tartary buckwheat tea and related health-care pills with new ionic liquid-based supramolecular hydrogels, *Food Control*, 2024, **155**, 110045.
97. G. Li, J. Li and Z. Yang, An electrochemical sensor based on graphene – chitosan – cyclodextrin modification for the detection of Staphylococcus aureus, *Carbon Lett.*, 2024, **34**, 495–504.
98. S. Guesmi, K. Moulaee, V. Bressi, H. Kahri, A. Khaskhoussi, C. Espro, H. Barhoumi and G. Neri, Non-enzymatic amperometric glucose sensing by novel Cu-MOF synthesized at room temperature, *Mater. Adv.*, 2024, **5**, 1160–1170.
99. D. Xiang, L. Zhao, Y. Wang, X. Li, X. Sun, J. Yang, H. Liu and C. Wan, A Highly Sensitive Non-Enzymatic Glucose Electrochemical Sensor Electrode Material of CuO Nanospheres/Activated Carbon Composites, *Russ. J. Electrochem.*, 2023, **59**, 1194–1205.
100. F. Khanmohammadi, B. M. Razavizadeh and E. Fooladi, Sensing and Bio-Sensing Research Application of electrochemical sensor modified by SBA-15/Fe_3O_4/polyaniline nanocomposite for determination of tyrosine in milk samples, *Sens. Bio-Sens. Res.*, 2023, **42**, 100602.
101. K. Postolović and Z. Stanić, Chitosan/TiO_2 nanoparticles modified carbon paste electrode as a sensitive voltammetric sensor for the determination of diclofenac sodium as an anti-inflammatory drug, *Mater. Today Commun.*, 2023, **37**, 107416.
102. S. Tang, A. Liang, M. Liu, W. Wang, F. Zhang and A. Luo, A glassy carbon electrode modified with a composite consisting of

electrodeposited chitosan and carboxylated multi - walled carbon nanotubes for simultaneous voltammetric determination of dopamine, serotonin and melatonin, *Carbon Lett.*, 2023, **33**, 2129–2139.

103. P. C. Ani, P. U. Nzereogu, A. C. Agbogu, F. I. Ezema and A. C. Nwanya, Cellulose from waste materials for electrochemical energy storage applications: A review, *Appl. Surf. Sci. Adv.*, 2022, **11**, 100298.
104. C. Bommier, R. Xu, W. Wang, X. Wang, D. Wen, J. Lu and X. Ji, Self-activation of cellulose: A new preparation methodology for activated carbon electrodes in electrochemical capacitors, *Nano Energy*, 2015, **13**, 709–717.
105. V. Molahalli, A. Sharma, K. Bijapur, G. Soman, N. Chattham and G. Hegde, Low-cost bio-waste carbon nanocomposites for sustainable electrochemical devices: A systematic review, *Mater. Today Commun.*, 2024, **38**, 108034.
106. D. Burhani, V. S. D. Voet, R. Folkersma, D. Maniar and K. Loos, Potential of Nanocellulose for Microplastic Removal: Perspective and Challenges, *Tetrahedron Green Chem.*, 2024, 100045.
107. M. Das, S. Lalsangi, S. Santra and R. Banerjee, Nanocellulose as a carrier for improved drug delivery: Progresses and innovation, *J. Drug Delivery Sci. Technol.*, 2024, **97**, 105743.
108. R. H. Althomali, K. A. Alamry, M. A. Hussein and R. M. Guedes, Hybrid PANI@dialdehyde carboxymethyl cellulose/ZnO nanocomposite modified glassy carbon electrode as a highly sensitive electrochemical sensor, *Diamond Relat. Mater.*, 2022, **122**, 108803.
109. L. Dong, S. Ren, X. Zhang, Y. Yang, Q. Wu and T. Lei, In-situ synthesis of Pt nanoparticles/reduced graphene oxide/cellulose nanohybrid for nonenzymatic glucose sensing, *Carbohydr. Polym.*, 2023, **303**, 120463.
110. A. Javed, B. Fatima, D. Hussain, S. E. Zahra Jawad, M. Subhan and M. Najam-ul-Haq, Effect of narcotic drugs on neurotransmitter: Electrochemical determination of heroin and dopamine by graphene oxide/carboxymethylcellulose/magnesium oxide nanohybrid membrane, *J. Mol. Liq.*, 2023, **384**, 122154.
111. Y. Zhang, C. Zhang, W. Chen and Z. Liu, One-step laser-induced Cu-embedded graphene for non-enzymatic glucose sensing in beverages, *J. Alloys Compd.*, 2024, **992**, 174563.
112. X. Zhang and Q. Zhang, A molecular imprinting electrochemical sensor for detection of anticancer drug amsacrine, *Sens. Rev.*, 2024, **3**, 388–394.
113. A. Y. El-Moghazy, N. Amaly, G. Istamboulie, N. Nitin and G. Sun, A signal-on electrochemical aptasensor based on silanized cellulose nanofibers for rapid point-of-use detection of ochratoxin A, *Microchim. Acta*, 2020, **187**(9), 535.
114. N. O. Gomes, E. Carrilho, S. A. S. Machado and L. F. Sgobbi, Bacterial cellulose-based electrochemical sensing platform: A smart material for miniaturized biosensors, *Electrochim. Acta*, 2020, **349**, 136341.
115. D. Büyüktaş, M. Ghaani, C. Rovera, D. Carullo, R. T. Olsson, F. Korel and S. Farris, A screen-printed electrode modified with gold

nanoparticles/cellulose nanocrystals for electrochemical detection of 4,4′-methylene diphenyl diamine, *Heliyon*, 2023, **9**(4), e15327.
116. X. Huang, Z. Ni, H. Su, Y. Shang, H. Liu, Y. He, H. Meng and Y. Dong, Cellulose nanocrystalline and sodium benzenesulfonate-doped polypyrrole nano-hydrogel/Au composites for ultrasensitive detection of carcinoembryonic antigen, *New J. Chem.*, 2021, **45**, 5551–5560.
117. S. Akhter, M. Shalauddin, W. J. Basirun, V. S. Lee, S. R. Ahmed, A. R. Rajabzadeh and S. Srinivasan, A highly selective bifunctional nanosensor based on nanocellulose and 3D polypyrrole decorated with silver-gold bimetallic alloy to simultaneously detect methotrexate and ciprofloxacin, *Sens. Actuators, B*, 2022, **373**, 132743.
118. J. Zhou, F. Lu and Z. Wu, Effects of a plasma jet on electrochemical properties of silk fibroin hydrogel doped with graphene oxide, *Polym. Polym. Compos.*, 2022, **30**, 1–12.
119. D. Molinnus, A. Drinic, H. Iken, N. Kröger, M. Zinser, R. Smeets, M. Köpf, A. Kopp and M. J. Schöning, Towards a flexible electrochemical biosensor fabricated from biocompatible Bombyx mori silk, *Biosens. Bioelectron.*, 2021, **183**, 1–9.
120. A. Seves, M. Romanò, T. Maifreni, S. Sora and O. Ciferri, The microbial degradation of silk: A laboratory investigation, *Int. Biodeterior. Biodegrad.*, 1998, **42**, 203–211.
121. Q. Chen, Y. Liu, K. Gu, J. Yao, Z. Shao and X. Chen, Silk-Based Electrochemical Sensor for the Detection of Glucose in Sweat, *Biomacromolecules*, 2022, **23**, 3928–3935.
122. Q. Wang, P. McArdle, S. L. Wang, R. L. Wilmington, Z. Xing, A. Greenwood, M. L. Cotten, M. M. Qazilbash and H. C. Schniepp, Protein secondary structure in spider silk nanofibrils, *Nat. Commun.*, 2022, **13**, 1–12.
123. S. Ling, C. Li, J. Adamcik, S. Wang, Z. Shao, X. Chen and R. Mezzenga, Directed growth of silk nanofibrils on graphene and their hybrid nanocomposites, *ACS Macro Lett.*, 2014, **3**, 146–152.
124. K. Satheesh and R. Jayavel, Synthesis and electrochemical properties of reduced graphene oxide via chemical reduction using thiourea as a reducing agent, *Mater. Lett.*, 2013, **113**, 5–8.
125. D. Kuang, W. Yu, J. Liu, Y. Yin and C. Wang, Flexible MWCNT/Silk fibroin film decorated with Pt NPs for electrochemical glucose sensors, *Microchem. J.*, 2023, **191**, 108760.
126. W. Chang, Y. Zhu, Y. Ma, Z. Zheng and C. Wang, Silk derived Fe/N-Doping porous carbon nanosheets for chloramphenicol electrochemical detection, *Curr. Anal. Chem.*, 2022, **18**, 1017–1028.
127. S. V. Selvi, A. Prasannan, H. Yu, V. Lincy and P.-D. Hong, Bio-mineralized tin/bismuth oxide nanoparticles with silk fibroins for efficient electrochemical detection of 2-nitroaniline in river water samples, *Environ. Res.*, 2023, **221**, 115285.
128. Y. Wang, J. Deng, H. Liu, R. Ran, T. Zhang and Z. Li, Enzyme-chitosan hydrogels for high sensitivity flexible silk-based electrochemical glucose sensor, *Microchem. J.*, 2024, **200**, 110306.

129. Q. Chen, Y. Liu, K. Gu, J. Yao, Z. Shao and X. Chen, Silk-Based Electrochemical Sensor for the Detection of Glucose in Sweat, *Biomacromolecules*, 2022, **23**, 3928–3935.
130. A. Noor, H. Khalid, M. Aslam, A. Hayat, A. F. Khan, M. Nasir, A. A. Chaudhry and M. H. Nawaz, Graphene oxide reinforced silk fibroin nanocomposite as an electroactive interface for the estimation of dopamine, *RSC Adv.*, 2022, **12**, 29319–29328.

CHAPTER 10

Role of Low-carbon Materials in Construction and Infrastructure

NEETU VISHWAKARMA,[a] POOJA SHARMA,[b] UPASANA BAGRI,[c] SHALINI SINGH CHHETRI,[c] MONIKA SINDHU,[d] MEENAKSHI CHOUDHARY[e] AND AJIT SHARMA*[c]

[a] Department of Physics, School of Chemical Engineering and Physical Sciences, Lovely Professional University, Punjab, 144001, India; [b] Department of Mechanical Engineering, CEMMPRE, University of Coimbra, R. Luis Reis dos Santos, 3030-788, Coimbra, Portugal; [c] Department of Chemistry, School of Chemical Engineering and Physical Science, Lovely Professional University, Punjab, 14411, India; [d] Department of Chemistry, University Centre for Research and Development, Chandigarh University, Chandigarh, Punjab, 140413, India; [e] Department of Research and Innovation, Maya Dcvi University, Dehradun, Uttarakhand, 248011, India

*Email: ajitsharma2003@gmail.com

10.1 Introduction

The manufacturing industry is the greatest single consumer of energy worldwide. Furthermore, the construction industries are the largest world-wide consumers of materials.[1,2] Accordingly, buildings account for 19% of the emissions of greenhouse gases worldwide. According to recent research, buildings provide the best short-term reduction prospects for lowering

Low Carbon Materials: Paving a Sustainable Path for a Greener Future
Edited by Santanu Patra, Sudheesh K. Shukla and Mika Sillanpää

Published by the Royal Society of Chemistry, www.rsc.org

emissions of greenhouse gases.[3] As a result, regulations demanding upgrades to building performance and fabric have been introduced by policymakers. One example of this is the Energy Performance of Buildings Directive from the European Union (EU).[3,4] These laws have mostly addressed the operational production of greenhouse gases linked to the consumption of energy for tasks such as lighting, space heating, and cooling. Regulators have not, however, included the greenhouse gases emissions connected to the initial manufacturing of structures.[5,7]

Embodied carbon can contribute to whole-life carbon emissions in any range from 2% to 80%, according to a recent study of building life cycle evaluations.[6] The exact ratio is determined by a variety of factors, such as the building's intended purpose, location, the type of materials used, and the service life and projected energy supply. Certain types of structures, such as commercial warehousing, tend to have higher proportions of embedded emissions, with up to 90% of the total coming from these sources.[7] Because of enhanced operational performance and decreased carbon intensity of the power supply, it is anticipated that the portion of life cycle emissions related to embedded carbon will increase more with decreases in operational emissions.[7–9] Conversely, with more effective buildings usually needing more material use and an anticipated increase in building activity, absolute rises in embodied carbon can be projected.

Considering the temporal distribution of emissions adds even more significance to embodied emissions. Because annual emissions are not the primary factor preventing unacceptably high levels of environmental change, some scientists have argued that in policy-making and economic evaluations, a higher weighting should be given to current emissions savings rather than emissions savings in the future.[8,9] Building assessments that take into account the temporal distribution of emissions have shown that, in the time horizon pertinent to adopted environmental mitigation objectives, the rapid release of emissions, or a "carbon spike," associated with emissions during the construction period can outweigh life cycle emissions. The same authors later questioned the value of developing additional projects as a way to mitigate climate change as a result of this conclusion. Still, new development is frequently inevitable, and in these situations, there needs to be a stronger emphasis on a reduction of embodied carbon.[10–12]

Including the effects of upstream power generation, the building and construction industries globally account for about 40% of energy-associated carbon dioxide emissions associated with the construction and operation of structures. The effects of embodied carbon in construction materials and goods are often not addressed by current building rules, but they do address operational energy. When combined across all industrial sectors, management of materials which includes both manufacturing and collection is responsible for over 50% of all greenhouse gas emissions.

These embodied consequences associated with the production of building materials grow more significant as building activities become more efficient.

The world's populations will double the quantity of building floor area between now and 2060, which is the same as creating the population of New York on a monthly basis for 40 years. Embodied carbon, or the emissions related to the production of building materials and construction, will account for a large portion of the carbon footprint of these new structures.

Embodied carbon emissions are those that are irreversibly released into the atmosphere as soon as a building is constructed, when compared to operational emissions of carbon dioxide, which can be decreased over time *via* energy-efficient building upgrades and the consumption of renewable energy.[13–15]

The Route Map for Low-carbon Green Construction published in 2013 highlighted that achieving a sustainable built environment requires integrating substantial reductions in operational emissions with a targeted 39% decrease in carbon dioxide emissions by 2050, relative to the 2010 baseline.[16,17] Prologis, the first for-profit corporation to tackle this problem, assessed and mitigated 110% of the embodied carbon linked to the construction of the distribution facility in Pineham in 2007. Starting in 2009, they standardized this methodology for all UK operations.[18,19] The other major clients, including British Land, the Crown Estate, and Land Securities, have followed suit and now demand an evaluation of the embodied carbon on all projects with significant value. Numerous businesses showcasing cost benefits *via* embedded carbon control have reinforced the commercial case.[20] Embedded carbon was included as a permissible solution under the previously established no carbon construction standards due to the efforts of a group of prominent practitioners in the UK because of this increasing demand.[21–23] Planning applications must already include a rudimentary estimate of embodied carbon, according to a number of local authorities. Additionally, whole-life carbon assessment has received extra credit in sustainability assessment programs in the UK like DGNB, BREEAM (Building Research Establishment Environmental Assessment Method), LEED (Leadership in Energy and Environmental Design), and Green Star. As of this writing, the publicly accessible database of embodied carbon, named WRAP, which was introduced in 2014, has evaluations for about 231 projects and is expected to grow further.[24,25] It is obvious that embodied carbon mitigation and assessment will soon be taken into regular account on a significant number of major construction projects.

The need for construction infrastructure and building has expanded significantly along with economic expansion, industrialization, and urbanization, particularly in the transportation and housing areas.[26,27] Chinese Ministry of Construction data show that between 1980 and 2008, the country's total building floor area increased from 10.2 billion square meters to 52.7 billion square meters.[28–30] Additionally, between 2005 and 2020, new structures with a total area of 30 billion square meters would be developed. Moreover, in reaction to the 2008 global financial crisis, the Chinese government approved a stimulus program expected to be worth four trillion RMB. The majority of the funds were allocated to different construction projects for infrastructure.[31,32] The global change in the environment has been increased by urbanization, which is the primary source of carbon dioxide emissions.[33] Building construction as well as the upkeep of railway and road networks has required substantial quantities of building supplies. These demands and actions have had a negative influence on the environment, particularly on the manufacture of cement and steel.[34,35] The continuous manufacture of steel and cement would increase the release of carbon dioxide and demand a large number of raw materials. Resources for basic materials are few, though. Only 25 billion tons of suitable limestone, which is needed to create cement, are reserved in China.[36,37]

Reducing raw material usage and emissions of carbon dioxide is vital and important to preserve the environment and slow down global warming. In this particular setting, the following questions must be addressed: In the future, how much cement and steel will be used to create structures as well as transportation infrastructure? What is the level of carbon dioxide emissions going to be? To contribute to the creation of a low-carbon, dematerialized society, how can we limit the use of cement and steel and minimize carbon dioxide emissions in the construction and infrastructure sectors? This study's primary goal is to evaluate future material flow stock and associated carbon dioxide emissions estimates for the construction of transport and building infrastructure. To achieve a low-carbon, dematerialized society, it will be imperative to extend the lifespan of transport infrastructure and buildings and to create a civilization that consumes fewer resources.[38,39] In a case study of buildings in Nagoya City, Japan, the consequences of extending a building's lifespan were previously calculated.[40] This study employed scenario analysis to assess the impact of extending the lifespan of buildings and infrastructure projects on lowering material consumption and greenhouse gas emissions. It also offered recommendations for achieving a low-carbon, dematerialization society (Table 10.1).[41,42]

10.2 Literature Review of Earlier Work

The obstacles to specific types of "environmentally friendly construction" or "beneficial building" have been the subject of several earlier studies. While some of these studies define sustainability broadly and include social and economic components, others have concentrated only on the environmental components of viability.[43–45] The embodied emissions of building materials have not been included in these studies, which have a tendency to focus only on operating emissions, the implementation of technologies that are energy-efficient, or the accomplishment of regulatory desires like zero-carbon homes.[46–48] Very few, if any, have concentrated particularly on the obstacles to selecting different materials in order to reduce carbon emissions that are embodied. As a result, the review that follows incorporates material from two different streams. Studies that shed light on the institutional and cultural hurdles impeding sustainable development in the construction sector are

Table 10.1 Different low-carbon materials applied in the construction and infrastructure sectors.

Material	Description	Benefits
Recycled materials	Materials recovecred from waste streams	Reduces waste, conserves resources, lowers carbon emissions
Bio-based materials	Materials derived from renewable biological sources	Reduces dependence on fossil fuels, improves indoor air quality, enhances sustainability
Geopolymer concrete	Concrete that uses alternative binders instead of Portland cement	Lower carbon emissions, enhanced durability
Cross-laminated timber (CLT)	Engineered wood product made of layers of wood glued together	High strength, durability, and fire resistance
Energy-efficient insulation	Materials that reduce heat transfer	Improves energy efficiency, lowers carbon emissions
Solar-powered building materials	Materials that integrate solar panels or other energy-harvesting technologies	Generates renewable energy, reduces reliance on grid power
Low volatile organic compounds (VOC) paints and coatings	Paints and coatings with low levels of VOCs	Improves indoor air quality, reduces environmental pollution
Green roofing materials	Materials that support vegetation on roofs	Improves insulation, reduces stormwater runoff, enhances biodiversity
Sustainable wood products	Wood products from sustainably managed forests	Supports forest conservation, reduces deforestation

included in the first stream.[49,50] The second stream includes in-depth research on the use of particular substitute materials, such as straw bale. In order to estimate the use of materials and emissions of carbon dioxide, we used comprehensive data on the material utilization intensity of infrastructure and buildings in this study.[51,52]

10.3 Innovation Challenges in the Construction Sector

The construction business is supplier-driven, extremely fragmented, and risk averse. The majority of construction companies are small-scale enterprises with limited resources for research, development, and innovation.[53,54] Few businesses are able to thoroughly evaluate every aspect of a new material, and the ability to capitalize on new technology often depends on the availability of skilled and experienced personnel. Because of the nature of the business, it is necessary to switch between short-term projects, many of which have a distinct identity and a constantly changing cast of stakeholders. As a result, professionals learn *via* each project they work on, gaining perspectives and abilities from their unique experiences.[55,56] This causes inventions to spread slowly and causes delayed incremental progress. The inadequate transfer of knowledge from academics to industry further impedes the procedure of knowledge creation.[57,58] This resistance to innovation is reinforced by a litigious industry climate, which makes it necessary for new building products to obtain acceptability and legitimacy through a large volume of pre-implementation proof.[59,60] A person's capacity to make comprehensive project judgments is frequently weakened by the typically fragmented, multi-party construction process. Therefore, often, particular stakeholders believe they can't implement long-term fixes "down the line."[61] It has long been recognized that an organized strategy to construction must replace traditional partisan ties.[62–64] Nevertheless, contractual arrangements frequently prevent the supply chain and design teams from integrating effectively. Unfortunately, the expertise of every stakeholder cannot be properly utilized unless there is more communication and early supply chain engagement. Decisions on projects are frequently made too late for practical or cost-effective implementation without this early engagement.[65,66]

The current financial situation has constrained demand for new low-carbon buildings, particularly recent market shifts. Clients often prioritize project cost, timeliness, performance, and aesthetics.[67–69] Consequently, a glaring dilemma exists. There must be a decrease in the amount of carbon that is embodied in emissions connected to the usage of building materials. This calls for the widespread use of substitute materials in an industry that has historically and legitimately shied away from innovation. This transformation must begin without significant client, financial, or regulatory incentives. Individuals' tenacity and dedication continue to be the primary drivers of advancements in the sector in this vacuum.[70,71]

10.4 Challenges Preventing the Building Sector from Using Substitute Materials

There are several factors that hinder the process of choosing novel materials. Specifically, there are four issues: there are no short- to medium-term economic benefits; there is inadequate marketing; professionals in engineering are not adequately informed about novel materials; and there are no comprehensive demonstration projects or supporting material performance information. The authors contend that in addition to effective marketing and involvement of stakeholders, this can be countered by including design guidelines.[72,73]

Material decisions are usually influenced by other objectives, even in situations when there is enough knowledge and example projects available. A global study of design teams carried out in 2012 by Arup on behalf of the World Business Council for Sustainable Development showed that, while many factors

Table 10.2 Typical obstacles to the adoption of less carbon-embodied building materials.

Institution and habitual	Commercial	Performance and technical related aspects	Understanding and views
The departmental locking in to known materials is the outcome of targeted training and recruiting	▪ Premium for uncertainty applied to new items ▪ High transaction expenses for further study and professional growth	• Absence of relevant performance data • Absence of large-scale demonstration initiatives	Ignorance of options by clients and unfavorable opinions held by practitioners
Appropriate material palette is encouraged by the culture of institutions and established procedures	▪ Exorbitant price of new goods ▪ Market externalization of integrated embedded emission costs	Absence of design guidelines, instruments and standards details	Practitioner's ignorance and lack of experience with alternatives
Due to time restrictions, favored designs are chosen over alternatives	Funds invested in pre-existing resources (for example, in supply chain connections, training, *etc.*)	Limitations and constraints imposed by policy and laws	Client's unfavorable opinions
Customary specialization and past actions of individual practitioners	Financing availability for medium-sized enterprises	Materials and technologies used are those available locally	Uncertainty in policy
Insufficient user–producer connections	Reluctance to assume the related economic risk	Ineffectively constructed supply networks	Selection is impeded by perceived worries about the sourcing of materials
Material choice is thought to be independent of the practitioner's influence	The financing of the project is incompatible with the timeline	Having trouble getting insurance for new and recycled materials	Simply viewed as low importance, with other factors taking prominence

contribute to substance choice, cost is the primary consideration, and sustainable material criteria are frequently less important than the project team's prior knowledge and experiences.[74] A previous analysis of this collection of work is produced as a list of several typical obstacles that are compiled in Table 10.2.[40]

A material's applicability and sustainability are largely determined on the spot and specific to the project variables. The option with the lowest embodied carbon will vary depending on the type of structure and the specific project requirements. The promotion of the best choice for each unique project must be the ultimate objective of legislators and low-carbon building promoters. Thus, it is crucial to simultaneously promote a wide range of material possibilities. This calls for the development of skills and regulations that acknowledge and encourage the wide range of choices available. As a result, while evaluating the obstacles to the adoption of specific materials is important, studies also need to find common points of leverage and strategies that promote a variety of solutions. This study uses a survey and interviews with top building professionals to further investigate these prevalent hurdles and opportunities for leverage.

10.5 Study Goals and Limitations

This study's main goal was to comprehend the practical, cultural, technological, and financial obstacles that keep building industry professionals from choosing among the many materials that are frequently recognized as having reduced embodied carbon content. It also aimed to comprehend how advocacy organizations, professional associations, and regulations may help remove these obstacles. In the building business, embodied carbon is still an uncommon subject despite recent advances in understanding. It will likely take many years before embedded carbon quantification and prevention are standard practice in all industries and organizations. Consequently, the purpose of this study was not to enlist individuals who would make up an accurate representation of the construction sector worldwide.[75] Rather, it was aimed at those who had a lot of experience with low-carbon materials.[76] The opinions and firsthand knowledge of these early-believers will ultimately influence the manufacturing sector's future embodied carbon reduction strategy. As a result, knowing their goals and backgrounds can help in creating policies and recommendations for the industry across all sectors.[77]

The supply chains in the construction business are usually complex and long term, encompassing many different professions. Numerous actors possess essentially distinct incentives and priorities. The goal was made to restrict the study's participants to experts in the planning, specification, and building processes. Developers, final consumers, or material producers were not the target audience for the research or interviews. Additional targeted research centered on these groups' perspectives would be very beneficial.

10.6 Methods

The study used a mixed-methods qualitative approach that included a survey and several semi-structured interviews. This methodology is widely employed in several fields and was chosen to offer the intended blend of comprehensiveness and depth.[78]

A sequential explanatory design was chosen, in which a survey was used to collect preliminary qualitative as well as quantitative information on the adoption hurdles, and then in-depth interviews were conducted to further explore the barriers that were found. A Qualtrics-hosted online survey was accessible from April 3rd to May 23rd, 2014. The most suitable format was determined to be an open discussion forum since it made it simple for practitioners to distribute the survey and increase the quantity and variety of replies. Links to the poll were sent to a specific group of people with a lot of experience, well-established contacts, LinkedIn groups, and many prominent industry mailing lists.

The main objectives of the study were to collect demographic information, determine the perceived impact and accountability of different professions on material choice and carbon footprint reduction, learn about the opinions of the participants using a variety of 24 low-carbon example materials, and investigate perceived obstacles and motivators for low-carbon material adoption.[79] Respondents were prompted to think about more broad hurdles and motivations to alternate materials in addition to the questions concerning individual materials.[80] Every time an observer was asked to select an item from a predetermined list, they were given the ability to add more choices and add comments.

10.7 Conclusions

This study's main goal was to comprehend the practical, cultural, technological, and financial obstacles that keep building industry professionals from choosing among the many materials that are frequently recognized as having reduced embodied carbon content. An analysis of earlier research evaluating operational mitigation of emissions and environmentally friendly practice adoption hurdles in the construction sector identified a shared set of institutional and cultural constraints. The findings in this chapter strongly imply that these obstacles also hinder the selection of substitute materials as a strategy for reducing embodied emissions of carbon. Many of these obstacles are similar for all materials, and they include low expertise and skill levels, opinions regarding high costs, insufficient time for design to consider novel alternatives, insufficient information from material manufacturers, and the incapacity to create an efficient or group chain of responsibility. The lack of readily available product and construction level carbon information as well as benchmarks are another issue that hinders design teams. By promoting earlier supply chain engagement making efficient use of whole-life pricing, and making adjustments to agreements and tender documentation, the construction sector might try to overcome these obstacles. Many practitioners need further training, and companies doing their first carbon footprint evaluations need to set up internal processes to make sure that knowledge is transferred from project to project. The industry needs to disseminate the body of knowledge regarding embodied carbon as well. To enable benchmarking, this entails posting data to shared repository systems, exchanging standardized submission forms, and publicly talking about their achievements and shortcomings. Instead of waiting for legislation to take action, the sector needs to keep strengthening the expanding business case and actively encourage clients to participate in assessments.

Professional organizations have a responsibility to support the exchange of knowledge between businesses and to create an embodied carbon community. Institutions can support important demonstration initiatives, offer training programs and advice, collect cost information on their own to support the business case, and assist in sharing the knowledge gained by early actors. However, ultimately, in order to build on the early efforts of moral leaders, control will also be necessary. The industry is beginning to acknowledge in several large nations that embodied carbon needs to be regulated in some way. By establishing a long-term course for this kind of legislation, governments can offer clarification.

Embodied carbon reduction has a clear chance to contribute significantly to many countries efforts to fulfill their increasingly demanding carbon finances. Prompt action and regulation could aid in the rapid creation of knowledge, the collection of data more quickly, and the expansion of a business with substantial export potential. Early participants have a great chance to take the lead globally in a sector that is expanding, creating jobs for skilled labor, expanding the market for substitute materials, and achieving notable GHG emission reductions.

Moreover, neither the impact of reducing the intensity of material usage nor the recycling of waste building materials was estimated in this study. A comprehensive program for the future should address the repurposing of waste materials for construction and the application of technological advancements to reduce the intensity of consumption of raw materials.

Low-carbon building and infrastructure materials have an excellent future. As climate change becomes more widely recognized, there is a growing need for sustainable solutions. The use of low-carbon materials will be fueled by developments in technology, legislation that is supportive of them, growing consumer awareness, and integration with the circular economy. The UN Sustainable Development Goals, which include cheap and clean energy, sustainable cities and communities, and climate action, may be greatly aided by these resources. We can design a constructed environment that is more robust and sustainable, benefiting both people and the environment, by utilizing the promise of low-carbon materials.

References

1. S. D. I. R. du Can and L. Price, Sectoral trends in global energy use and greenhouse gas emissions, *Energy Policy*, 2008, **36**(4), 1386–1403.

2. F. Krausmann, S. Gingrich, N. Eisenmenger, K. H. Erb, H. Haberl and M. Fischer-Kowalski, Growth in global materials use, GDP and population during the 20th century, *Ecol. Econ.*, 2009, **68**(10), 2696–2705.
3. Change, I. C., Synthesis Report. Contribution of working groups I. II and III to the fifth assessment report of the intergovernmental panel on climate change, *151*(10.1017), 2014.
4. T. Nauclér and P. A. Enkvist, Pathways to A Low-Carbon Economy: Version 2 of the Global Greenhouse Gas Abatement Cost Curve, 2009.
5. En, C. E. N., 15978: 2011 Sustainability of Construction Works—assessment of environmental performance of buildings—calculation method, CEN, Brussels, Belgium, 2011.
6. T. Ibn-Mohammed, R. Greenough, S. Taylor, L. Ozawa-Meida and A. Acquaye, Operational *vs.* embodied emissions in buildings—A review of current trends, *Energy Build.*, 2013, **66**, 232–245.
7. S. Sturgis and G. Roberts, *Redefining zero: Carbon profiling as a solution to whole life carbon emission measurement in buildings*, RICS Research, London, UK, 2010.
8. H. D. Matthews, S. Solomon and R. Pierrehumbert, Cumulative carbon as a policy framework for achieving climate stabilization, *Philos. Trans. R. Soc., A*, 2012, **370**(1974), 4365–4379.
9. J. Rhys, *Cumulative Carbon Emissions and Climate Change: Has the Economics of Climate Policies Lost Contact with the Physics?* Oxford Institute for Energy Studies, 2011.
10. J. Heinonen, A. Säynäjoki and S. Junnila, A longitudinal study on the carbon emissions of a new residential development, *Sustainability*, 2011, 3(8), 1170–1189.
11. A. Säynäjoki, J. Heinonen and S. Junnila, A scenario analysis of the life cycle greenhouse gas emissions of a new residential area, *Environ. Res. Lett.*, 2012, 7(3), 034037.
12. The Green Construction Board, Low Carbon Routemap for the UK Built Environment, 2013.
13. J. Giesekam, J. R. Barrett and P. Taylor, Construction sector views on low carbon building materials, *Build. Res. Inform.*, 2016, **44**(4), 423–444.
14. J. Giesekam, J. Barrett, P. Taylor and A. Owen, The greenhouse gas emissions and mitigation options for materials used in UK construction, *Energy Build.*, 2014, **78**, 202–214.
15. Union, I., Communication from the Commission to the European Parliament, the Council, the European Economic and Social Committee and the Committee of the Regions, Brussels, 2014, http://ww.w.xploit-eu.com/pdfs/Europe,202020, 20.
16. N. Watson, P. Walker, A. Wylie and C. Way, Evaluating the barriers to entry for non-conventional building materials, in *IABSE 2012 Spring Conference on Global Thinking in Structural Engineering: Recent Achievements*, International Association for Bridge and Structural Engineering (IABSE), 2012, pp. 254–255

17. H. Bergsdal, R. A. Bohne and H. Brattebø, Projection of construction and demolition waste in Norway, *J. Ind. Ecol.*, 2007, **11**(3), 27–39.
18. Y. Ma and S. M. Bao, Status quo of building energy conservation and green building in China, China Construction Newspaper, 2006, vol. 29, p. 1.
19. W. M. Morrison, *China's Economic Conditions*, DIANE Publishing, 2010.
20. N. B. Grimm, S. H. Faeth, N. E. Golubiewski, C. L. Redman, J. Wu, X. Bai and J. M. Briggs, Global change and the ecology of cities, *Science*, 2008, **319**(5864), 756–760.
21. Y. Q. Han, Thinking about the circle economy for Chinese cement industry, *Build. Mater. Dev.*, 2008, **1**, 14–20.
22. H. Okamoto, *Change to stock society: Infrastructure construction in a period of long lifetime*, Kajima Press, 2006
23. H. Tanikawa, A. Onishi, Y. Takahira, S. Hashimoto, O. Higashi, H. Shirakawa and H. Imura, Four-dimensional mapping of building material stock and energy consumption for stock-type and low carbon society: a case study of buildings in Nagoya city, *J. Life Cycle Assess., Jpn.*, 2010, **6**(2), 92–101.
24. K. Williams and C. Dair, What is stopping sustainable building in England? Barriers experienced by stakeholders in delivering sustainable developments, *Sustainable Dev.*, 2007, **15**(3), 135–147.
25. T. Kershaw and S. Simm, Thoughts of a design team: Barriers to low carbon school design, *Sustainable Cities Soc.*, 2014, **11**, 40–47.
26. J. Pinkse and M. Dommisse, Overcoming barriers to sustainability: an explanation of residential builders' reluctance to adopt clean technologies, *Business Strategy Environ.*, 2009, **18**(8), 515–527.
27. M. Osmani and A. O'Reilly, Feasibility of zero carbon homes in England by 2016: A house builder's perspective, *Build. Environ.*, 2009, **44**(9), 1917–1924.
28. S. K. Arora, R. W. Foley, J. Youtie, P. Shapira and A. Wiek, Drivers of technology adoption—the case of nanomaterials in building construction, *Technol. Forecase. Soc.*, 2014, **87**, 232–244.
29. M. Osmani and A. O'Reilly, Feasibility of zero carbon homes in England by 2016: A house builder's perspective, *Build. Environ.*, 2009, **44**(9), 1917–1924.
30. J. Pinkse and M. Dommisse, Overcoming barriers to sustainability: an explanation of residential builders' reluctance to adopt clean technologies, *Business Strategy Environ.*, 2009, **18**(8), 515–527.
31. A. Moncaster, D. Hinds, H. Cruickshank, P. M. Guthrie, N. Crishna, K. Baker *et al.*, Knowledge exchange between academia and industry, in *Proceedings of the Institution of Civil Engineers-Engineering Sustainability*, Thomas Telford Ltd, 2010, vol. 163, no. 3, pp. 167-174.
32. J. Egan, *Rethinking construction*, Published by Department of Trade and Industry, URN 03/951, London, 1998.

33. Royal Academy of Engineering, *Engineering a low carbon built environment. The discipline of Building Engineering Physics*, 2010.
34. T. Kershaw and S. Simm, Thoughts of a design team: Barriers to low carbon school design, *Sustainable Cities Soc.*, 2014, **11**, 40–47.
35. J. Persson and S. Grönkvist, Drivers for and barriers to low-energy buildings in Sweden, *J. Cleaner Prod.*, 2015, **109**, 296–304.
36. L. Wang, A. Toppinen and H. Juslin, Use of wood in green building: a study of expert perspectives from the UK, *J. Cleaner Prod.*, 2014, **65**, 350–361.
37. S. Gold and F. Rubik, Consumer attitudes towards timber as a construction material and towards timber frame houses–selected findings of a representative survey among the German population, *J. Cleaner Prod.*, 2009, **17**(2), 303–309.
38. C. Zhang and L. Canning, Application of non-conventional materials in construction, *Proc. Instit. Civil Eng.-Constr. Mater.*, 2011, **164**(4), 165–172.
39. Arup & World Business Council for Sustainable Development (WBCSD), *Material choice for green buildings*, 2012, 201201.
40. J. Giesekam, J. Barrett, P. Taylor and A. Owen, The greenhouse gas emissions and mitigation options for materials used in UK construction, *Energy Build.*, 2014, **78**, 202–214.
41. A. Tashakkori and C. Teddlie, *Handbook of mixed methods in social & behavioral research*, Sage Publications, 2003, p. 0761920730.
42. N. Watson, P. Walker, A. Wylie and C. Way, Evaluating the barriers to entry for non-conventional building materials, in *IABSE 2012 Spring Conference on Global Thinking in Structural Engineering: Recent Achievements*, International Association for Bridge and Structural Engineering (IABSE), 2012, pp. 254-255.
43. Arup & World Business Council for Sustainable Development (WBCSD), Material choice for green buildings, 2012, 20120.
44. J. Giesekam, J. R. Barrett and P. Taylor, Construction sector views on low carbon building materials, *Build. Res. Inform.*, 2016, **44**(4), 423–444.
45. F. Shi, T. Huang, H. Tanikawa, J. Han, S. Hashimoto and Y. Moriguchi, Toward a low carbon–dematerialization society: Measuring the materials demand and CO_2 emissions of building and transport infrastructure construction in China, *J. Ind. Ecol.*, 2012, **16**(4), 493–505.
46. G. Churkina, A. Organschi, C. P. Reyer, A. Ruff, K. Vinke and Z. Liu, *et al.*, Buildings as a global carbon sink, *Nat. Sustainability*, 2020, **3**(4), 269–276.
47. J. Hildebrandt, N. Hagemann and D. Thrän, The contribution of wood-based construction materials for leveraging a low carbon building sector in Europe, *Sustainable Cities Soc.*, 2017, **34**, 405–418.
48. T. C. Ling and M. Tiong, Developing carbon-neutral construction materials using wastes as carbon sink, in *ISRM International*

Symposium-Asian Rock Mechanics Symposium, ISRM, 2021, p. ISRM-ARMS11.
49. A. Amiri, J. Ottelin, J. Sorvari and S. Junnila, Cities as carbon sinks—classification of wooden buildings, *Environ. Res. Lett.*, 2020, **15**(9), 094076.
50. S. A. Miller, E. Van Roijen, P. Cunningham and A. Kim, Opportunities and challenges for engineering construction materials as carbon sinks, *RILEM Technical Lett.*, 2021, **6**, 105–118.
51. N. Alaux, M. R. M. Saade, E. Hoxha, B. Truger and A. Passer, Future trends in materials manufacturing for low carbon building stocks: A prospective macro-scale analysis at the provincial level, *J. Cleaner Prod.*, 2023, **382**, 135278.
52. A. Laveglia, N. Ukrainczyk, N. De Belie and E. A. Koenders, From Quarry to Carbon Sink: Process-based LCA Modelling of Lime-based Construction Materials for Net-Zero and Carbon-Negative Transformation, *Green Chem.*, 2024, 6584–6600.
53. S. Uzan, *Design of a low carbon building: Case study of an architectural competition*, 2019.
54. F. P. Campo and M. Grosso, Lime Based Construction Materials as a Carbon Sink, *Key Eng. Mater.*, 2022, **922**, 139–145.
55. L. R. Caldas, A. B. Saraiva, A. F. Lucena, M. H. Y. Da Gloria, A. S. Santos and R. D. Toledo Filho, Building materials in a circular economy: The case of wood waste as CO_2-sink in bio concrete, *Resour., Conserv. Recycl.*, 2021, **166**, 105346.
56. R. Guo, Y. Zhao, Y. Shi, F. Li, J. Hu and H. Yang, Low carbon development and local sustainability from a carbon balance perspective, *Resour., Conserv. Recycl.*, 2017, **122**, 270–279.
57. S. Ramakrishna, M. Pervaiz, J. Tjong, P. Ghisellini and M. M. Sain, Low-carbon materials: genesis, thoughts, case study, and perspectives, *Circ. Econ. Sustainability*, 2021, 1–16.
58. X. Liu, Low-carbon utilization of coal gangue under the carbon neutralization strategy: a short review, *J. Mater. Cycles Waste Manage.*, 2023, **25**(4), 1978–1987.
59. A. M. Omer, Focus on low carbon technologies: The positive solution, *Renewable Sustainable Energy Rev.*, 2008, **12**(9), 2331–2357.
60. F. G. Li, C. Bataille, S. Pye and A. O'Sullivan, Prospects for energy economy modelling with big data: Hype, eliminating blind spots, or revolutionising the state of the art?, *Appl. Energy*, 2019, **239**, 991–1002.
61. X. Wang, G. Zhao, C. He, X. Wang and W. Peng, Low-carbon neighborhood planning technology and indicator system, *Renewable Sustainable Energy Rev.*, 2016, **57**, 1066–1076.
62. S. Gupta, A. Kashani, A. H. Mahmood and T. Han, Carbon sequestration in cementitious composites using biochar and fly ash–Effect on mechanical and durability properties, *Constr. Build. Mater.*, 2021, **291**, 123363.

63. T. Zhou, X. Luo, X. Liu, X. Zhai, Y. Sun and G. Liu, *et al.*, The green and low-carbon development pathways in the urban and rural building sector in Shaanxi Province, China, *Energy Build.*, 2024, **306**, 113952.
64. E. Paloheimo and O. Salmi, Evaluating the carbon emissions of the low carbon city: A novel approach for consumer based allocation, *Cities*, 2013, **30**, 233–239.
65. Y. Fan and F. Wei, Contributions of natural carbon sink capacity and carbon neutrality in the context of net-zero carbon cities: A case study of Hangzhou, *Sustainability*, 2022, **14**(5), 2680.
66. K. Saddul, *What are the Prospects for Reducing Embodied Carbon, or Creating Carbon Sinks, in the Materials from Which New University Buildings are Constructed?*, 2021.
67. S. Lehmann, Low carbon construction systems using prefabricated engineered solid wood panels for urban infill to significantly reduce greenhouse gas emissions, *Sustainable Cities Soc.*, 2013, **6**, 57–67.
68. Z. Liu, D. Guan, D. Crawford-Brown, Q. Zhang, K. He and J. Liu, A low-carbon road map for China, *Nature*, 2013, **500**(7461), 143–145.
69. T. Luo, Y. Tan, C. Langston and X. Xue, Mapping the knowledge roadmap of low carbon building: A scientometric analysis, *Energy Build.*, 2019, **194**, 163–176.
70. N. Tripathi, C. D. Hills, R. S. Singh and C. J. Atkinson, Biomass waste utilisation in low-carbon products: harnessing a major potential resource, *npj Clim. Atmos. Sci.*, 2019, **2**(1), 35.
71. Y. Zhang and J. Wang, Low-carbon green space construction in urban communities, *Nat., Environ. Pollut. Technol.*, 2014, **13**(3), 571.
72. M. Cao, F. Wang, S. Ma, H. Geng and K. Sun, Recent advances on greenhouse gas emissions from wetlands: mechanism, global warming potential, and environmental drivers, *Environ. Pollut.*, 2024, 124204.
73. J. K. Choi, B. R. Bakshi and T. Haab, Effects of a carbon price in the US on economic sectors, resource use, and emissions: An input–output approach, *Energy Policy*, 2010, **38**(7), 3527–3536.
74. W. L. Shang and Z. Lv, Low carbon technology for carbon neutrality in sustainable cities: A survey, *Sustainable Cities Soc.*, 2023, **92**, 104489.
75. D. Ioannidou, G. Sonnemann and S. Suh, Do we have enough natural sand for low-carbon infrastructure?, *J. Ind. Ecol.*, 2020, **24**(5), 1004–1015.
76. B. V. Venkatarama Reddy, Sustainable materials for low carbon buildings, *Int. J. Low-Carbon Technol.*, 2009, **4**(3), 175–181.
77. F. Orsini and P. Marrone, Approaches for a low-carbon production of building materials: A review, *J. Cleaner Prod.*, 2019, **241**, 118380.
78. R. Agrawal, S. Bhagia, A. Satlewal and A. J. Ragauskas, Urban mining from biomass, brine, sewage sludge, phosphogypsum and e-waste for reducing the environmental pollution: Current status of availability,

potential, and technologies with a focus on LCA and TEA, *Environ. Res.*, 2023, **224**, 115523.
79. S. Foxell, *A carbon primer for the built environment*, Routledge, 2014.
80. Y. A. Alli, A. Bamisaye, M. O. Bamidele, N. O. Etafo, S. Chkirida and A. Lawal, *et al.*, Transforming waste to wealth: Harnessing carbon dioxide for sustainable solutions, *Results Surf. Interfaces*, 2024, 100321.

CHAPTER 11

Low-carbon Materials for Environmental Monitoring

T. TASIĆ, V. MILANKOVIĆ AND T. LAZAREVIĆ-PAŠTI*

University of Belgrade, VINČA Institute of Nuclear Sciences – National Institute of the Republic of Serbia, Serbia
*Email: lazarevictlj@yahoo.com

11.1 Introduction

In an era marked by unprecedented environmental challenges, the significance of environmental monitoring has risen to the forefront of global discourse. As our planet struggles with the far-reaching consequences of climate change, escalating pollution levels, and the degradation of fragile ecosystems, the need for comprehensive monitoring mechanisms has become more critical than ever before. The growing global awareness of the profound impact of human activities on the environment has sparked a significant change in how we think as a society. From the local to the international stage, societies are recognizing the urgent call for sustainable solutions that can mitigate and, where possible, reverse the environmental damage created by decades of unrestrained practices.

The urgency to mitigate climate change has underscored the need for eco-conscious practices, and a crucial path for achieving this lies in the incorporation of low-carbon materials (LCMs) within environmental monitoring technologies.[1] The significance of materials science in contributing to sustainable practices cannot be overstated. Traditional monitoring devices, often reliant on materials with substantial carbon footprints,[2] are now being modified to minimize environmental harm. The investigation of LCMs for environmental applications is not merely an exploration of novel substances; it is a strategic attempt to revolutionize the basics of environmental

Low Carbon Materials: Paving a Sustainable Path for a Greener Future
Edited by Santanu Patra, Sudheesh K. Shukla and Mika Sillanpää

Published by the Royal Society of Chemistry, www.rsc.org

monitoring. By prioritizing materials that offer improved sensitivity, selectivity, and durability, more efficient and environmentally responsible monitoring infrastructure is enabled.

The integration of LCMs into environmental monitoring systems is not only a technological advancement but also an environmental necessity. Traditional monitoring devices often rely on materials and processes that contribute significantly to greenhouse gas emissions and environmental pollution. In contrast, LCMs offer a promising alternative, particularly those derived from sustainable sources or through eco-friendly processes. These materials help reduce the overall carbon footprint of monitoring technologies, making them more compatible with global efforts to mitigate climate change. Additionally, utilizing LCMs in sensors and detectors enhances their performance while promoting sustainable practices.

This chapter contributes to the ongoing discourse on sustainable practices in environmental monitoring, with a particular focus on low-carbon nanomaterials. These materials are emphasized because they are predominantly used in environmental monitoring applications due to their high sensitivity, tunability, and efficiency in detecting various contaminants. Carbon-based nanomaterials, in particular, offer a compelling combination of low environmental impact, ease of synthesis from renewable sources, and excellent performance in real-world monitoring scenarios. Given the focus of this chapter, we have concentrated on carbon-based materials as they present one of the most effective solutions in terms of balancing sustainability and functionality. By reviewing various LCMs, exploring their synthesis methods, and defining their unique properties, we aim to provide a comprehensive resource for researchers, practitioners, and policymakers invested in sustainable technologies. The chapter goes beyond a mere exploration of materials; it actively speaks up for the adoption of low-carbon alternatives in environmental sensors, detectors, and monitoring devices. It highlights the benefits of such materials, emphasizing improved sensitivity, selectivity, and durability. Furthermore, the investigation extends into the challenges and future prospects of integrating LCMs into environmental monitoring systems. Through this discussion, the chapter raises awareness of potential hurdles and provides a roadmap for overcoming them, facilitating the practical implementation of sustainable monitoring practices. In essence, this chapter is a call to action, advocating for sustainability in environmental monitoring. By embracing LCMs and advancing materials science, we not only contribute to the resolution of contemporary environmental challenges but also actively participate in the creation of a more sustainable and resilient future.

11.2 Principles of Low-carbon Materials

LCMs refer to substances or compounds that possess a significantly reduced carbon footprint compared to traditional materials when utilized in various applications, including manufacturing, construction, and technology.[3,4]

These materials are chosen for their ability to contribute to environmental sustainability by minimizing greenhouse gas emissions, energy consumption, and overall ecological impact throughout their life cycle.[4] The most important characteristics of LCMs are listed in Table 11.1.

The characteristics of LCMs are intertwined with sustainability principles, energy efficiency, and environmental responsibility. Their selection and application contribute to a holistic approach to addressing contemporary challenges associated with climate change, pollution, and resource depletion.[5–7] Still, their efficiency for the intended purpose should be optimal.[3]

Table 11.1 The most important characteristics of LCMs.

Characteristic	Description
Reduced carbon footprint	The primary characteristic of LCMs is their inherently lower carbon footprint. This reduction is achieved by selecting materials and manufacturing processes that prioritize energy efficiency, resource conservation, and emissions reduction.
Environmentally friendly synthesis	LCMs often exhibit characteristics associated with environmentally friendly synthesis methods. This includes processes that consume fewer resources, generate less waste, and minimize the use of hazardous substances.
Renewable or recyclable components	Many LCMs incorporate renewable resources or are designed for easy recycling. This characteristic contributes to the circular economy, promoting sustainable practices by reducing reliance on finite resources and minimizing end-of-life environmental impact.
Energy efficiency	LCMs are frequently chosen for their energy-efficient properties. This can involve materials that require less energy for production, contribute to the efficiency of devices they are incorporated into (such as sensors or detectors), or exhibit favorable thermal properties.
Longevity and durability	Sustainability is often enhanced through durable materials that have a longer lifespan. LCMs may possess properties that make them more resistant to wear and degradation, reducing the frequency of replacements and associated environmental costs.
Biodegradability	Some LCMs are designed to be biodegradable, meaning they can naturally break down over time without causing harm to the environment. This characteristic is especially relevant in applications where disposal is a concern.
Enhanced performance	Beyond their environmental benefits, LCMs are chosen for their ability to provide equivalent or improved performance compared to traditional counterparts. This characteristic ensures that sustainability efforts do not compromise functionality.
Versatility	LCMs are often versatile and adaptable to a range of applications. This characteristic allows for their incorporation into various industries and technologies, providing a broad impact on reducing the overall environmental footprint across diverse sectors.

While carbon-based materials are commonly associated with low-carbon initiatives due to their widespread use and potential for carbon sequestration, non-carbon materials can also contribute to sustainability goals in different ways. Nanoparticles and metal oxide materials can also be considered LCMs, depending on their production methods, environmental impact, and applications.[8] Moreover, several types of composites are considered LCMs due to their reduced environmental impact compared to traditional composite materials.[9] These low-carbon composites are often made from renewable or recycled materials and are designed to minimize energy consumption, emissions, and resource depletion throughout their life cycle. Examples of such materials include natural fiber composites,[10,11] recycled composites,[12,13] wood–plastic composites,[14–16] bio-based composites,[17,18] and hybrid composites.[19] This chapter will focus on carbon-based materials, considering their important role in environmental monitoring.

The structure of carbon-based LCMs is not very different from that of traditional carbon materials.[20–24] By definition, "Low-carbon materials are the ones that simultaneously lower embodied and operational carbon emissions, reduce carbon footprint, and enhance materials' circularity, compared to the conventional material choices without compromising the end product's functional requirements and performance properties."[3,4] The carbon footprint includes all the greenhouse gas emissions produced throughout a product's life.[3] LCMs aim to reduce this impact, making our environment healthier and promoting fairness through a circular economy. These materials, like waste biomass (such as sawdust and agricultural residue), not only absorb CO_2 while growing but actually have a negative CO_2 footprint.[3] It's important to understand that LCMs are just the beginning of sustainable practices because materials are the basic building blocks for products, which, in turn, provide various essential services.[3]

LCMs come in various dimensionalities, ranging from zero-dimensional (0D) nanoparticles to two-dimensional (2D) sheets and three-dimensional (3D) structures. Examples of LCMs with different dimensionalitics are given in Table 11.2.

In technical terms, 0D materials formally have all three dimensions under 100 nm, but common 0D carbon materials typically have much smaller dimensions, often just a few nanometers.[25] Examples include fullerenes,[26,27] graphene quantum dots (GQDs),[28,29] carbon quantum dots (CQDs),[30,31] and nanodiamonds.[32] The small size ensures that a significant portion of carbon

Table 11.2 Examples of LCMs with different dimensionalities.

Dimensionality	Example
Zero-dimensional (0D)	Carbon nanoparticles, carbon dots, and fullerenes
One-dimensional (1D)	Carbon nanotubes (CNTs), single-walled (SWCNTs) or multi-walled (MWCNTs)
Two-dimensional (2D)	Graphene
Three-dimensional (3D)	Graphite foam, mesoporous carbon materials

atoms is exposed on the material's surface, resulting in an exceptionally high surface-to-volume ratio.

Moving to 1D carbon nanomaterials, such as single-walled and multi-walled carbon nanotubes (SWCNTs and MWCNTs), their dimensions, particularly in terms of length, are typically larger than 100 nm.[33] These materials have a well-defined structure, described by the folding scheme of graphene sheets, maintaining a constant diameter along the entire nanotube. Carbon nanotubes are readily available commercially, and their qualities are reported through a standardized procedure.

Within the sphere of 2D carbon materials, graphene is a good example.[34] Graphene, an infinite sheet of carbon organized in a hexagonal lattice, exhibits extraordinary thermal and mechanical properties.[35] Its hexagonal structure is dictated by the sp^2 hybridization of carbon atoms, resulting in in-plane chemical bonds.[35] The π electron cloud extends above and below the graphene plane, rendering it a zero-gap semi-metal with remarkable chemical inertness.[36] The chemical reactivity of graphene primarily stems from defects and edges.[37,38] The determination of the specific surface area (SSA) of graphene poses a geometrical challenge, and for pristine graphene, the SSA reaches 2630 $m^2\ g^{-1}$. In practical terms, graphene synthesis employs both bottom-up and top-down approaches.[39]

Graphite, a member of the carbon allotropes family, is distinguished by its hexagonal lattice structure, which is comprised of graphene sheets stacked through weak van der Waals forces. In its three-dimensional (3D) form, graphite showcases anisotropic conductivity, with its exceptional electron mobility and electrical conductivity primarily demonstrated within the (0001) plane of the hexagonal layers. The layered arrangement allows for easy slippage between the graphene sheets, contributing to graphite's lubricating properties.[25]

Mesoporous carbon materials represent a class of carbonaceous substances characterized by a well-defined porous structure with pore sizes typically in the mesoporous range (2–50 nm).[40] The unique architecture of mesoporous carbon materials is advantageous for various applications. Mesoporous carbon materials feature a network of interconnected pores, providing a large surface area and high pore volume. This porous architecture contributes to their exceptional adsorption capacity, making them valuable in gas storage, environmental remediation, and sensing applications.[41]

In this chapter, we focus on carbon-based nanomaterials due to their significant advantages in terms of sustainability, reduced CO_2 footprint, and versatility. These materials stand out in the field of low-carbon technologies for several key reasons. First, carbon-based nanomaterials can be synthesized from abundant, renewable precursor materials like biomass and biowaste. This reduces the reliance on non-renewable resources and minimizes the environmental impact associated with mining or energy-intensive production processes typically required for alternative materials. Using biowaste as a feedstock addresses waste management challenges and aligns with circular economy principles by transforming waste into valuable resources. Additionally, the energy efficiency of carbon-based nanomaterial production processes offers a significant advantage. For instance, synthesis techniques such as hydrothermal carbonization and pyrolysis can be optimized to consume less energy than manufacturing other materials like metal oxides or ceramics. These processes can operate at relatively lower temperatures and use renewable feedstocks, further contributing to a lower overall energy demand and reduced carbon emissions. Another important aspect is end-of-life management. Carbon-based nanomaterials offer potential for recyclability and reuse, contributing to their sustainability profile. Unlike certain metal or polymer-based materials, which may pose environmental risks due to persistence or difficulty in recycling, carbon-based materials tend to degrade more easily or can be repurposed in subsequent applications. This lowers their environmental persistence and supports a more sustainable lifecycle, from synthesis to disposal. Together, these attributes make carbon-based nanomaterials a highly suitable option for low-carbon technologies, particularly in applications such as environmental monitoring, where both performance and sustainability are crucial considerations.

11.3 Low-carbon Materials for Environmental Sensors

The utilization of LCMs in monitoring devices yields numerous environmental benefits, significantly reducing the ecological footprint associated with traditional materials.[42] Primarily, incorporating these materials decreases carbon emissions substantially throughout the entire life cycle, from raw material extraction to end-of-life disposal. The energy efficiency inherent in many LCMs further contributes to overall energy conservation, aligning with sustainability objectives. Moreover, the use of renewable resources or materials designed for easy recycling promotes resource conservation and supports a circular economy, minimizing the environmental impact of raw material extraction. Durable and long-lasting low-carbon materials extend the lifespan of monitoring devices, reducing the frequency of replacements and diminishing the need for additional raw materials and energy. Additionally, the selection of low-carbon materials emphasizes improved environmental compatibility, minimizing harm to ecosystems, and promoting sustainable practices within the technology industry.[42]

LCMs and traditional materials exhibit severe differences in their carbon footprints. As mentioned, LCMs are specifically chosen or designed to minimize greenhouse gas emissions throughout their life cycle, incorporating energy-efficient production processes, renewable resources, and recyclability. The emphasis on a circular economy and the durability of these materials further contributes to reduced carbon emissions. In contrast, traditional materials may have higher carbon footprints due to less efficient production processes, limited recyclability, and dependence on non-renewable resources. The extraction and processing of traditional materials often result in increased carbon emissions, worsened by their shorter lifespan and the need for more frequent replacements. The choice between low-carbon and traditional materials hinges on specific environmental goals and considerations. LCMs stand out for their intentional efforts to mitigate the environmental impact associated with materials used in various applications.[42]

The principles guiding the selection and development of LCMs are vital in the search for sustainable environmental monitoring technologies. One of the key considerations is the intentional effort to minimize the overall carbon footprint throughout the life cycle of the materials.[42] It involves prioritizing materials with reduced emissions during extraction, manufacturing, transportation, and disposal. Energy efficiency emerges as a critical principle, advocating for the use of materials that require less energy for production and operation. By incorporating energy-efficient materials, the environmental impact associated with monitoring devices is significantly diminished. Another fundamental principle is the utilization of renewable resources in the development of LCMs. This principle aligns with the broader goal of reducing dependency on finite resources and promoting a more sustainable sourcing strategy. By incorporating materials derived from renewable sources, such as bio-based or recycled content, the environmental impact of raw material extraction is mitigated, contributing to a circular and regenerative approach. Further, recyclability stands out as an important principle in selecting and developing LCMs.[42] Materials designed for easy recycling facilitate the creation of a closed-loop system, minimizing waste and reducing the demand for new raw materials. This principle not only aligns with environmental sustainability but also addresses the escalating issue of electronic waste associated with monitoring devices. Durability and longevity are also essential considerations. Opting for materials with a prolonged lifespan reduces the frequency of replacements, subsequently lowering the demand for additional raw materials and the associated carbon emissions from manufacturing. This principle supports the idea of creating robust and resilient monitoring devices that can withstand the test of time, aligning with the ethos of sustainability. Moreover, a critical principle involves assessing the entire life cycle of the materials, from cradle to grave. Comprehensive life cycle assessments enable a holistic understanding of the environmental impact of a material, guiding decision-makers in selecting materials that offer the most significant reduction in overall carbon emissions.

Besides sustainability, the performance of the selected material within the environmental sensor is a critical aspect in the selection and development of LCMs. The material's ability to meet the specific requirements of sensor applications, such as sensitivity, selectivity, stability, resolution, linearity, range, repeatability, reproducibility, and response time, plays a key role in the overall efficacy of the monitoring system.[43] Striking a balance between environmental responsibility and functionality is crucial to ensure that the adoption of LCMs does not compromise the performance of the monitoring devices. The selected materials must exhibit characteristics that enhance the sensor's overall sensitivity to environmental changes. It involves careful consideration of the material's electrical, mechanical, and chemical properties to optimize its responsiveness to targeted stimuli. The material's capability to detect and accurately measure environmental parameters, such as pollutants or climate indicators, directly influences the reliability and precision of the monitoring system. Furthermore, the durability and stability of the chosen material in the environmental sensor are critical factors. Environmental sensors are often deployed in harsh conditions, exposed to varying temperatures, humidity levels, and potentially corrosive elements. The material must withstand these environmental challenges without compromising its structural integrity or sensor performance over time.

Balancing environmental concerns with the need for high-performance materials is a complex task. Advances in materials science play a pivotal role in identifying or engineering materials that not only meet sustainability criteria but also excel in the demanding operational conditions of environmental sensors. Rigorous testing and validation procedures are essential to ensure that the selected materials align with both eco-conscious goals and the stringent performance requirements of modern monitoring technologies. Ultimately, the most important principle guiding the selection and development of LCMs for environmental monitoring is the holistic consideration of sustainability and efficiency. Achieving a harmonious integration of environmental responsibility and optimal sensor performance is key to developing monitoring devices that contribute to a sustainable future and provide accurate, reliable, and efficient data for informed environmental decision-making.

11.4 Synthesis and Fabrication Methods of Low-carbon Materials

Green chemistry exploits natural sources such as plants, microorganisms, or their by-products to synthesize desired LCMs.[42] Strict regulations mandate reducing or eliminating poisonous or carcinogenic compounds during synthesis procedures. Various templates can be employed in the preparation of LCMs, offering flexibility in material structure. Metals and metal oxides do not sound very green but are abundant in the Earth's crust and work well as starting materials for low-carbon synthesis. Ethylene or natural gas reserves can be potential sources of carbon.[42] LCMs, such as nanotubes and carbon dots, are great for making sensors that detect harmful environmental pollutants. To make these materials work even better as sensors, we can modify their surface with different groups, a process called functionalization.[42] There are two ways to do this: non-covalent methods use catalysts like metallic oxides or nanoparticles, while covalent methods attach chemical groups directly to the surface of LCMs.

Table 11.3 The common synthesis techniques for low-carbon materials.

Method	Material
Chemical vapor deposition	Graphene
Liquid-phase exfoliation	Graphene, graphene oxide
Hydrothermal and solvothermal methods	Graphene, carbon nanotubes
Arc discharge and laser ablation	Carbon nanotubes, carbon dots
Electrochemical deposition	Carbon nanomaterials
Mechanical exfoliation	Graphene
Template-assisted synthesis	Carbon nanotubes

Different synthesis techniques offer diverse approaches to producing LCMs with tailored properties, enabling their application in various fields. The choice of a specific synthesis method depends on the desired properties, scalability, and intended applications of the low-carbon material. An overview of common synthesis techniques for LCMs is presented in Table 11.3.

Chemical vapor deposition (CVD) is a widely used method for synthesizing LCMs like graphene. In this process, a precursor gas containing carbon is introduced into a high-temperature reactor. The carbon in the gas is deposited onto a substrate, forming a thin layer of the desired LCM. CVD allows for precise control over the growth and quality of the material.[44]

Liquid-phase exfoliation is a technique commonly employed for producing LCMs such as graphene and graphene oxide. In this method, bulk materials (*e.g.*, graphite) are dispersed in a liquid medium and mechanical or ultrasonic forces are applied to break down the material into individual layers. The resulting dispersion can be used to obtain LCMs with varying thicknesses.[45]

Hydrothermal and solvothermal methods involve the reaction of precursor materials in high-temperature, high-pressure aqueous or solvent environments. These techniques are versatile and can be used to synthesize various LCMs, including carbon nanotubes and certain forms of graphene. The controlled conditions influence the structure and properties of the resulting materials.[46]

Arc discharge and laser ablation are methods commonly employed for the synthesis of carbon nanotubes. In arc discharge, a high electric current is passed through graphite electrodes in an inert gas atmosphere, leading to the formation of carbon nanotubes. Laser ablation involves vaporizing a carbon target using a laser, resulting in the production of carbon nanotubes.[47,48]

Electrochemical deposition is a method for depositing LCMs onto a substrate through the use of an electric current. It is commonly used for the synthesis of materials like reduced graphene oxide and certain carbon-based nanomaterials. The process offers control over the material's thickness and morphology.[49,50]

Mechanical exfoliation, often referred to as the "Scotch tape" method, involves peeling layers of a material, such as graphite, to obtain thin sheets of LCMs like graphene. While simple, this method is not suitable for large-scale production and is primarily used for research and laboratory purposes.[51]

Template-assisted synthesis involves using templates or molds to guide the growth of LCMs. This method allows for the precise control of the material's structure and dimensions. Various templates, including porous membranes or sacrificial materials, can be employed to synthesize materials like carbon nanotubes.[52,53]

The scalability and reproducibility in large-scale production of LCMs are of utmost importance to ensure practical viability and widespread adoption. Several key factors must be addressed to scale up production processes effectively while maintaining consistency and quality. The optimization of material synthesis routes and fabrication techniques to enhance efficiency and yield is essential.[42] It involves fine-tuning reaction parameters such as temperature, pressure, and reaction time to achieve optimal results on a larger scale.[54] Next, a reliable and sustainable supply chain for raw materials must be secured, ensuring consistency in material composition and quality.[55] Quality control measures to monitor raw material purity and integrity must be implemented, minimizing variability in the final product. Investing in scalable manufacturing infrastructure and equipment capable of accommodating increased production volumes is important, necessitating upgrades to reactors, purification systems, and processing facilities while maintaining operational efficiency. Robust process monitoring and control systems should be implemented to track key parameters and variables throughout production, enabling real-time adjustments to optimize product quality and minimize deviations. Standardized protocols and procedures for production must be established to ensure consistency and reproducibility across batches, accompanied by rigorous quality assurance protocols to validate product quality and compliance with industry standards. Environmental sustainability and safety considerations should be integrated into production processes to minimize resource consumption, waste generation, and environmental impact, adhering to regulatory guidelines and best practices to ensure worker safety and compliance.[56] Additionally, striving for cost-efficiency and economies of scale in production processes is essential to maximize profitability and competitiveness, with continuous investment in research and development initiatives facilitating innovation and improvement in production technologies and methodologies.[57] By addressing these considerations, manufacturers can overcome challenges associated with scalability and reproducibility, paving the way for the successful large-scale production of LCMs with consistent quality and performance.

Evaluating the environmental impact during the synthesis process of LCMs is another thing that should be considered to ensure sustainability and minimize adverse effects on ecosystems and human health. While there isn't a one-size-fits-all procedure, several methodologies and approaches can be employed to assess environmental impacts. Conducting a comprehensive life cycle assessment (LCA) of the synthesis process, from raw material extraction to end-of-life disposal, is imperative, as it quantifies environmental impacts across all stages of the product life cycle, including resource depletion, greenhouse gas emissions, energy consumption, and waste generation.[58] Additionally, calculating the environmental footprint of the synthesis process provides insights into environmental burdens and opportunities for improvement.[59–61] Material flow analysis helps identify inefficiencies and potential environmental hotspots, guiding efforts to optimize resource utilization and minimize waste generation. Assessing ecotoxicity and potential human health risks associated with raw materials and by-products is crucial, as is utilizing carbon and environmental footprint calculators to estimate environmental impacts. Implementing environmental management systems, such as ISO 14001,[62] establishes procedures for monitoring and improving environmental performance, while the development of sustainability metrics facilitates benchmarking and comparison with industry standards. Engaging stakeholders in discussions about environmental impact assessment fosters transparency and collaboration, driving collective action toward sustainability goals. By systematically employing these procedures and methodologies, manufacturers can effectively evaluate and mitigate the environmental impact of LCM synthesis, contributing to the transition toward more sustainable and environmentally responsible production practices.

11.5 Applications of Low-carbon Materials in Environmental Monitoring

Numerous materials derived from biomass have proven effective in detecting a wide range of environmental pollutants, including metal ions,[63] pesticides,[64] pharmaceuticals,[65] and dyes.[66] However, by reviewing the literature, it is easy to find that quantum dots (QDs) are most often used for this purpose. They are often sourced from graphene and graphene oxide in an environmentally friendly manner. Moreover, they can also be synthesized from biowaste and biomass, which makes them even more suitable in terms of environmental protection (see Figure 11.1).

GQDs have gathered significant scientific interest owing to their fluorescence, excellent biocompatibility, low toxicity, and attractive physicochemical characteristics. Abbas *et al.* present an eco-friendly, acid-free, and sustainable approach for synthesizing GQDs using a cost-effective and renewable carbon source: biomass waste.[67] This method mitigates the risk of contamination from strong acids and reduces expenses associated with expensive precursors. Blue fluorescent GQDs were successfully synthesized from spent tea leaves, a green biomass waste. Figure 11.2 illustrates the schematic process of GQD preparation from spent tea leaves.

The results reveal that the GQDs show a size distribution ranging from 1 to 5 nm, with an average size of approximately 3 ± 0.4 nm and a thickness of about 1 nm, comprising one to three graphene layers. The as-prepared GQDs exhibit interesting size-dependent optical properties and significant surface

Figure 11.1 Common natural sources for low-carbon QDs synthesis.

functionalization. These GQDs were applied as fluorescence probes for ferric ion detection, leveraging their unique optical characteristics. A highly sensitive sensor is developed with a detection limit as low as 0.29 μM (see Figure 11.3).

Saud *et al.* conducted a study on synthesizing GQDs using ethanolic extracts of eucalyptus tree leaves *via* a hydrothermal treatment technique (see Figure 11.4).[68] Various combinations of heat treatment times and temperatures were employed during the process. Comprehensive analyses were conducted to assess the optical, morphological, and compositional properties of the GQDs synthesized *via* this green method.

The study revealed that the highest yield of GQDs was achieved at a reaction temperature of 300 °C. The optimal treatment period of 480 minutes also resulted in a maximum product yield of approximately 44.34%. The quantum yields of the GQDs synthesized under these conditions (referred to as GQD/300) were measured at 0.069. Moreover, the D/G ratio of GQD/300 was found to be 0.532, indicating the presence of a nano-crystalline graphite

Figure 11.2 Schematic illustration of the development of GQDs from spent tea leaves. Reproduced from ref. 67, https://doi.org/10.3390/nano12203696, under the terms of the CC BY 4.0 license, https://creativecommons.org/licenses/by/4.0/.

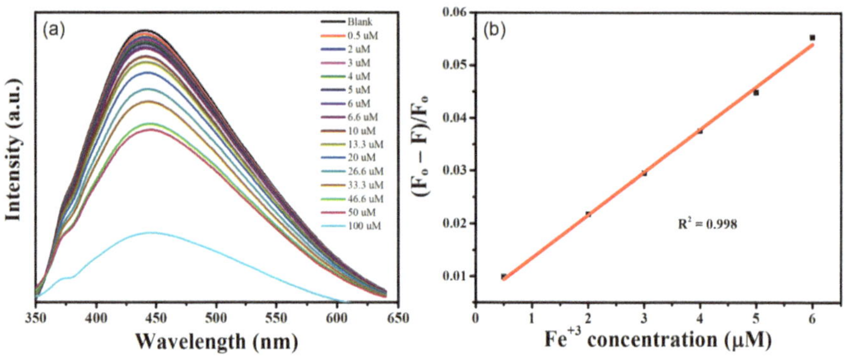

Figure 11.3 (a) Fluorescence spectra of GQDs (10 mg mL^{-1}) with Fe^{3+} concentrations ranging from 0 to 100 μM, with an excitation wavelength of 340 nm and (b) linear regression plot of fluorescence intensity and Fe^{3+} concentration in the low concentration range of 0.5–6 μM. Reproduced from ref. 67, https://doi.org/10.3390/nano12203696, under the terms of the CC BY 4.0 license, https://creativecommons.org/licenses/by/4.0/.

structure. Transmission electron microscopy (TEM) images illustrated the formation of GQD/300 with particle sizes ranging from 2.0 to 5.0 nm. Importantly, it was observed that GQD/300 exhibited selective detection capabilities toward Fe^{3+} ions, underscoring its potential utility in metal ion sensing applications.

Carbon quantum dots (CQDs) have recently gained significant attention for their applications in detecting pesticides, metal ions, pharmaceuticals, and other environmental contaminants. They are nanometer-sized carbon-based materials with unique optical and electrochemical properties. One of the notable features of CQDs is their versatile synthesis from a wide range of precursor materials. Researchers have successfully synthesized CQDs from

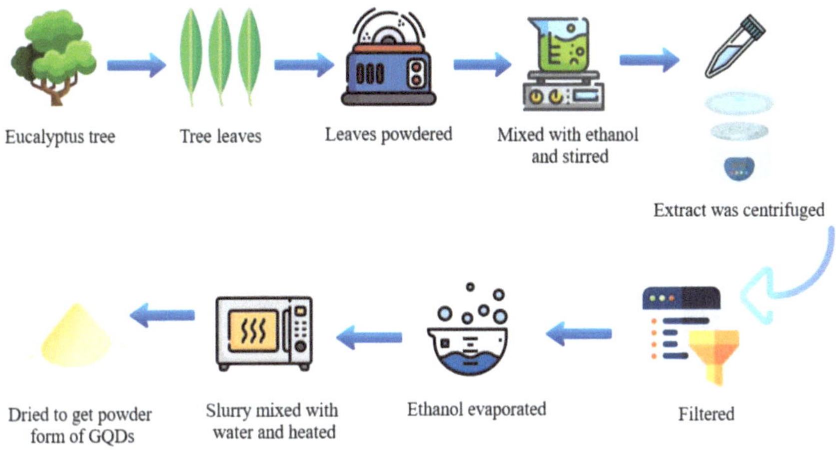

Figure 11.4 Graphical diagram for the preparation of GQDs from eucalyptus tree leaves. Reproduced from ref. 68, https://doi.org/10.3390/nano13010148, under the terms of the CC BY 4.0 license, https://creativecommons.org/licenses/by/4.0/.

various natural sources, including cauliflower,[69] banana peel,[70] tomatoes,[71] rice,[72] broccoli,[73] cabbage,[74] cucumber,[75] and many others. Each of these precursors offers distinct chemical compositions and structural characteristics, which can influence the properties of the resulting CQDs. The synthesis of CQDs from natural sources often involves simple and cost-effective methods, making them attractive for large-scale production. These methods typically include carbonization, hydrothermal treatment, or microwave-assisted synthesis. By utilizing sustainable and abundant biomass as precursors, researchers can reduce environmental impact and promote the utilization of renewable resources. Furthermore, the unique properties of CQDs, such as their size-dependent fluorescence, surface functionalization, and chemical stability, make them ideal candidates for environmental sensing applications. They can be employed as fluorescent probes for detecting and quantifying various contaminants in water, soil, air, and food samples with high sensitivity and selectivity. While optical detection methods are more commonly employed with CQDs due to their excellent photoluminescence properties, electrochemical sensors utilizing CQDs have also shown promise.

Examples of CQDs from natural sources for the detection of environmental contaminants are listed in Table 11.4.

11.6 Challenges and Future Prospects

Integrating LCMs into monitoring systems presents complex challenges that demand multi-layered solutions. The primary challenge is to ensure compatibility between LCMs and existing monitoring system components, such

Table 11.4 Examples of CQDs from natural sources for the detection of environmental contaminants.

Analyte	Type of precursor	Method of detection	Linearity range	Limit of detection	Real samples	Ref.
Pesticides						
Amicarbazone, diazinon, and glyphosat	Cauliflower juice	Optical		0.5 ng mL^{-1}, 0.25 ng mL^{-1} and 2 ng mL^{-1}	Cherry tomato	76
Diazinon	Rose flowers	Optical	0.02 to 1 μM	0.01 μM	Tap water and river water	77
Isoprothiolane	*Calotropis procera* leaves	Optical	1.0 mM to 0.05 μM	11.58 nM	Fruit, rice	78
Fenitrothion, dithianon, and dinoseb	*Ginkgo biloba* leaves	Optical	1.0 to 2000 nM, 1.0 to 1000 nM, and 1.0 to 3000 nM	0.36 nM, 0.28 nM and 0.66 nM	Rice	79
Paraoxon-ethyl	Sucrose	Optical	Up to 5.80 mM	0.22 μM	Vegetables	80
Chlorpyrifos and quinalphos	Tagetes erecta	Optical	0.05 to 100.0 μM and 0.01 to 50.0 μM	2.1 ng mL^{-1} and 1.7 ng mL^{-1}	Fruit, rice	81
Hydrazine	Potato pulp	Optical	0.15 to 1.5 μM	0.15 μM	Water bodies	82
Dimethoate	Pork rib bones	Optical	0.15 to 5.0 μM	0.064 μM	River water and farm water	83
Isocarbophos	Leek	Optical	1.0×10^{-9} to 1.0×10^{-3} M	3.0×10^{-10} M	Broccoli	84
Chlorpyrifos	Burning ash of waste paper	Optical	0.01 to 1.0 μg mL^{-1}	3 ng mL^{-1}	Water	85
Isocarbophos	Soybean	Optical	1.0×10^{-9} to 1.0×10^{-4} M	3.0×10^{-10} M	Broccoli	86
Dichlorvos	Feather	Optical	6.0×10^{-9} to 6.0×10^{-8} M	3.8×10^{-9} M	Cabbage and fruit juice	87
Glyphosate	Wool	Optical	0.025 to 2.5 μg mL^{-1}	12 ng mL^{-1}	Cereal	88
Glyphosate	*Litchi chinensis*	Optical		8 ng mL^{-1}	Pearl river water, tea, and soil	89
Thiabendazole	Rosemary leaves	Optical	0.03 to 1.73 μg mL^{-1}	8 ng mL^{-1}	Apple, orange, and tomato juices	90
Chlorpyrifos	*Psidium guajava* leaves	Electrochemical	0.01 to 1 μM	1.5 nM		91

Atrazine, pretilchlor, dichlorvos, alachlor, carbofuran and metolachlor	Orange peel	Optical	10 to 200 ng mL^{-1}	4.7 ng mL^{-1}	Hawthorn samples	92
Metals						
Co^{2+}	Pigskin	Optical	1 to 100 μM	0.68 μM		93
Cr^{6+}	Lemon peel waste	Optical	2.5 to 50 μM	0.073 μM		94
Cu^{2+}	Bamboo leaves	Optical	0.333 to 66.6 μM	0.115 μM	River water	95
Cu^{2+}	*Acacia concinna* seeds	Optical	0.01 to 10 μM	0.0043 μM	Water	96
Cu^{2+}	Grass	Optical	0 to 50 μM	0.001	Water	97
Fe^{3+}	Honey	Optical	0.005 to 100 μM	0.0017 μM	Human blood	98
Fe^{3+}	Wool	Optical	0.10 to 10 μM	0.01 μM		99
Fe^{3+}	Coriander leaves	Optical	0 to 60 μM	0.4 μM		100
Hg^{2+}	*Tamarindus indica* leaves	Optical	0 to 0.1 μM	0.006 μM	Pond water	101
Fe^{3+}	Papaya	Optical	1 to 10 μM and 25 to 700 μM (water soluble), 1 to 8 μM and 100 to 800 μM (ethanol soluble)	0.48 μM (water soluble), 0.29 μM (ethanol soluble)		102
Hg^{2+}	Pomelo peel	Optical	0.0005 to 0.01 μM	0.00023 μM	Lake water	103
Hg^{2+}	Flour	Optical	0.0005to 0.01 μM	0.0005 μM	Lake water	104
Hg^{2+}	Edible mushrooms	Optical	0 to 0.1 μM	0.00413 μM		105
Fe^{3+}	*Prunus avium* fruit	Optical	0 to 100 μM	0.96 μM		106
Fe^{3+}	Onion waste	Optical	0 to 20 μM	0.31 μM	Water	107
Fe^{3+}	Black tea	Optical	0.25 to 60 μM	0.25 μM	Human serum	108

Table 11.4 (*Continued*)

Analyte	Type of precursor	Method of detection	Linearity range	Limit of detection	Real samples	Ref.
Hg^{2+}	Hair	Optical	0 to 75 µM	0.01 µM	Water	109
Fe^{3+}	*Citrus sinensis*	Optical	0.01 to 100 µM	0.003 µM	Environmental water, urine, blood	110
Hg^{2+}	Tea	Optical	0 to 100 µM	0.001 µM	Lake water	111
Hg^{2+} and Cu^{2+}	Urine	Optical	0 to 100 µM	1.8 µM (Hg^{2+}) and 1.7 µM (Cu^{2+})		112
Hg^{2+}	Strawberry juice	Optical	0.01 to 50 µM	0.003 µM	Water	113
Pb^{2+}	*Lantana camara* berries	Optical	0 to 0.2 µM	0.00964 µM	Tap water, pond water, serum, urine, and wastewater	114
Cr^{6+}	Mushroom	Optical	10 to 50 µM	0.73 µM	Carbon dot-coated filter paper with Cr^{6+}	115
Dyes						
Alizarin red S	Cornstalk	Optical	10 to 150 µM	2.65 µM		116
Methylene blue	Lychee seeds	Optical		50 mM		117
Tartrazine	Citrus lemon peel	Optical	0.6 to 23.5 µM	60.0 nM		110
Pharmaceuticals						
Tetracycline	Waste tea residue	Optical	0.06 to 8.32 µM	0.04 µM	Urine and pharmaceutical samples	118

Tetracycline	Carp roe	Optical	0.1 to 50 μM	41.7 nM	River water, serum, and milk	119
Tetracycline	Waste tobacco leaves	Optical		0.52 μM	River water and tetracycline tablet	120
Oxytetracycline	Sweet potato peels	Optical	50 to 900 ng mL^{-1}	15.3 ng mL^{-1}		121
Tetracycline, terramycin and chlortetracycline	Rice	Optical	3.32 to 32.26 μM	0.2367, 0.3739 and 0.2791 μM	River water and tap water	122
Tetracycline, oxytetracycline and chlortetracycline	Tobacco	Optical	6×10^{-6} to 4×10^{-9} M, 2×10^{-6} to 2×10^{-8} M, and 2×10^{-7} to 2×10^{-8} M	5.18 nM, 6.06 nM, and 14 nM		123
Tetracycline and fluoroquinolones	*Curcuma amada*	Optical	0 to 16 μM	33 nM and 2 nM		124
Tetracycline, chlortetracycline and oxytetracycline	Industrial hemps	Optical	0 μM to 50 μM	0.0330 μM, 0.27 μM and 0.28 μM		125
Amoxicillin	*Sophorajaponica L.* flower	Optical	5 to 80 μM	3.40 μM	The human plasma sample	126
Ciprofloxacin	*Cinnamomum tamala (tejpata)* leaves	Optical	1 to 100 μM	6.06 μM		127

as sensors, detectors, and data-processing units. Addressing this issue requires careful attention to materials' physical and chemical properties to ensure integration without compromising the sensor's performance.

Additionally, optimizing the performance of monitoring systems using LCMs presents a complex issue, necessitating extensive research and development efforts to enhance sensitivity, selectivity, durability, and reliability while minimizing energy consumption and environmental impact. Moreover, the initial cost of integrating LCMs into monitoring systems may exceed that of conventional materials, presenting a significant barrier to adoption. Scaling up production to meet demand while maintaining affordability and consistency further complicates this challenge. Technological limitations, including limited material availability and processing techniques, add another layer of complexity to the integration process.

Nevertheless, despite these challenges, there are many promising future prospects. Continued research and development efforts are expected to drive technological innovation, leading to the development of advanced LCMs with superior performance, functionality, and cost-effectiveness tailored to the specific needs of monitoring systems. Interdisciplinary collaboration between materials scientists, engineers, and environmental scientists holds enormous potential for accelerating progress in this field. Embracing principles of the circular economy, such as resource efficiency, recycling, and waste reduction, can further enhance the sustainability of monitoring systems by promoting the use of recycled and renewable materials.

Linking the integration of LCMs to the UN's Sustainable Development Goals (SDGs) underscores their broader significance. For instance, Goal 9 (Industry, Innovation, and Infrastructure) is supported by advancing monitoring technologies that foster sustainable industrialization. Goal 12 (Responsible Consumption and Production) aligns with using recycled and renewable materials in monitoring systems, promoting more sustainable consumption patterns. Furthermore, Goal 13 (Climate Action) is directly addressed by reducing the carbon footprint of monitoring technologies, contributing to global efforts to combat climate change.

While integrating LCMs into monitoring systems presents a difficult challenge, it also offers significant opportunities for technological advancement, environmental sustainability, and public well-being. By addressing these challenges and leveraging future prospects, humanity can drive progress toward building more resilient, efficient, and sustainable monitoring systems capable of addressing the complex challenges of the 21st century.

Declaration of Generative AI and AI-assisted Technologies in the Writing Process

During the preparation of this work, the authors used ChatGPT 3.5 in order to refine certain sections of the text and enhance its clarity. After using this tool/service, the authors reviewed and edited the content as needed and take full responsibility for the content of the publication.

Acknowledgements

The authors acknowledge the support provided by the Serbian Ministry of Science, Technological Development, and Innovations (contract number: 451-03-66/2024-03/200017).

References

1. Ş. Kılkış, G. Krajačić, N. Duić, M. A. Rosen and M. dA. Al-Nimr, *Energy Convers. Manage.*, 2022, **269**, 116146.
2. S. McAlister, F. McGain, M. Breth-Petersen, D. Story, K. Charlesworth, G. Ison and A. Barratt, *Lancet Reg. Health – West. Pac.*, 2022, **24**, 100459.
3. O. Das, Á. Restás, V. Shanmugam, G. Sas, M. Försth, Q. Xu, L. Jiang, M. S. Hedenqvist and S. Ramakrishna, *Mater. Circ. Econ.*, 2021, **3**, 26.
4. K. Liu, M. Tebyetekerwa, D. Ji and S. Ramakrishna, *Matter*, 2020, **3**, 590–593.
5. V. Milanković, T. Tasić, M. Pejčić, I. Pašti and T. Lazarević-Pašti, *Foods*, 2023, **12**, 2397.
6. Đ Katnić, S. J. Porobić, T. Lazarević-Pašti, M. Kojić, T. Tasić, M. Marinović-Cincović and D. Živojinović, *J. Water Process Eng.*, 2023, **56**, 104487.
7. B. Katnić, S. J. Porobić, I. Vujčić, M. M. Kojić, T. Lazarević-Pašti, V. Milanković, M. Marinović-Cincović and D. Z. Živojinović, *Radiat. Phys. Chem.*, 2024, **214**, 111277.
8. M. S. Chavali and M. P. Nikolova, *SN Appl. Sci.*, 2019, **1**, 607.
9. S. Maiti, M. R. Islam, M. A. Uddin, S. Afroj, S. J. Eichhorn and N. Karim, *Adv. Sustainable Syst.*, 2022, **6**, 2200258.
10. A. Gholampour and T. Ozbakkaloglu, *J. Mater. Sci.*, 2020, **55**, 829–892.
11. I. Elfaleh, F. Abbassi, M. Habibi, F. Ahmad, M. Guedri, M. Nasri and C. Garnier, *Results Eng.*, 2023, **19**, 101271.
12. W. Ballout, N. Sallem-Idrissi, M. Sclavons, C. Doneux, C. Bailly, T. Pardoen and P. Van Velthem, *Sci. Rep.*, 2022, **12**, 5928.
13. Y. Yang, R. Boom, B. Irion, D.-J. van Heerden, P. Kuiper and H. de Wit, *Chem. Eng. Process.*, 2012, **51**, 53–68.
14. D. J. Gardner, Y. Han and L. Wang, *Curr. For. Rep.*, 2015, **1**, 139–150.
15. A. Wechsler and S. Hiziroglu, *Build. Environ.*, 2007, **42**, 2637–2644.
16. A. Ashori, *Bioresour. Technol.*, 2008, **99**, 4661–4667.
17. J. J. Andrew and H. N. Dhakal, *Compos., Part C: Open Access*, 2022, **7**, 100220.
18. B. Madsen, P. Brøndsted and T. L. Andersen, in *Advances in Wind Turbine Blade Design and Materials*, ed. P. Brøndsted and R. P. L. Nijssen, Woodhead Publishing, 2013, pp. 363-386.

19. M. Ö. Seydibeyoğlu, A. Dogru, J. Wang, M. Rencheck, Y. Han, L. Wang, E. A. Seydibeyoğlu, X. Zhao, K. Ong, J. A. Shatkin, S. Shams Es-haghi, S. Bhandari, S. Ozcan and D. J. Gardner, *Polymers*, 2023, **15**, 984.
20. A. Jocić, S. Breitenbach, D. Bajuk-Bogdanović, I. A. Pašti, C. Unterweger, C. Fürst and T. Lazarević-Pašti, *Molecules*, 2022, **27**, 1477.
21. A. Jocić, S. Breitenbach, I. A. Pašti, C. Unterweger, C. Fürst and T. Lazarević-Pašti, *Environ. Sci. Pollut. Res.*, 2022, **29**, 35138–35149.
22. V. Anićijević, T. Tasić, V. Milanković, S. Breitenbach, C. Unterweger, C. Fürst, D. Bajuk-Bogdanović, I. A. Pašti and T. Lazarević-Pašti, *Int. J. Environ. Res. Public Health*, 2023, **20**, 4553.
23. T. Tasić, V. Milanković, K. Batalović, S. Breitenbach, C. Unterweger, C. Fürst, I. A. Pašti and T. Lazarević-Pašti, *Foods*, 2023, **12**, 2362.
24. T. Lazarević-Pašti, A. Jocić, V. Milanković, T. Tasić, K. Batalović, S. Breitenbach, C. Unterweger, C. Fürst and I. A. Pašti, *C*, 2023, **9**, 103.
25. T. Lazarević-Pašti, *Chemosensors*, 2023, **11**, 93.
26. A. Barberis, Y. Spissu, A. Fadda, E. Azara, G. Bazzu, S. Marceddu, A. Angioni, D. Sanna, M. Schirra and P. A. Serra, *Biosens. Bioelectron.*, 2015, **67**, 214–223.
27. C. Zhang, J. He, Y. Zhang, J. Chen, Y. Zhao, Y. Niu and C. Yu, *Biosens. Bioelectron.*, 2018, **102**, 94–100.
28. L. Lu, L. Zhou, J. Chen, F. Yan, J. Liu, X. Dong, F. Xi and P. Chen, *ACS Nano*, 2018, **12**, 12673–12681.
29. Z. S. Qian, X. Y. Shan, L. J. Chai, J. J. Ma, J. R. Chen and H. Feng, *Biosens. Bioelectron.*, 2014, **60**, 64–70.
30. H. Li, X. Yan, S. Qiao, G. Lu and X. Su, *ACS Appl. Mater. Interfaces*, 2018, **10**, 7737–7744.
31. M. Zheng, S. Ruan, S. Liu, T. Sun, D. Qu, H. Zhao, Z. Xie, H. Gao, X. Jing and Z. Sun, *ACS Nano*, 2015, **9**, 11455–11461.
32. Z. Wang, T. Hu, R. Liang and M. Wei, *Front. Chem.*, 2020, **8**, DOI: 10.3389/fchem.2020.00320.
33. G. Ćirić-Marjanović, I. Pašti and S. Mentus, *Prog. Mater. Sci.*, 2015, **69**, 61–182.
34. Y. Zhu, S. Murali, W. Cai, X. Li, J. W. Suk, J. R. Potts and R. S. Ruoff, *Adv. Mater.*, 2010, **22**, 3906–3924.
35. A. C. Ferrari, F. Bonaccorso, V. Fal'ko, K. S. Novoselov, S. Roche, P. Bøggild, S. Borini, F. H. L. Koppens, V. Palermo, N. Pugno, J. A. Garrido, R. Sordan, A. Bianco, L. Ballerini, M. Prato, E. Lidorikis, J. Kivioja, C. Marinelli, T. Ryhänen, A. Morpurgo, J. N. Coleman, V. Nicolosi, L. Colombo, A. Fert, M. Garcia-Hernandez, A. Bachtold, G. F. Schneider, F. Guinea, C. Dekker, M. Barbone, Z. Sun, C. Galiotis, A. N. Grigorenko, G. Konstantatos, A. Kis, M. Katsnelson, L. Vandersypen, A. Loiseau, V. Morandi, D. Neumaier, E. Treossi, V. Pellegrini, M. Polini, A. Tredicucci, G. M. Williams, B. Hee Hong, J.-H. Ahn, J. Min Kim, H. Zirath, B. J. van Wees, H. van der Zant, L. Occhipinti, A. Di Matteo, I. A. Kinloch, T. Seyller, E. Quesnel, X. Feng, K. Teo, N. Rupesinghe, P. Hakonen, S. R. T. Neil, Q. Tannock, T. Löfwander and J. Kinaret, *Nanoscale*, 2015, **7**, 4598–4810.

36. I. A. Pašti, A. Jovanović, A. S. Dobrota, S. V. Mentus, B. Johansson and N. V. Skorodumova, *Appl. Surf. Sci.*, 2018, **436**, 433–440.
37. I. A. Pašti, A. Jovanović, A. S. Dobrota, S. V. Mentus, B. Johansson and N. V. Skorodumova, *Phys. Chem. Chem. Phys.*, 2018, **20**, 858–865.
38. W. Yuan, Y. Zhou, Y. Li, C. Li, H. Peng, J. Zhang, Z. Liu, L. Dai and G. Shi, *Sci. Rep.*, 2248, 20133.
39. P. Avouris and C. Dimitrakopoulos, *Mater. Today*, 2012, **15**, 86–97.
40. S. Mehdipour-Ataei and E. Aram, *Catalysts*, 2023, **13**, 2.
41. A. J. Schwanke, R. Balzer and S. Pergher, in *Handbook of Ecomaterials*, ed. L. M. T. Martínez, O. V. Kharissova and B. I. Kharisov, Springer International Publishing, Cham, 2017, pp. 1–22.
42. S. Chawla, P. Rai, T. Garain, S. Uday and C. M. Hussain, *Trends Environ. Anal. Chem.*, 2022, **33**, e00156.
43. O. Kanoun, T. Lazarević-Pašti, I. Pašti, S. Nasraoui, M. Talbi, A. Brahem, A. Adiraju, E. Sheremet, R. D. Rodriguez, M. Ben Ali and A. Al-Hamry, *Sensors*, 2021, **21**, 4131.
44. M. Saeed, Y. Alshammari, S. A. Majeed and E. Al-Nasrallah, *Molecules*, 2020, **25**, 3856.
45. Y. Xu, H. Cao, Y. Xue, B. Li and W. Cai, *Nanomaterials*, 2018, **8**, 942.
46. L. Ndlwana, N. Raleie, K. M. Dimpe, H. F. Ogutu, E. O. Oseghe, M. M. Motsa, T. A. M. Msagati and B. B. Mamba, *Materials*, 2021, **14**, 5094.
47. A. Szabó, C. Perri, A. Csató, G. Giordano, D. Vuono and J. B. Nagy, *Materials*, 2010, **3**, 3092–3140.
48. R. Das, Z. Shahnavaz, M. E. Ali, M. M. Islam and S. B. Abd Hamid, *Nanoscale Res. Lett.*, 2016, **11**, 510.
49. L. Hu, Y. Song, J. Ge, J. Zhu, Z. Han and S. Jiao, *J. Mater. Chem. A*, 2017, **5**, 6219–6225.
50. D. Y. Ozherelkov, S. A. Eremin, V. N. Anikin, S. V. Chernyshikhin, A. Y. Nalivaiko and A. A. Gromov, *Mater. Chem. Phys.*, 2021, **267**, 124673.
51. I. Imae, *Coatings*, 2021, **11**, 297.
52. V. Pavlenko, S. Khosravi H, S. Żółtowska, A. B. Haruna, M. Zahid, Z. Mansurov, Z. Supiyeva, A. Galal, K. I. Ozoemena, Q. Abbas and T. Jesionowski, *Mater. Sci. Eng., R*, 2022, **149**, 100682.
53. Z.-G. Shi, Y.-Q. Feng, L. Xu, S.-L. Da and M. Zhang, *Carbon*, 2004, **42**, 1677–1682.
54. L. Ge, C. Zhao, M. Zuo, J. Tang, W. Ye, X. Wang, Y. Zhang and C. Xu, *J. Anal. Appl. Pyrolysis*, 2022, **168**, 105747.
55. C. Ma, M. Zhang, Y. Ding, Y. Xue, H. Wang, P. Li and D. Wu, *Nanomaterials*, 2023, **13**, 1768.
56. M. A. Rosen and H. A. Kishawy, *Sustainability*, 2012, **4**, 154–174.
57. X. Qi, C. Qin, W. Zhong, C. Au, X. Ye and Y. Du, *Materials*, 2010, **3**, 4142–4174.
58. A. Santos, A. Barbosa-Póvoa and A. Carvalho, *Curr. Opin. Chem. Eng.*, 2019, **26**, 139–147.
59. D. Pandey, M. Agrawal and J. S. Pandey, *Environ. Monit. Assess.*, 2011, **178**, 135–160.

60. E. Loyarte-López, M. Barral and J. C. Morla, *Sustainability*, 2020, **12**, 1629.
61. R. S. Treptow, *J. Chem. Educ.*, 2010, **87**, 168–171.
62. International Organization for Standardization (ISO). ISO 14001:2015 Environmental management systems — Requirements with guidance for use. ISO, Geneva, Switzerland, 2015.
63. S. Bilge, L. Karadurmus, A. Sınağ and S. A. Ozkan, *TrAC, Trends Anal. Chem.*, 2021, **145**, 116473.
64. I. Sadok, A. Krzyszczak-Turczyn, B. Czech, P. Parlakidis and Z. Vryzas, *Food Chem.*, 2025, **492**, 145467.
65. C. Onfray and A. Thiam, *Micromachines*, 2023, **14**, 1688.
66. S. J. Malode, M. M. Shanbhag, R. Kumari, D. S. Dkhar, P. Chandra and N. P. Shetti, *J. Pharm. Biomed. Anal.*, 2023, **222**, 115102.
67. A. Abbas, Q. Liang, S. Abbas, M. Liaqat, S. Rubab and T. A. Tabish, *Nanomaterials*, 2022, **12**, 3696.
68. A. Saud, H. Saleem, N. Munira, A. A. Shahab, H. Rahman Siddiqui and S. J. Zaidi, *Nanomaterials*, 2023, **13**, 148.
69. V. C. Hoang, L. H. Nguyen and V. G. Gomes, *J. Electroanal. Chem.*, 2019, **832**, 87–96.
70. R. Atchudan, T. N. J. I. Edison, S. Perumal, N. Muthuchamy and Y. R. Lee, *Fuel*, 2020, **275**, 117821.
71. S. Rodríguez-Varillas, T. Fontanil, Á. J. Obaya, A. Fernández-González, C. Murru and R. Badía-Laíño, *Appl. Sci.*, 2022, **12**, 773.
72. V. K. Jothi, K. Ganesan, A. Natarajan and A. Rajaram, *J. Fluoresc.*, 2021, **31**, 427–436.
73. H. Man, M. Chaima, X. Wang, L. Xiu, L. Yang and J. Huang, *Arabian J. Sci. Eng.*, 2023, **48**, 8315–8324.
74. N. Sharma, G. S. Das and K. Yun, *Appl. Microbiol. Biotechnol.*, 2020, **104**, 7187–7200.
75. A. Z. Alanazi, K. Alhazzani, A. M. Mostafa, J. Barker, M. M. El-Wekil and A.-M. B. H. Ali, *Microchem. J.*, 2024, **197**, 109800.
76. F. Ashrafi Tafreshi, Z. Fatahi, S. F. Ghasemi, A. Taherian and N. Esfandiari, *PLoS One*, 2020, **15**, e0230646.
77. Z. Shekarbeygi, N. Farhadian, S. Khani, S. Moradi and M. Shahlaei, *Microchem. J.*, 2020, **158**, 105232.
78. S. Ghosh, A. R. Gul, C. Y. Park, P. Xu, S. H. Baek, J. R. Bhamore, M. W. Kim, M. Lee, S. K. Kailasa and T. J. Park, *Microchem. J.*, 2021, **167**, 106272.
79. W.-K. Li, J.-T. Feng and Z.-Q. Ma, *Carbon*, 2020, **161**, 685–693.
80. M. M. F. Chang, I. R. Ginjom and S. M. Ng, *Sens. Actuators, B*, 2017, **242**, 1050–1056.
81. S. Ghosh, A. R. Gul, C. Y. Park, M. W. Kim, P. Xu, S. H. Baek, J. R. Bhamore, S. K. Kailasa and T. J. Park, *Chemosphere*, 2021, **279**, 130515.
82. S. D. Hiremath, B. Priyadarshi, M. Banerjee and A. Chatterjee, *J. Photochem. Photobiol., A*, 2020, **389**, 112258.
83. H. Liu, J. Ding, L. Chen and L. Ding, *J. Photochem. Photobiol., A*, 2020, **400**, 112724.

84. Y. Hu, J. Li and X. Li, *Anal. Bioanal. Chem.*, 2019, **411**, 7879–7887.
85. B. Lin, Y. Yan, M. Guo, Y. Cao, Y. Yu, T. Zhang, Y. Huang and D. Wu, *Food Chem.*, 2018, **245**, 1176–1182.
86. Y. Hu, Z. Chen, F. Lai and J. Li, *J. Mater. Sci.*, 2019, **54**, 8627–8639.
87. J. Hou, G. Dong, Z. Tian, J. Lu, Q. Wang, S. Ai and M. Wang, *Food Chem.*, 2016, **202**, 81–87.
88. L. Wang, Y. Bi, J. Hou, H. Li, Y. Xu, B. Wang, H. Ding and L. Ding, *Talanta*, 2016, **160**, 268–275.
89. D. Tang, Y. Lin and Q. Zhou, *Mikrochim. Acta*, 2018, **185**, 476.
90. N. Kazemifard, A. A. Ensafi and B. Rezaei, *Food Chem.*, 2020, **310**, 125812.
91. S. R. Dash, S. S. Bag and A. K. Golder, *Electroanalysis*, 2022, **34**, 1141–1149.
92. L. Fu, Y. Hou, H. Guo, C. Liu, Y. Li, B. Jia, L. Zhou and W. Kong, *Microchem. J.*, 2022, **176**, 107253.
93. X. Wen, L. Shi, G. Wen, Y. Li, C. Dong, J. Yang and S. Shuang, *Sens. Actuators, B*, 2016, **235**, 179–187.
94. A. Tyagi, K. M. Tripathi, N. Singh, S. Choudhary and R. K. Gupta, *RSC Adv.*, 2016, **6**, 72423–72432.
95. Y. Liu, Y. Zhao and Y. Zhang, *Sens. Actuators, B*, 2014, **196**, 647–652.
96. J. R. Bhamore, S. Jha, T. J. Park and S. K. Kailasa, *Sens. Actuators, B*, 2018, **277**, 47–54.
97. S. Liu, J. Tian, L. Wang, Y. Zhang, X. Qin, Y. Luo, A. M. Asiri, A. O. Al-Youbi and X. Sun, *Adv. Mater.*, 2012, **24**, 2037–2041.
98. X. Yang, Y. Zhuo, S. Zhu, Y. Luo, Y. Feng and Y. Dou, *Biosens. Bioelectron.*, 2014, **60**, 292–298.
99. L. Shi, B. Zhao, X. Li, G. Zhang, Y. Zhang, C. Dong and S. Shuang, *Sens. Actuators, B*, 2016, **235**, 316–324.
100. A. Sachdev and P. Gopinath, *Analyst*, 2015, **140**, 4260–4269.
101. D. Bano, V. Kumar, V. K. Singh and S. H. Hasan, *New J. Chem.*, 2018, **42**, 5814–5821.
102. N. Wang, Y. Wang, T. Guo, T. Yang, M. Chen and J. Wang, *Biosens. Bioelectron.*, 2016, **85**, 68–75.
103. W. Lu, X. Qin, S. Liu, G. Chang, Y. Zhang, Y. Luo, A. M. Asiri, A. O. Al-Youbi and X. Sun, *Anal. Chem.*, 2012, **84**, 5351–5357.
104. X. C. Qin, Z. Y. Guo, Z. M. Liu, W. Zhang, M. M. Wan and B. W. Yang, *J. Photochem. Photobiol., B*, 2013, **120**, 156–162.
105. S. Venkateswarlu, B. Viswanath, A. S. Reddy and M. Yoon, *Sens. Actuators, B*, 2018, **258**, 172–183.
106. T. N. J. I. Edison, R. Atchudan, J.-J. Shim, S. Kalimuthu, B.-C. Ahn and Y. R. Lee, *J. Photochem. Photobiol., B*, 2016, **158**, 235–242.
107. R. Bandi, B. R. Gangapuram, R. Dadigala, R. Eslavath, S. S. Singh and V. Guttena, *RSC Adv.*, 2016, **6**, 28633–28639.
108. P. Song, L. Zhang, H. Long, M. Meng, T. Liu, Y. Yin and R. Xi, *RSC Adv.*, 2017, **7**, 28637–28646.
109. Y. Guo, L. Zhang, F. Cao and Y. Leng, *Sci. Rep.*, 2016, **6**, 35795.

110. T. Chatzimitakos, A. Kasouni, L. Sygellou, A. Avgeropoulos, A. Troganis and C. Stalikas, *Talanta*, 2017, **175**, 305–312.
111. J. Wei, B. Liu and P. Yin, *RSC Adv.*, 2014, **4**, 63414–63419.
112. J. B. Essner, C. H. Laber, S. Ravula, L. Polo-Parada and G. A. Baker, *Green Chem.*, 2016, **18**, 243–250.
113. H. Huang, J.-J. Lv, D.-L. Zhou, N. Bao, Y. Xu, A.-J. Wang and J.-J. Feng, *RSC Adv.*, 2013, **3**, 21691–21696.
114. R. Bandi, R. Dadigala, B. R. Gangapuram and V. Guttena, *J. Photochem. Photobiol., B*, 2018, **178**, 330–338.
115. M. R. Pacquiao, M. D. G. de Luna, N. Thongsai, S. Kladsomboon and P. Paoprasert, *Appl. Surf. Sci.*, 2018, **453**, 192–203.
116. Z. Li, Q. Wang, Z. Zhou, S. Zhao, S. Zhong, L. Xu, Y. Gao and X. Cui, *Microchem. J.*, 2021, **166**, 106250.
117. M. Xue, M. Zou, J. Zhao, Z. Zhan and S. Zhao, *J. Mater. Chem. B*, 2015, **3**, 6783–6789.
118. D. B. Gunjal, Y. M. Gurav, A. H. Gore, V. M. Naik, R. D. Waghmare, C. S. Patil, D. Sohn, P. V. Anbhule, R. V. Shejwal and G. B. Kolekar, *Opt. Mater.*, 2019, **98**, 109484.
119. X. Tang, L. Wang, H. Ye, H. Zhao and L. Zhao, *J. Photochem. Photobiol., A*, 2022, **424**, 113653.
120. Y.-M. Liang, H. Yang, B. Zhou, Y. Chen, M. Yang, K.-S. Wei, X.-F. Yan and C. Kang, *Anal. Chim. Acta*, 2022, **1191**, 339269.
121. H. Liu, L. Ding, L. Chen, Y. Chen, T. Zhou, H. Li, Y. Xu, L. Zhao and N. Huang, *J. Ind. Eng. Chem.*, 2019, **69**, 455–463.
122. H. Qi, M. Teng, M. Liu, S. Liu, J. Li, H. Yu, C. Teng, Z. Huang, H. Liu, Q. Shao, A. Umar, T. Ding, Q. Gao and Z. Guo, *J. Colloid Interface Sci.*, 2019, **539**, 332–341.
123. H. Miao, Y. Wang and X. Yang, *Nanoscale*, 2018, **10**, 8139–8145.
124. B. K. Korah, A. R. Chacko, S. Mathew, B. K. John, T. Abraham and B. Mathew, *Anal. Bioanal. Chem.*, 2022, **414**, 4935–4951.
125. H. Qi, C. Liu, J. Jing, T. Jing, X. Zhang, J. Li, C. Luo, L. Qiu and Q. Li, *Dyes Pigm.*, 2022, **206**, 110555.
126. X.-R. Guo, Y.-M. Dong, X.-Y. Chen and J. Chen, *Spectrochim. Acta, Part A*, 2022, **282**, 121703.
127. N. Chaudhary, D. Verma, J. Gopal Sharma and P. R. Solanki, *Mater. Lett.*, 2022, **308**, 131090.

CHAPTER 12

Low-carbon Materials: Transportation and Automotive Applications

ATUL SHARMA,[a] RASHMI PUNDEER,[b] NEHA DUA,[c] HARISH KUMAR,[d] D. K. SHARMA,[e] SAIF ALI CHAUDHRY[a] AND BINDU MANGLA*[f]

[a] Department of Chemistry, Jamia Millia Islamia, New Delhi 110025, India; [b] Department of Chemistry, Indira Gandhi University, Meerpur 122502, Rewari, Haryana, India; [c] Department of Chemistry, Indian Institute of Technology, Roorkee 247667, Haridwar, Uttarakhand, India; [d] Department of Chemistry, School of Chemical Sciences, Central University of Haryana, Mahendergarh, 123031, India; [e] Department of Sciences, Manav Rachna University, Faridabad-121004, Haryana, India; [f] Department of Chemistry, J.C. Bose University of Science and Technology, YMCA, Faridabad 121006, Haryana, India

*Email: bindumangla@gmail.com

12.1 Introduction

The search for sustainable development and climate change responses has led to a focus on low-carbon products in many sectors. Characterized by the reduction of carbon emissions during production, use, and disposal, this material plays a key role in the transition to a low-carbon economy. Climate change has brought unprecedented challenges to ecosystems, economies, and human societies worldwide. All emerging sectors need to reach a rapid and collaborative consensus on the role of anthropogenic greenhouse gas emissions in driving climate change. Among these, the materials industry

Low Carbon Materials: Paving a Sustainable Path for a Greener Future
Edited by Santanu Patra, Sudheesh K. Shukla and Mika Sillanpää

Published by the Royal Society of Chemistry, www.rsc.org

stands as both a significant contributor to emissions and a potential catalyst for transformative change.[1]

Materials are the fundamental elements of contemporary civilization, appearing in everything from technology to infrastructure.[2] Nonetheless, traditional manufacturing techniques and materials – like steel, cement, and plastics – contribute significantly to carbon emissions. Achieving the global climate targets requires addressing the carbon intensity of materials production and use. Switching to low-carbon materials is not without its difficulties. Numerous obstacles like a lack of technical knowledge, scalability and performance restrictions, cost, market demand, and competition, are making this shift difficult.

Despite the abovementioned challenges, there are infinite opportunities in this sector, but carbon footprint reduction is not a deal to be tackled by individual efforts alone. It can be achieved *via* collaborative means by connecting interdisciplinary research with materials science. A well planned cooperative strategy mechanism covering business, education, government, and civil society should share best practices, resources, and knowledge, in order to switch to low-carbon materials. However, it should be a matter of policy interventions. There is a need to set up policies requiring emissions reductions, carbon pricing schemes, and subsidies for sustainable alternatives to foster an atmosphere favorable to innovation. This chapter discusses low-carbon material prospects and problems, while laying out a roadmap for achieving climate neutrality.

12.2 Low-carbon Materials in Vehicle Design

The automotive sector is at a critical juncture as it faces the need to prevent climate change and reduce its ecological impact. The sector is one of the largest producers of carbon emissions worldwide and must use effective strategies to ensure sustainable development.[3]

This section examines the use of low-carbon materials in vehicle manufacturing and how it can transform the transportation industry and lead to future prosperity. The burning of fossil fuels in automobiles is a major source that contributes to the world's carbon emissions. The automotive industry is increasing the use of alternative technologies to support the urgent need to reduce its carbon footprint and mitigate climate change.[4]

The design of the conventional car has put a great reliance on energy-intensive materials like steel, aluminum, and plastic, all serving as a major contributor to greenhouse gas emissions. However, with the use of low-carbon substitutes these GHG emissions can be reduced by proper use of materials science and engineering. Many low-carbon materials such as recycled materials, bio-based materials, heavy metals, and composites have been investigated for use in automotive applications.[5]

This knowledge has the opportunity to reduce environmental impact, increase fuel efficiency, and reduce vehicle weight. Another important strategy to improve vehicle economy and reduce emissions is weight.

By replacing traditional materials with lighter alternatives such as aluminum alloys or carbon fiber-reinforced polymer (CFRP), automakers can save weight without sacrificing performance or safety.

As we move toward electric vehicles (EVs), the new equipment presents unique opportunities and challenges. The primary energy storage for EVs is lithium-ion batteries, which are based on materials such as nickel, cobalt, and lithium. For EVs to become widespread, ethical and safety measures must be taken, and battery use and recycling must be improved. Life cycle assessment (LCA) is essential to assess the environmental impact of a vehicle throughout its operating life. This process is important for reducing emissions and identifying improvement opportunities.

A framework for comparing sustainability performance and promoting continual improvement is provided by environmental certification programs, such as LEED (leadership in energy and environmental design) for automobiles. The promise of low-carbon materials is not without difficulty, yet there are still a number of obstacles to overcome.[6] These consist of consumer acceptability, supply chain restrictions, cost considerations, and technology limitations.

Continued innovation and cooperation to create and apply sustainable technology and low-carbon materials will shape vehicle design in the future. To accelerate the transition to a more sustainable transportation system, funding for R&D, legislation that supports it, and consumer education initiatives will be crucial.

12.3 Advancements in Energy-storage Materials

12.3.1 Electric Vehicle Battery Materials

The rapid electrification of the automotive industry has brought EVs to the forefront of transportation innovation. Central to the success and proliferation of EVs are the batteries that power them. Here, the intricate world of electric vehicle battery materials, exploring the advancements, challenges, and future prospects shaping the evolution of this critical technology are discussed.

With growing concerns over climate change and air pollution, electric vehicles have emerged as a promising solution to decarbonize transportation. EVs offer the potential to reduce greenhouse gas emissions, improve air quality, and decrease dependence on fossil fuels, making them a key driver of sustainable mobility. At the heart of every electric vehicle lies a rechargeable battery pack that stores and delivers electrical energy to power the vehicle's electric motor.[7] The performance, energy density, durability, and cost of these batteries are crucial factors influencing the adoption and acceptance of electric vehicles on a mass scale. Electric vehicle batteries typically have several main components, including cathodes, anodes, electrolytes, and separators. Each of these components plays an important role in the battery's overall functionality, performance, and safety. Battery product development has led to progress in improving energy density, life cycle, charging, and safety. Innovations in cathode materials such as lithium metal phosphate (LFP), nickel manganese cobalt oxide (NMC), and nickel cobalt aluminum (NCA) have given the electric car more energy and longer journeys. Despite its success, EV battery technology still faces a number of challenges and limitations. Issues such as limited availability of raw materials, supply chain constraints, cost considerations, and environmental concerns are hindering EVs' widespread adoption and growth.[7]

Extraction, processing, and disposal of battery products can impact the environment and society, including environmental damage, water pollution, and human rights violations. Designing of strong and closed-loop recycling systems can help save valuable information and reduce waste. New technologies such as solid-state, lithium–sulfur, and sodium-ion batteries offer potential performance and safety-enhancing alternatives to traditional lithium-ion batteries and are being designed and installed.

12.3.2 High-efficiency Engine Components

EVs have emerged as a promising tool for reducing carbon emissions and reducing dependence on fossil fuels. At the heart of every electric vehicle is the drive train, which is made up of many components that work together to convert electricity into energy.

This section discusses the field of high-performance electronic devices, reviewing new developments, integration strategies, and optimization techniques that have enabled success in this important vehicle.

As concerns about climate change and air pollution grow, the automotive industry is undergoing major changes; electric vehicles are gaining importance as cleaner and more efficient than internal combustion engines. Electric vehicles can reduce greenhouse gas emissions, improve air quality, and improve safety.[8] An electric car drive system consists of several components, each of which plays a key role in converting electrical energy stored in the battery into mechanical energy to move the vehicle forward.

Electric motors, the primary driver of electric vehicles, convert electrical energy into all forms of energy. Advances in motor design include the use of permanent magnet (PMSYM) materials, improved performance, power output, and electronic components. The electric motor and motor control the flow of electrical power between the battery pack and the motor, making power transfer efficient and enabling the motor to operate. Innovations in semiconductor technologies, such as silicon carbide (SiC) and gallium nitride (GaN) power components, enable higher switching frequencies, lower costs, and increased overall efficiency.

The transmission and drivetrain components of an electric vehicle play a key role in transferring power from the engine to the wheels, ensuring power distribution and efficiency. Single-speed transmissions, multi-speed transmissions, and direct-drive transmissions are common transmission options used in electric vehicles, and offer performance, convenience, and cost advantages in electric vehicles. Good thermal management is essential to maintaining the temperature of EV equipment, increasing efficiency and ensuring long-term reliability. Liquid cooling systems such as heat pumps, and thermal generators are used to capture the heat generated by the engine, electric generator, and battery pack.

Regenerative braking systems allow an electric vehicle to recover kinetic energy during deceleration and braking and convert it back into electrical energy to charge the battery. Regenerative braking systems increase overall efficiency and extend driving by harnessing energy that would otherwise be dissipated as heat. Rapid electric development and infrastructure development, continuous research and development, as well as supportive policies and incentives, are crucial to overcoming these challenges and enabling the world to adopt electric vehicles.

12.3.3 Fuel Cell and Hydrogen Storage Materials

As the world strives to transition to cleaner, more sustainable energy sources, hydrogen has emerged as a promising alternative with the potential to revolutionize many areas, including transportation, business, and power generation. At the heart of hydrogen-based technologies are fuel cells and hydrogen storage systems, both of which rely on advanced materials for energy conversion and storage. In this section, we explore recent developments in fuel cells and hydrogen storage materials, and highlight their role in unlocking the potential of the hydrogen industry.

Hydrogen is abundant, versatile, and emits only water vapour when used as a fuel, making it an attractive option for reducing greenhouse gas emissions and combating climate change. The hydrogen industry aims to use hydrogen as a clean energy source to replace fossil fuels and help transition to a low-carbon future.

Fuel cells are electrochemical systems that only produce water as a byproduct while converting hydrogen and oxygen into heat and power. Fuel cells' electrolyte, electrodes, and catalysts are essential parts that depend on cutting-edge materials to facilitate effective energy conversion.[9] Solid oxide fuel cells (SOFCs) and proton exchange membrane fuel cells (PEMFCs) are two of the most promising fuel cell technologies for a range of uses, such as portable electronics, stationary power generation, and transportation. Fuel cell material development has advanced significantly, with an emphasis on enhancing cost-effectiveness, performance, and durability. Platinum-group metals (PGMs) and non-precious metal catalysts are examples of catalyst materials that are essential for promoting the electrochemical reactions that take place inside fuel cells.[10]

One of the key challenges in realizing the full potential of hydrogen-based technologies is the safe and efficient storage of hydrogen. Hydrogen storage materials, such as metal hydrides, carbon-based materials, and chemical hydrides, offer different storage mechanisms and capacities depending on the application requirements. Advances in material design, synthesis, and characterization are essential for developing high-capacity, lightweight, and cost-effective hydrogen storage solutions.

Metal hydrides are among the most promising hydrogen storage materials, offering high storage densities, reversible hydrogen uptake/release, and excellent safety characteristics.[11] Transition metal hydrides, including complex metal hydrides and alanates, exhibit favorable hydrogen storage properties and have been extensively studied for automotive, stationary, and portable applications.

Chemical hydrides, such as ammonia borane, sodium borohydride, and formic acid, offer reversible hydrogen storage through chemical reactions that release hydrogen upon demand. While chemical hydrides typically have lower gravimetric and volumetric storage densities compared to other storage methods, their ease of handling and potential for onboard hydrogen generation make them suitable for certain applications, such as portable electronics and backup power systems.

It is necessary to carefully evaluate system design, efficiency, safety, and cost when integrating fuel cells and hydrogen storage systems into real-world applications. System-level optimization is necessary to maximize the performance and dependability of hydrogen-based technologies in a variety of applications. This includes heat management, pressure regulation, and hydrogen distribution.

12.4 Low-carbon Materials in Interior and Exterior Components

The low-carbon transportation infrastructure aims at decreasing carbon emissions throughout its life cycle. Technologies focus on solid waste utilization, green construction, renewable energy, and low-carbon preventive maintenance such as using industrial waste in roadbeds and applying energy-efficient construction techniques. These efforts aim to reduce the environmental impact and promote sustainability.

12.4.1 Sustainable Interior Materials

The evaluation of the dynamic characteristics of non-woven composites is crucial. This study focuses on composites including flax, hemp, glass fibers, and kenaf combined with polypropylene. It examines how factors like reinforcement, fiber to matrix weight ratio, fiber orientation, and porosity content influence the damping behavior. These superior damping properties make nonwoven composites particularly appealing for automotive applications where effective vibration dissipation and prevention of fatigue fractures is essential. The use of natural fibers as reinforcement in composite materials has gained attention, especially within the automotive industry, primarily due to their favorable mechanical properties compared to synthetic fibers.[12,13]

Nonwoven composites, engineered for specific mechanical properties, have garnered interest for automotive interior parts.[14–16] Martin *et al.* reported that nonwoven composites show good mechanical properties which depend on various parameters including their manufacturing.[17]

Roussière *et al.*[18] examined the compressive and tensile behavior of nonwoven composites consisting of randomly oriented flax fibers reinforcing various types of matrix materials at a consistent fiber content. Their findings demonstrated that flax–polylactic acid (PLA) composite exhibited tensile properties comparable to those of glass–polyester composite. Additionally, Merotte *et al.*[19,20] emphasized the advantageous acoustical properties of randomly dispersed nonwoven flax–polypropylene (PP) composites, achievable by regulating the porosity content within the composite. Optimal acoustical properties were observed with a porosity content of 60%.

Numerous investigations have examined the dynamic characteristics of composites reinforced with synthetic fibers, concentrating on elements such as fiber angles and geometric ratios in laminated composites, as well as the impact of nanoplates and nanotubes as reinforcing elements.[21–23] Regarding the damping qualities of hybrid composites with polyamide nonwoven fabrics and compression-molded nonwoven mats based on different natural fibers, certain investigations have produced encouraging findings.[24]

Using DMA, research has been conducted on the dynamic characteristics of composites reinforced with flax, carbon, and glass fibers that are unidirectional and twill. According to their findings, flax-based composites offer better damping qualities than carbon and glass composites.[25] In this study, free–free flexural vibration experiments were used to evaluate the damping capabilities of four nonwoven composites: flax–PP, hemp–PP, kenaf–PP, and glass–PP. The findings showed that composites made of hemp, kenaf, and flax fibers had better damping qualities than glass–PP composites. Additionally, it was discovered that the porosity content and fiber/matrix weight ratio had a substantial impact on the flax–PP nonwoven composite's damping capacity. The energy dissipation and damping properties of the composite were improved by increasing the weight ratio of polypropylene.

Additionally, higher porosity content notably improved the damping properties of flax–PP composite. A predictive model for the loss factor of flax–PP composite based on porosity content was proposed, offering valuable insights for optimizing damping performance in automotive interior components.

The use of sustainable low-carbon material in the interior of vehicles containing cockpit, seat door, side trim, floor trim, headliner, and luggage compartment is the demand of the time and can be encouraged through consistent innovation strategies.[26] Natural fibers, hemp, kenaf, and flax are preferably used as environment-friendly alternatives to glass fibers because of lower cost, lower weight, lower density, higher load-bearing capacity, and higher strength. Natural fibers are used as reinforcement in interior parts like door panels because of their high strength. Kenaf has one of the best CO_2-absorption properties among plants.[27]

Natural fiber-reinforced plastic does not create sharp edges when breaking. Due to the high performance and low cost, the composites based on natural fibers are used for thermoplastics, thermosets, and elastomers in automotive interiors, for example, doors made up of flax/sisal fibers, and binders like polypropylene or polyurethane.[28] Lightweight and ecologically compatible interior surfaces can be obtained by using biocomposite materials. Natural fiber-reinforced polypropylene is used in the kenaf door trim for the BMW i3 electric vehicle. The supplier, Draxlmaier, made kenaf-fiber panels coated with transparent plastic.[29,30] Fibers of the tropical mallow plant kenaf are used in invisible door beams and instrument panel covers of BMW i3 to make a high-quality surface with a high degree of fineness. Yanfeng Automotive interiors presented recyclable and sustainable middle parts of the door panel. The supplier, International Automotive components, worked on replacing the classic material with the natural fiber product "EcoMatHot" consisting of 70% renewable material in the mounting frame of the vehicle roof lining with sliding roofs.[31]

12.4.2 Low-emission Materials: Transportation Infrastructure Carbon Emission Reduction Strategies

Transportation infrastructure contributes significantly to global carbon emissions, while road construction, rehabilitation, and maintenance result in approximately 72% of these emissions.[32] The assessment of transportation infrastructure's carbon emissions is a crucial first step in controlling significant environmental problems, as it provides empirical support for the implementation of green technology and policies.[33]

It has been discovered that, as compared to densely graded asphalt pavement during the use phase, greenhouse gas (GHG) emissions are reduced by 46.7% in permeable asphalt pavements. Numerous steps have been taken to reduce carbon emissions in the transportation system. In China, for example, the use of warm-mix asphalt and recycled aggregate has shown amazing results in lowering carbon emissions. The California Department of Transportation in the United States is demanding Environmental Product Declarations to quantify emissions from materials used in building projects and is deploying advanced pavement management systems.[34]

Vijayan *et al.* have presented a comprehensive writing on the use of carbon fiber-reinforced polymer (CFRP) composites in building lightweight structures owing to the high strength-to-weight ratio as compared to traditional building materials such as concrete and steel.[35] CFRP composites are being used in concrete bridges and beams to strengthen them, and also to increase their flexural and tensile strength in order to enhance their overall durability. The corrosion resistance of CFRP composites makes them a valuable material to be used under drastic environmental conditions of moisture such as in marine structures.[36] The composites, being highly adaptable, can be interwoven in different patterns as per the requirement of the structure.

Structures made up of CFRP composites possess high fatigue resistance, which is a favorable quality when the structure is to be used for heavy loads and constant movement takes place, for example, bridges.[37]

12.4.3 Comparison of CO_2 Emissions between ICEVs and EVs for Better Insights into Low Carbon Emissions

A case study in the US was carried out by Farzaneh and Jung on a comparison of the carbon footprints of internal combustion engine vehicles (ICEVs) and electric vehicles (EVs). The C-footprints of the two types of vehicles were investigated from cradle to grave. The carbon emission of EVs (363.2 gCO_2-eq km^{-1}) was found to be much lower than that of ICEVs (469.1 gCO_2-eq km^{-1}).[38] The municipal corporations of Ibague and Venadillo were chosen as urban and rural areas, respectively, from the Colombian region of Tolima. Global warming potentials (GWPs) of electric, diesel, and gasoline cars were obtained. Urban and rural areas differ in terms of C-emissions from private transportation due to factors like daily trips, daily distance travelled, average occupancy, fleet size, and vehicle specifications. In Ibague, daily carbon emissions from automobiles were found to be 460% higher than from motorcycles. On average, emissions from Ibague (140 g CO_2-eq p km^{-1}) were 23% higher than the rural area Venadillo (110 g CO_2-eq p km^{-1}). Also, average daily carbon emissions per passenger were 120% higher (2000 g CO_2-eq p km^{-1}) in the case of the urban area compared with the rural one (see Figure 12.1).[39]

Greenhouse gas emissions of an ICEV, *i.e.* a motorcycle, were compared with an electric motorcycle in Thailand. The motorcycle with an internal combustion engine released 43.6 g of CO_2 km^{-1}, an electric motorcycle using coal-generated electricity released 20.7 g of CO_2 km^{-1} and an electric motorcycle using natural-gas-generated electricity released as low as 10.7 g of CO_2 km^{-1}.[40] Morfeldt *et al.*[41] have studied the tailpipe CO_2 emission of IECVs w.r.t. four-wheel vehicles, and the value was found to be 173 g of CO_2 km^{-1}.[41] With an average one-passenger and life-span of 15 000 km, CO_2 emission of an electric bike is 22 g of CO_2-eq km^{-1} as compared to a conventional motorbike having 16 g of CO_2.eq km^{-1}.[42]

12.5 Challenges of Utilizing Materials with a Lower CO_2 Footprint in the Automotive Industry

12.5.1 Cost and Regulatory Certification Challenges

Cost plays an important role in the effective working of any policy. It is widely evident that many sustainable materials are slightly more expensive than traditional ones. Moreover, the scalability of the low-footprint carbon materials is in its infancy.

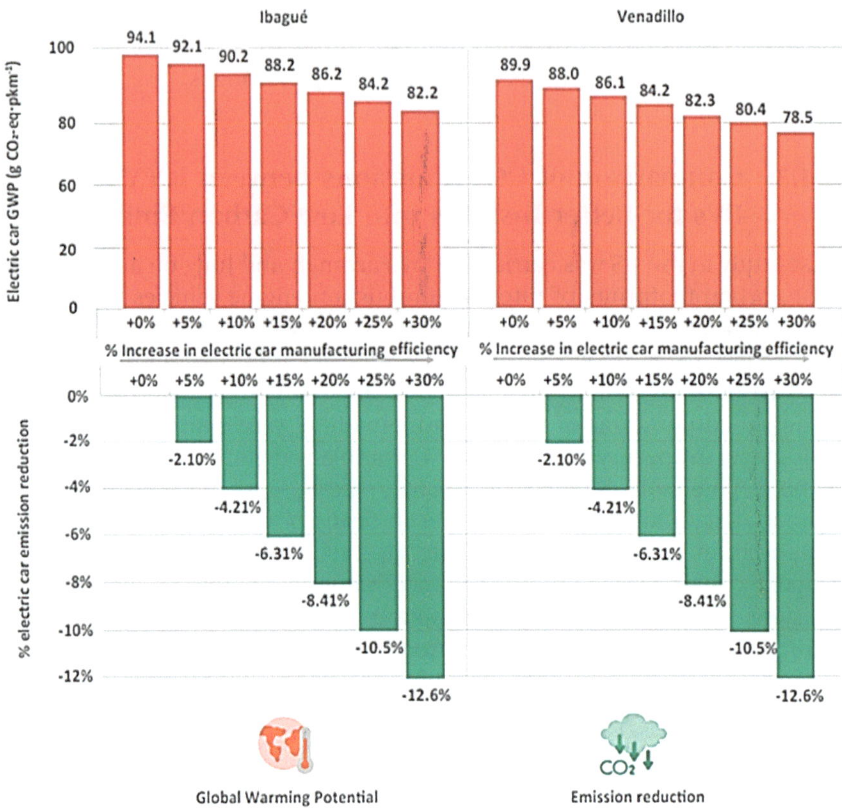

Figure 12.1 Sensitivity analysis of EV global warming potential with EV manufacturing efficiency. Reproduced from ref. 39, https://doi.org/10.1016/j.scs.2023.104658, under the terms of the CC BY 4.0 license, https://creativecommons.org/licenses/by/4.0/.

When the material is new, the issues regarding the material's performance and safety can be somewhat complex, which ultimately poses challenges in its certification.

12.5.2 Performance and Durability

Low-CO_2 materials may not offer the same strength or durability as offered by traditional materials and therefore this is also a major challenge with these materials. Moreover, the long-term durability of sustainable materials may be uncertain as bio-based materials can degrade faster when exposed to weathering and other atmospheric conditions.

Hybrid electric vehicles (HEVs) are seen as transitional because of their limited electricity replenishment opportunities, while fuel cell electric vehicles (FCEVs) face challenges related to high hydrogen production costs and safety concerns. As a result, the focus of the discussion leans toward

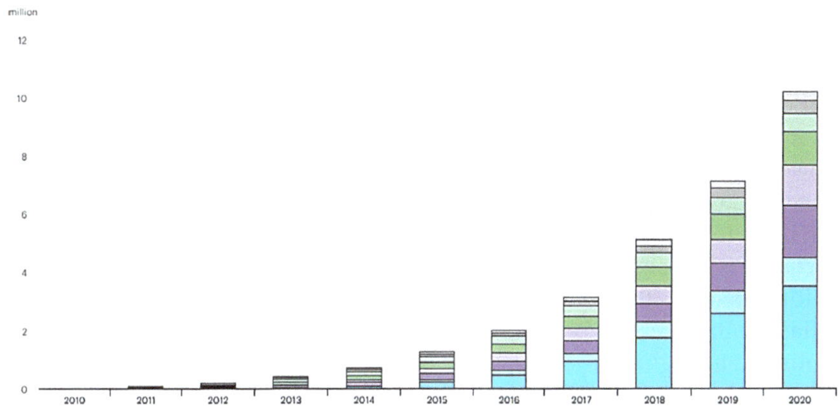

Figure 12.2 Worldwide electric car stock by region and type (International Energy Agency, 2021. Electric Vehicles. Research Report, Paris, France). Reproduced from ref. 55, https://www.iea.org/reports/global-ev-outlook-2021, under the terms of the CC BY 4.0 license, https://creativecommons.org/licenses/by/4.0/.

battery electric vehicles (BEVs), which are more widely adopted and supported in the current market context (see Figure 12.2).

12.5.3 Technological Barrier

Low-CO_2 materials require new manufacturing technologies which are quite difficult to implement with existing production facilities, along with the fact that material integration will also be problematic as it can require adjustments in vehicle design and assembly.

12.5.3.1 Cost and Supply Chain Considerations

While reformulated fuels—particularly drop-in variants—can generally operate within current transportation fleets, achieving meaningful lifecycle reductions in greenhouse gas emissions requires substantial financial commitments from the refining sector. This effort often involves both the modernization of pre-existing processing systems and the establishment of advanced production units capable of supporting low-emission fuel technologies. Such technological advancement presents a major economic constraint, reflecting the high capital and operational expenditures necessary to meet stringent environmental and performance standards.

In view of the cost-effectiveness of low-carbon fuels, consideration of the total cost of ownership is essential, which includes environmental benefits, maintenance costs, and vehicle efficiency in its overall lifecycle. In case of non-drop-in fuels, the capital cost and feasibility of developing infrastructure depends on the region and transport sector. In the US, there are approximately 150 000 fuel stations, out of which about 95% are owned by independent retailers.

The large number of stations and fragmented ownership seem to be the major obstacle for retail deployment of fuels, and emphasizes the interest in drop-in fuels for road transportation.

Government policies on subsidizing low-carbon fuels can be very supportive in incentivizing the widespread adoption of these fuels. One of the challenges in using non-drop-in fuels is the lengthy and costly processes associated with certifying new fuels, as is the case in the marine and aviation sectors. International codes for ships using gases and other low-flashpoint fuels (IGF code) for liquefied natural gas (LNG) are well established and guidelines for methanol have also been adopted recently, while ammonia and hydrogen are yet to be formally established within the code. For the aviation industry, the drop-in sustainable aviation fuel (SAF) is considered to be a direct replacement of fossil jet fuel as it is devoid of any requirement for equipment or infrastructure. However, the different jet fuels need to go through a three-phase qualification process (ASTM D4054).[43] Investment in the infrastructure required for production, distribution, and storage such as refueling stations and processing units is essential for their scalability. With the increase in the production volumes, unit costs will decrease, making the fuels more cost-competitive. Advancements in research and development can help to improve the efficiency of low-carbon fuel and thus reduce production costs also. Considering non-drop-in fuels, many supply chains need to be established or scaled. The supply chain begins with the feedstock sourcing which includes captured CO_2, biomass, renewable electricity, and waste materials. Some value chains based on biomass have the limitation of the availability of raw feedstocks. However, the value chains involving captured carbon dioxide and renewable energy have fewer limitations.[43,44]

In order to optimize production efficiency, biofuel refineries or renewable energy plants which are critical elements of supply chain must be located strategically.

For the distribution of low-carbon fuel from the production site to end users (airports, fueling stations), an efficient network of logistics is essential to reduce supply chain bottlenecks. A resilient supply chain built by establishing an effective transportation system and diversifying feedstock sources can help to ensure continuous availability of the fuel and mitigate disruptions.[45]

12.5.3.2 Regulatory and Safety Standard Compliance

Policymakers are urged to establish market mechanisms for charging infrastructures and grid services to facilitate integration into grid operations. Additionally, more than 20 countries have initiated plans to phase out combustion engine vehicles, offering incentives to promote electric vehicle production and usage. Ambitious governmental policies, such as China's

New Energy Automobile Industry Plan and the European Union's Sustainable and Smart Mobility Strategy, aim to accelerate electric vehicle development and adoption, with a focus on low-carbon transformation in transportation.

These parameters include increasing the deployment of alternative and clean energy-powered vehicles, electrifying urban transit and heavy-freight vehicles, and setting targets for reducing carbon emissions intensity and energy consumption in transportation by 2030. Globally, there has been significant growth in the adoption of electric buses and trucks, with China leading the way, followed by Europe, India, and Latin America. In California, efforts to achieve carbon neutrality by 2045 include measures such as the Low Carbon Fuel Standard and transitioning to zero-emission vehicles. However, achieving carbon neutrality in transportation requires addressing emissions throughout the entire well-to-wheel process, including energy production, distribution, and storage.

12.6 Impact of Low-carbon Materials on the Transportation Sector

12.6.1 Reduction of Greenhouse Gas Emissions

Many studies develop strategies for low-carbon transportation policies through scenario analysis, often concentrating on specific aspects like fuel consumption policies or carbon taxation.[46] Many of the low-carbon policies highlight the transportation system and emission of carbon.

Different circumstances and demands play crucial roles in selecting low-carbon transportation policies. However, urban transportation involves many stakeholders and factors, and it is difficult to examine only one integrated approach to low-carbon policies. Therefore, this chapter adopts a multi-actor multivariate analysis (MAMCA) with an additional optimization model based on social analysis (SNA) to evaluate low-carbon assessment. This method considers stakeholder opinions and their capacity to acquire resources and information, ensuring a more comprehensive and accurate evaluation of alternatives. Utilizing MAMCA with SNA proves effective and offers valuable insights for the development of effective policies regarding low-carbon transport, as demonstrated in the case of Tianjin.[47] Many scientists have summarized the policies and discussed which type of policy is more beneficial. A summary is provided in Table 12.1.

These low-carbon policies are all supposed to have short-term or medium-term effects and are dedicated to the operation stage within the traffic system. The details are provided in Table 12.2.

The MAMCA method is used to develop in-depth analysis and evaluate the various transport policies. Multi-criteria analysis (MCA)[56] and multi-criteria decision analysis (MCDA)[57] have been increasingly used in transportation policy evaluation to address multiple objectives effectively. However, as the number of stakeholders involved grows, traditional MCA methods become

Table 12.1 Government agencies program on urban low-carbon transport policies.

Best possible low-carbon transport policies	Government agencies
Different operation management methods	China
Traffic management	China
Technologies	United States
Planning and design	United States
Finance and investment	United States
Marketing and promotion	China
Tax adjustment	China
Public participation	United States
Safety	United Kingdom, United States, Canada

Table 12.2 Details of the selected low-carbon policies.

Policy measures	Description
Tax adjustment	Policies that are used to adjust taxes and related to transportation
Environmental measures	Publicity regarding environmental protection regulations
Traffic management	Policies that redistribute the demand for travel time by influencing people's behavior
Price adjustment mechanism	Policies that adjust public transportation prices
Multi-operation mechanism	Policies that introduce a variety of operating mechanisms for public transport infrastructure

less practical due to challenges in facilitating dialogue and achieving consensus. To overcome these limitations, Macharis proposed the multi-actor multi-criteria analysis (MAMCA) method, which explicitly incorporates the viewpoints of different stakeholders. MAMCA has been applied in various transportation decision problems, offering a systematic and comprehensive approach. This study combines MAMCA with MCDA, considering stakeholder opinions to analyze and evaluate transportation policies. While MAMCA is effective, determining stakeholder weights is crucial and can be addressed using social network analysis (SNA).[48] For example, research using MAMCA shows that most respondents believe that government funding and subsidies are the best way to achieve low-cost transportation, excluding education, government control, and traffic management. While the technology division seeks to maximize revenues from such initiatives, end users and the planning department support top-down economic policies due to their wider influence. Governmental oversight bodies and scholars, on the other hand, have more expansive goals. They challenge the efficacy and efficiency of public funds and subsidies and support traffic demand management as a quicker way to control traffic and safeguard the environment. As a result, the analysis emphasizes the variety of stakeholder preferences. Through comprehension of these diverse goals and standards, decision-makers can modify and execute selected policy plans appropriately. For example, governments should provide more convenience and preferential measures, including lower public transportation costs, subsidies for eco-friendly vehicles, and more public transportation options, to win over end consumers.

Government departments are primarily concerned about air pollution and gasoline reliance among all traffic-related environmental issues. Stakeholders believe that price adjustment systems, environmental awareness campaigns, and governmental funding and subsidies are generally helpful in addressing the campaign for improved air quality. In particular, tax adjustment strategies are thought to be a better option for lowering reliance on fuel.[49]

It has been observed that key elements such as increasing efficiency, flexibility, energy saving, and renewable energy injection play a significant role in managing the 30% of carbon emissions from transportation.

Many studies have also focused on life cycle analysis (LCA) to assess carbon releases within the railroad sector.[50] Previous studies focused on infrastructure and operation phases in railway LCA. However, it is worth noting that the contribution of vehicles to the life of the railway should also be taken into account because their lifespan is short compared to domestic equipment. Currently, there is a scarcity of research addressing the entire life cycle carbon emissions of railroad vehicles. The authors examined the complete life cycle emissions of Beijing–Shanghai trains using various LCA methods. Nonetheless, variations in evaluation boundaries and methodologies between these investigations highlight the unique life cycles of each transportation mode, resulting in divergent evaluation outcomes. Therefore, it is imperious to analyze both high-speed rail (HSR) trains and road transport vehicles within a unified statistical framework and evaluation scope to comprehensively elucidate the carbon emission reduction benefits of rail transportation.[51]

Results show that HSR significantly reduces emissions compared to private vehicles and buses, emitting only 24–32% of private vehicle emissions and 47–89% of bus emissions. Predictive modelling suggests potential emission reduction through energy structure optimization. HSR's lifecycle emissions mainly stem from the operational phase. Electric energy usage notably reduces emissions during operation, with BEVs and electric buses (EBs) showing significant reductions compared to ICEVs and diesel electric buses (DEBs). Choosing HSR over road travel can lead to a significant emissions reduction. Predictions indicate that deeper energy structure optimization can further reduce carbon emissions from HSR transportation.[52]

12.6.2 Societal and Economic Benefits

The survival of humans and the sustainable development of cities have been very difficult due to city expansion, greenhouse gas emission, energy consumption, and increased vehicle population. The atmospheric environment of cities is diminished due to air pollution. The advancement of a low-carbon economy has become a foremost part of the global need of the hour.

Low-carbon transport parameters were divided into three parts by Kazuki Nakamura and Yoshitsugu Hayashi, *i.e.* avoidance strategies, change strategies, and development strategies. The impact of these three strategies was on reducing carbon dioxide through various methods, such as avoidance strategies to reduce the need for travel, glass replacement strategies to reduce traffic dependence on vehicles, and development strategies to reduce the use of emissions.[53]

Use of biomass as a carbon-negative fuel for transportation is significant in terms of reducing dependency on fossil fuels and in the mitigation of GHG emissions. Jasper *et al.*[54] carried out a study which involved the building of Aspen Plus Process and GaBi life cycle assessment models for a biomass gasification refinery with (a) Fischer–Tropsch synthesis of synthetic diesel, (b) H-production, and (c) electricity. The products were meant for passenger cars. The overall efficiency range for biomass-to-liquids is 20–40%, for biomass-to-hydrogen 58–61%, and for electricity it is only 25–26%. The production costs were also determined for all the products, being $3.171–6.409 gal for FT diesel configurations, $1.895–1.929 per kg H_2 for hydrogen, and $0.133 $kW^{-1} h^{-1}$ for electricity production. As per the market rates at the time of the study, the most profitable product for the biorefinery was hydrogen. FT diesel is also profitable but with a lower profit margin. However, as prices of oil, hydrogen, and electricity change, the choice of product will change also.[54]

12.7 Conclusion and Future Prospects

This chapter has described the importance of low-carbon materials in revolutionizing transportation, with a special focus on their applications in the automotive sector. The chapter plays a critical role in reducing carbon footprints in the transportation sector. However, an advancement in the automotive technologies serves as a challenge too in the path of carbon emission reduction. However, considering the challenges, there are infinite opportunities also in the path such as innovations in material sciences, sustainable environment, economic implications, circular economy, integration of regular supports, *etc.* Overall, focusing on lifecycle sustainability and collaborative efforts, industries can address current challenges and unlock new possibilities for reducing carbon emissions and enhancing vehicle performance.

Acknowledgements

The authors would like to express their thanks for support from J.C. Bose University, YMCA Faridabad; Jamia Millia Islamia, New Delhi, Indira Gandhi University, Meerpur, and IIT Roorkee.

References

1. Y. Guo, L. Luo, T. Liu, L. Hao, Y. Li, P. Liu and T. Zhu, *J. Environ. Sci.*, 2024, **136**, 682–697.

2. S. L. Jarvenpaa and A. Essén, *Inf. Organ.*, 2023, **33**, 100449.
3. L. Chen, M. Ma and X. Xiang, *Sustainable Cities Soc.*, 2023, **96**, 104654.
4. K. Shahzad and I. I. Cheema, *J. Power Sources*, 2024, **591**, 233888.
5. R. Kumar, K. Sadeghi, J. Jang and J. Seo, *Sci. Total Environ.*, 2023, **882**, 163446.
6. E. Fontana, *J. Clean. Prod.*, 2019, **236**, 117675.
7. S. Cheng, C. Li, F. Chai and H. Gong, *Energy Proc. Part A*, 2012, **17**, 249–257.
8. M. Bharathidasan, V. Indragandhi, V. Suresh, M. Jasiński and Z. Leonowicz, *Energy Rep.*, 2022, **8**, 9662–9685.
9. J. Han, J. Feng, P. Chen, Y. Liu and X. Peng, *Energy Convers. Manag.*, 2022, **15**, 100265.
10. N. A. Szaro, S. C. Ammal, F. Chen and A. Heyden, *J. Power Sources*, 2024, **603**, 234411.
11. A. Bishnoi, S. Pati and P. Sharma, *J. Power Sources*, 2024, **608**, 234609.
12. C. Alves, P. M. Ferrão, A. J. Silva, L. G. Reis, M. Freitas, L. B. Rodrigues and D. E. Alves, *J. Clean. Prod.*, 2010, **18**(4), 313–327.
13. M. Khalfallah, B. Abbès, F. Abbès, Y. Q. Guo, V. Marcel, A. Duval, F. Vanfleteren and F. Rousseau, *Mater. Des.*, 2014, **64**, 116–126.
14. S. V. Joshi, L. T. Drzal, A. K. Mohanty and S. Arora, *Composites, Part A*, 2004, **35**(3), 371–376.
15. K. Rao and K. M. Rao, *Compos. Struct.*, 2007, **77**(3), 288–295.
16. P. Wambua, J. Ivens and I. Verpoest, *Compos. Sci. Technol.*, 2003, **63**(9), 1259–1264.
17. N. Martin, P. Davies and C. Baley, *Ind. Crops Prod.*, 2016, **83**, 194–205.
18. F. Roussière, C. Baley, G. Godard and D. Burr, *Appl. Compos. Mater.*, 2012, **19**, 171–188.
19. J. Merotte, A. Le Duigou, A. Bourmaud, K. Behlouli and C. Baley, *Polym. Test.*, 2016, **51**, 174–180.
20. J. Merotte, A. Le Duigou, A. Kervoelen, A. Bourmaud, K. Behlouli, O. Sire and C. Baley, *Polym. Test.*, 2018, **66**, 303–311.
21. G. Thilagavathi, E. Pradeep, T. Kannaian and L. Sasikala, *J. Ind. Text.*, 2010, **39**(3), 267–278.
22. D. V. Parikh, Y. Chen and L. Sun, *Text. Res. J.*, 2006, **76**(11), 813–820.
23. H. R. Gu, S. J. Kim and H. A. Kim, *Procedia Eng.*, 2017, **200**, 45–52.
24. M. Assarar, A. E. l Mahi and J. Berthelot, *Int. J. Mater. Sci.*, 2013, **3**(2), 61–71.
25. F. Duc, P. E. Bourban, C. J. G. Plummer and J. A. Månson, *Composites, Part A*, 2014, **64**, 115–123.
26. J. E. Dölle, *Lieferantenmanagement in der Automobilindustrie*, Springer-Verlag, 2013, DOI: 10.1007/978-3-8349-4043-8.
27. R. Dunne, D. Desai, R. Sadiku and J. Jayaramudu, *J. Reinf. Plast. Compos.*, 2016, **35**(13), 1041–1050.
28. A. Bjurenstedt and F. Lärneklint, *Engelska*, 2004, **245**, 1402–1617.
29. T. Bröker and D. Ostner, *Automobil Ind.*, 2017, **62**(8), 94–95.

30. S. Gelowicz, T. Günnel, H. Hammer and C. Otto, *Automobil Ind.*, 2017, **62**(5), 34–37.
31. W. Wellbrock, D. Ludin, L. Röhrle and W. Gerstlberger, *Int. J. Corporate Soc. Responsibility*, 2020, **5**, 1–11.
32. Y. Wang, H. Li, A. Abdelhady and J. Harvey, *Int. J. Transp. Sci. Technol.*, 2018, **7**(3), 169–178.
33. J. Liu, H. Li, Y. Wang and H. Zhang, *Transp. Res. Part D: Transport Environ.*, 2020, **85**, 102381.
34. J. Xiao, J. Wang, Z. Sun and J. Li, *Gonglu Jiaotong Keji/J. Highway Transp. Res. Dev.*, 2005, 52–55.
35. D. Vijayan, A. Sivasuriyan, P. Devarajan, A. Stefańska, L. Wodzyński and E. Koda, *Buildings*, 2023, **13**(6), 1509.
36. M. Ananthkumar, K. M. Mini, C. Prakash, S. V. Sharma and A. C. B. Krishnaa, *Mater. Sci. Eng.*, 2020, **872**(1), 012137.
37. A. Chalot, L. Michel and E. Ferrier, *Composites, Part B*, 2019, **177**, 107255.
38. F. Farzaneh and S. Jung, *J. Cleaner Prod.*, 2023, **390**, 136111.
39. J. Montoya-Torres, O. Akizu-Gardoki and M. Iturrondobeitia, *Sustainable Cities Soc.*, 2023, **96**, 104658.
40. K. Tuayharn, P. Kaewtatip, K. Ruangjirakit, P. Limthongkul, SAE Technical Paper, 2015.
41. J. Morfeldt, S. Davidsson Kurland and D. J. A. Johansson, *Transp. Res. Part D: Transport Environ.*, 2021, **95**, 102807.
42. I. Philips, J. Anable and T. Chatterton, *Transport Policy*, 2022, **116**, 11–23.
43. C. Panoutsou and K. Maniatis, *Concawe Report RED II Annex IX A/B*, 2021.
44. M. M. F. Hasan, L. M. Rossi, D. P. Debecker, K. C. Leonard, Z. Li, B. C. E. Makhubela, C. Zhao and A. Kleij, *ACS Sustainable Chem. Eng.*, 2021, **9**, 12427–12430.
45. K. Morganti, K. Moljord, R. Pearson, M. Vermeire, M. Traver, P. Scorletti, T. de Melo, Y. Wang, P. China and J. Repasky, *Energy Environ. Sci.*, 2024, **17**, 531–568.
46. E. R. Sarabi, *Procedia Eng.*, 2011, **21**, 989–996.
47. W. R. Morrow, K. S. Gallagher, G. Collantes and H. Lee, *Energy Policy*, 2010, **38**(3), 1305–1320.
48. C. A. Bana e Costa, *J. Multi-Criteria Decision Anal.*, 2001, **10**(2), 111–125.
49. C. Banville, M. Landry, J. Martel and C. Boulaire, *J. Int. Federation Syst. Res.*, 1998, **15**(1), 15–32.
50. L. Bilgili, S. L. Kuzu, A. Y. Çetinkaya and P. Kumar, *Sustainable Cities Soc.*, 2019, **49**, 101635.
51. D. Candelaresi, A. Valente, E. Bargiacchi and G. Spazzafumo, *Hybrid Technologies for Power Generation*, Elsevier, 2022, pp. 475–495.
52. J. A. G. Sánchez, J. M. L. Martínez, J. L. Martín, M. N. F. Holgado and H. A. Morales, *Energy Convers. Manage.*, 2013, **74**, 332–343.
53. K. Nakamura and Y. Hayashi, *Transport Policy*, 2013, **29**, 264–274.
54. M. Jasper, N. Rafati, K. Schimmel, A. Shahbazi, F. Li, M. Mba-Wright and L. Wang, *Energy Convers. Manag.: X*, 2022, **14**, 100208.

55. E. M. Bibra, E. Connelly, M. Gorner, C. Lowans, L. Paoli, J. Tattini, J. Teter, C. LeCroy, O. MacDonnell, D. Welch, R. Palmer, D. Sharma and C. Xu, *Global EV Outlook 2021*, International Energy Agency, 2021, Available from: https://www.iea.org/reports/global-ev-outlook-2021.
56. D. A. Tsamboulas, *Transport Policy*, 2007, **14**(1), 11–26.
57. C. Macharis, L. Milan and S. Verlinde, *Research in Transportation Business & Management*, 2014, **11**, 75–84.

CHAPTER 13

Low-carbon Materials for Reducing Air Pollutants

ABHISHEK SINGH[a] AND TANVIR ARFIN*[a,b]

[a] Air Resource, Environmental Resource Planning and Management, CSIR-National Environmental Engineering Research Institute (CSIR-NEERI), Nehru Marg, Nagpur-440020, India; [b] Academy of Scientific and Innovative Research (AcSIR), Ghaziabad-21002, India
*Email: tanvirarfin@gmail.com

13.1 Introduction

The industrial age, the advent of internal combustion engines, and breakthroughs in engineering materials dramatically propelled human progress.[1] This resulted in the extensive production of items in factories and industries, establishing a new benchmark.[2] Embracing a linear economy model governed the cradle-to-grave lifetime of materials,[3] encompassing extraction, production,[4] consumption, and disposal, motivated by swift economic growth through consumption.[5] It is essential to acknowledge the hazards associated with carbon dioxide (CO_2) emissions from fossil fuels, the limited availability of these energy sources, and the environment's incapacity to manage the overproduction of pollutants.[6]

Ensuring the well-being of individuals and society while preserving the environment for future generations is crucial, making sustainable development essential.[7] This can be accomplished by shifting to a sustainable economy that emphasizes decarbonization.[8] A low-carbon economy can be realized by adopting materials, goods, and facilities that minimize greenhouse gas emissions and waste generation.[9]

Sustainable development emphasizes the responsible use of materials to reduce environmental harm and ensure long-term resource availability.[10]

Low Carbon Materials: Paving a Sustainable Path for a Greener Future
Edited by Santanu Patra, Sudheesh K. Shukla and Mika Sillanpää

Published by the Royal Society of Chemistry, www.rsc.org

Choosing sustainable materials is vital for minimizing the carbon footprint of goods and structures, thereby supporting broader sustainability objectives.[11] Life cycle assessment (LCA) is a helpful technique used to examine the environmental impacts of materials from their inception to disposal, aiding in selecting sustainable options.[12] Sustainable materials need less energy to manufacture, resulting in reduced environmental harm compared to traditional materials.[13]

Biomimicry draws inspiration from nature and can create sustainable materials that mimic natural processes and structures.[14] Innovations in sustainable materials, such as bio-based polymers and green composites, offer alternatives to traditional materials.[15] The circular economy advocates for material reuse, recycling, and repurposing to minimize waste and foster sustainable resource utilization.[16] Implementing sustainable material procurement strategies, such as using locally available materials and minimizing transportation distances, can effectively reduce carbon emissions and support local economies.[17]

The extraction and development of environmentally friendly assets must exert low or no detrimental impacts on the environment and society.[18] Ongoing research and development in sustainable materials lead to innovative advancements, providing new solutions for industries to achieve sustainability goals.[19] Incorporating sustainable materials into production and manufacturing processes can encourage sustainable development and cultivate a more environmentally conscious lifestyle.[20]

Nanoparticles have attracted significant interest in recent years across multiple fields, such as engineering,[21] medicine,[22] and electronics.[23] These compounds exhibit unique physicochemical characteristics owing to their peculiar atomic and electrical configurations and a higher density of surface atoms compared to more extensive materials.[24] Nanoparticles exhibit greater reactivity due to their enhanced surface area-to-volume ratio[25] and the development of new features suitable for various applications.[26] Nanoscale medicine has seen notable progress in using biocompatible materials for testing and therapeutics.[27] Nanotechnology has significantly improved imaging techniques by introducing contrast agents made from biodegradable polymer-based nanogels.[28] These nanogels incorporate various materials such as chitosan,[29] gelatin,[30] protein,[31] alginate,[32] starch,[33] carbon nanotubes,[34] nanoceramics,[35] pullulan,[36] polysaccharides,[37] cellulose,[38] and dendrimers.[39] The use of nanoparticles in addressing environmental contamination shows promising potential.[40] These materials exhibit strong reactivity, efficiently adsorb contaminants, and are economically advantageous relative to conventional methods.[41] Nanoparticles have shown their capacity to assimilate heavy metals, reducing heavy metal pollution.[42] Integrating nanotechnology continually expands the boundaries of scientific research and opens new avenues for innovation.[43] Nanotechnology can revolutionize multiple industries, enhance living standards, and tackle urgent global issues.[44]

Persistent organic pollutants (POPs) are hazardous substances that persist in the environment, accumulate in living organisms, and pose significant threats to health and ecosystems.[45] These pollutants can travel great distances through air and water, leading to widespread global contamination.[46] POPs include compounds such as dioxins, polychlorinated biphenyls (PCBs), and certain pesticides, all of which have been linked to adverse effects on human health and wildlife.[47] The Stockholm Convention is a significant international treaty aimed at eradicating or limiting the production and utilization of dangerous substances to protect human health and the environment.[48]

The advancement of materials with low carbon emissions significantly influences climate change and is a critical factor in sustainability assessments. Surprisingly, there is a current need for specific guidelines set by the International Organization for Standardization (ISO) or national standards for producing low-carbon materials. This study emphasizes the increasing enthusiasm for developing and marketing low-carbon materials and their capacity to mitigate air pollution. It is important to clarify that when we mention "carbon" in the context of low-carbon materials, we highlight their ability to reduce carbon dioxide emissions rather than referring to elemental carbon.

13.2 Low-carbon Materials

Carbon compounds demonstrate remarkable adaptability and can exist in various forms known as allotropes. These forms comprise pure or mixed hybridization orbitals ($sp^1/sp^2/sp^3$). A deeper understanding of their fundamental structures and associated features has the potential to drive the development of innovative carbon materials. Preliminary investigations into carbon materials and their orbital hybridization are now being evaluated.[49]

Carbon materials provide significant value owing to their capacity to adsorb contaminants in flue gases, providing a cost-effective, versatile,[50] and chemically stable solution with robust adsorption properties.[51] Friction materials commonly integrate carbon materials like graphite and coke, with graphite consistently displaying a flake-like structure, mainly observed under a microscope.[52]

Low-carbon materials are a crucial solution in the fight against climate change and air pollution. These materials are designed to minimize carbon emissions throughout their life cycle, from manufacturing to disposal. Low-carbon materials are crucial for fostering a sustainable and environmentally responsible future by employing sophisticated methods to mitigate greenhouse gas (GHG) emissions and sequester them.[53]

These materials are essential in the battle against air pollution. They work to reduce the release of GHGs and other harmful substances throughout their lifecycle, from manufacturing to usage and disposal. Enhancing air quality and alleviating the impacts of climate change are of utmost importance.[54] Their reduced carbon footprint distinguishes them from conventional materials, accomplished by sustainable sourcing, optimized production techniques, improved recyclability, and the creation of novel formulations.

Table 13.1 The progression of materials development through time.

Natural materials age	Stone age	3300 BCE
	Copper age	
	Bronze age	1200 BCE
Pre-materials age	Iron age	600 BCE
	Rubber age	
	Steel age	100 AD
Functional materials age	Silicon age	1980 AD
	Molecular age	Modern age
	Low-carbon age	

Embracing low-carbon materials is pivotal in advancing sustainability and curbing environmental damage. These materials are crucial in minimizing the carbon footprint and combating climate change. This study introduces the concept of low-carbon materials and incorporates them into the historical chronology of materials development (see Table 13.1). Using low-carbon materials is essential for the worldwide endeavour toward an environmentally sustainable, carbon-neutral future.

13.3 Essential Characteristics of Low-carbon Materials

The points below highlight crucial aspects of low carbon materials:

- **Sustainable sourcing**: Materials are responsibly sourced from renewable resources or methods with minimal environmental impact.
- **Efficient manufacturing**: Production processes are optimized to minimize energy consumption and CO_2 emissions.
- **Recyclability**: Materials are either easily recyclable or made from recycled content.
- **Longevity and durability**: Using long-lasting materials reduces the need for frequent replacements, thereby decreasing overall carbon emissions.
- **Innovation in composition**: Developing new materials or composites with an inherently lower carbon footprint promotes innovation in material composition.
- **Innovation in composition**: Development of new materials or composites that inherently have a lower carbon footprint.

13.4 Drive and Uncertainty

The international initiative for a low-carbon future and sustainable economic development is guided by the United Nations Framework Convention on Climate Change (UNFCCC), created in 1992 and the significant Paris Agreement of 2016, ratified by 189 nations. These pivotal agreements call for nations to adopt policies aimed at stabilizing atmospheric GHG

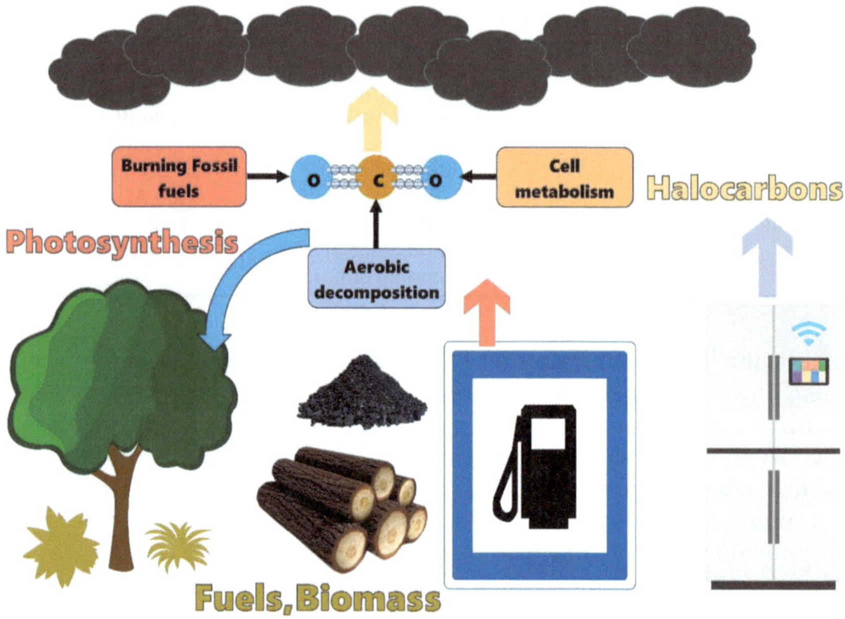

Figure 13.1 Human activities, particularly in materials engineering, play a significant role in generating GHG emissions, with CO_2 being a major contributor. These emissions have a long-lasting impact on the atmosphere, as represented by the size of the arrows.

concentrations to combat the adverse impacts of climate change. It is imperative to acknowledge that both industrialized and developing nations have equal accountability in mitigating emissions and averting actions that destabilize the Earth's climate.

Human activities, notably the burning of fossil fuels and materials engineering, have significantly altered the chemical configuration of the Earth's environment, leading to grave climate change repercussions. The rising levels of CO_2 and N_2O, chiefly attributable to anthropogenic activities like combustion and materials engineering, present a significant hazard and will remain in the upper atmosphere for a prolonged duration (see Figure 13.1).

Notwithstanding international climate accords, emissions persist in escalating, caused by an urgent amplification of endeavours to locate and exploit fossil fuel supplies globally. At the current emission rate, irreversible global warming is on the horizon. It is essential to acknowledge that this warming is a consequence of accumulated carbon emissions throughout human history, highlighting the importance of a comprehensive approach to emissions reduction.[55]

13.5 Manufacturing

In manufacturing, it is crucial to acknowledge that low-carbon materials are currently limited to certain elements on the periodic table. This underscores the pressing need to raise awareness within the broader materials community. This essay aims to underscore the importance of developing a thorough explanation of low-carbon materials, products, and associated services.

A carbon footprint encapsulates the total GHG emissions, both direct and indirect, generated throughout the life process of an item or material. The primary objective of low-carbon materials is to elevate the quality of the living environment and foster social equality through a low-carbon eco-innovation, all while safeguarding Earth's assets and environment for future generations. Low-carbon biomass assets include agricultural residue, sawdust from paper mills and forestry, and similar materials. These materials absorb CO_2 as they grow, resulting in a negative CO_2 footprint. Understanding the need for low-carbon materials is but one aspect of the overarching sustainability concept, as materials provide essential elements for products that support diverse services.

The production of materials relies on fossil fuels to varying degrees. Although low-carbon materials like metals contain little carbon, their production and use heavily depend on fossil fuels, leading to significant CO_2 emissions. It is crucial to reduce this dependency to mitigate overall CO_2 emissions. Conversely, wood products are carbon-intensive, sequestering substantial quantities of carbon absorbed from ecosystems and necessitating reduced manufacturing energy.

Preserving forests as sustainable carbon reservoirs is crucial for attaining a carbon-neutral environment by compensating for CO_2 emissions. Carbon capture and storage (CCS) approaches, which store CO_2 in subsurface water storage facilities or as carbonated minerals, also play a crucial role in this effort.[56] Responsible forest management, encompassing regular harvesting and replacement of trees, guarantees that carbon from harvested trees is sequestered in wood products for prolonged durations.[57]

Increasing reliance on carbon sequestration, rather than reducing its usage, is necessary to remove significant quantities of CO_2 derived from the Earth's atmosphere and return it to the Earth. Therefore, terms like "low-carbon economy"[58] or "post-carbon economy" can be misleading.[59] This study investigates the principal sources of CO_2 emissions associated with the manufacturing and utilization of materials with differing carbon contents and offers recommendations for reducing these emissions.

13.6 Environmental Pollution

Air pollution presents a critical global challenge with profound implications for the well-being of people, ecological systems, and the world's climate. It encompasses various chemical substances and atmospheric phenomena that may profoundly affect human health, comfort, the food supply chain, asset valuations, and the overall health of ecosystems. These effects may be transient or enduring.

Airborne contaminants exist in various forms – solid, gas, or liquid particles – varying in size from large to minuscule. They arise from natural occurrences like volcanic eruptions and forest fires, as well as from human activities such as manufacturing, transportation, and energy production. Despite comprising only 0.01% of the Earth's atmospheric emissions, anthropogenic sources can still profoundly affect the natural world, the environment, and humanity.[60]

The impact of air pollution on ecosystems and human health is profound and should never be disregarded. Even slight alternations in the atmospheric chemical equilibrium can profoundly disturb climate patterns, harm ecosystems, and threaten the existence of countless species.[61] These effects include acid rain, ground-level ozone, and photochemical smog. These pollutants, whether visible or not, consist of particles and gases not naturally present.

While natural phenomena like pollen, dust storms, and wildfires contribute to these pollutants, the predominant and more significant damage stems from human activity.[62] The combustion of biomass, which includes burning coal, wood, and other resources in automobiles, residences, and workplaces, emits contaminants that deteriorate the air quality.[63]

13.6.1 Particulate Matter

Particulate matter (PM) is a significant air pollutant classified by the US Environmental Protection Agency (USEPA). It comprises particles released into the atmosphere from various sources such as fossil fuel combustion, industrial discharges, fog, dust, and smoke (see Figure 13.2). The dimensions of these particles vary from sub-micrometres to 10 micrometres or smaller.[64]

PM poses a significant health risk due to its ability to deeply infiltrate the respiratory system, causing infections and worsening heart and lung conditions. Furthermore, PM impacts the climate system by modifying cloud formation, altering the Earth's radiation balance, and contributing to stratospheric air pollution. The dimensions of PM correlate directly with its detrimental effects, as PM_{10} (particles with a diameter of less than 10 micrometres) presents the greatest hazard by potentially entering the lungs and bloodstream, resulting in respiratory and cardiovascular complications.

Extensive scientific research has established a clear connection between exposure to particle pollution and various health conditions, including bronchitis, asthma, respiratory carcinoma, and untimely mortality. Certain demographics, such as young children, older adults, and persons with existing cardiovascular or pulmonary ailments, exhibit heightened vulnerability to the detrimental impacts of particulate pollution.[65]

Incorporating carbon nanostructures into air particle monitoring equipment can vastly improve PM capture. Using r-GO, g-C_3N'NSs, and CNTs as nanostructures in filters holds great potential for enhancing air quality and combating the detrimental impacts of air pollution. This work examines the

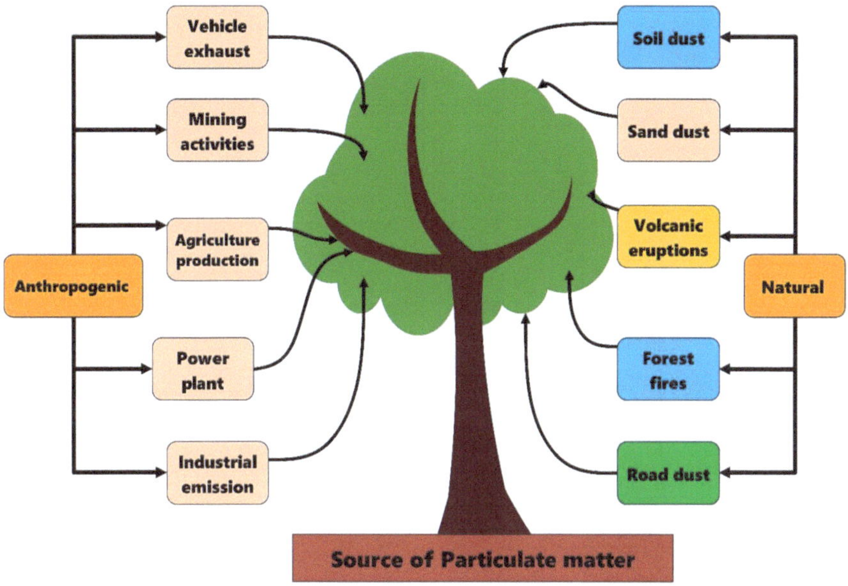

Figure 13.2 A diagram depicting the origins of particulate matter (PM).

influence of these nanomaterials on filter efficacy and potential application of Laser-induced Breakdown Spectroscopy (LIBS) for accurate monitoring of heavy metals. The findings of this study will be crucial in developing novel strategies to mitigate air pollution and protect human health and the environment.[66]

Zhong's group has devised an innovative technique for directly synthesizing one-dimensional (1D) carbon nanotubes (CNTs) directly on a porous silicon carbide (SiC) membrane substrate using chemical vapor deposition (CVD). This approach eliminates the need for a sacrificial or transition surface. They have developed three unique designs of CNT-coated SiC membranes: interweaving CNTs/SiC membrane (I-CNTs/SiC), spiral CNTs-coated SiC membrane (S-CNTs/SiC), and vertical CNTs/SiC membrane (V-CNTs/SiC). The S-CNTs/SiC arrangement exhibited remarkable performance, achieving an ultrafine PM filtering efficiency of 99.48% and a gas permeance of 450 $m^3 m^{-2} h^{-1} kPa^{-1}$.[67]

13.6.2 Volatile Organic Compounds

Volatile organic compounds (VOCs) are potent organic chemicals in many products that swiftly evaporate into the environment. Due to their significant volatility and mobility, they endure without degradation and can traverse considerable distances in the environment.[68] VOCs stem from both natural and human activities (see Figure 13.3).

VOCs have a detrimental effect on the environment and human health. These chemicals volatilize under ambient pressure and room temperature,

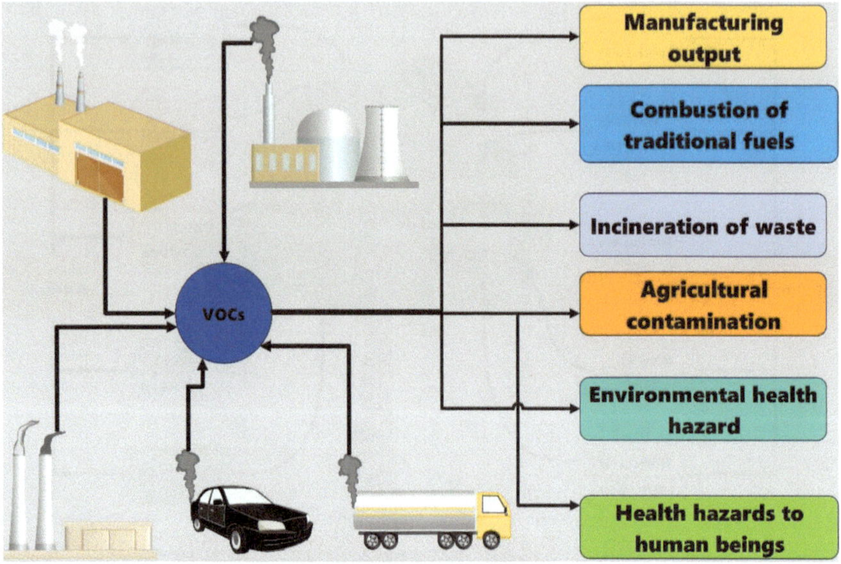

Figure 13.3 The possible origins of volatile organic compounds (VOCs).

Table 13.2 Classification of VOCs as contaminants.

VOC classification	Examples	Boiling point range °C
Very volatile organic compounds	C_3H_8 C_4H_{10} CH_3Cl	0 to 50–100
Volatile organic compounds	CH_2O C_7H_8 C_3H_6O C_3H_8O	50–100 to 240–260
Semi-volatile organic compounds	Pesticides (chlorodane, DDT) $C_{14}H_9Cl_5$ Plasticizers (phthalates)	240–260 to 380–400

and may be present in both enclosed and open environments. The classification of VOCs based on their volatile properties is presented in Table 13.2.

VOCs have a significant impact on human health and the environment. They raise concerns because of their toxicity, carcinogenicity, and detrimental impacts, and their role in material degradation and the generation of tropospheric photochemical oxidants. VOCs contribute to stratospheric ozone depletion, climate change, and the emission of odor. Various nanomaterials have been engineered to mitigate VOCs, with their efficacy contingent upon aspects like porosity, dimensions, electrostatic attraction, surface characteristics, and chemical composition. Studies have demonstrated the efficacy of nanomaterials in reducing VOCs and addressing environmental pollution.[69] Graphene and graphene oxide (GO) are particularly noteworthy for their ability to adsorb gaseous pollutants due to their ease of acquisition, large surface area, unique pore structure, chemical stability, and thermal stability.[70] TiO_2/graphene nanocomposites, synthesized through a hydrothermal response of GO and TiO_2 in C_2H_5OH and H_2O, exhibited superior stability and efficiency in the photocatalytic deterioration of C_6H_6 in the gas stage compared to pure TiO_2.[71] Graphene oxide and reduced graphene oxide have demonstrated efficacy in toluene removal,[72] utilizing their capacity to establish π–π interactions and hydrophobic and electrostatic associations with the contaminant, thereby promoting adsorption on their surfaces.

13.6.3 Carbon Dioxide

The increasing CO_2 emissions from various sources, including volcanic eruptions, fossil fuel use, deforestation, and industrialization, are intensifying the greenhouse effect,[73] causing global warming and environmental challenges such as ocean acidification.[74] Notwithstanding considerable technological advancements and ample raw materials, fossil fuels remain the primary option. Therefore, it is imperative to devise strategies to mitigate or eradicate carbon dioxide promptly.[75] The essential characteristics of an ideal CO_2 adsorbent are:

- Fast reaction rate
- Rejuvenation
- Thermal capture
- High selectivity
- Attainability

A variety of carbon-based materials meet these requirements and can be efficient CO_2 adsorbents, including:

- Carbon aerogels
- Carbon nanotubes
- Activated carbon
- Graphene
- Biochar
- Polymer-based carbon
- Other carbons

Their capacity to adapt their shape, expansive surface area, functional properties, numerous pores, exceptional gas storage ability, reliable duplicability, and electronic characteristics all contribute to their effectiveness.[76]

Carbon nanotubes (CNTs) possess remarkable properties such as a hollow tubular configuration, robust mechanical strength, water resistance, and elevated surface area, making them excellent for capturing CO_2 under high pressure.[77] However, their adsorption capacity at low pressure is constrained by the predominance of micropores and functional groups.[78] To mitigate this constraint, the incorporation of chemical changes or the fabrication of composites can substantially improve the CO_2 adsorption efficiency of CNTs. For instance, researchers discovered that tetraethylenepentamine (TEPA)-modified carbon reduced the activation energy for CO_2 desorption to 39.9 $kJ\,mol^{-1}$, indicating potential cost savings in CO_2 capture.[79]

Moreover, to tackle the limited surface area of polyaspartamide (PAA) differentiated from poly succinimide (PSI), multi-walled carbon nanotubes (MWCNTs) were integrated, resulting in a PAA-MWCNT adsorbent with significantly increased pore volume and surface area by 41 and 31 times, respectively. The application of PAA on MWCNTs resulted in the creation of carbon-containing defects and reduced graphitized structures, hence facilitating enhanced carbon dioxide absorption (70 $mg\,g^{-1}$) at a temperature of 25 °C.[80]

13.6.4 Carbon Monoxide

It is crucial to separate carbon monoxide from carbon dioxide, methane, hydrogen, and nitrogen for environmental preservation and chemical synthesis. Carbon monoxide (CO) is a harmful gas associated with climate change, acid rain, and other environmental issues.[81]

CO is a significant byproduct of numerous processes, frequently utilized to produce CO_2 for disposal and energy recovery.[82] It is also an essential precursor for numerous organic compounds and can function as an independent product. However, the challenge lies in separating CO from other gases such as nitrogen (N_2), CO_2, methane (CH_4), hydrogen sulphide (H_2S), hydrogen (H_2), and other sulfur compounds, which requires significant energy and expense – separating CO from N_2 is particularly challenging due to their similar physical properties, hindering conventional separation techniques. Consequently, the reutilization of carbon monoxide as a resource is limited, and CO-containing mixtures are typically burned for their calorific value, resulting in significant CO_2 emissions. Creating economically feasible separation technologies for recycling CO byproducts is crucial for generating valuable raw materials and mitigating CO_2 emissions, while this endeavor presents significant challenges CO can be used in a number of industrial processes such as:

- Hydroformylation
- Koch reactions
- Water gas shift reactions
- Fischer–Tropsch reactions

and to form industrially important products including:

- Metal carbonyls
- Phosgene

- Methanol
- Acetic acid
- Oxo-alcohols

The World Health Organization (WHO) has identified concentrations exceeding six parts per million (ppm) over extended periods as potentially hazardous. Abnormal levels of carboxyhemoglobin (COHb) exceeding 2% in nonsmokers and 10% in smokers can trigger symptoms, according to the WHO.[83] Recent research indicates that carbon monoxide poisoning results in tissue hypoxia and generates direct cellular alternations through several mechanisms, resulting in immunological or inflammatory dysfunction.[84] The consequences arise from disturbances in normal signalling mechanisms caused by endogenous CO, a physiological gas generated by the enzymatic breakdown of heme.[85] Endogenous CO is presently being studied in preclinical and phase 1 trials as a prospective treatment for specific disorders.[86]

The synthetic adsorbents have shown remarkable efficacy in removing carbon monoxide, with traditional chicken eggshells achieving the greatest CO removal rate of 59.32%. The adsorption kinetics aligned with the pseudo-first-order equation, and the adsorption isotherms conformed to the Langmuir equation.[87] In our case study, AC-Fe was the most efficient CO adsorbent. After adsorption, the carbon content rose dramatically from 62.25% to 89.67%, demonstrating the efficient capture of carbon components, likely from CO or CO_2 present in the biogas. Nevertheless, there was a significant reduction in CO concentration.[88]

13.7 Low-carbon Materials and CO_2 Footprint

Low-carbon materials are intrinsically important for sustainable development as well as climate change mitigation. Through substantial lifecycle GHG emissions reduction, these materials offer a pathway to a more sustainable future. This ecological footprint is achieved *via* a combination of sustainable sourcing and energy-efficient manufacturing in conjunction with advanced carbon capture technologies. Bio-based polymers, such as polylactic acid (PLA) which is derived from renewable materials like corn or sugarcane, could produce up to 60% less CO_2 than fossil fuel-based plastics. This significant reduction arises from the carbon absorbed during the growth of biomass and the low energy requirements for their production.[89]

Nanocomposites are also durable and light and, hence, have become the best sustainable alternative to synthetic composites. Additionally, green composites containing natural fibers such as jute, hemp, or bamboo have emerged as robust eco-friendly substitutes to synthetic composites. Research shows that these green materials use up to 50% less energy to produce while providing comparable mechanical properties. The implementation of hemp fibers would greatly reduce the carbon footprint due to reduced replacement by energy-intensive materials like glass fibers, and causes an excellent reduction in lifecycle CO_2 emissions.[90]

Concrete, in particular geopolymer concrete, is an emerging low-carbon alternative that is worth celebrating as it is a substantial advancement in cement material innovation that has been shrouded in darkness. Cements based on Portland cement can be substituted with industrial byproducts like fly ash and slag, thereby ensuring a whopping 80% reduction in CO_2 emissions. This is important because traditional cement production funding alone is reported to have a global contribution of approximately 8% to CO_2 emissions. Furthermore, Carbon Capture and Utilization (CCU) embedded in construction materials aids in sequestering greenhouse gases while enhancing the material performance. Biocarbon solutions through the embedding of CO_2 into calcium carbonate minerals have another advantage in that they boost the performance of materials.[91]

The highlights of LCA that emphasize low-carbon materials include economic and ecological benefits. For instance, timber that is sustainably managed can store about 1.8 tons of CO_2 per cubic meter and also utilizes 60% less energy to process when compared to steel. This has a dual role, minimizing emissions during the production phase while acting as long-term carbon sinks.[92]

Low-carbon materials have to be integrated in construction industries because they help tackle climate change over time and also improve air quality. They do this by changing how resources are used and encouraging better methods. These changes can help reduce structural, packaging, and industrial waste. This puts industries in a stronger position to move toward a carbon neutral economy.[93]

13.8 Future Perspectives

The potential of low-carbon materials to mitigate air pollutants is emerging and provides diverse ways to address this environmental issue.[94] With the advancement of technology, there is an increasing focus on creating low-carbon materials that reduce air pollution while fostering sustainability across all sectors. A vital area of investigation entails optimizing manufacturing processes to reduce carbon emissions while improving the performance and durability of materials. Moreover, there is a significant transition toward incorporating renewable energy sources into material production, hence diminishing the carbon footprint of the manufacturing industry.

The development of innovative materials with improved adsorption capabilities for absorbing air pollutants, including VOCs, PM, and GHGs, is another promising avenue. These materials, comprising carbon-based adsorbents and nanomaterials, provide very effective solutions for the elimination of atmospheric pollutants.

Furthermore, advancements in nanotechnology facilitate the precise engineering of materials at the molecular level, resulting in customized features for targeted pollutant removal applications. The incorporation of low-carbon materials in infrastructure and urban planning can significantly mitigate air pollution in urban environments. Incorporating green roofs, permeable pavements, and other sustainable building materials can alleviate pollution and enhance urban air quality.

Moreover, low-carbon materials offer the potential to improve indoor air quality by creating sustainable construction materials and air purification systems. Cooperative endeavours among researchers, policymakers, manufacturers, and the public are essential for maximising the efficacy of low-carbon materials in mitigating air pollution. Public awareness and educational initiatives can facilitate the widespread adoption of low-carbon materials and sustainable behaviours. Moreover, funding research and development programs targeting low-carbon materials is essential for expediting innovation and fostering technological progress in pollution control.

The extensive utilization of low-carbon materials can substantially aid global efforts to address climate change and improve air quality for present and future generations.

13.9 Conclusion

Adopting low-carbon materials offers a viable approach to addressing air pollution and promoting environmental sustainability. Prioritizing materials with a reduced carbon footprint over their whole lifecycle can mitigate significant pollutants such as particulate matter, VOCs, CO_2, and CO.

Low-carbon materials are meticulously engineered to reduce adverse environmental effects throughout their lifecycle, from production to disposal. Incorporating these materials in construction and industrial settings directly mitigates particulate matter and VOC emissions, significantly improving air quality and public health. Many low-carbon materials are designed to emit reduced levels of VOCs, which are detrimental to the environment and human health. The decrease in VOCs enhances indoor air quality and diminishes the formation of ground-level ozone, a harmful air pollutant.

It employs efficient production processes to manufacture low-carbon products that substantially leverage modern technologies to reduce CO_2 and CO emissions. By incorporating renewable resources and enhancing production processes, these materials significantly diminish greenhouse gas emissions, contributing to climate change mitigation. The durability and longevity of low-carbon materials are essential. Enhanced durability diminishes the necessity for frequent replacements and repairs, thus resulting in reduced carbon emissions and resource use.

Ongoing innovations in creating new composites and technologies are essential for improving the environmental advantages of low-carbon materials. These innovative advancements are broadening the horizons of materials science and providing highly efficient strategies for mitigating air pollutants. Adopting and promoting low-carbon materials can result in substantial progress toward cleaner air and a more sustainable future for industries and communities.

Low-carbon materials are essential in mitigating air pollution. Their capacity to reduce emissions of particulate matter, VOCs, CO_2, and CO highlights their significance in environmental preservation and sustainable development.

Low-carbon materials and their related products are engineered to reduce environmental impact while ensuring optimal performance. The emphasis is on promoting low-carbon materials by designing them at the atomic and molecular levels, optimizing the processing of raw materials, employing renewable and recycled products, and modifying designs to minimize waste. The difficulties of sampling and inquiry pose substantial hurdles in identifying and mitigating environmental contaminants.

Subsequent studies should focus on augmenting efficiency, particularly in visible light, to improve the removal of indoor pollutants. Incorporating low-carbon materials into product design methods can substitute high GHG-emitting components with low-emission alternatives. Sustainable products can be incorporated into an innovative company strategy that retains ownership while providing clients with cyclical solutions. Attaining a sustainable future necessitates developing and using low-carbon materials, products, and services through integrating innovative technology with socially responsible rules and ethical concerns along the biogeochemical spectrum. It is essential to allocate more resources to exploring and innovating low-carbon materials and developing a proficient workforce.

Acknowledgements

The authors are grateful to the director, CSIR-NEERI, for providing the necessary facilities for this work. The manuscript has been checked for plagiarism using the licensed version of iThenticate software and recorded in the Knowledge Resource Centre, CSIR-NEERI, Nagpur for anti-plagiarism (KRC No.: CSIR-NEERI/KRC/2024/MAY/AR/2).

References

1. T. Arfin, in *Biomaterials in food packaging*, ed. M. Yusuf and S. A. Khan, Jenny Stanford Publishing, Singapore, 2022, ch. 2, pp. 49–81.
2. S. Narkhede, E. Phillips, A. Singhai, A. Dadsena, R. Sahu, T. Arfin, A. Shakya and P. Parkhey, in *Catalytic applications of biochar for environmental remediation: sustainable strategies towards a circular economy*, ed. R. T. Kapoor, M. Sillanpaa and M. Rafatullah, ACS Publications, Washington DC, 2024, ch. 5, vol. 2, pp. 95–140.
3. O. Aquatar, P. R. Mogarkar, S. Rangari and T. Arfin, in *Caspases: roles and significance in cell death and disease*, ed. J. Warren, Nova Science Publishers, New York, 2016, ch. 2, pp. 1–18.

4. P. R. Mogarkar, O. Aquatar, S. Rangari and T. Arfin, in *Caspases: roles and significance in cell death and disease*, ed. J. Warren, Nova Science Publishers, New York, 2016, ch. 2, pp. 19–38.
5. T. Arfin, A. Singh, N. Mathew and A. Tirpude, *Innovation Chem. Mater. Sustainability*, 2024, **1**, 66–75.
6. T. Arfin and S. Wanjari, in *Advances in chemistry research*, ed. J. C. Taylor, Nova Science Publishers, New York, ch. 12, vol. 26, 2015, pp. 213–239.
7. S. A. Mandavgane, S. Rayulu, A. A. Chilkalwar, S. M. Anjankar, A. A. Moinuddin, T. Arfin and A. A. Mandavgane, *Rasayan J. Chem.*, 2023, **16**, 2239–2248.
8. N. Malik, A. U. Khan, S. Naqvi and T. Arfin, *J. Mol. Liq.*, 2016, **221**, 12–18.
9. V. Shanmugam, R. A. Mensah and M. Forsth, *et al.*, *Compos., Part C: Open Access*, 2021, **27**, 100138.
10. S. S. Waghmare, D. H. Lataye, T. Arfin and S. Rayalu, *Int. J. Innovative Res. Sci., Eng. Technol.*, 2015, **4**, 11998–12010.
11. T. Arfin, Md. O. Aquatar and S. S. Waghmare, *Int. J. Sci. Res. Dev.*, 2015, **3**, 908–917.
12. T. Arfin, N. Mathew and P. Mondal, in *Biobased packaging materials: Sustainable alternative to conventional packaging materials*, ed. S. Ahmed, Springer, Singapore, 2024, ch. 11, pp. 279–311.
13. F. Mohammad, T. Arfin and H. A. Al-Lohedan, in *Graphene-based nanotechnologies for energy and environmental applications: Micro and nano technologies*, ed. M. Jawaid, A. Ahmad and D. Lokhat, Elsevier, Netherland, 2019, 1st edn, ch. 8, pp. 145–154.
14. N. Malik, T. Arfin and A. U. Khan, in *Nanomaterials for drug delivery and Therapy*, ed. A. M. Grumezescu, Elsevier, Netherland, 2019, ch. 13, pp. 373–402.
15. T. Arfin, A. Tarannum and K. Sonawane, in *Green and sustainable advanced materials: processing and characterization*, ed. S. Ahmed and C. M. Hussain, Scrivener Publishing LLC, Beverly, 2018, ch. 1, vol. 1, pp. 1–34.
16. T. Arfin and S. Athar, in *Nanomaterials: biomedical, environmental, and engineering applications*, ed. S. Kanchi, S. Ahmed, M. I. Sabela and C. M. Hussain, Scrivener Publishing LLC, Beverly, 2018, ch. 3, pp. 93–103.
17. T. Arfin and K. Sonawane, in *Green and sustainable advanced packaging materials*, ed. S. Ahmed, Springer, Switzerland, 2018, ch. 1, pp. 1–32.
18. T. Arfin and A. Tarannum, in *Handbook of nanomaterials for industrial applications*, ed. C. M. Hussain, Elsevier, Netherlands, 2018, ch. 6, pp. 127–134.
19. T. Arfin, in *Chitosan: derivatives, composites and applications*, ed. S. Ahmed and S. Ikram, Scrivener Publishing, 2017, ch. 5, pp. 115–150.
20. F. Mohammad, T. Arfin and H. A. Al-Lohedan, *Mater. Sci. Eng., C*, 2017, **71**, 735–743.
21. T. Arfin and Rafiuddin, *Desalination*, 2012, **284**, 100–105.
22. T. Arfin and F. Mohammad, *Adv. Ind. Eng. Manag.*, 2014, **3**, 25–38.

23. F. Mohammad, L. du Plessis and T. Arfin, in *X-ray diffraction, structure, principles and applications*, ed. K. Shih, Nova Science Publishers, New York, 2013, ch. 7, pp. 161–182.
24. P. K. Obulapuram, T. Arfin, F. Mohammad, K. Kumari, S. K. Khiste, H. A. Al-Lohedan and M. Chavali, *ACS Omega*, 2021, **6**, 33614–33626.
25. S. Bansod, T. Arfin, R. Singh, S. Ahmad and K. Neeti, in *Chemically modified carbon nanotubes for commercial applications*, ed. J. Aslam, C. M. Hussain and R. Aslam, Wiley-VCH GmbH, Germany, 2023, ch. 12, pp. 271–298.
26. T. Arfin, in *Functionalized nanomaterials-based devices for environmental applications: a volume in micro and nano technologies*, ed. C. M. Hussain, S. K. Shukla and G. M. Joshi, Elsevier, Netherland, 2021, ch. 9, pp. 199–218.
27. T. Arfin, in *Nanomaterials in diagnostic tools and devices*, ed. S. Kanchi and D. Sharma, Elsevier, Netherland, 2020, ch. 3, pp. 85–112.
28. T. Arfin, in *Sustainable nanocellulose and nanohydrogels from natural sources: micro and nano technologies*, ed. F. Mohammad, H. A. Al-Lohedan and M. Jawaid, Elsevier, Netherland, 2020, ch. 12, pp. 255–274.
29. T. Arfin, in *Handbook of biopolymers: advances and multifaceted applications*, ed. S. Ahmed, S. Kanchi and G. Kumar, Jenny Stanford Publishing, Singapore, 2018, ch. 7, pp. 167–182.
30. S. Athar and T. Arfin, in *Natural polymers: Derivatives, blends and composites*, ed. S. Ikram and S. Ahmed, Nova Science Publishers, New York, 2017, ch. 11, vol. 2, pp. 199–216.
31. T. Arfin, S. Athar and S. Rangari, in *Handbook of biopolymers: advances and multifaceted applications*, ed. S. Ahmed, S. Kanchi and G. Kumar, Pan Stanford Publishing, Singapore, 2019, 1st edn, ch. 4, pp. 75–93.
32. T. Arfin and K. Sonawane, in *Alginates: applications in the biomedical and food industries*, ed. S. Ahmed, John Wiley, New York, 2019, ch. 3, pp. 45–58.
33. F. Mohammad, T. Arfin, I. B. Bwatanglang and H. A. Al-Lohedan, in *Bio-based polymers and nanocomposites: preparation, processing, properties & performance*, ed. M. L. Sanyang and M. Jawaid, Springer Nature, Switzerland, 2019, ch. 8, pp. 157–181.
34. T. Arfin, in *Environmental applications of carbon nanomaterials-based devices*, ed. S. Mallakpour and M. Hussain, Wiley-VCH Verlag, Germany, 2021, ch. 13, pp. 313–329.
35. T. Arfin, in *Handbook of consumer Nanoproducts*, ed. S. Mallakpour and C. M. Hussain, Springer, Singapore, 2022, ch. 14, pp. 337–354.
36. T. Arfin, in *Pullulan: processing, properties, and applications*, ed. S. Ahmed, A. Soundararajan, Jenny Standford Publishing, Singapore, 2021, ch. 2, pp. 35–58.
37. T. Arfin, in *Marine polysaccharides: an overview*, ed. S. Ahmed and A. Soundararajan, Jenny Stanford Publishing, Singapore, 2019, ch. 1, pp. 1–11.

38. F. Mohammad, T. Arfin, N. Saba, M. Jawaid and H. A. Al-Lohedan, in *Electrically conductive polymers and polymer composites: from synthesis to biomedical applications*, ed. A. Khan, M. Jawai, A. A. P. Khan and A. M. Asiri, Wiley-VCH, Germany, 2018, ch. 9, pp. 181–197.
39. T. Arfin and F. Mohammad, in *Advances in materials science research Vol.21*, ed. M. C. Wythers, Nova Science Publishers, New York, 2015, ch. 7, pp. 157–174.
40. S. Ahmad, R. Singh, T. Arfin and K. Neeti, *Environ. Sci.: Adv.*, 2022, **1**, 620–661.
41. P. K. Obulapuram, T. Arfin, F. Mohammad, S. K. Khiste, M. Chavali, A. N. Albalawi and H. A. Al-Lohedan, *Polymers*, 2021, **13**, 3490.
42. T. Arfin and A. Tarannum, *J. Environ. Chem. Eng.*, 2019, **7**, 102811.
43. T. Arfin and Rafiuddin, *Electrochim. Acta*, 2009, **54**, 6928–6934.
44. T. Arfin and Rafiuddin, *J. Electroanal. Chem.*, 2009, **636**, 113–122.
45. T. Arfin, P. Kate, A. Singh and K. Kumari, in *Persistent organic pollutants: Gaps in management and associated challenges*, ed. K. Kumari, CRC Press, Boca Raton, 2021, ch. 8, pp. 201–226.
46. K. Kumari and T. Arfin, in *Pollutants of global concern: a comprehensive overview of persistent organic pollutants*, ed. K. Kumari, Springer, Cham, 2024, ch. 14, pp. 195–204.
47. A. Somanathan, N. Mathew, A. M. Pillai, P. Mondal and T. Arfin, in *Bioplastics for sustainability: Manufacture, technologies and environment*, ed. A. K. Mishra and C. M. Hussain, Elsevier, Netherland, 2024, ch. 13, pp. 313–354.
48. A. Somanathan, N. Mathew, A. M. Pillai and T. Arfin, in *Handbook of nanofillers*, ed. S. Mallakpour, C. M. Hussain, Springer Nature, Singapore, 2024, ch. 30, pp. 1–40.
49. J. Li, D. Yin and Y. Qin, *Manuf. Rev.*, 2023, **10**, 13.
50. T. Arfin, R. Bushra and F. Mohammad, *Graphene Technol.*, 2016, **1**, 1–15.
51. F. Mohammad, T. Arfin and H. A. Al-Lohedan, *Mater. Chem. Phys.*, 2019, **229**, 117–123.
52. Y. Hou, Y. Chen, X. He, F. Wang, Q. Cai and B. Shen, *Chem. Eng. J.*, 2014, **490**, 151424.
53. C. R. Iddon and S. K. Firth, *Energy Build.*, 2013, **67**, 479–488.
54. B. V. V. Reddy, *Int. J. Low Carbon Technol.*, 2009, **4**, 175–181.
55. J. Hansen, P. Kharecha and M. Sato, *et al.*, *PLoS One*, 2013, **8**, e81648.
56. A. Sanna, M. Dri, M. R. Hall and M. Maroto-Valer, *Appl. Energy*, 2012, **99**, 545–554.
57. J. F. Bastin, Y. Finegold, C. Garcia, D. Mollicone, M. Rezende, D. Routh, C. M. Zohner and T. W. Crowther, *Science*, 2019, **365**, 77–79.
58. R. Richels and J. Edmonds, *Energy Policy*, 1995, **23**, 373–378.
59. J. Wiseman, T. Edwards and K. Luckins, *Environ. Innov. Soc. Transit.*, 2013, **8**, 76–93.
60. A. M. Pillai and T. Arfin, in *Electrocatalytic materials for renewable energy*, ed. S. K. Shukla, C. M. Hussain, S. Patra and M. Choudhary, Scrivener Publishing LLC, Beverly, 2024, ch. 11, pp. 303–331.

61. N. Mathew, A. Tirpude and A. M. Pillai, in *Bioremediation technologies: for wastewater and sustainable circular bioeconomy*, ed. R. T. Kapoor and M. Rafatullah, De walter de Gruyte gmbh, Berlin, 2023, ch. 14, pp. 235–274.
62. N. Mathew, A. Somanathan, A. Tirpude and T. Arfin, *Environ. Pollut. Manag.*, 2024, **1**, 1–14.
63. T. Arfin, A. M. Pillai, N. Mathew, A. Tirpude, R. Bang and P. Mondal, *Environ. Sci. Pollut. Res.*, 2023, **30**, 125347–125369.
64. R. Espinosa, *et al.*, *Appl. Soft Comput.*, 2021, **113**, 107850.
65. H. Saleem, S. J. Zaidi, A. F. Ismail and P. S. Goh, *Chemosphere*, 2022, **287**, 132083.
66. Y. A. Attia, A. E. Ezet and S. Saeed, *et al.*, *Sci. Rep.*, 2024, **14**, 621.
67. K. Yuan, S. Feng and F. Zheng, *et al.*, *Ind. Eng. Chem. Res.*, 2020, **59**, 19680–19688.
68. P. Pandey and R. Yadav, *Int. J. Plant Environ.*, 2018, **4**, 14–26.
69. F. D. Guerra, M. F. Attia and D. C. Whitehead, *et al.*, *Molecules*, 2018, **23**, 1760.
70. T. Arfin and S. N. Rangari, *Anal. Methods*, 2018, **10**, 347–358.
71. Y. Zhang, Z. R. Tang, X. Fu and Y. J. Xu, *ACS Nano*, 2010, **4**, 7303–7314.
72. V. Kumar, Y. S. Lee, J. W. Shin, K. H. Kim, D. Kukkar and Y. F. Tsang, *Environ. Int.*, 2020, **135**, 105356.
73. R. Waheed, D. Chang, S. Sarwar and W. Chen, *J. Clean. Prod.*, 2018, **172**, 4231–4238.
74. L. Wang, F. Sun and F. Hao, *et al.*, *Chem. Eng. J.*, 2020, **383**, 123205.
75. X. Gao, J. Li and M. Zheng, *et al.*, *Process Saf. Environ.*, 2021, **156**, 598–616.
76. S. S. Waghmare and T. Arfin, *Int. J. Modern Trends Eng. Res.*, 2015, **2**, 355–361.
77. T. Arfin, in *Handbook of bionanocomposites: Green and sustainable materials*, ed. S. Ahmed and S. Kanchi, Jenny Stanford Publishing, Singapore, 2018, ch. 9, pp. 235–246.
78. G. Singh, R. Bahadur and J. M. Lee, *et al.*, *Chem. Eng. J.*, 2021, **406**, 126787.
79. M. Irani, A. T. Jacobson and K. A. M. Gasem, *et al.*, *Fuel*, 2017, **206**, 10–18.
80. J. M. Ngoy, N. Wagner, L. Riboldi and O. Bolland, *Energy Procedia*, 2014, **63**, 2230–2248.
81. N. Mozaffari, S. Solaymani and A. Achour, *et al.*, *J. Phys. Chem. C*, 2020, **124**, 3692–3701.
82. S. Hou, C. Chen and C. Chang, *et al.*, *Energy Convers. Manag.*, 2011, **52**, 2758–2767.
83. S. D. Brown and C. A. Piantadosi, *Adv. Exp. Med. Biol.*, 1989, **248**, 747–754.
84. L. K. Weaver, *N. Engl. J. Med.*, 2009, **369**, 1217–1225.
85. R. F. Coburn, *Annu. Rev. Med.*, 1973, **24**, 241–250.
86. R. Motterlini and L. E. Otterbein, *Nat. Rev. Drug Discovery*, 2010, **9**, 728–743.
87. A. J. Adewale, J. A. Sonibare and J. A. Adeniran, *et al.*, *Next Mater.*, 2024, **2**, 100100.

88. K. R. Gbangbo, A. R. Kouakou and A. D. Ehouman, *Cast Stud. Chem. Environ. Eng.*, 2023, **7**, 100337.
89. L. K. Turner and F. G. Collins, *Constr. Build. Mater.*, 2013, **43**, 125–130.
90. R. Sathre and J. O'Connor, *Environ. Sci. Policy*, 2010, **13**, 104–114.
91. M. Jamshidian, E. A. Tehrany and M. Imran, *et al.*, *Compr. Rev. Food Sci. Food Saf.*, 2010, **9**, 552–571.
92. S. V. Joshi, L. T. Drzal, A. K. Mohanty and S. Arora, *Composites, Part A*, 2004, **35**, 371–376.
93. K. L. Scrivener, V. M. John and E. M. Gartner, *Cem. Concr. Res.*, 2018, **114**, 2–26.
94. A. Somanathan, N. Mathew and T. Arfin, in *Waste-derived nanoparticles: Synthesis, applications, and sustainability*, ed. J. Aslam, R. Aslam and C. M. Hussain, Elsevier, Netherland, 2024, ch. 11, pp. 281–318.

CHAPTER 14

Recycling of Low-carbon Materials

AHMED I. OSMAN,*[a,†] CHENHAO ZHOU,[b,†] ZHONGHAO CHEN,[c,d] MAHMOUD NASR,*[e] BASEM E. KESHTA,[f] DAVID W. ROONEY[g] AND POW-SENG YAP*[b,h,i]

[a] School of Engineering, Technology, and Design, Canterbury Christ Church University, Canterbury, CT1 1QU, UK; [b] Department of Civil Engineering, Xi'an Jiaotong-Liverpool University, Suzhou, 215123, China, Email: times27zhou@163.com; [c] College of Environmental and Resource Sciences, Zhejiang University, Hangzhou 310058, Zhejiang, China; [d] School of Engineering, Westlake University, Hangzhou, Zhejiang 310030, China; [e] Nanocomposite Catalysts Lab., Chemistry Department, Faculty of Science at Qena, South Valley University, Qena 83523, Egypt; [f] Chemistry Department, Faculty of Science, Tanta University, Tanta 31527, Egypt; [g] School of Chemistry and Chemical Engineering, Queen's University Belfast, Belfast BT9 5AG, Northern Ireland, UK; [h] School of Engineering, Faculty of Innovation and Technology, Taylor's University, 47500 Subang Jaya, Selangor, Malaysia; [i] Centre for Sustainable Societies, Taylor's University, 47500 Subang Jaya, Selangor, Malaysia
*Emails: ahmed.osman@canterbury.ac.uk; mahmoud.nasr62@sci.svu.edu.eg; PowSeng.Yap@taylors.edu.my

[†] Co-first authors.

Low Carbon Materials: Paving a Sustainable Path for a Greener Future
Edited by Santanu Patra, Sudheesh K. Shukla and Mika Sillanpää

Published by the Royal Society of Chemistry, www.rsc.org

14.1 Introduction

Carbon neutrality, the goal of offsetting carbon dioxide emissions by removing it from the atmosphere, has emerged as a key global environmental objective in the 21st century.[1] Various countries, including the UK, Japan, and China, have committed to achieving carbon neutrality by specific target years.[1–3] However, a major constraint in realizing this goal is the limited focus on the material outputs of economies in existing strategies and research.

While efforts have primarily concentrated on transitioning energy sources from fossil fuels to renewables, insufficient attention has been given to reducing emissions from fossil fuel usage in energy-intensive industries and material production.[4,5] These industries, which include iron and steel, aluminum, chemicals, cement, and forest products, contribute significantly to greenhouse gas emissions. Additionally, buildings and vehicles that rely on these materials contribute to emissions as well.[1,6]

Materials vary in their reliance on fossil fuels during manufacturing. Low-carbon materials, such as metals, have a low-carbon content but emit substantial CO_2 during production and use.[7] On the other hand, high-carbon materials like wood products sequester carbon from the atmosphere and require less energy for production.[8] Forests, along with carbon capture and storage technologies, play a vital role in offsetting CO_2 emissions and achieving carbon neutrality.[9]

The International Energy Agency (IEA) incorporates material efficiency and recycling strategies into its global scenarios to reduce CO_2 emissions and energy consumption.[10] Recycling and efficient material usage can significantly decrease CO_2 emissions from the industrial sector, with potential savings of millions of tons of CO_2 equivalent per year.[11] However, the cost-effectiveness of recycling and its interaction with other CO_2 reduction measures in the energy system requires further exploration.[12]

Addressing emissions from material production is crucial as it influences energy demand.[13,14] By integrating the analysis of materials and energy, a better understanding of future CO_2 emission trends can be gained.[15] Unfortunately, only a small percentage of journal papers focus on materials, and these publications are spread across multiple scientific literatures with limited interaction.[1,16] To bridge this gap and contribute to the carbon neutrality literature, this chapter aims to demonstrate how recycling of low-carbon materials and usage trends can either facilitate or delay the achievement of carbon neutrality.

The chapter herein presents an intricate examination of the spectrum of recycling methodologies tailored for low-carbon materials, distinguishing between mechanical, chemical, and biological approaches. Each method is presented not just in terms of its technical aspects but also through the lens of its environmental impact, efficiency, and contribution to sustainable development goals. Mechanical techniques, characterized by processes such as shredding and sorting, offer a direct path to material recovery.[17] Chemical recycling, including advanced methods like pyrolysis and gasification, transforms materials at the molecular level, providing innovative routes to resource utilization.[18,19] Biological strategies, encompassing composting and anaerobic digestion, highlight the role of natural processes in waste management and material recovery.[20]

This chapter also investigates challenges inherent in the recycling of low-carbon materials, from technological barriers and economic constraints to policy and regulatory gaps. However, it also illuminates the potential for overcoming these obstacles through innovation, collaboration, and the persistent pursuit of efficiency and sustainability in recycling practices. By providing a comprehensive overview of the current state of play, this chapter aims to foster a deeper understanding among readers of the critical role recycling plays in our collective journey toward a more sustainable future, encouraging active participation and support for sustainable practices across society.

14.2 Recycling Methods for Low-carbon Materials

Sustainability in reducing carbon emissions and adapting to constantly changing climate is an urgent need at present,[21,22] and rethinking the way we handle material flows may bring significant improvements in curbing the threats of environmental degradation and global warming.[23] Based on global recognition and strong environmental awareness, people have conducted extensive explorations on sustainable measures and alternative uses of materials.[24–26] At present, the recycling technology of low-carbon materials is a research hotspot. Figure 14.1 and Table 14.1 summarize the methods of recovering low-carbon materials from different aspects (mechanical, chemical, and biological) and specifically elaborate on the effectiveness of these recycling technologies.

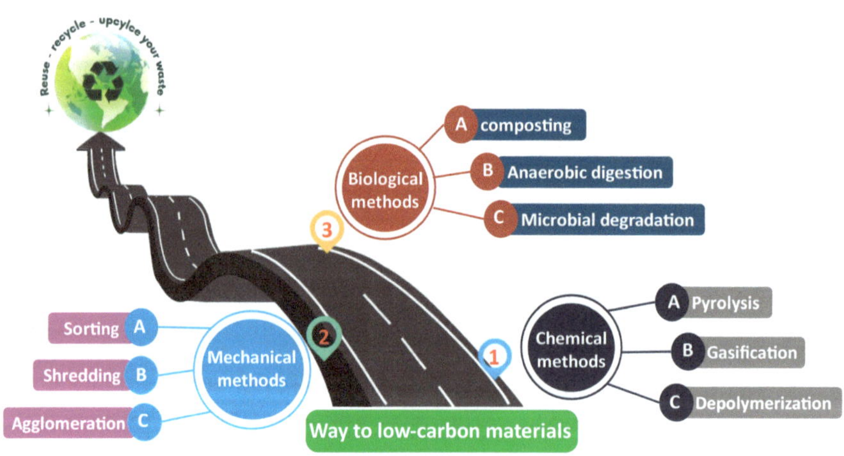

Figure 14.1 Schematic illustration of the various recycling methods for low-carbon materials.

Table 14.1 Comparison of recycling methods for low-carbon materials.

Recycling aspects	Recycling method	Technical feature	Benefit	Ref.
Mechanical	Shredding and size reduction	Controllable and targeted reduction of material size	Environmental protection, low cost and easy operation	27 and 28
	Sorting and separation technologies	Direct classification and indirect classification based on material properties	Efficient and diverse	29–31
	Agglomeration and pelletizing	Transformed into particles and used for subsequent molding	Environmental protection, high reduction degree of products	32 and 33
Chemical	Pyrolysis and gasification	Decomposition of materials by high-temperature heating and gas injection	Strong stability and low cost	34–36
	Depolymerization and solvent-based processes	Chemical depolymerization and enhanced recovery efficiency through catalyst	A wide range of applications	37 and 38
	Hydrothermal and thermochemical conversion	Apply high-temperature and high-pressure conditions in a closed reaction vessel	Environmental protection and cost-effective	39–41
Biological	Composting and anaerobic digestion	Decomposition of materials by microorganisms and bacteria	Environmental protection and high efficiency	42 and 43
	Microbial degradation and bioremediation	Closely related to the catalytic activity of enzymes and microbial activities	Wide application, environmental protection, and cost-effective	44 and 45
	Fermentation and bioconversion	Using recycled materials as microbial substrate	Generating more renewable value-added products	46–48

14.2.1 Mechanical Recycling Techniques

14.2.1.1 Shredding and Size Reduction

Reducing size through crushing, grinding, or milling processes is a mechanical method for recovering low-carbon materials. Typically, a crusher can be used to complete large-scale initial crushing.[49] Then, a hammer mill can be used to reduce size further.[50] In addition, commonly used tools include swing hammer crushers, crocodile scissors, circular mills, impact crushers, and others.[51] This method has significant environmental and cost advantages. On the one hand, the process of chopping and reducing size does not produce harmful gases that can cause air pollution or chemical wastewater pollution.[52] On the other hand, the equipment used is often easy to operate and inexpensive.

14.2.1.2 Sorting and Separation Technologies

Sorting and separation technologies can efficiently achieve material recycling and utilization.[53] These technologies can be divided into two types: direct classification and indirect classification. Direct classification refers to the separation work based on the inherent characteristics of materials such as size, density, conductivity, and magnetism.[54] Common direct separation techniques include shaking screening, hydraulic cyclone, and magnetic density separation.[51] At the same time, indirect classification is an automated robot classification activity based on sensor detection, such as laser-induced breakdown spectroscopy (LIBS), spectral imaging-based sorting, and optical-based sorting.[51]

14.2.1.3 Agglomeration and Pelletizing

Granulation is a process of aggregation,[55] which refers to the processing of materials into particles.[56] The generated particles can be used for subsequent molding and processing, and the resulting products often have performance similar to that of the original products. Agglomeration and granulation processes have developed into the main agglomeration techniques for particles and have strong recycling and utilization effects.[57] The final formed particles often obtain different external characteristics based on the type of equipment and process selected. Common granulation techniques mainly include agitation, comparison, and globulation.[58]

14.2.2 Chemical Recycling Approaches

14.2.2.1 Pyrolysis and Gasification

Chemical recovery technology based on pyrolysis and gasification is in a leading position due to its stability and cost-effectiveness.[59] Pyrolysis refers to the high-temperature heating of materials in an anaerobic environment, which leads to the decomposition of materials into high-, medium-, and low-molecular-weight products,[60] such as gases and liquids, which can often be used as fuels or for the manufacture of other compounds.[61] Gasification refers to injecting gas into a reactor, decomposing materials by reacting with gasification agents such as water and oxygen, and finally forming synthetic gas together with solid, liquid, and other by-products.[61] The composition and yield of synthetic gas are often determined by various factors such as gasification agents and temperature.[62]

14.2.2.2 Depolymerization and Solvent-based Processes

The material can be recovered through chemical depolymerization through dissolution, and the recovered monomers or reaction intermediates can be used for the synthesis of new compounds and theoretically achieve unlimited recovery work.[63] To improve the rate and yield of depolymerization products, different types of catalysts are widely used, such as bases, zeolites, and nanoparticles.[64] In addition, the recovery framework of solvent-based processes typically consists of three parts: impurity removal, dissolution, and re-precipitation.[65] To selectively recover or remove corresponding compound components, it is often necessary to find the optimal solvent based on solubility rules and other factors.

14.2.2.3 Hydrothermal and Thermochemical Conversion

Hydrothermal and thermochemical conversion are effective ways for material recycling and utilization.[66] Hydrothermal technology refers to the thermochemical conversion process under high-temperature (200–374 °C) and high-pressure (4–22 MPa) conditions applied in a closed reaction vessel.[67] This technology has significant advantages in environmental protection and cost-effectiveness. Through the thermochemical conversion route, the materials that need to be recovered can be converted into heat energy through combustion and liquid fuels and chemical products, thereby achieving higher value utilization of material resources.[66,68]

14.2.3 Biological Recycling Methods

14.2.3.1 Composting and Anaerobic Digestion

Composting and anaerobic digestion are effective in material recovery.[69,70] Composting can be divided into a high-speed initial stage and a secondary low-interest rate stage.[71] The high temperature generated by bacteria during the initial high-speed stage will accelerate the decomposition of compounds, followed by the gradual death of microorganisms in the secondary low-speed stage. As the temperature decreases, the decomposition process tends to stabilize. Anaerobic digestion refers to the process in which bacteria decompose materials and form various recyclable resources through natural biological processes under relatively anaerobic conditions.[71]

14.2.3.2 Microbial Degradation and Bioremediation

Microbial degradation and bioremediation are widely used recovery technologies for environmental pollution treatment.[72] Biodegradation involves a series of steps involving the use of enzymes,[73] and the catalytic activity of enzymes can significantly improve the rate of microbial degradation. Using enzymes for biodegradation is a green, environmentally friendly, and cost-effective recycling method.[74] Besides, bioremediation includes biological enhancement and biological stimulation.[75] Bio-enhancement refers to the integration of exogenous microorganisms, while bio-stimulation refers to stimulating the degradation ability of one's microorganisms by increasing nutrients.

14.2.3.3 Fermentation and Bioconversion

Fermentation and biotransformation are considered suitable methods for converting recycled materials into value-added products such as active compounds and fuels.[76] Fermentation often uses recycled materials as the substrate for microorganisms while allowing them to grow in environments with low water content or no free water, ultimately generating various recyclable resources in the form of biotransformation.[77] Many factors affect the fermentation and biotransformation processes, such as pH, temperature, water activity, and particle size.[76]

14.3 Benefits and Impacts of Recycling Low-carbon Materials

At present, relevant documents aimed at supporting low-carbon products have been passed at both the international and national levels, encouraging economic development to transition from traditional routes to green circular routes.[78] Low-carbon development can achieve stable economic and social development while addressing global climate change.[1,79] Low-carbon material recycling is an important component of low-carbon development. This section outlines the benefits and impacts of recycling low-carbon materials, specifically related to reducing greenhouse gas emissions, protecting natural resources, energy conservation, reducing carbon footprint, circular economy, and waste management solutions.

14.3.1 Reduction of Greenhouse Gas Emissions

With the continuous development of global industrialization, the issue of global warming caused by increasing greenhouse gas emissions is a concern for countries around the world. Energy conservation and emissions reduction are crucial for curbing global warming.[80] The recycling of low-carbon materials often reduces the net emissions of greenhouse gases. The process of transforming recycled materials into new products can significantly replace the production of these main products, which often require a large amount of energy and raw material input,[81] ultimately achieving the goal of minimizing the impact of resource utilization on greenhouse gas emissions through the highest quality utilization of recycled low-carbon materials.[82] In addition, taking the application of low-carbon materials in the transportation sector as an example, while decarbonizing transportation will significantly reduce the dependency on geopolitically sensitive fossil fuels, it will also have a series of economic impacts on society and the economy, including oil price fluctuations.[83]

To better quantify the reduction of greenhouse gas emissions caused by low-carbon material recycling, decision-makers can use the emission coefficient of material recycling to evaluate greenhouse gas emissions. This coefficient is usually developed based on life cycle assessment (LCA), which is a mature and internationally recognized method commonly used to quantify the emissions of products throughout their entire life cycle.[81] Taking LCA's evaluation of the cement manufacturing process as an example, when 100% of fossil fuels are replaced by wood waste, greenhouse gas emissions will be significantly reduced by 40% on the original basis.[84]

14.3.2 Conservation of Natural Resources

The extraction of raw materials is crucial for economic activities, but it will also increasingly exacerbate the damage to the natural environment.[85] Therefore, it is necessary to transition to a low-carbon economy, and the recycling and utilization of many low-carbon materials will play a role in protecting natural resources.

Fossil energy is a relatively important area for emission reduction. Taking the consumption growth of fossil fuels in China as an example, although China has achieved many economic and social achievements, it has also faced problems such as resource constraints and ecosystem pollution. Strengthening low-carbon material recycling and controlling fossil energy consumption have a significant impact on resource conservation and environmental protection.[86] However, expanding the production of low-carbon energy materials may require a large amount of resources (land, water, *etc.*), which can also lead to adverse environmental consequences such as deforestation and water scarcity.[87]

Sand and gravel are the most commonly used raw materials in the construction industry, so the consumption of natural resources often accounts for a significant proportion of the construction industry. The use of secondary recycled materials represented by recycled aggregate (RA) in concrete can significantly reduce the demand for mineral extraction.[88] The use of RA can make outstanding contributions to natural resource conservation and sustainable development and can also produce sustainable green concrete with carbon efficiency and cost-effectiveness.[88]

14.3.3 Energy Savings and Carbon Footprint Reduction

To significantly reduce gas emissions, it is important to go beyond well-known choices and seek new emission reduction opportunities.[89] The current focus of managers is mainly on reducing energy demand by improving energy efficiency and focusing on efforts to decarbonize emissions.[90] The recycling and sustainable management of low-carbon materials is a reliable strategy to reduce the environmental and resource footprint of buildings.[91–93] However, in the transition to low carbon, it is often necessary to consider the availability of suitable technology and energy resources. Low-carbon materials can bring significant challenges in terms of production and energy availability to a certain extent.[94]

Using the Chinese city of Shanghai as the research object, the energy-saving and carbon footprint reduction capabilities of recycled materials were evaluated. Based on the current recycling system in Shanghai, it can reduce 8.7 million tons of standard coal and 168 100 tons of carbon dioxide.[95] The recycling of scrap steel and non-ferrous metals is the main factor, accounting for 44% and 42% of energy savings, as well as 60% and 33% of carbon dioxide emissions reductions.[95] In addition, taking the recycling and utilization of wood waste as an example, this measure can often reduce the physical emissions of buildings by about 3–23%.[90] From the above emission reduction measures, it can be found that, in the future, emphasis should be placed on the outstanding contribution of low-carbon material recycling strategies to achieving carbon savings.

14.3.4 Circular Economy and Waste Management Solutions

The recycling of low-carbon materials is crucial for achieving a low-carbon and circular economy. The recycling and utilization of low-carbon materials can be used to produce high-value-added materials, and these materials can be used as accessories in the main production processes of various industries; despite the energy consumption and environmental pollution issues during the production process, this approach often leads to a doubling of economic benefits.[96,97] A circular economy is an important pillar supporting global sustainable development. The circular economy can effectively alleviate the supply pressure of resources, preserve and optimize the original value of products and materials by combining the recyclability and repairability of materials,[98] and help address the challenges brought about by global resource scarcity and climate change.[99]

Circular waste management can not only achieve savings in energy, fuel, labor, and maintenance costs but also ultimately achieve the goal of reducing greenhouse gas emissions.[100] The recyclability demonstrated by circular waste management in the construction industry has a significant impact on reducing landfill waste. On the one hand, it can significantly reduce the frequency of mining activities such as gravel and limestone. On the other hand, it has a significant effect on reducing greenhouse gas emissions during cement production.[98,101] Therefore, the circular economy and waste management achieved through the recycling of low-carbon materials will play an important role in mitigating climate change and managing resource utilization.

14.4 Challenges in Recycling Low-carbon Materials

In recent years, research on recycling solutions for low-carbon materials has attracted widespread attention.[102] The recycling and utilization of

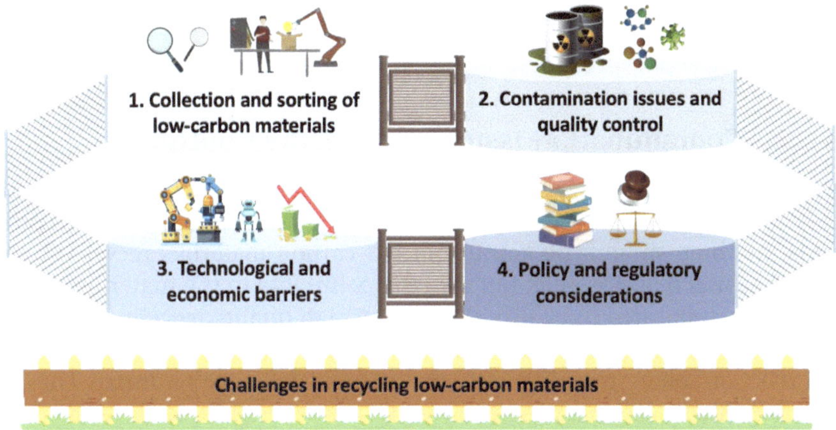

Figure 14.2 An illustration of the challenges faced by low-carbon material recycling.

low-carbon materials have significant advantages in the economic, social, and environmental fields. However, not all low-carbon materials can be functionally recycled, and materials that need to be recycled may be dispersed or lost due to incineration or landfills. This not only leads to irreversible resource loss but also has related adverse effects on the environment.[103] This section provides an overview of the challenges faced in the process of recycling low-carbon materials (Figure 14.2), specifically related to the collection and sorting of low-carbon materials, contamination issues and quality control, technological and economic barriers, as well as policy and regulatory considerations.

14.4.1 Collection and Sorting of Low-carbon Materials

In the process of low-carbon material recycling, although recyclable compound components dominate in low-carbon materials, the final amount of recovered compounds is unsatisfactory.[104] The collection and classification of low-carbon materials are based on diverse infrastructure, which often varies significantly due to regional, national, and cultural differences.[105] Lack of a systematic and standardized material collection environment, such as a lack of trained operators and inadequate infrastructure, can pose serious challenges to low-carbon material recycling.[106]

In addition, sorting, as a key link in the material recycling process, can usually be operated manually or by automated machines. However, this process may also be limited due to uncertainty caused by low manual efficiency and sensor recognition errors in automated machines.[106] In the screening process of low-carbon materials, the material is first classified based on its quality and then used for subsequent extraction and processing based on its properties. It is not difficult to find that any errors in the screening process will have a significant adverse impact on the recovery of low-carbon materials.

14.4.2 Contamination Issues and Quality Control

The strategy of recycling low-carbon materials requires finding a balance between the material recovery rate and material recovery quality.[107] In other words, not only should the amount of low-carbon materials recycled be increased as much as possible, but also the pollutants in the materials should be carefully separated to ensure the goal of high-quality recycling. However, in the process of recycling low-carbon materials, it is often inevitable to face the risk of poor recycling of toxic substances to humans or the environment.[108]

The mechanical and chemical recycling methods used in the process of recycling low-carbon materials may lead to the generation and release of harmful particles in the materials. For example, recycling methods such as crushing and pyrolysis may release harmful substances into the environment in smaller particles. When harmful particles are released into water, they can seriously pollute surface water and groundwater, and these particles are likely to be ingested by fish and other organisms, ultimately entering the food chain and causing serious negative impacts on ecosystems and human health. Similarly, when harmful exudates enter the soil, soil pollution may occur due to an increase in heavy metal concentration and other indices.[109]

14.4.3 Technological and Economic Barriers

The recycling market for low-carbon materials will face limitations in terms of technology and economic barriers.[110] We need to pay special attention to the role of technological innovation. The use of advanced recycling technologies has a significant impact on the recycling of low-carbon materials. The backwardness of recycling technologies often leads to products that cannot become value-added products that meet market requirements.[106] In addition, if the recycling market operates under stable economic conditions and policy measures are introduced for material recycling management to reflect its social costs, material recycling will be at its optimal level.[110]

The combination of technological progress and a better economic environment can effectively promote the recycling status of low-carbon materials.[111] The importance of conducting technological research and innovation in the recycling industry is self-evident in continuously improving the recycling efficiency and quality of low-carbon materials. The simultaneous development of new recycling technologies and the purchase of new recycling equipment often rely on high economic support and cooperation. On the contrary, if faced with limitations such as lack of technology and insufficient budget, the recycling industry often experiences adverse situations such as chaotic disposal and secondary pollution, ultimately leading to low added value of recycled products and poor standardization of the recycling process.

14.4.4 Policy and Regulatory Considerations

The lack of a national policy framework and extensive supervision and management is a major challenge for carrying out low-carbon material recycling work. The legislative framework related to low-carbon material recycling is often complex, as it includes policies related to circular economy, waste management, and product safety.[112] By comparing the economic benefits, distribution impacts, and legal limitations of various policies, it can be found that it is not easy to achieve clear recognition of a particular policy or a combination of policies.[113]

Managers often face three challenges when selecting and deploying various recycling policies,[113] namely (1) addressing the inefficiency of the recycling market to improve the role that the market should play, (2) designing policy tools in the absence of complete regulation of waste disposal behavior, and (3) conducting supervision and management work through price or quantity adjustments. In addition, the threats faced by policies and regulatory considerations in the low-carbon process are also reflected in the government's failure to provide economic support to recyclers through tax reductions or capital subsidies for industry investment, a lack of expert systems to provide reference suggestions for management departments, and insufficient evaluation reference data.[106]

14.5 Future Trends and Innovations in Low-carbon Material Recycling

The establishment of a new development model for a low-carbon economy has been widely recognized in the face of the threat posed by global climate change.[114] Reducing energy consumption has a significant impact on reducing global energy demand and carbon emissions.[115,116] At the same time, low-carbon innovation and development with sustainable development goals such as carbon reduction have become a research hotspot.[117] Therefore, further research and development work is still needed for the recycling of low-carbon materials. An overview of the future trends and innovations in low-carbon material recycling is shown in Figure 14.3, including advances in recycling technology, integration of digitalization and artificial intelligence, emerging circular business models, as well as policy and industry initiatives.

14.5.1 Advancements in Recycling Technologies

People use recycling technology to convert discarded low-carbon materials into environmentally friendly and recyclable resources, thereby alleviating resource shortages to a certain extent and promoting sustainable

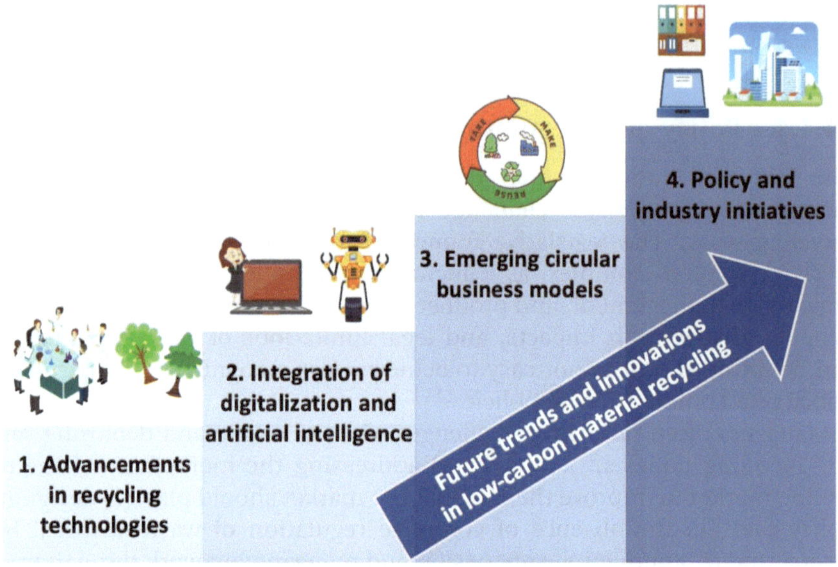

Figure 14.3 Systematic illustration of future trends and innovations in low-carbon material recycling.

development. Technological innovation, as a fundamental driving force for sustainable development, plays a decisive role in providing technical support for low-carbon material recycling.[118] The advancement of recycling technology can significantly improve material recycling efficiency, optimize recycling processes, and protect natural resources and the ecological environment. This will not only have a substantial positive impact on low-carbon development but also greatly help achieve sustainable development goals.[119] Therefore, innovative progress in recycling technology is an important direction for the sustainable development of the low-carbon material recycling industry in the future.

Technological innovation with strong development potential can coordinate economic growth and environmental quality. Technological innovation plays an extremely important role in both economic growth and environmental quality protection.[119] Taking some innovative low-carbon material recycling technologies as an example, the use of geographic information systems in the collection process can significantly improve the collection efficiency of low-carbon materials.[120] The use of biomass gasification and mixed fermentation technology can often convert low-carbon materials into various clean energy sources while significantly reducing the cost of recycling and treatment.[121]

14.5.2 Integration of Digitalization and Artificial Intelligence

With the continuous improvement of the global technological level, the application and demand of intelligent solutions in low-carbon material recycling management will also rapidly develop in major markets in the future.[122] Digitization is the driving force for decarbonization; therefore, we need to establish a synergistic effect between digitalization and low-carbon material recycling.[123] Applying hybrid digital technology based on artificial intelligence to the recycling of low-carbon materials will greatly optimize the management process of low-carbon material recycling. For example, this technology can continuously evaluate and optimize the collection process of low-carbon materials through machine learning capabilities and graph theory methods.[124]

In addition, the widespread use of digitalization in the recycling industry will maximize energy consumption while significantly controlling greenhouse gas emissions.[123] If intelligent solutions are deployed in the production, transportation, and construction sectors, it is expected that by 2030, digital technology can reduce greenhouse gas emissions to 15%.[123] The combination of digitization and artificial intelligence has the potential to significantly improve data availability and integrity while also continuously optimizing business process management in the recycling industry.[122]

14.5.3 Emerging Circular Business Models

Sustainable consumption practices and material recycling are decisive factors driving changes in current socio-economic patterns.[125] Integrating low-carbon material recycling into the circular economy is an emerging paradigm that can help achieve sustainable development goals.[126] In recent years, the development of the circular economy has been rapid. Circular economy usually refers to the collection of valuable materials from waste logistics through the recyclability and repairability of products. The circular business model is one of the main components of the circular economy, which is usually defined as a business model of circulation, expansion, strengthening, and energy circulation.[127]

The recycling and utilization of resources are often a circular economy strategy that preserves material value.[128] Low-carbon materials will be recycled and used in the same value chain at the end of their life cycle, generating by-products through recycling and then being transferred to the same process for input. This circular model can ensure the integrity and reusability of resources, thereby achieving maximum resource utilization and minimum pollutant emissions. Circular business models can often be combined with clean production, comprehensive resource utilization, sustainable consumption, and other aspects.

14.5.4 Policy and Industry Initiatives

Policy is the action guide for government agencies. On the one hand, it will safeguard business strategies with economic and environmental performance. On the other hand, it can induce technological innovation and promote coordinated cooperation among different industries. At the economic level, the recycling and utilization process of low-carbon materials can be regulated and optimized through the use of taxes, sanctions, and incentive measures.[129] For example, by saving tax-free funds, some enterprises can be better sanctioned, while considering the scarcity of public funds,[130] funds should be properly allocated to support the research and development of recycling technologies.

In terms of industry initiatives, the recycling of low-carbon materials is often stimulated by market pressure, company conditions, emission alternatives, and cost advantages. The market pressure mainly includes customer demand for sustainable products, pricing policies, company procurement, and technological innovation support.[129] If faced with improper policies and industry measures, such as a lack of clear and guiding recycling management standards, loss of legitimacy of market barriers, and unequal competition supported by management departments, it will have a significant negative impact on the efficient recycling and utilization of low-carbon materials.

14.6 Conclusion

This work highlights the crucial role of recycling in achieving environmental sustainability. It synthesizes the complex interplay of techniques, challenges, and opportunities into a coherent narrative that underscores the transformative potential of recycling within the broader context of global sustainability and climate action. It also investigates the holistic approach to recycling low-carbon materials, including revealing the technical, environmental, and economic dimensions of various recycling methodologies. Furthermore, it has highlighted the innovation and ingenuity driving the field forward while also acknowledging the significant hurdles that must be overcome. This brings to light the multifaceted nature of recycling, encompassing not just the technical challenges of material recovery and reprocessing but also the broader societal, economic, and policy frameworks that shape the recycling ecosystem.

The chapter suggests a future where advances in recycling technology, coupled with robust policy support and a reinforced collective environmental consciousness, come together to fully harness the benefits of low-carbon material recycling. It envisions a world where recycling is not just an aspect of waste management but a cornerstone of a sustainable, circular economy that minimizes waste, reduces resource extraction, and mitigates climate change. This future is predicated on the principles of innovation, collaboration, and sustainability, calling for concerted efforts from governments, industry, communities, and individuals to realize the full potential of recycling in paving the way for a sustainable and resilient future.

In this light, the chapter serves not just as an exposition of the current state of recycling methods but as a clarion call to action. It invites all stakeholders to engage with the challenges and opportunities of recycling with renewed vigor, emphasizing that the path to sustainability is a shared journey that requires the collective will, innovation, and persistence of society at large. Through this expanded narrative, the chapter aims to inspire, inform, and galvanize a global movement toward sustainable resource management and environmental stewardship, marking a pivotal step in our ongoing quest for a sustainable future.

References

1. A. Grainger and G. Smith, The role of low carbon and high carbon materials in carbon neutrality science and carbon economics, *Curr. Opin. Environ. Sustainability*, 2021, **49**, 164–189.
2. Z. Wu, X. Huang, R. Chen, X. Mao and X. Qi, The United States and China on the paths and policies to carbon neutrality, *J. Environ. Manage.*, 2022, **320**, 115785.
3. L. Xin, H. Sun and X. Xia, Renewable energy technology innovation and inclusive low-carbon development from the perspective of spatio-temporal consistency, *Environ. Sci. Pollut. Res. Int.*, 2023, **30**(8), 20490–20513.
4. O. dQ. F. Araújo, I. B. Boa Morte, C. L. T. Borges, C. R. V. Morgado and J. L. de Medeiros, Beyond clean and affordable transition pathways: A review of issues and strategies to sustainable energy supply, *Int. J. Electr. Power Energy Syst.*, 2024, **155**, 109544.
5. B. Wu, Z. Wang, Y. Tian and S. Zheng, The impact of industrial transformation and upgrading on fossil energy elasticity in China, *J. Cleaner Prod.*, 2024, **434**, 140287.
6. A. Sanna, M. Dri, M. R. Hall and M. Maroto-Valer, Waste materials for carbon capture and storage by mineralisation (CCSM) – A UK perspective, *Appl. Energy*, 2012, **99**, 545–554.
7. D. Raabe, The Materials Science behind Sustainable Metals and Alloys, *Chem. Rev.*, 2023, **123**(5), 2436–2608.
8. M. Kuittinen, C. Zernicke, S. Slabik and A. Hafner, How can carbon be stored in the built environment? A review of potential options, *Archit. Sci. Rev.*, 2021, **66**(2), 91–107.
9. A. Grainger, L. R. Iverson, G. H. Marland and A. Prasad, Comment on "The global tree restoration potential", *Science*, 2019, **366**(6463), eaay8334.
10. A. Elshkaki, The implications of material and energy efficiencies for the climate change mitigation potential of global energy transition scenarios, *Energy*, 2023, **267**, 126596.
11. A. I. Osman, M. Nasr, M. Farghali, A. K. Rashwan, A. Abdelkader, A. H. Al-Muhtaseb, I. Ihara and D. W. Rooney, Optimizing biodiesel production from waste with computational chemistry, machine learning and policy insights: a review, *Environ. Chem. Lett.*, 2024, 1–67.
12. F. Kullmann, P. Markewitz, L. Kotzur and D. Stolten, The value of recycling for low-carbon energy systems - A case study of Germany's energy transition, *Energy*, 2022, **256**, 124660.

13. M. Nasr, S. A. Halawy, S. El-Nahas, A. Abdelkader and A. I. Osman, Direct and easily prepared nanocomposite impurity-free hydroxyapatite derived from CKD as an effective catalyst for trans-2-butene production, *Appl. Catal., A*, 2023, **652**, 119039.
14. M. Nasr, A. Abdelkader, S. El-Nahas, A. I. Osman, A. Abdelhaleem, H. A. El Nazer, D. W. Rooney and S. A. Halawy, Utilizing Undissolved Portion (UNP) of Cement Kiln Dust as a Versatile Multicomponent Catalyst for Bioethylene Production from Bioethanol: An Innovative Approach to Address the Energy Crisis, *ACS Omega*, 2023, **9**(1), 1962–1976.
15. N. Alaux, M. Ruschi Mendes Saade, E. Hoxha, B. Truger and A. Passer, Future trends in materials manufacturing for low carbon building stocks: A prospective macro-scale analysis at the provincial level, *J. Cleaner Prod.*, 2023, **382**, 135278.
16. R. W. Wimbadi and R. Djalante, From decarbonization to low carbon development and transition: A systematic literature review of the conceptualization of moving toward net-zero carbon dioxide emission (1995–2019), *J. Cleaner Prod.*, 2020, **256**, 120307.
17. M. A. Martín-Lara, J. A. Moreno, G. Garcia-Garcia, S. Arjandas and M. Calero, Life cycle assessment of mechanical recycling of post-consumer polyethylene flexible films based on a real case in Spain, *J. Cleaner Prod.*, 2022, **365**, 132625.
18. M. Klotz, C. Oberschelp, C. Salah, L. Subal and S. Hellweg, The role of chemical and solvent-based recycling within a sustainable circular economy for plastics, *Sci. Total Environ.*, 2024, **906**, 167586.
19. A. I. Osman, Z. Y. Lai, M. Farghali, C. L. Yiin, A. M. Elgarahy, A. Hammad, I. Ihara, A. S. Al-Fatesh, D. W. Rooney and P. S. Yap, Optimizing biomass pathways to bioenergy and biochar application in electricity generation, biodiesel production, and biohydrogen production, *Environ. Chem. Lett.*, 2023, **21**(5), 2639–2705.
20. K. Archana, A. S. Visckram, P. Senthil Kumar, S. Manikandan, A. Saravanan and L. Natrayan, A review on recent technological breakthroughs in anaerobic digestion of organic biowaste for biogas generation: Challenges towards sustainable development goals, *Fuel*, 2024, **358**, 130298.
21. G. Seyfang, Community action for sustainable housing: Building a low-carbon future, *Energy Policy*, 2010, **38**(12), 7624–7633.
22. M. Farghali, A. I. Osman, I. M. A. Mohamed, Z. Chen, L. Chen, I. Ihara, P.-S. Yap and D. W. Rooney, Strategies to save energy in the context of the energy crisis: a review, *Environ. Chem. Lett.*, 2023, **21**(4), 2003–2039.
23. S. Lehmann, Low carbon construction systems using prefabricated engineered solid wood panels for urban infill to significantly reduce greenhouse gas emissions, *Sustainable Cities Soc.*, 2013, **6**, 57–67.
24. A. Mohajerani, L. Burnett, J. V. Smith, S. Markovski, G. Rodwell, M. T. Rahman, H. Kurmus, M. Mirzababaei, A. Arulrajah, S. Horpibulsuk and F. Maghool, Recycling waste rubber tyres in construction materials

and associated environmental considerations: A review, *Resour., Conserv. Recycl.*, 2020, **155**, 104679.
25. M. Yang, L. Chen, J. Wang, G. Msigwa, A. I. Osman, S. Fawzy, D. W. Rooney and P.-S. Yap, Circular economy strategies for combating climate change and other environmental issues, *Environ. Chem. Lett.*, 2023, **21**(1), 55–80.
26. L. Chen, Y. Zhang, Z. Chen, Y. Dong, Y. Jiang, J. Hua, Y. Liu, A. I. Osman, M. Farghali, L. Huang, D. W. Rooney and P.-S. Yap, Biomaterials technology and policies in the building sector: a review, *Environ. Chem. Lett.*, 2024, **22**, 715–750.
27. C. Tian, G. Zhou, J. Zhang and C. Zhang, Optimization of cutting parameters considering tool wear conditions in low-carbon manufacturing environment, *J. Cleaner Prod.*, 2019, **226**, 706–719.
28. J. Lee, K. Kim, H. Cho, J. Ok and S. Kim, Shredding and liberation characteristics of refrigerators and small appliances, *Waste Manage.*, 2017, **59**, 409–421.
29. J. Pluskal, R. Šomplák, V. Nevrlý, V. Smejkalová and M. Pavlas, Strategic decisions leading to sustainable waste management: Separation, sorting and recycling possibilities, *J. Cleaner Prod.*, 2021, **278**, 123359.
30. S. P. Gundupalli, S. Hait and A. Thakur, A review on automated sorting of source-separated municipal solid waste for recycling, *Waste Manage.*, 2017, **60**, 56–74.
31. J.-P. Lange, Managing plastic waste– sorting, recycling, disposal, and product redesign, *ACS Sustainable Chem. Eng.*, 2021, **9**(47), 15722–15738.
32. J. Pal, Innovative development on agglomeration of iron ore fines and iron oxide wastes, *Miner. Process. Extr. Metall. Rev.*, 2019, **40**(4), 248–264.
33. S. G. Şerban and I. Kiss, Identifying the possibilities for superior recovery by pelletization of industry related small and powdery iron containing waste, *Acta Polytechnica Hungarica*, 2021, **18**(4), 79–104.
34. M. Policella, Z. Wang, K. G. Burra and A. K. Gupta, Characteristics of syngas from pyrolysis and CO_2-assisted gasification of waste tires, *Appl. Energy*, 2019, **254**, 113678.
35. J. Dong, Y. Tang, A. Nzihou and Y. Chi, Key factors influencing the environmental performance of pyrolysis, gasification and incineration Waste-to-Energy technologies, *Energy Convers. Manage.*, 2019, **196**, 497–512.
36. P. Prasertcharoensuk, S. J. Bull and A. N. Phan, Gasification of waste biomass for hydrogen production: Effects of pyrolysis parameters, *Renewable Energy*, 2019, **143**, 112–120.
37. Y.-B. Zhao, X.-D. Lv and H.-G. Ni, Solvent-based separation and recycling of waste plastics: A review, *Chemosphere*, 2018, **209**, 707–720.
38. M. Konarova, N. Batalha, G. Fraga, M. H. Ahmed, S. Pratt and B. Laycock, Integrating PET chemical recycling with pyrolysis of mixed plastic waste *via* pressureless alkaline depolymerization in a hydrocarbon solvent, *Waste Manage.*, 2024, **174**, 24–30.

39. G. Sztancs, A. Kovacs, A. J. Toth, P. Mizsey, P. Billen and D. Fozer, Catalytic hydrothermal carbonization of microalgae biomass for low-carbon emission power generation: the environmental impacts of hydrochar co-firing, *Fuel*, 2021, **300**, 120927.
40. J. Lee, K.-Y. A. Lin, S. Jung and E. E. Kwon, Hybrid renewable energy systems involving thermochemical conversion process for waste-to-energy strategy, *Chem. Eng. J.*, 2023, **452**, 139218.
41. F. Cheng, A. A. Small and L. M. Colosi, The levelized cost of negative CO_2 emissions from thermochemical conversion of biomass coupled with carbon capture and storage, *Energy Convers. Manage.*, 2021, **237**, 114115.
42. A. Le Pera, M. Sellaro, E. Bencivenni and F. D'Amico, Environmental sustainability of an integrate anaerobic digestion-composting treatment of food waste: Analysis of an Italian plant in the circular bioeconomy strategy, *Waste Manage.*, 2022, **139**, 341–351.
43. J. Zhang, H. Tian, X. Wang and Y. W. Tong, Effects of activated carbon on mesophilic and thermophilic anaerobic digestion of food waste: Process performance and life cycle assessment, *Chem. Eng. J.*, 2020, **399**, 125757.
44. S. Verma and A. Kuila, Bioremediation of heavy metals by microbial process, *Environ. Technol. Innovation*, 2019, **14**, 100369.
45. A. Saravanan, P. S. Kumar, P. A. Duc and G. Rangasamy, Strategies for microbial bioremediation of environmental pollutants from industrial wastewater: A sustainable approach, *Chemosphere*, 2023, **313**, 137323.
46. M. Xu, M. Yang, H. Sun, M. Gao, Q. Wang and C. Wu, Bioconversion of biowaste into renewable energy and resources: A sustainable strategy, *Environ. Res.*, 2022, **214**, 113929.
47. K. Amulya and S. V. Mohan, Green hydrogen based succinic acid and biopolymer production in a biorefinery: Adding value to CO_2 from acidogenic fermentation, *Chem. Eng. J.*, 2022, **429**, 132163.
48. S. F. Ahmed, M. Kabir, A. Mehjabin, F. T. Z. Oishi, S. Ahmed, S. Mannan, M. Mofijur, F. Almomani, I. A. Badruddin and S. Kamangar, Waste biorefinery to produce renewable energy: Bioconversion process and circular bioeconomy, *Energy Rep.*, 2023, **10**, 3073–3091.
49. D. García, I. Vegas and I. Cacho, Mechanical recycling of GFRP waste as short-fiber reinforcements in microconcrete, *Constr. Build. Mater.*, 2014, **64**, 293–300.
50. X. Li, R. B. Bai and J. McKechnie, Environmental and financial performance of mechanical recycling of carbon fibre reinforced polymers and comparison with conventional disposal routes, *J. Cleaner Prod.*, 2016, **127**, 451–460.
51. S. P. Gundupalli, S. Hait and A. Thakur, A review on automated sorting of source-separated municipal solid waste for recycling, *Waste Manage.*, 2017, **60**, 56–74.
52. A. C. Meira Castro, J. P. Carvalho, M. C. S. Ribeiro, J. P. Meixedo, F. J. G. Silva, A. Fiúza and M. L. Dinis, An integrated recycling approach for

GFRP pultrusion wastes: recycling and reuse assessment into new composite materials using Fuzzy Boolean Nets, *J. Cleaner Prod.*, 2014, **66**, 420–430.
53. P. Sajeesh and A. K. Sen, Particle separation and sorting in microfluidic devices: a review, *Microfluid. Nanofluid.*, 2014, **17**(1), 1–52.
54. G. Gaustad, E. Olivetti and R. Kirchain, Improving aluminum recycling: A survey of sorting and impurity removal technologies, *Resour., Conserv. Recycl.*, 2012, **58**, 79–87.
55. J. Pal, Innovative Development on Agglomeration of Iron Ore Fines and Iron Oxide Wastes, *Miner. Process. Extr. Metall. Rev.*, 2018, **40**(4), 248–264.
56. K. Ragaert, L. Delva and K. Van Geem, Mechanical and chemical recycling of solid plastic waste, *Waste Manage.*, 2017, **69**, 24–58.
57. Y. Rama Murthy, G. U. Kapure, S. K. Tripathy and G. P. Sahu, Recycling of ferromanganese gas cleaning plant (GCP) sludge by novel agglomeration, *Waste Manage.*, 2018, **80**, 457–465.
58. S. Muley, T. Nandgude and S. Poddar, Extrusion-spheronization a promising pelletization technique: In-depth review, *Asian J. Pharm. Sci.*, 2016, **11**(6), 684–699.
59. O. Dogu, M. Pelucchi, R. Van de Vijver, P. H. M. Van Steenberge, D. R. D'hooge, A. Cuoci, M. Mehl, A. Frassoldati, T. Faravelli and K. M. Van Geem, The chemistry of chemical recycling of solid plastic waste *via* pyrolysis and gasification: State-of-the-art, challenges, and future directions, *Prog. Energy Combust. Sci.*, 2021, **84**, 100901.
60. F. A. López, O. Rodríguez, F. J. Alguacil, I. García-Díaz, T. A. Centeno, J. L. García-Fierro and C. González, Recovery of carbon fibres by the thermolysis and gasification of waste prepreg, *J. Anal. Appl. Pyrolysis*, 2013, **104**, 675–683.
61. J. Dong, Y. J. Tang, A. Nzihou and Y. Chi, Key factors influencing the environmental performance of pyrolysis, gasification and incineration Waste-to-Energy technologies, *Energy Convers. Manage.*, 2019, **196**, 497–512.
62. N. Déparrois, P. Singh, K. G. Burra and A. K. Gupta, Syngas production from co-pyrolysis and co-gasification of polystyrene and paper with CO_2, *Appl. Energy*, 2019, **246**, 1–10.
63. A. Carniel, V. A. Waldow and A. M. Castro, A comprehensive and critical review on key elements to implement enzymatic PET depolymerization for recycling purposes, *Biotechnol. Adv.*, 2021, **52**, 107811.
64. Z. Laldinpuii, S. Lalhmangaihzuala, Z. Pachuau and K. Vanlaldinpuia, Depolymerization of poly(ethylene terephthalate) waste with biomass-waste derived recyclable heterogeneous catalyst, *Waste Manage.*, 2021, **126**, 1–10.
65. Y. B. Zhao, X. D. Lv and H. G. Ni, Solvent-based separation and recycling of waste plastics: A review, *Chemosphere*, 2018, **209**, 707–720.
66. Y. J. Wu, H. Y. Wang, H. Y. Li, X. Han, M. Y. Zhang, Y. Sun, X. D. Fan, R. Tu, Y. M. Zeng, C. C. Xu and X. W. Xu, Applications of catalysts in

thermochemical conversion of biomass (pyrolysis, hydrothermal liquefaction and gasification): A critical review, *Renewable Energy*, 2022, **196**, 462–481.
67. L. Zhan, L. Jiang, Y. L. Zhang, B. Gao and Z. M. Xu, Reduction, detoxification and recycling of solid waste by hydrothermal technology: A review, *Chem. Eng. J.*, 2020, **390**, 124651.
68. Z. Chen, L. Chen, K. S. Khoo, V. K. Gupta, M. Sharma, P. L. Show and P.-S. Yap, Exploitation of lignocellulosic-based biomass biorefinery: A critical review of renewable bioresource, sustainability and economic views, *Biotechnol. Adv.*, 2023, **69**, 108265.
69. M. A. Bustamante, A. P. Restrepo, J. A. Alburquerque, M. D. Pérez-Murcia, C. Paredes, R. Moral and M. P. Bernal, Recycling of anaerobic digestates by composting: effect of the bulking agent used, *J. Cleaner Prod.*, 2013, **47**, 61–69.
70. Q. C. Meng, H. B. Liu, H. D. Zhang, S. Y. Xu, E. Lichtfouse and Y. B. Yun, Anaerobic digestion and recycling of kitchen waste: a review, *Environ. Chem. Lett.*, 2022, **20**(3), 1745–1762.
71. D. Briassoulis, A. Pikasi and M. Hiskakis, Organic recycling of post-consumer/industrial bio-based plastics through industrial aerobic composting and anaerobic digestion - Techno-economic sustainability criteria and indicators, *Polym. Degrad. Stab.*, 2021, **190**, 109642.
72. S. J. Varjani, Microbial degradation of petroleum hydrocarbons, *Bioresour. Technol.*, 2017, **223**, 277–286.
73. F. Abbasian, R. Lockington, M. Mallavarapu and R. Naidu, A Comprehensive Review of Aliphatic Hydrocarbon Biodegradation by Bacteria, *Appl. Biochem. Biotechnol.*, 2015, **176**(3), 670–699.
74. R. Koshti, L. Mehta and N. Samarth, Biological Recycling of Polyethylene Terephthalate: A Mini-Review, *J. Polym. Environ.*, 2018, **26**(8), 3520–3529.
75. M. Wu, W. Li, W. A. Dick, X. Ye, K. Chen, D. Kost and L. Chen, Bioremediation of hydrocarbon degradation in a petroleum-contaminated soil and microbial population and activity determination, *Chemosphere*, 2017, **169**, 124–130.
76. C. R. Chilakamarry, A. M. Mimi Sakinah, A. W. Zularisam, R. Sirohi, I. A. Khilji, N. Ahmad and A. Pandey, Advances in solid-state fermentation for bioconversion of agricultural wastes to value-added products: Opportunities and challenges, *Bioresour. Technol.*, 2022, **343**, 126065.
77. M. Kopke and S. D. Simpson, Pollution to products: recycling of 'above ground' carbon by gas fermentation, *Curr. Opin. Biotechnol*, 2020, **65**, 180–189.
78. F. Orsini and P. Marrone, Approaches for a low-carbon production of building materials: A review, *J. Cleaner Prod.*, 2019, **241**, 118380.
79. A. I. Osman, M. Farghali, Y. Dong, J. Kong, M. Yousry, A. K. Rashwan, Z. Chen, A. Al-Fatesh, D. W. Rooney and P.-S. Yap, Reducing the carbon footprint of buildings using biochar-based bricks and insulating materials: a review, *Environ. Chem. Lett.*, 2024, **22**(1), 71–104.

80. X. Y. Zhang, K. X. Jiao, J. L. Zhang and Z. Y. Guo, A review on low carbon emissions projects of steel industry in the World, *J. Cleaner Prod.*, 2021, **306**, 127259.
81. D. A. Turner, I. D. Williams and S. Kemp, Greenhouse gas emission factors for recycling of source-segregated waste materials, *Resour., Conserv. Recycl.*, 2015, **105**, 186–197.
82. M. Corsten, E. Worrell, M. Rouw and A. van Duin, The potential contribution of sustainable waste management to energy use and greenhouse gas emission reduction in the Netherlands, *Resour., Conserv. Recycl.*, 2013, **77**, 13–21.
83. K. Shahzad and I. Iqbal Cheema, Low-carbon technologies in automotive industry and decarbonizing transport, *J. Power Sources*, 2024, **591**, 233888.
84. L. H. Zhang and W. E. Mabee, Comparative study on the life-cycle greenhouse gas emissions of the utilization of potential low carbon fuels for the cement industry, *J. Cleaner Prod.*, 2016, **122**, 102–112.
85. R. Arendt, V. Bach and M. Finkbeiner, Environmental costs of abiotic resource demand for the EU's low-carbon development, *Resour., Conserv. Recycl.*, 2022, **180**, 106057.
86. J. K. He, Global low-carbon transition and China's response strategies, *Adv. Clim. Change Res.*, 2016, 7(4), 204–212.
87. J. Benajes, A. García, J. Monsalve-Serrano and M. Guzmán-Mendoza, A review on low carbon fuels for road vehicles: The good, the bad and the energy potential for the transport sector, *Fuel*, 2024, **361**, 130647.
88. S. C. Bostanci, M. Limbachiya and H. Kew, Use of recycled aggregates for low carbon and cost effective concrete construction, *J. Cleaner Prod.*, 2018, **189**, 176–196.
89. K. Schanes, S. Giljum and E. Hertwich, Low carbon lifestyles: A framework to structure consumption strategies and options to reduce carbon footprints, *J. Cleaner Prod.*, 2016, **139**, 1033–1043.
90. M. Norouzi, A. N. Haddad, L. Jimenez, S. Hoseinzadeh and D. Boer, Carbon footprint of low-energy buildings in the United Kingdom: Effects of mitigating technological pathways and decarbonization strategies, *Sci. Total Environ*, 2023, **882**, 163490.
91. A. I. Osman, Z. Y. Lai, M. Farghali, C. L. Yiin, A. M. Elgarahy, A. Hammad, I. Ihara, A. S. Al-Fatesh, D. W. Rooney and P.-S. Yap, Optimizing biomass pathways to bioenergy and biochar application in electricity generation, biodiesel production, and biohydrogen production, *Environ. Chem. Lett.*, 2023, **21**(5), 2639–2705.
92. L. Chen, Z. Chen, Z. Xie, L. Wei, J. Hua, L. Huang and P.-S. Yap, Recent developments on natural fiber concrete: A review of properties, sustainability, applications, barriers, and opportunities, *Dev. Built Environ.*, 2023, **16**, 100255.
93. J. Kong, Y. Dong, A. Poshnath, B. Rismanchi and P.-S. Yap, Application of Building Integrated Photovoltaic (BIPV) in Net-Zero Energy Buildings (NZEBs), *Energies*, 2023, **16**(17), 6401.

94. S. Darda, T. Papalas and A. Zabaniotou, Biofuels journey in Europe: Currently the way to low carbon economy sustainability is still a challenge, *J. Cleaner Prod.*, 2019, **208**, 575–588.
95. H. J. Dong, Y. Geng, X. M. Yu and J. J. Li, Uncovering energy saving and carbon reduction potential from recycling wastes: A case of Shanghai in China, *J. Cleaner Prod.*, 2018, **205**, 27–35.
96. R. AliAkbari, M. H. Ghasemi, N. Neekzad, E. Kowsari, S. Ramakrishna, M. Mehrali and Y. Marfavi, High value add bio-based low-carbon materials: Conversion processes and circular economy, *J. Cleaner Prod.*, 2021, **293**, 126101.
97. G. Lan, J. Yang, R. P. Ye, Y. Boyjoo, J. Liang, X. Liu, Y. Li, J. Liu and K. Qian, Sustainable carbon materials toward emerging applications, *Small Methods*, 2021, **5**(5), 2001250.
98. M. Möslinger, G. Ulpiani and N. Vetters, Circular economy and waste management to empower a climate-neutral urban future, *J. Cleaner Prod.*, 2023, **421**, 138454.
99. J. L. Xie, Z. Q. Xia, X. Tian and Y. W. Liu, Nexus and synergy between the low-carbon economy and circular economy: A systematic and critical review, *Environ. Impact Assess. Rev.*, 2023, **100**, 107077.
100. B. Fang, J. Yu, Z. Chen, A. I. Osman, M. Farghali, I. Ihara, E. H. Hamza, D. W. Rooney and P.-S. Yap, Artificial intelligence for waste management in smart cities: a review, *Environ. Chem. Lett.*, 2023, **21**(4), 1959–1989.
101. X. Peng, Y. Jiang, Z. Chen, A. I. Osman, M. Farghali, D. W. Rooney and P.-S. Yap, Recycling municipal, agricultural and industrial waste into energy, fertilizers, food and construction materials, and economic feasibility: a review, *Environ. Chem. Lett.*, 2023, **21**(2), 765–801.
102. Q. Q. Chen, Y. Gu, Z. Y. Tang, W. Wei and Y. H. Sun, Assessment of low-carbon iron and steel production with CO recycling and utilization technologies: A case study in China, *Appl. Energy*, 2018, **220**, 192–207.
103. M. Andersson, M. L. Söderman and B. A. Sandén, Challenges of recycling multiple scarce metals: The case of Swedish ELV and WEEE recycling, *Resour. Policy*, 2019, **63**, 101403.
104. Y. Moh and L. Abd Manaf, Solid waste management transformation and future challenges of source separation and recycling practice in Malaysia, *Resour., Conserv. Recycl.*, 2017, **116**, 1–14.
105. J. N. Hahladakis and E. Iacovidou, An overview of the challenges and trade-offs in closing the loop of post-consumer plastic waste (PCPW): Focus on recycling, *J. Hazard. Mater.*, 2019, **380**, 120887.
106. A. D. N. Gunarathne, T. P. Y. C. Tennakoon and J. R. Weragoda, Challenges and opportunities for the recycling industry in developing countries: the case of Sri Lanka, *J. Mater. Cycles Waste Manage.*, 2019, **21**(1), 181–190.
107. P. H. Brunner and H. Rechberger, Waste to energy–key element for sustainable waste management, *Waste Manage.*, 2015, **37**, 3–12.

108. J. Knapp, A. Allesch, W. Muller and A. Bockreis, Methods to estimate the transfer of contaminants into recycling products - A case study from Austria, *Waste Manage.*, 2017, **69**, 88–100.
109. S. Kumar, H. Dhar, V. V. Nair, Rena, J. Govani, S. Arya, J. K. Bhattacharya, A. N. Vaidya and A. B. Akolkar, Environmental quality monitoring and impact assessment of solid waste dumpsites in high altitude sub-tropical regions, *J. Environ. Manage.*, 2019, **252**, 109681.
110. F. Nicolli, N. Johnstone and P. Söderholm, Resolving failures in recycling markets: the role of technological innovation, *Environ. Econ. Policy Stud.*, 2012, **14**(3), 261–288.
111. N. Nayak and D. Apelian, Opportunities and Barriers to Resource Recovery and Recycling from Shredder Residue in the United States, *JOM*, 2014, **66**(11), 2367–2376.
112. M. S. Qureshi, A. Oasmaa, H. Pihkola, I. Deviatkin, A. Tenhunen, J. Mannila, H. Minkkinen, M. Pohjakallio and J. Laine-Ylijoki, Pyrolysis of plastic waste: Opportunities and challenges, *J. Anal. Appl. Pyrolysis*, 2020, **152**, 104804.
113. P. Söderholm and T. Ekvall, Metal markets and recycling policies: impacts and challenges, *Miner. Econ.*, 2019, **33**(1–2), 257–272.
114. T. D. Moshood, G. Nawanir, F. Mahmud, S. Sorooshian and A. Q. Adeleke, Green and low carbon matters: A systematic review of the past, today, and future on sustainability supply chain management practices among manufacturing industry, *Cleaner Eng. Technol.*, 2021, **4**, 100144.
115. X. Y. Wang, G. Y. Wang, T. Y. Chen, Z. W. Zeng and C. K. Heng, Low-carbon city and its future research trends: A bibliometric analysis and systematic review, *Sustainable Cities Soc.*, 2023, **90**, 104381.
116. L. Chen, G. Msigwa, M. Yang, A. I. Osman, S. Fawzy, D. W. Rooney and P.-S. Yap, Strategies to achieve a carbon neutral society: a review, *Environ. Chem. Lett.*, 2022, **20**(4), 2277–2310.
117. H. Yin, J. Y. Zhao, X. Xi and Y. H. Zhang, Evolution of regional low-carbon innovation systems with sustainable development: An empirical study with big-data, *J. Cleaner Prod.*, 2019, **209**, 1545–1563.
118. A. Goutam Mukherjee, U. Ramesh Wanjari, R. Chakraborty, K. Renu, B. Vellingiri, A. George, R. S. C and A. Valsala Gopalakrishnan, A review on modern and smart technologies for efficient waste disposal and management, *J. Environ. Manage.*, 2021, **297**, 113347.
119. Z. Quan, X. Xu, W. Wang, J. Jiang and S. Gao, Do industrial solid waste recycling and technological innovation promote low-carbon development in China? New insights from NARDL approach, *Sci. Total Environ*, 2024, **916**, 170446.
120. S. A. Hassan, N. G. M. Jameel and B. Şekeroğlu, Smart solid waste monitoring and collection system, *Int. J.*, 2016, **6**(10), 1–12.
121. H. I. D. I. Muri and D. R. Hjelme, Sensor Technology Options for Municipal Solid Waste Characterization for Optimal Operation of Waste-to-Energy Plants, *Energies*, 2022, **15**(3), 1105.

122. C. Burger, M. Kalverkamp and A. Pehlken, Decision making and software solutions with regard to waste management, *J. Cleaner Prod.*, 2018, **205**, 210–225.
123. T. A. Kurniawan, M. H. D. Othman, X. Liang, H. H. Goh, P. Gikas, T. D. Kusworo, A. Anouzla and K. W. Chew, Decarbonization in waste recycling industry using digitalization to promote net-zero emissions and its implications on sustainability, *J. Environ. Manage.*, 2023, **338**, 117765.
124. K. H. Yu, Y. Zhang, D. N. Li, C. E. Montenegro-Marin and P. M. Kumar, Environmental planning based on reduce, reuse, recycle and recover using artificial intelligence, *Environ. Impact Assess. Rev.*, 2021, **86**, 106492.
125. N. Suchek, C. I. Fernandes, S. Kraus, M. Filser and H. Sjögrén, Innovation and the circular economy: A systematic literature review, *Business Strategy Environ.*, 2021, **30**(8), 3686–3702.
126. U. Awan and R. Sroufe, Sustainability in the Circular Economy: Insights and Dynamics of Designing Circular Business Models, *Appl. Sci.*, 2022, **12**(3), 1521.
127. M. T. Islam, U. Iyer-Raniga and S. Trewick, Recycling Perspectives of Circular Business Models: A Review, *Recycling*, 2022, **7**(5), 79.
128. E. Hultberg and R. Pal, Lessons on business model scalability for circular economy in the fashion retail value chain: Towards a conceptual model, *Sustainable Prod. Consumption*, 2021, **28**, 686–698.
129. F. Murakami, A. Sulzbach, G. M. Pereira, M. Borchardt and M. A. Sellitto, How the Brazilian government can use public policies to induce recycling and still save money?, *J. Cleaner Prod.*, 2015, **96**, 94–101.
130. N. F. da Cruz, P. Simões and R. C. Marques, Economic cost recovery in the recycling of packaging waste: the case of Portugal, *J. Cleaner Prod.*, 2012, **37**, 8–18.

CHAPTER 15

Toward a Greener Future with Low-carbon Materials

ASHIMA THAKUR,[a] SAIMA JAN,[a] SHREYA SEMWAL,[a] NANDINI DHUPAR,[a] PHẠM THANH ĐẠNG,[b] SUDHEESH KUMAR SHUKLA[c,d] AND AJIT SHARMA*[a]

[a] Department of Chemistry, School of Chemical Engineering and Physical Science, Lovely Professional University, Punjab, 14411, India; [b] VNU Key Laboratory of Advanced Material for Green Growth, Faculty of Chemistry, University of Science, Vietnam National University, Vietnam; [c] Department of Chemical Sciences, University of Johannesburg, Doornfontein Campus, P.O. Box 17011, Johannesburg, 2008, South Africa; [d] Department of Biosciences, Graphic Era (Deemed to be University), Dehradun 248002, Uttarakhand, India

*Email: ajitsharma2003@gmail.com

15.1 Introduction

Global industrial expansion has resulted in a significant increase in greenhouse gas emissions, with CO_2 being the main greenhouse gas.[1] In 2019, burning fossil fuels was responsible for 86% of greenhouse gas emissions. The idea that human-produced CO_2 is the main cause of the recent global warming is supported by the decrease in carbon emissions seen in 2020 during the COVID-19 pandemic lockdown,[2] as well as the rise in greenhouse gases and trends in global temperature.[3,4] Certain industries have the most effects on atmospheric carbon emissions, including power plants, cement, steel and iron, and glass producers.[5] During an era when sustainable solutions are essential and environmental concerns continue to rise, it is now critical that they use low-CO_2 emission materials.[6] Low-carbon materials are

Low Carbon Materials: Paving a Sustainable Path for a Greener Future
Edited by Santanu Patra, Sudheesh K. Shukla and Mika Sillanpää

Published by the Royal Society of Chemistry, www.rsc.org

essential to sustainability because they enable industries to lessen their impact on the environment and fight climate change.[7] From manufacture to disposal, these materials are intended to produce fewer carbon emissions. Low-carbon materials are essential to relive the stress faced by the world. Thus we should adapt low-carbon materials for energy, healthcare, and for the environment and society while examining the impact and future of low-carbon materials.[8]

15.1.1 Low-carbon Materials for Greener Energy

In order to reduce the environmental impact of energy production and consumption, low-carbon materials play an essential role in the advancement of greener energy solutions. Renewable energy technologies including solar panels, wind turbines, and energy-storage systems are made using materials like recycled metals, innovative polymers, and sustainable composites to decrease their carbon footprint. For example, durable and lightweight materials extend the durability and efficiency of renewable energy installations while consuming less energy overall during production, cutting back on the use of fossil fuels for cooling and heating. Low-carbon materials speed the transition, encourage sustainable resource use, and reduce greenhouse gas emissions by assisting in the development and application of renewable energy technology. Low-carbon materials have many uses in the energy field, by which greenhouse gas emissions are decreased, making greener energy for the world, including:

- Fuel cell materials
- Battery materials
- Energy-harvesting materials
- Supercapacitor materials

15.1.2 Batteries with Low-carbon Materials

In the present period of rapid technological advancement and industrialization, batteries are becoming increasingly necessary for energy-storage solutions. As seen in Figure 15.1, they create a number of environmental issues despite their extensive usage. For example, lithium-ion batteries are valued for their high energy density, but their highly resource-intensive extraction processes degrade the environment.[9,10] Similarly, nickel-cadmium batteries are problematic because of cadmium, a dangerous heavy metal, while lead-acid batteries, despite being cheap, offer major toxicity hazards because of their lead content. By reducing their environmental impact, developments in low-carbon battery materials present an appropriate solution to these problems.[11] Since electric vehicles (EVs) generate significantly less carbon than vehicles powered by internal combustion engines, low-carbon materials are especially important in the development of EV batteries.[12,13] Policies encouraging companies to embrace sustainable practices have assisted in this change globally, accelerating the transformation to a low-carbon economy.[14,15] The use of fossil fuels during mining, refining, and

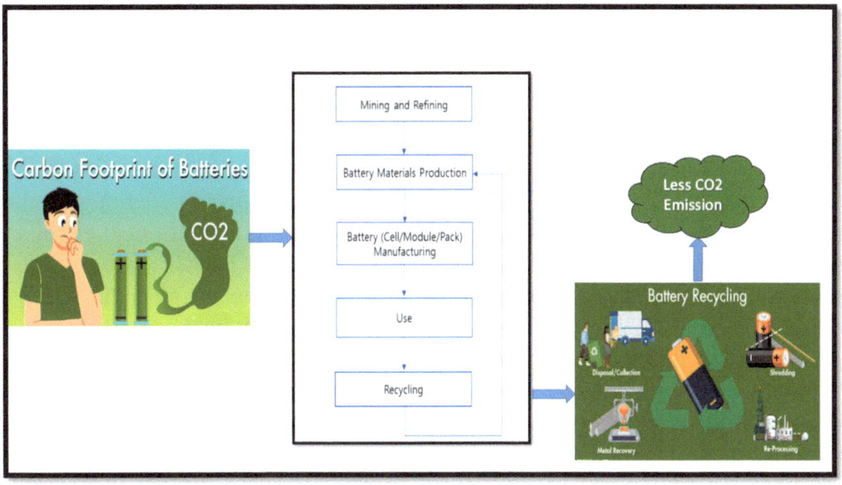

Figure 15.1 Carbon footprint of batteries.

material processing is the main source of CO_2 emissions in battery production, according to life cycle studies.[16,17] In order to solve this, traditional energy inputs in these phases must be replaced by renewable energy sources. Emissions can be further decreased by using low-carbon substitutes for cathode and anode materials. By lowering dependency on virgin resources and the environmental impact of extraction and processing, recycling end-of-life batteries is also essential.[18,19] Together, these developments open the door for environmentally friendly battery technologies and greatly contribute to a more sustainable future.

15.1.3 Low-carbon Fuel Cells (FCs)

An innovating kind of electrochemical energy converter, fuel cells (FCs) provide a sustainable and effective substitute for traditional systems by directly converting chemical energy into electrical energy without the need for combustion.[20] Under a variety of pH levels, hydrogen (H_2) is oxidized at the anode in these devices to produce electrons, protons (H^+), or water (H_2O). The chemical cycle is completed when the electrons reach the cathode *via* an external circuit, where oxygen is reduced to produce water or hydroxide ions (OH^-).[21] Fuel cell performance is greatly impacted by the selection of electrode and electrolyte materials since they must satisfy strict thermal, electrochemical, and physio-chemical specifications that call for sophisticated synthesis methods for optimization.[22] Compared to internal combustion engines, fuel cells offer a number of advantages. As the main fuel, hydrogen provides an abundant and almost limitless source of energy.[23] It is an exceptional energy carrier with an energy density that is two to three times higher than that of petrol and second only to nuclear energy. When compared to the lowest limits of internal combustion engines running on the Carnot cycle, hydrogen fuel cells have high conversion efficiencies, reaching up to 83% theoretically and about 65% real efficiency.[24–26] Due to their low carbon emissions and just water emissions as a consequence, they are totally sustainable. Fuel cells' modular design, ease of maintenance, affordability, and long service life are all influenced by their dependable performance, which includes steady power output, quick dynamic response, and few mechanical components. Solid oxide fuel cells (SOFCs) and proton exchange membrane fuel cells (PEMFCs) stand out among the different kinds of fuel cells.[27] PEMFCs are perfect for transportation applications, but SOFCs are best suited for stationary power generation because of their high operating temperatures. PEMFCs are ideal for automobiles since they require high-purity hydrogen, have quick start-up times, and function effectively at temperatures lower than 100 °C.[28] Fuel cell technology advancement is essential to creating a low-carbon, safe, and effective energy system. By offering high-performance, sustainable solutions, these technologies have the potential to completely transform the energy sector and accelerate the shift to a more environmentally friendly future.

15.2 Low-carbon Supercapacitors

Supercapacitors have drawn a lot of interest and are frequently used in electronic device memory backup systems due to their faster energy storage than batteries. Their significance has increased even more with the introduction of hybrid cars, which has advanced their development to unprecedented levels.[29,30] Supercapacitors are essential for applications like vehicle braking and acceleration because of their high power density, which shields batteries from high-frequency cycles of charging and discharging.[31] They are also perfect as emergency door power sources for airplanes due to their safe and dependable operation.[32] Because of its high specific surface area, chemical stability, and superior electrical conductivity, carbon is the most often used electrode material for electric double-layer supercapacitors.[33]

15.2.1 Activated Carbon

Because of its high surface area and porosity, which aid in the adsorption of electrolyte ions, activated carbon is a material that is frequently utilized in supercapacitor electrodes and hydrogen storage.[34] It is environmentally friendly since it may be produced from a variety of carbon-rich precursors, including wood, coconut shells, and agricultural waste, as shown in Figure 15.2.[35]

15.2.2 Carbon Nanotubes

The low density, stability, low cost, and superior electrical conductivity of carbon, as well as its resistance to corrosion, make it a popular material for supercapacitor electrodes.[37] Graphene, carbon nanotubes (CNTs), graphitic arrays, and amorphous carbon have all demonstrated exceptional electrocatalytic activity for oxygen reduction processes.[38,39] These substances are

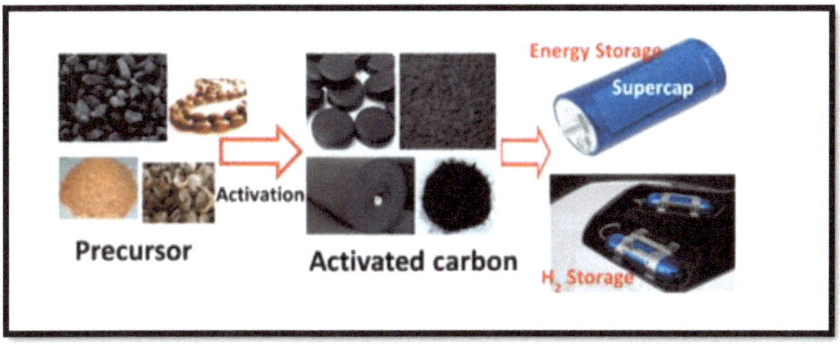

Figure 15.2 Activated carbon used for H_2 and energy storage. Reproduced from ref. 36 with permission from the Royal Society of Chemistry.

also essential for use in energy systems, supercapacitors, and electrochemical sensing.[40] In particular, CNTs are remarkable for their electrical conductivity and mechanical strength, which makes them a perfect option for supercapacitor electrodes. Improvements in synthesis techniques have made it possible to use more environmentally friendly and sustainable manufacturing procedures, whereas the production of CNTs in the past was energy-intensive and costly.[41,42]

15.2.3 Low-carbon Energy-harvesting Materials

Over time, the energy consumption of portable devices, including wearable electronics and smartwatches, has rapidly increased. These gadgets depend on fossil fuel-free continuous charging methods.[43] Still, battery life and charging time are significant issues that call for creative ways to power these gadgets. Given that solar energy is a potential renewable energy source, it has become an effective and sustainable way to increase the operating time of portable devices.[44,45] Fiber-shaped batteries have drawn interest as adaptable and flexible solutions in energy textiles to address this.[46]Aqueous zinc-ion (Zn-ion) batteries, which can be charged effectively in a variety of illumination situations, are the basis for photo-powered energy textiles. Because of their special qualities, semiconducting materials like MoS_2@TiO_2@Ti are employed for such batteries. The well-known semiconductor MoS_2 exhibits Zn-ion storage capacity by producing electrons and holes in sunlight. These holes and electrons move to TiO_2 and Ti, respectively, where photo-generated holes strengthen molybdenum's (Mo) oxidation state. At the Zn anode, photo-generated electrons convert Zn^{2+} to Zn, whereas this process releases Zn^{2+} ions into the electrolyte.[47,48] The coiling procedure used to create the fiber-shaped aqueous Zn-ion battery increases light exposure for the photo-powered cathode and boosts energy efficiency. Under simulated solar light and different environmental conditions, the constructed battery shows typical electrochemical performance with a photo-conversion efficiency of about 1%. As a result, it is a low-carbon and effective energy-harvesting material.[49]

15.3 Low-carbon Materials for Greener Healthcare

Globally, greenhouse gas emissions from healthcare institutions are on the rise and account for 5% of CO_2 emissions.[50] Hospitals and other healthcare facilities must reduce their use of nonrenewable energy sources and their direct emissions. This involves adopting renewable energy, installing battery management systems, and providing subsidies for solar panels to reduce electricity consumption. The employment of biological sensors in conjunction with green technology is one remarkable development.[51,52] Since biomarker concentrations can be extremely low in the early stages of disease, sensors in the biomedical field should be extremely sensitive and effective for early diagnosis.

15.3.1 Graphene as a Biosensor

The amount of trash plastic is increasing with the increase in urbanization and industry. Biochar from organic matter, including plastic waste, can be used to create graphene, a substance made of graphitic carbon.[53] The use of graphene as a biomedical sensor, such as fluorescent, label-free optical, and electrochemical sensors, is made possible by this technique, which also helps to lower greenhouse gas emissions. The optical characteristics of graphene make it an excellent choice for fluorescence detection. Graphene and its matching fluorescence resonance energy transfer (FRET) pair interact more or less when a target molecule is present, which causes noticeable variations in fluorescence. In two-way biosensing, graphene and its derivatives can act as either donors or acceptors. Graphene nanoparticles' FRET capabilities make them perfect for fluorescent biosensors, offering a more environmentally friendly technology that lowers the biosensor industry's carbon footprint.

15.4 Low-carbon Materials for a Greener Environment

Among the many innovations driving this movement, photocatalysts are particularly notable for their ability to truly accelerate change. Unique materials known as photocatalysts utilize light, mostly sunlight, to catalyze chemical reactions. They can absorb photons and start the electron transfer process because of their semiconductor-based characteristics. Photocatalysis is a phenomenon that has enormous potential for use in sustainable manufacturing, water purification, environmental remediation, and the generation of renewable energy.[55] Photocatalysts provide an alternative to traditional catalytic processes that mostly rely on fossil fuels by harnessing solar energy. This change lessens the negative effects of using fossil fuels on the environment, such as greenhouse gas emissions and air pollution.[56] Energy efficiency and environmental sustainability are also addressed by methods like the photocatalytic reduction of carbon dioxide (CO_2) into useful chemicals like carbon monoxide (CO), methane (CH_4), methanol (CH_3OH), and higher hydrocarbons (C^{2+}), as well as the high-density production of green hydrogen through water splitting and photoreforming. Zinc oxide (ZnO)-based materials have shown great promise as photocatalysts for processes triggered by sunlight. Their effectiveness is hampered by issues including high overpotential, limited absorption of visible light, fast charge recombination, and poor electron–hole pair separation.[58] By immobilizing ZnO photocatalysts with dopants, the overall efficiency of photoreduction processes is increased, charge migration and separation are improved, and CO_2 adsorption capacity is increased.[59] Because of their adaptability to a wide range of uses, photocatalysts are an essential part of low-carbon technology. A few examples follow.

15.4.1 Pollution Remediation

Through the process of photo-oxidation, which efficiently breaks down organic contaminants in air and water, photocatalysts are essential to pollution remediation. This procedure significantly reduces the carbon emissions connected with conventional pollution control technologies by providing an energy-efficient and chemical-free substitute.[60]

15.4.2 Water Treatment

When exposed to UV light, photocatalysts in water treatment purify the water by decomposing organic contaminants and even getting rid of germs and viruses. This technology also reduces the CO_2 footprint by reducing the need for carbon-intensive disinfection techniques like ozonation or chlorination.[61]

15.4.3 Conversion of CO_2 into Fuel

The conversion of CO_2 into fuel is also made possible by photocatalysts like ZnO and TiO_2, which absorb light energy and produce photo-induced electron–hole pairs. As shown in Figure 15.3, these pairings drive redox processes that convert CO_2 into fuels that can be used.[57,62,63]

15.5 Low-carbon Materials for Greener Construction

With almost 4 gigatons of cement used in 2020, the demand for building materials has significantly increased due to the world's population growth and fast urbanization. Cement production is a significant source of CO_2 emissions, with 0.9 tons of CO_2 produced with every ton of cement. Low-carbon cement-based composites (LCCs) are crucial for lowering the carbon footprint in order to address this environmental impact.[64–66] A sustainable substitute for conventional cement production techniques is the use of renewable resources and industrial wastes. However, it can be difficult to provide strength and endurance in adverse weather circumstances when combining these elements with regular

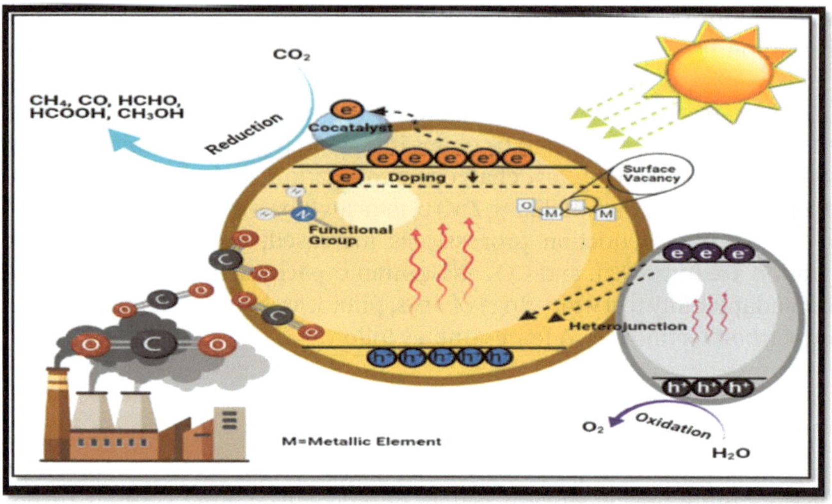

Figure 15.3 Conversion of CO_2 into fuel with the help of a photocatalyst. Reproduced from ref. 54 with permission from Elsevier, Copyright 2021.

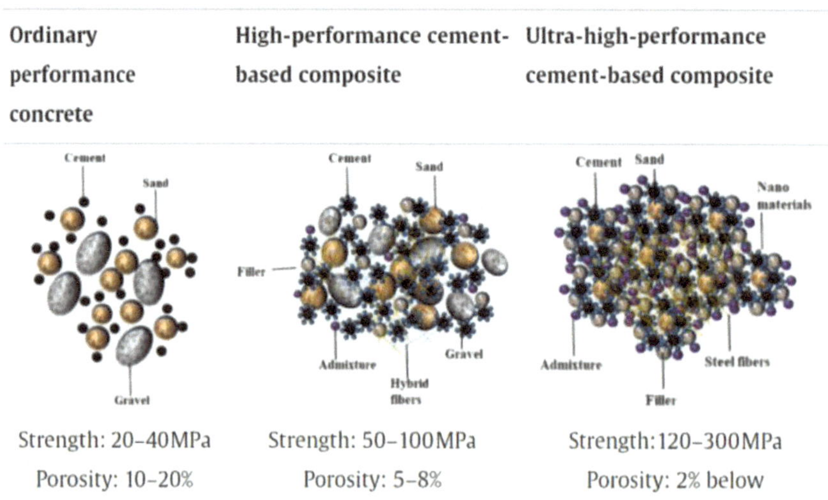

Figure 15.4 Comparative composition and mechanical properties of cement-based composites across performance grades. Reproduced from ref. 72, https://doi.org/10.1016/j.mtsust.2024.100712, under the terms of the CC BY 4.0 license, https://creativecommons.org/licenses/by/4.0/.

cement.[67,68] To overcome these problems, low-carbon, high-performance cement-based composites (LCHPCCs) must be developed.

When compared to ordinary concrete, ultra-high-performance and high-performance cementitious composites have stronger compression and less porosity, as indicated in Figure 15.4. By combining materials like powders, aggregates, and fibers – including steel and hybrid fibers – structural strength and durability are increased.[69] These cutting-edge composites demonstrate the building industry's dedication to sustainable development and are highly sought after for durable, load-bearing structures. By adding eco-friendly components like sustainable cement additives, aggregates, and fibers, researchers are actively enhancing these mixtures and lessening their negative environmental effects.[70] Developing environmentally friendly construction materials requires optimizing these composites' life cycle assessment and design (Table 15.1).[71,73]

Using low-embodied carbon materials is one of the creative ways to reduce carbon footprints in the construction industry. Emissions can be greatly decreased by making minor adjustments, such as choosing exposed concrete for floors and ceilings.

Table 15.1 Construction materials categorized by climate impact based on their global warming potential (GWP). Materials range from high-carbon emitters to climate-negative options, highlighting the benefits of bio-based alternatives. (GWP = global warming potential in kg $CO_2e\ kg^{-1}$.)

Category	Material	GWP (kg $CO_2e\ kg^{-1}$)
High-carbon (net GWP > 1 kg $CO_2e\ kg^{-1}$)	Insulated triple glazing	99.89
	Wood-aluminum window frame	18.43
	PVC window frame	8.77
	Wood window frame	4.18
	Waterproofing membrane (polyethylene)	2.70
	Steel	2.22
Low-carbon (net GWP < 1 kg $CO_2e\ kg^{-1}$)	Mineral plaster	1.00
	Solid wood softwood	1.05
	Ceramic tiles	0.87
	Bamboo cladding	0.75
	Bamboo flooring	0.63
	Wood – hardwood for finishing	0.53
	Gypsum plasterboard	0.39
	CLT/GLT (softwood)	0.36
	OSB (oriented strand board) 10/22	0.29
	OSB 3/10/18	0.27
	Cement screed	0.16
	Concrete deep foundations	0.13
	Clay plaster	0.09
Climate-negative (net GWP < 0 kg $CO_2e\ kg^{-1}$)	Solid wood – hardwood for structure	−0.14
	GLB 10/16	−0.15
	CLB 16/10	−0.16
	Reed mats	−0.17
	CLB and GLB	−0.18
	Straw	−0.20
	Hemp fibers	−0.21

15.5.1 Addition of Recycled Aggregates

Adding recycled aggregates (RAs) to concrete reduces its carbon footprint; material characteristics are maintained by using about 20% RA. Reducing transportation-related emissions requires local sourcing.

15.5.2 Adopting Recyclable or Low-embodied Carbon

When compared to conventional concrete bricks, low-rise structures can reduce their embodied carbon emissions by roughly 30% by using external walls made of low-embodied carbon blocks like clay or Polycare.

15.5.3 Discovering New Concrete

Carbon impacts are further decreased by using engineered bamboo for structural or decorative parts and creative concrete solutions like wood concrete and ultra-low carbon concrete. Sustainable substitutes are provided by insulation materials such as foam glass with a high recycled content or recycled polystyrene.

15.5.4 Embracing Recycling

Environmental footprints can be significantly reduced by using recycled aluminum rather than raw aluminum, and there are even CO_2-absorbing possibilities available with modern plasterboards like Fermacell or Ecosmart.

15.5.5 Considering Composite Window Frames

Sustainable substitutes for steel and aluminum are provided by composite window frames and timber-based products, like cross-laminated timber (CLT), which meet both structural and acoustic needs.

15.6 Low-carbon Materials in Transportation and the Automotive Industry

The economy, social structures, and ecology are all significantly impacted by transportation. Public transportation systems are no longer sustainable due to the dependence on fossil fuel-powered vehicles, especially private automobiles, which also contribute to the production of CO_2, CO, volatile organic compounds, and harmful air pollutants including sulfur and nitrogen oxides. Carbon black, heavy metals, and micro- and nanoplastics are examples of fine particles that worsen air quality problems and provide serious health hazards, such as asthma attacks, heart attacks, and cardiac arrhythmias.[74] In order to tackle these issues, transportation planning is putting more of an emphasis on lowering its detrimental effects on the environment, climate, and human health while enhancing energy efficiency in both freight and passenger transportation.[75]

Significant traffic congestion brought on by urbanization and the rise in personal automobile use has also resulted in delays, higher fuel consumption, and higher emissions of pollutants.[76] Additionally, relying on fossil fuels for transportation has significant geopolitical and economic ramifications, as it makes countries dependent on oil producers and exposes their economy to volatile oil prices.[77] Reducing these problems and fostering a more sustainable future depend heavily on the development of sustainable and ecologically friendly transportation options.[78]

Research initiatives on green transportation seek to solve critical energy and environmental issues. This strategy combines improved public transit infrastructure, economical conventional fuel consumption, electric vehicle technology breakthroughs, and biogas as a sustainable bus fuel. Reduced traffic, better energy and resource sustainability, less pollution, increased safety and security, fewer accidents, and faster travel times are all possible outcomes of an efficient green transportation system (Table 15.2).[79]

The production of fuel cell vehicles, electric vehicle battery technology, the creation of effective charging infrastructure, the use of lightweight materials in vehicle design, the integration of smart grid systems, and the incorporation of renewable energy sources into transportation systems are some of the major advancements, as shown in Table 15.1.[81]

Battery electric vehicles (BEVs), hydrogen fuel cell vehicles, and hybrid systems are examples of low-carbon technologies that are becoming competitive substitutes for conventional gasoline and diesel automobiles. The development of these technologies is being accelerated by significant expenditures made by major manufacturers in order to address the growing demand for environmentally sustainable transportation. As a transitional phase toward complete electrification, hybrid cars are more fuel-efficient and emit fewer pollutants than their traditional counterparts. In the meantime, BEVs, which are exclusively powered by rechargeable batteries, provide significant decreases in greenhouse gas emissions and air pollution.[82–84]

Table 15.2 The efficacy of green transportation.[80]

Main contribution	Description	Associated outcomes
Contribution to economic growth	High productivity in green transport systems	- Increase investment in transport and vehicle efficiency - Increase public transport usage
Contribution to welfare society	Safe and secure transport systems	- Increase private investment - Less health risk
Contribution to improvement of quality environment	Energy efficiency and low air emissions	- Travel and vehicle efficiency - Reduced cost on fuel and energy

15.7 Steps Toward a Greener World with Carbon Neutrality

Carbon dioxide (CO_2) and other greenhouse gases released from commercial activities raise the global concern about climate change.[85] This climate change risk led research toward controlling climate change, thus stressing the goal of achieving carbon neutrality (CN) because carbon emissions are the forefront cause of this.[86] Carbon neutrality means zero CO_2 emissions from different approaches such as transportation, batteries, and construction, and making the future greener and sustainable for living. The world should be carbon neutral by 2050. For achieving carbon neutrality below are the 12 steps which need to be followed.[87]

1. **The methods of capturing and utilizing carbon dioxide** (CCU) will create a novel carbon cycle, referred to as "atmosphere to atmosphere," which holds significant promise for reducing carbon emissions indirectly. Moving forward, CCU is poised to play a crucial role in achieving carbon neutrality.
2. **Replacement of raw materials.** Fossil fuels are used as a raw material for energy production, electricity, and heat, but for the carbon neutrality clean energy and fuels low-carbon materials have drawn attention and will also play an important role in carbon neutrality.[88] The raw material act as a main material used for the production of the product and their application, for example, petrol and diesel are used as a source of energy in the vehicle, and are nonrenewable sources of energy with a very high carbon footprint material and should be replaced by low-carbon materials using renewable sources and cleaner materials for the production of products.
3. **Cleaner production low-carbon**, sustainable development and industrial activities are connected to each other by cleaner production and emphasize technology development.[89] This cleaner production mainly focuses on the saving of raw materials, reduction of energy, and reduced use of toxic and harmful substances, thus reducing pollutant and carbon dioxide emissions.[90] This cleaner production mainly focuses on eliminating high energy consumption and adopting advance production technologies.
4. **Consumption waste recycling** – consumption waste refers to waste materials generated after the products are consumed, for example, plastic bottles, waste clothes, *etc.* Recycling of the waste consumption material and its use as a raw material for the production of new materials reduces the emission of carbon. The emission of carbon will be lowered by 51.8% when waste materials are recycled and used as raw materials for the production of new materials.[91]
5. **Consumption waste recycling and raw material substitution are two different strategies.** The former emphasises reusing or repurposing

materials produced from consumption waste, while the latter focusses on substituting primary raw materials with sustainable or alternative sources.

6. **Industrial product and waste exchange** – for the development of a sustainable world infrastructure, solid waste is a vital element. For carbon neutrality, products and waste infrastructure are equally important. The collection of solid waste and treatment, sorting, and recycling before incineration in landfills has a carbon emission reduction effect.[92] For example, fly ash and recycling cinders used to produce bricks and cement can not only achieve the goal of carbon emission reductions but also provide great economic benefits.[93]
7. **Improving energy infrastructure**, including energy exchange systems, significantly contributes to society's goal of achieving carbon neutrality by reducing carbon emissions.[94,95] Currently, a significant portion of total carbon emissions comes from energy usage, which includes both direct (Scope 1) and indirect (Scope 2) emissions.[96] Scope 1 emissions are directly from sources owned or controlled, such as stationary and mobile combustion, production processes, and accidental leaks, while Scope 2 emissions are from outsourced electricity, steam, heating, and cooling. Energy infrastructure enables centralized heating, cooling, and power supply systems, decreasing heat and electricity consumption rates.[97] Additionally, it promotes efficient and staged energy utilization. Across various energy infrastructure projects (EIPs), comprehensive energy infrastructure is consistently observed.
8. **The water infrastructure** within society comprises facilities such as water supply plants, various wastewater treatment plants, reclaimed water facilities, and the network of pipes for recycled water distribution. Through a centralized water supply, gradual utilization, centralized wastewater treatment, and discharge, reductions in both fresh water consumption and chemical oxygen demand (COD) emissions can be realized.[97] In industry, waste, water, and energy are all intertwined, creating a complicated and interdependent system.[98] Solid wastes like sludge and straw can serve as energy sources, with steam generated through boiler combustion being utilized for energy purposes. Furthermore, when the steam cools down, it can be condensed into water, thus serving as a valuable water resource.
9. **Buildings and transportation** play crucial roles in achieving low-carbon goals. Low-carbon buildings are constructed using environmentally friendly materials and technologies aimed at reducing carbon emissions. These materials, like recycled slag and fly ash, have extended lifespans and streamline production processes.[99] Transportation-related carbon emissions encompass those generated during the movement of raw materials and goods. Strategies to mitigate these emissions include adopting cleaner fuels and enhancing transport efficiency[100]

10. **Service infrastructure**, particularly encompassing the information systems and organizational structures, plays a key role in carbon reduction efforts. The information system aspect focuses on optimizing energy usage through the utilization of big data platforms, thereby reducing carbon emissions.
11. **Land use change involves** transitioning from one land type to another[101] such as converting cropland to industrial use or landscaping. This change directly impacts carbon storage in soil and vegetation. Unused land can be transformed into landscaped areas, incorporating features like recreational hiking and walking trails.[102] Landscaping serves to not only enhance visual appeal but also sequester carbon dioxide, thereby advancing carbon-neutrality goals.
12. **Carbon storage** involves capturing carbon dioxide from major stationary sources like power plants and chemical facilities and transporting it to either oceanic or geological sites for long-term storage.[103] These technologies are potent tools for promoting the clean and efficient utilization of coal and addressing global climate change, making substantial contributions to achieving carbon neutrality over the coming decades.

15.8 Conclusion

Finally, the use of low-carbon materials in the transition to a greener future is critical to reducing the negative environmental effects of industrial growth. Greenhouse gas emissions from conventional energy sources can be greatly decreased by using sustainable materials in energy-storage devices like fuel cells and low-carbon batteries. Utilizing green technologies in healthcare, such as graphene biosensors, can reduce carbon footprints and increase efficiency. Low-carbon materials have the potential to drive sustainable practices in a number of environmental applications, such as the use of photocatalysts for water treatment and pollution remediation. Furthermore, novel building methods, recycled materials, and high-performance, low-carbon composites can be very advantageous to the construction industry. It will take coordinated efforts from multiple sectors, with a focus on waste recycling, cleaner industry, better energy and water infrastructure, and creative approaches to land use and carbon storage, to reach carbon neutrality by 2050. We can build a resilient and sustainable future that effectively fights climate change and encourages environmental care by adopting these strategies.

15.9 Future Prospective

The development of a more sustainable and greener future depends on low-carbon materials. Materials like green concrete, bio-based polymers, and recycled metals are examples of creative, workable solutions that reduce the carbon emissions that are frequently connected with conventional building and manufacturing. By adopting these alternatives, we're actively tackling some of the most difficult environmental issues of our day, in addition to improving the way things are created. It's a method of reevaluating our construction and use, opening the door to a healthier planet for all. Globally, decarbonization has emerged as a crucial area of study: adopting the sustainable and renewable sources to decrease the carbon footprint thus decreasing the greenhouse gas emission. To achieve decarbonization goals, many technologies are being studied and adopted, for example, use of electric vehicles, and reusable batteries. Using low-carbon materials to adapt to a cleaner future means going with eco-friendly solutions like bamboo, low-emission concrete, and recycled steel. Carbon footprints can be further decreased by implementing energy-efficient designs, passive construction methods, and on-site renewable energy systems like solar panels. Reusing and recycling resources is ensured by putting an emphasis on lifespan assessments and advancing a circular economy, which reduces waste. Sustainable practices are promoted by updated building codes, financial incentives, and green certifications. Green building efforts are supported and understood by the general public when they are accompanied by public awareness campaigns and professional education and training. Achieving decarbonization goals requires constant improvement and close monitoring of environmental effects by industrial stakeholders. In addition to changing sectors, low-carbon materials are encouraging us to reconsider how we use and recycle resources. These materials promote a circular economy, in which nothing is wasted and everything has a purpose, by emphasizing reusability and waste minimization. Consider recycling building trash into new building materials or converting agricultural waste into bioplastics. These methods lessen the need to mine the planet for new resources while also keeping garbage out of landfills. It's about changing our thinking to live in balance with the environment, not merely reducing emissions.

References

1. I. Chakraborty and P. Maity, COVID-19 outbreak: Migration, effects on society, global environment and prevention, *Sci. Total Environ.*, 2020, **728**, 138882.
2. P. M. Forster, H. I. Forster, M. J. Evans, M. J. Gidden, C. D. Jones and C. A. Keller, *et al.*, Current and future global climate impacts resulting from COVID-19, *Nat. Clim. Change*, 2020, **10**(10), 913–919.
3. Z. Liu, P. Ciais, Z. Deng, R. Lei, S. J. Davis and S. Feng, *et al.*, Near-real-time monitoring of global CO_2 emissions reveals the effects of the COVID-19 pandemic, *Nat. Commun.*, 2020, **11**(1), 5172.
4. R. Li and S. Li, Carbon emission post-coronavirus: continual decline or rebound?, *Struct. Change Econ. Dyn.*, 2021, **57**, 57–67.
5. T. Huo, Y. Ma, W. Cai, B. Liu and L. Mu, Will the urbanization process influence the peak of carbon emissions in the building sector? A dynamic scenario simulation, *Energy Build.*, 2021, **232**, 110590.

6. P. Yang, Y. F. Yao, Z. Mi, Y. F. Cao, H. Liao and B. Y. Yu, *et al.*, Social cost of carbon under shared socioeconomic pathways, *Global Environ. Change*, 2018, **53**, 225–232.
7. E. Papadis and G. Tsatsaronis, Challenges in the decarbonization of the energy sector, *Energy*, 2020, **205**, 118025.
8. Z. Zhang, S. Y. Pan, H. Li, J. Cai, A. G. Olabi, E. J. Anthony and V. Manovic, Recent advances in carbon dioxide utilization, *Renewable Sustainable Energy Rev.*, 2020, **125**, 109799.
9. M. M. Thackeray, C. Wolverton and E. D. Isaacs, Electrical energy storage for transportation—approaching the limits of, and going beyond, lithium-ion batteries, *Energy Environ. Sci.*, 2012, **5**(7), 7854–7863.
10. X. Xiao, Z. Zhang, W. Yu, W. Shang, Y. Ma and P. Tan, Achieving a high-specific-energy lithium–carbon dioxide battery by implementing a bi-side-diffusion structure, *Appl. Energy*, 2022, **328**, 120186.
11. B. Nykvist, F. Sprei and M. Nilsson, Assessing the progress toward lower priced long range battery electric vehicles, *Energy Policy*, 2019, **124**, 144–155.
12. T. Nagaura, Lithium ion rechargeable battery, *Prog. Batteries Sol. Cells*, 1990, **9**, 209.
13. A. Sarkar, V. R. Dharmaraj, C. H. Yi, K. Iputera, S. Y. Huang and R. J. Chung, *et al.*, Recent Advances in Rechargeable Metal–CO_2 Batteries with Nonaqueous Electrolytes, *Chem. Rev.*, 2023, **123**(15), 9497–9564.
14. ACEA, CO_2 emissions from heavy-duty vehicles preliminary CO_2 baseline, Assoc. Des Constr. Eur. *d'Automobile (March)*, 2020
15. Z. P. Cano, D. Banham, S. Ye, A. Hintennach, J. Lu, M. Fowler and Z. Chen, Batteries and fuel cells for emerging electric vehicle markets, *Nat. Energy*, 2018, **3**(4), 279–289.
16. Y. S. Najjar, Hydrogen safety: The road toward green technology, *Int. J. Hydrogen Energy*, 2013, **38**(25), 10716–10728.
17. Y. Qiao, J. Yi, S. Wu, Y. Liu, S. Yang, P. He and H. Zhou, Li–CO_2 electrochemistry: a new strategy for CO_2 fixation and energy storage, *Joule*, 2017, **1**(2), 359–370.
18. Y. Liu, R. Wang, Y. Lyu, H. Li and L. Chen, Rechargeable Li/CO_2–O_2 (2 : 1) battery and Li/CO_2 battery, *Energy Environ. Sci.*, 2014, 7(2), 677–681.
19. C. Zamfirescu and I. Dincer, Ammonia as a green fuel and hydrogen source for vehicular applications, *Fuel Process. Technol.*, 2009, **90**(5), 729–737.
20. D. D. T. Ferraren-De Cagalitan and M. L. S. Abundo, A review of biohydrogen production technology for application towards hydrogen fuel cells, *Renewable Sustainable Energy Rev.*, 2021, **151**, 111413.
21. E. Çabukoglu, G. Georges, L. Küng, G. Pareschi and K. Boulouchos, Fuel cell electric vehicles: An option to decarbonize heavy-duty transport? Results from a Swiss case-study, *Transp. Res. Part D: Transport Environ.*, 2019, **70**, 35–48.
22. E. Çabukoglu, G. Georges, L. Küng, G. Pareschi and K. Boulouchos, Fuel cell electric vehicles: An option to decarbonize heavy-duty

transport? Results from a Swiss case-study, *Transp. Res. Part D: Transport Environ.*, 2019, **70**, 35–48.
23. S. Sharma and B. G. Pollet, Support materials for PEMFC and DMFC electrocatalysts—A review, *J. Power Sources*, 2012, **208**, 96–119.
24. A. Chandan, M. Hattenberger, A. El-Kharouf, S. Du, A. Dhir and V. Self, *et al.*, High temperature (HT) polymer electrolyte membrane fuel cells (PEMFC)–A review, *J. Power Sources*, 2013, **231**, 264–278.
25. Y. Liu, W. Lehnert, H. Janßen, R. C. Samsun and D. Stolten, A review of high-temperature polymer electrolyte membrane fuel-cell (HT-PEMFC)-based auxiliary power units for diesel-powered road vehicles, *J. Power Sources*, 2016, **311**, 91–102.
26. S. Ramesohl and F. Merten, Energy system aspects of hydrogen as an alternative fuel in transport, *Energy Policy*, 2006, **34**(11), 1251–1259.
27. A. Arvay, J. French, J. C. Wang, X. H. Peng and A. M. Kannan, Nature inspired flow field designs for proton exchange membrane fuel cell, *Int. J. Hydrogen Energy*, 2013, **38**(9), 3717–3726.
28. G. Wang, C. Chen, B. A. Beshiwork and B. Lin, Developing a low-carbon hybrid of ammonia fuel cell and internal combustion engine for carbon neutrality, *Appl. Energy Combust. Sci.*, 2023, **16**, 100214.
29. C. Lin, T. Wu, X. Ou, Q. Zhang, X. Zhang and X. Zhang, Life-cycle private costs of hybrid electric vehicles in the current Chinese market, *Energy Policy*, 2013, **55**, 501–510.
30. Y. S. Lee, Y. A. Kumar, S. Sambasivam, S. A. Hira, K. Zeb and W. Uddin, *et al.*, $CoCu_2O_4$ nanoflowers architecture as an electrode material for battery type supercapacitor with improved electrochemical performance, *Nano-Struct. Nano-Obj.*, 2020, **24**, 100618.
31. C. Zhong, Y. Deng, W. Hu, J. Qiao, L. Zhang and J. Zhang, A review of electrolyte materials and compositions for electrochemical supercapacitors, *Chem. Soc. Rev.*, 2015, **44**(21), 7484–7539.
32. Z. Zhai, L. Zhang, T. Du, B. Ren, Y. Xu and S. Wang, *et al.*, A review of carbon materials for supercapacitors, *Mater. Des.*, 2022, **221**, 111017.
33. Y. Wang, L. Zhang, H. Hou, W. Xu, G. Duan and S. He, *et al.*, Recent progress in carbon-based materials for supercapacitor electrodes: a review, *J. Mater. Sci.*, 2021, **56**, 173–200.
34. L. Weinstein and R. Dash, Supercapacitor carbons, *Mater. Today*, 2013, **10**(16), 356–357.
35. D. Qu, Studies of the activated carbons used in double-layer supercapacitors, *J. Power Sources*, 2002, **109**(2), 403–411.
36. M. Sevilla and R. Mokaya, Energy storage applications of activated carbons: supercapacitors and hydrogen storage, *Energy Environ. Sci.*, 2014, **7**(4), 1250–1280.
37. M. T. T. Moghadam, M. Seifi, F. Jamali, S. Azizi and M. B. Askari, $ZnWO_4$-CNT as a superior electrode material for ultra-high capacitance supercapacitor, *Surf. Interfaces*, 2022, **32**, 102134.
38. H. Meskher, S. B. Belhaouari, K. Deshmukh, C. M. Hussain and F. Sharifianjazi, A magnetite composite of molecularly imprinted

polymer and reduced graphene oxide for sensitive and selective electrochemical detection of catechol in water and milk samples: An artificial neural network (ANN) application, *J. Electrochem. Soc.*, 2023, **170**(4), 047502.
39. S. Karthikeyan, B. Narenthiran, A. Sivanantham, L. D. Bhatlu and T. Maridurai, Supercapacitor: Evolution and review, *Mater. Today: Proc.*, 2021, **46**, 3984.
40. T. T. Steffen, L. C. Fontana, P. Hammer and D. Becker, Amine functionalization of carbon nanotubes with solid urea using different plasma treatments, *Appl. Surf. Sci.*, 2022, **583**, 152493.
41. H. Meskher, H. C. Mustansar, A. K. Thakur, R. Sathyamurthy, I. Lynch and P. Singh, *et al.*, Recent trends in carbon nanotube (CNT)-based biosensors for the fast and sensitive detection of human viruses: a critical review, *Nanoscale Adv.*, 2023, **5**(4), 992–1010.
42. M. Haghgoo, R. Ansari and M. K. Hassanzadeh-Aghdam, Predicting effective electrical resistivity and conductivity of carbon nanotube/carbon black-filled polymer matrix hybrid nanocomposites, *J. Phys. Chem. Solids*, 2022, **161**, 110444.
43. T. Xiong, X. Zhou, Y. Wang, T. Zhou, R. Huang and H. Zhong, *et al.*, Photo-powered all-in-one energy harvesting and storage fibers towards low-carbon smart wearables, *Energy Storage Mater.*, 2024, **65**, 103146.
44. D. Lau, N. Song, C. Hall, Y. Jiang, S. Lim and I. Perez-Wurfl, *et al.*, Hybrid solar energy harvesting and storage devices: The promises and challenges, *Mater. Today Energy*, 2019, **13**, 22–44.
45. H. Ryu, H. J. Yoon and S. W. Kim, Hybrid energy harvesters: toward sustainable energy harvesting, *Adv. Mater.*, 2019, **31**(34), 1802898.
46. Y. W. Chong, W. Ismail, K. Ko and C. Y. Lee, Energy harvesting for wearable devices: A review, *IEEE Sens. J.*, 2019, **19**(20), 9047–9062.
47. H. Akbari, M. C. Browne, A. Ortega, M. J. Huang, N. J. Hewitt, B. Norton and S. J. McCormack, Efficient energy storage technologies for photovoltaic systems, *Sol. Energy*, 2019, **192**, 144–168.
48. S. Seal, S. Rajaraman and J. Thomas, Wearable Devices: Fiber-Type Solar Cells, Nanogenerators, Batteries, and Supercapacitors for Wearable Applications, *Adv. Sci.*, 2018, **5**(9).
49. Sreekanth J. Varma, *et al.* Fiber-type solar cells, nanogenerators, batteries, and supercapacitors for wearable applications, *Advanced Science*, 2018, **5**(9), 1800340.
50. IEA, Global Energy Review: CO_2 Emissions in 2021—Analysis, 2021
51. S. Priyan, R. Matahen, D. Priyanshu and M. Mouqdadi, Environmental strategies for a healthcare system with green technology investment and pandemic effects, *Innov. Green Dev.*, 2024, **3**(1), 100113.
52. S. Priyan, Effect of green investment to reduce carbon emissions in an imperfect production system, *J. Climate Finance*, 2023, **2**, 100007.
53. Z. Zhang, X. G. Wang, X. Zhang, Z. Xie, Y. N. Chen and L. Ma, *et al.*, Verifying the rechargeability of Li–CO_2 batteries on working cathodes of Ni nanoparticles highly dispersed on N-doped graphene, *Adv. Sci.*, 2018, **5**(2), 1700567.

54. A. Bhattacharya and A. Selvaraj, Photocatalytic conversion of CO_2 into beneficial fuels and chemicals–a new horizon in atmospheric CO_2 mitigation, *Process Saf. Environ. Prot.*, 2021, **156**, 256–287.
55. S. Ali, J. Lee, H. Kim, Y. Hwang, A. Razzaq and J. W. Jung, *et al.*, Sustained, photocatalytic CO_2 reduction to CH_4 in a continuous flow reactor by earth-abundant materials: Reduced titania–Cu_2O Z-scheme heterostructures, *Appl. Catal., B*, 2020, **279**, 119344.
56. W. D. Jones, Carbon capture and conversion, *J. Am. Chem. Soc.*, 2020, **142**(11), 4955–4957.
57. D. Adekoya, M. Tahir and N. A. S. Amin, Recent trends in photocatalytic materials for reduction of carbon dioxide to methanol, *Renewable Sustainable Energy Rev.*, 2019, **116**, 109389.
58. H. Deng, F. Xu, B. Cheng, J. Yu and W. Ho, Photocatalytic CO_2 reduction of C/ZnO nanofibers enhanced by an Ni–NiS cocatalyst, *Nanoscale*, 2020, **12**(13), 7206–7213.
59. S. Chakrabortty, J. Nayak, B. Ruj, P. Pal, R. Kumar and S. Banerjee, *et al.*, Photocatalytic conversion of CO_2 to methanol using membrane-integrated green approach: a review on capture, conversion and purification, *J. Environ. Chem. Eng.*, 2020, **8**(4), 103935.
60. J. Huang, M. Lu, P. Wei, Y. Xie, H. Xie and M. Liu, *et al.*, Photocatalytic degradation of organic pollutants in sewage using CeO_2/ZnO 3D nanoflowers, *J. Environ. Chem. Eng.*, 2024, 113074.
61. S. N. Ahmed and W. Haider, Heterogeneous photocatalysis and its potential applications in water and wastewater treatment: a review, *Nanotechnology*, 2018, **29**(34), 342001.
62. S. N. Habisreutinger, L. Schmidt-Mende and J. K. Stolarczyk, Photocatalytic reduction of CO_2 on TiO_2 and other semiconductors, *Angew. Chem., Int. Ed.*, 2013, **52**(29), 7372–7408.
63. Y. Gao, K. Qian, B. Xu, Z. Li, J. Zheng and S. Zhao, *et al.*, Recent advances in visible-light-driven conversion of CO_2 by photocatalysts into fuels or value-added chemicals, *Carbon Resour. Convers.*, 2020, **3**, 46–59.
64. F. Althoey, W. S. Ansari, M. Sufian and A. F. Deifalla, Advancements in low-carbon concrete as a construction material for the sustainable built environment, *Dev. Built Environ.*, 2023, 100284.
65. F. Althoey, W. S. Ansari, M. Sufian and A. F. Deifalla, Advancements in low-carbon concrete as a construction material for the sustainable built environment, *Dev. Built Environ.*, 2023, 100284.
66. S. Ahmad, R. A. Assaggaf, M. Maslehuddin, O. S. B. Al-Amoudi, S. K. Adekunle and S. I. Ali, Effects of carbonation pressure and duration on strength evolution of concrete subjected to accelerated carbonation curing, *Constr. Build. Mater.*, 2017, **136**, 565–573.
67. A. Adesina, Recent advances in the concrete industry to reduce its carbon dioxide emissions, *Environ. Challenges*, 2020, **1**, 100004.
68. F. Wang, *et al.* Technologies and perspectives for achieving carbon neutrality, *The innovation*, 2021, **2**(4), 100180.

69. K. G. Kuder and S. P. Shah, Processing of high-performance fiber-reinforced cement-based composites, *Constr. Build. Mater.*, 2010, **24**(2), 181–186.
70. M. Hasani, F. M. Nejad, J. Sobhani and M. Chini, Mechanical and durability properties of fiber reinforced concrete overlay: Experimental results and numerical simulation, *Constr. Build. Mater.*, 2021, **268**, 121083.
71. R. T. Abd El-Hakim, G. M. Elgendy, S. M. El-Badawy and M. Amin, Performance evaluation of steel slag high performance concrete for sustainable pavements, *Int. J. Pavement Eng.*, 2022, **23**(11), 3819–3837.
72. M. Yang, L. Chen, J. Lai, A. I. Osman, M. Farghali, D. W. Rooney and P. S. Yap, Advancing environmental sustainability in construction through innovative low-carbon, high-performance cement-based composites: A review, *Mater. Today Sustainability*, 2024, 100712.
73. https://www.greengage-env.com/embodied-carbon/
74. Y. Van Fan, S. Perry, J. J. Klemeš and C. T. Lee, A review on air emissions assessment: Transportation, *J. Cleaner Prod.*, 2018, **194**, 673–684.
75. D. Sun, S. Zeng, H. Lin, X. Meng and B. Yu, Can transportation infrastructure pave a green way? A city-level examination in China, *J. Cleaner Prod.*, 2019, **226**, 669–678.
76. R. Zhang, J. Zhang, Y. Long, W. Wu, J. Liu and Y. Jiang, Long-term implications of electric vehicle penetration in urban decarbonization scenarios: An integrated land use–transport–energy model, *Sustainable Cities Soc.*, 2021, **68**, 102800.
77. P. Ahmadi, Environmental impacts and behavioral drivers of deep decarbonization for transportation through electric vehicles, *J. Cleaner Prod.*, 2019, **225**, 1209–1219.
78. L. Schipper, Automobile use, fuel economy and CO_2 emissions in industrialized countries: Encouraging trends through 2008?, *Transport Policy*, 2011, **18**(2), 358–372.
79. C. T. Lee, H. Hashim, C. S. Ho, Y. Van Fan and J. J. Klemeš, Sustaining the low-carbon emission development in Asia and beyond: Sustainable energy, water, transportation and low-carbon emission technology, *J. Cleaner Prod.*, 2017, **146**, 1–13.
80. K. J. Shah, S. Y. Pan, I. Lee, H. Kim, Z. You, J. M. Zheng and P. C. Chiang, Green transportation for sustainability: Review of current barriers, strategies, and innovative technologies, *J. Cleaner Prod.*, 2021, **326**, 129392.
81. K. Shahzad and I. I. Cheema, Low-carbon technologies in automotive industry and decarbonizing transport, *J. Power Sources*, 2024, **591**, 233888.
82. A. Kilic, B. Oral, D. Eroglu and R. Yildirim, Machine learning for beyond Li-ion batteries: powering the research, *J. Energy Storage*, 2023, **73**, 109057.
83. J. M. Rodríguez-Rego, A. Macías-García, L. Mendoza-Cerezo, A. Díaz-Parralejo and J. P. Carrasco-Amador, Design, machining and

characterization of the components required for the manufacture of a supercapacitor, *J. Energy Storage*, 2023, **73**, 109110.
84. S. Verma, S. Mishra, A. Gaur, S. Chowdhury, S. Mohapatra, G. Dwivedi and P. Verma, A comprehensive review on energy storage in hybrid electric vehicle, *J. Traffic Transp. Eng. (Engl. Ed.)*, 2021, **8**(5), 621–637.
85. L. Zhang, Y. Song, J. Shi, Q. Shen, D. Hu and Q. Gao, *et al.*, Frontiers of CO_2 capture and utilization (CCU) towards carbon neutrality, *Adv. Atmos. Sci.*, 2022, **39**(8), 1252–1270.
86. H. Pan, J. Page, R. Shi, C. Cong, Z. Cai and S. Barthel, *et al.*, Contribution of prioritized urban nature-based solutions allocation to carbon neutrality, *Nat. Clim. Change*, 2023, **13**(8), 862–870.
87. Y. Sun, Y. Li, J. Ning, H. Fu, F. Liu and Z. Feng, *et al.*, Twelve pathways of carbon neutrality for industrial parks, *J. Cleaner Prod.*, 2024, 140753.
88. B. Lin and Z. Li, Towards world's low carbon development: The role of clean energy, *Appl. Energy*, 2022, **307**, 118160.
89. Y. Zhou and L. Zhao, Impact analysis of the implementation of cleaner production for achieving the low-carbon transition for SMEs in the Inner Mongolian coal industry, *J. Cleaner Prod.*, 2016, **127**, 418–424.
90. G. C. de Oliveira Neto, J. M. F. Correia, P. C. Silva, A. G. de Oliveira Sanches and W. C. Lucato, Cleaner Production in the textile industry and its relationship to sustainable development goals, *J. Cleaner Prod.*, 2019, **228**, 1514–1525.
91. Q. Chen, X. Lai, H. Gu, X. Tang, F. Gao, X. Han and Y. Zheng, Investigating carbon footprint and carbon reduction potential using a cradle-to-cradle LCA approach on lithium-ion batteries for electric vehicles in China, *J. Cleaner Prod.*, 2022, **369**, 133342.
92. W. Xiao, T. Liu and X. Tong, Assessing the carbon reduction potential of municipal solid waste management transition: Effects of incineration, technology and sorting in Chinese cities, *Resour., Conserv. Recycl.*, 2023, **188**, 106713.
93. R. Branson, Re-constructing Kalundborg: the reality of bilateral symbiosis and other insights, *J. Cleaner Prod.*, 2016, **112**, 4344–4352.
94. Y. Guo, J. Tian and L. Chen, Managing energy infrastructure to decarbonize industrial parks in China, *Nat. Commun.*, 2020, **11**(1), 981.
95. Y. Guo, J. Tian and L. Chen, Water-energy nexus in China's industrial parks, *Resour., Conserv. Recycl.*, 2020, **153**, 104551.
96. X. Yu, H. Zheng, L. Sun and Y. Shan, An emissions accounting framework for industrial parks in China, *J. Cleaner Prod.*, 2020, **244**, 118712.
97. Y. Li and C. Ma, Circular economy of a papermaking park in China: a case study, *J. Cleaner Prod.*, 2015, **92**, 65–74.
98. J. M. Núñez-López, E. Rubio-Castro, M. M. El-Halwagi and J. M. Ponce-Ortega, Optimal design of total integrated residential complexes involving water-energy-waste nexus, *Clean Technol. Environ. Policy*, 2018, **20**, 1061–1085.
99. S. H. Cho and C. U. Chae, A study on life cycle CO_2 emissions of low-carbon building in South Korea, *Sustainability*, 2016, **8**(6), 1–19.

100. X. Wang, Y. Zhou, Q. Bi, Z. Cao and B. Wang, Research on the low-carbon development path and policy options of China's transportation under the background of dual carbon goals, *Front. Environ. Sci.*, 2022, **10**, 905037.
101. R. A. Houghton, Interactions between land-use change and climate-carbon cycle feedbacks, *Curr. Climate Change Rep.*, 2018, **4**, 115–127.
102. R. P. Côté and C. Liu, Strategies for reducing greenhouse gas emissions at an industrial park level: a case study of Debert Air Industrial Park, Nova Scotia, *J. Cleaner Prod.*, 2016, **114**, 352–361.
103. A. Mualim, H. Huda, A. Altway, J. P. Sutikno and R. Handogo, Evaluation of multiple time carbon capture and storage network with capital-carbon trade-off, *J. Cleaner Prod.*, 2021, **291**, 125710.

Subject Index

1-(3-dimethylaminopropyl)-3-ethylcarbodiimide hydrochloride (EDC), 208
1-ethyl-3-methylimidazolium tetrafluoroborate, 57

absorption coefficient, 107
ACs. *See* active carbons (ACs)
action potentials (APs), 198
active carbons (ACs), 51, 79, 229, 362
AFCs. *See* alkaline fuel cells (AFCs)
agglomeration, 338
agricultural residues, 7
agricultural waste, 47, 49
air pollution, 297, 316
alginate-based hydrogels, 80
alkaline batteries, 26
alkaline fuel cells (AFCs), 66, 77–79, 84, 87
allotropes, 316
alternating current voltammetry (ACV), 191
aluminum alloys, 297
Alzheimer's disease, 168
amperometric biosensors, 162–163
anaerobic digestion, 339
analyte, 179, 180
anion exchange membranes, 82
anode materials, 31–32
anodes, 74, 76, 132
antibody-based biosensors, 187
  structure of, 187–188
  technologies of, 189
  types of, 188–189
antimicrobial peptide-based biosensors, 199–200
antimicrobial peptides (AMPs), 199, 200
anti-reflective coating, 104
aptamers, 190, 191
arc discharge, 278
artificial intelligence (AI), 216, 346–347
artificial or synthetic graphite, 148
atomic force microscopy (AFM), 241
automotive industry, 368–369

bacterial infections, 192
bacteriophage-based biosensors, 199
bandgap, 107
band structure engineering, 113
barium titanate, 115
batteries, 23, 46, 230, 360–361
  challenges and opportunities, 35
    economic benefits, 39–40
    recent advancements, 38–39
    recycling and environmental impact, 35–36
    regulations, 36–37
    sustainable sources, 37–38
  characteristics of, 29
  low-carbon materials in, 29
    anode materials, 31–32
    cathode materials, 29–31
    electrolytes, 32–35
  recycling system for, 25
  types of, 26–27
  working principle of, 27–29

battery electric vehicles (BEVs), 305, 369
Beijing–Shanghai trains, 309
bio-based biosensors, 184
  antibody-based biosensors, 187–189
  antimicrobial peptide-based biosensors, 199–200
  bacteriophage-based biosensors, 199
  cell-based biosensors, 196–199
  enzyme-based biosensors, 184–187
  nucleic acid aptamer-based biosensors, 189–196
bio-based polymers, 13, 315, 325
biocatalytic processes, 79
biocellular sensors, 197
biocompatibility, 178, 211
bioconversion, 340
biodegradability, 178, 211
biodegradable polymers, 116, 119
bio-derived polymers, 57
biodiversity, 102
bioelectrocatalysis, 80
bioelectronic noses, 198
biofuel cells, 79–82
biogenic synthesis, 59
bio-inspired materials, 125
biological recycling methods, 339
  composting and anaerobic digestion, 339
  fermentation and bioconversion, 340
  microbial degradation and bioremediation, 339–340
biomass, 230, 310
biomass waste, 138–140
biomedicine, 231
bio-methanol, 131
bioplastics, 8
biopolymers, 79
bioreceptor, 179
bioremediation, 339–340
biosensors, 146, 171, 178, 230
  bio-based biosensors, 184
    antibody-based biosensors, 187–189
    antimicrobial peptide-based biosensors, 199–200
    bacteriophage-based biosensors, 199
    cell-based biosensors, 196–199
    enzyme-based biosensors, 184–187
    nucleic acid aptamer-based biosensors, 189–196
  carbon-based biosensors, 211–216
  challenges and future trends, 216–217
  characteristics of, 180
  different types of, 180
    electrochemical biosensors, 180
    gravimetric biosensors, 182
    optical biosensors, 180–182
    thermal biosensors, 182–183
  elements contained within, 179–180
  paper-based biosensors, 200–207
  silk-based biosensors, 207–211
Bombyx mori, 210
bottom-up method, 152, 153
building-integrated photovoltaics (BIPV), 121
Building Research Establishment Environmental Assessment Method (BREEAM), 256

capacitive biosensors, 164
capacitors, 46
carbon, 227
  spike, 255
  storage, 372

carbon-based biosensors, 211
types of carbon nanomaterials and their applications, 211
carbon dots, 213–214
carbon nanotubes, 213
fullerenes, 214
graphene, 212
graphene oxide, reduced graphene oxide, and their derivatives, 213
carbon-based electrodes, 228
carbon-based materials, 27, 146, 163, 166, 211, 226, 273, 299
carbon-based nanomaterials (CNMs), 157, 274
carbon black, 153
fabrication routes and specifications of, 153–154
carbon capture and storage (CCS), 319
carbon capture and utilization (CCU), 326
carbon-coated magnetic nanoparticles (ccMNPs), 80
carbon dioxide (CO_2), 314, 323–324, 370
carbon dots (CDs), 156–157, 161, 211, 213–214, 229
carbon fiber-reinforced polymer (CFRP), 297, 302
carbon fibers (CFs), 155, 229
carbon footprint reduction, 296, 341–342
carbonization method, 231
carbon monoxide (CO), 324–325, 365
carbon nanofibers, 168
carbon nanotubes (CNTs), 31, 51, 79, 148, 157, 166, 211, 213, 274, 315, 323, 362–363
fabrication routes, 149–150
key features for biosensing applications, 150
carbon neutrality (CN), 16–17, 370
carbon quantum dots (CQDs), 156, 273, 282, 283
carboxyhemoglobin (COHb), 325
carboxymethyl cellulose, 57
carrier concentration optimization, 113
carrier mobility, 107
catalysts, 137
cathode, 74, 76
cathode materials, 29–31
CDs. *See* carbon dots (CDs)
cell-based biosensors, 196
applications of, 197–199
methods of cell immobilization in, 196–197
cell–cell interactions, 197
cell–drug interactions, 197
cellulose, 170, 235–240
cellulose acetate (CA), 238
cellulose-based aerogel, 235
cellulose-based materials, 118
cellulose derivatives, 118
ceramics, 115, 228, 275
chemical binding, 184
chemical energy, 95
chemical functionalization, 119
chemical hydrides, 300
chemical luminescence biosensor, 182
chemical recycling approaches, 335, 338
depolymerization and solvent-based processes, 339
hydrothermal and thermochemical conversion, 339
pyrolysis and gasification, 338–339
chemical vapor deposition (CVD), 136, 149, 150, 278, 321
chemiluminescence, 181, 191
chitosan, 232–235, 315
chromium, 136
ciprofloxacin (CIF), 238
circular business models, 347
circular economy, 15–16, 55, 125, 310, 342
Clark electrode, 178
climate change, 59, 295, 297, 316
CNTs. *See* carbon nanotubes (CNTs)

colorimetric methods, 191, 201
combined heat and power (CHP), 67
complementary determining regions (CDRs), 187, 189
composite materials, 111
composting, 339
compressed air energy storage (CAES), 28
conducting polymers, 82, 110–111, 162
conductometric biosensors, 163–164
construction sector
 challenges preventing building sector from using substitute materials, 259–261
 innovation challenges in, 258–259
 methods, 261–262
 study goals and limitations, 261
consumption waste recycling, 370
contact electrification, 118
co-precipitation method, 50
cost reduction, 3
 energy savings and reduced utility bills, 4
 initial costs *versus* lifecycle costs, 3–4
CQDs. *See* carbon quantum dots (CQDs)
cross-laminated timber (CLT), 368
current collectors, 58
CV. *See* cyclic voltammetry (CV)
CVD. *See* chemical vapor deposition (CVD)
cyclic voltammetry (CV), 163, 191, 239

Debye equation, 148
depolymerization, 339
differential pulse voltammetry (DPV), 191, 233
digitalization, 346–347
Dirac points, 152
direct e^- transfer (DET), 80
direct piezoelectric effect, 114
disease diagnosis, 201–203
doping, 112
double-layer supercapacitor systems, 58

economic benefits, 39–40
ecosystems, 7
electric double-layer capacitors (EDLCs), 47
electric vehicles (EVs), 39, 297, 298, 303, 360
 battery materials, 297–298
electrocatalysts, 137
electrochemical biomedical sensors, 225
 properties and importance of low-carbon materials, 227–232
 utilizing low-carbon material for development of, 232
  cellulose, 235–240
  chitosan, 232–235
  silk, 241
electrochemical biosensors, 162, 180
 amperometric and voltammetric biosensors, 162–163
electrochemical deposition, 278
electrochemical energy, 46
electrochemical impedance spectroscopy (EIS) technique, 164
electrochemical processes, 135
electrochemical reactions, 29
electrochemical sensors, 226, 243
electrochemical water-splitting process, 131
 electrocatalysts
  from biomass waste, 138–140
  from electronic waste, 140–141
  from natural minerals, 138
 low-carbon materials for water-splitting process, 132–137
electrodes, 52, 79, 81
 configuration, 119
 materials, 49, 57

electrolytes, 32–35
electromagnetic energy, 95
electromagnetic induction, 97
electronic waste, 47, 53, 55, 140–141
embodied carbon, 255, 256
encapsulation, 104
energy consumption, 280
energy devices
role of materials in, 24
types of, 65–66
energy-dispersive X-ray spectroscopy (EDX), 233
energy efficiency, 276
energy harvesting, 117
advantages of using low-carbon materials in, 100
compliance with regulations and standards, 101
economic benefits, 100
enhanced sustainability of energy systems, 100
improved energy efficiency, 100
positive social impact, 101
reduced environmental impact, 100
applications of low-carbon energy-harvesting materials, 120
green building technologies, 121–122
Internet of Things (IoT) devices, 121
sustainable energy solutions for developing regions, 122–123
wearable electronics, 120
challenges and limitations of, 101
economic challenges, 101
environmental and social impacts, 102
regulatory and policy challenges, 102–103
supply chain issues, 101–102
technical challenges, 101
technological barriers, 102
definition and scope of, 94–96
emerging trends and future prospects, 125
advanced material design, 125
multi-functional materials, 125
sustainability and circular economy, 125
fundamentals of, 96
explanation of ideas and principles of, 96–97
role of materials in energy conversion, 98–99
types of energy sources for, 98
importance of, 93–94
importance of continued research and development in field, 125–126
integration with existing energy infrastructure, 124
compatibility with existing systems, 124
grid integration, 124
reliability and maintenance, 124
materials, 96
potential impact of low-carbon energy-harvesting materials on global sustainability, 126
properties of low-carbon materials, 99–100
scalability and commercialization challenges, 123
economic barriers, 123–124
regulatory and standardization issues, 124
technical challenges, 123

energy harvesting (*continued*)
types of low-carbon energy-harvesting materials, 103
photovoltaic materials, 103–108
piezoelectric materials, 114–117
thermoelectric materials, 108–114
triboelectric materials, 118–120
energy infrastructure projects (EIPs), 371
energy savings, 341–342
energy storage, 125
energy-storage devices, 23, 25–26, 27, 34, 37, 40, 145
energy transfer mechanisms, 97
enthalpy (ΔH), 182
environmental adaptability, 206
environmental impact, 84–85, 279
environmentally friendly synthesis methods, 170
environmental monitoring, 121, 146, 205–206, 270
applications of low-carbon materials in, 280–283
environmental pollution, 200, 205, 319
carbon dioxide, 323–324
carbon monoxide, 324–325
particulate matter, 320–321
volatile organic compounds, 321–323
environmental sensors, 277
enzyme-based biosensors, 184
applications of, 185–186
development of, 186–187
methods of enzyme immobilization, 184–185
enzyme electrode, 184
enzymes, 56, 79, 80, 184
Escherichia coli, 197
ethyl cellulose (EC), 238
ethylene, 277
ethylene glycol (EG), 140
evanescent wave (EW), 181
Faradaic efficiency, 137
Faradaic processes, 47, 55
FCs. *See* fuel cells (FCs)
fermentation, 340
Fermi energy, 151
field emission scanning electron microscopy (FESEM), 239
Fischer–Tropsch synthesis, 310
flexible and stretchable electronics, 120
flow batteries, 82, 83
fluorescence, 161, 181
fluorescence resonance energy transfer (FRET), 161, 162, 364
fluorescent biosensors, 161–162
fluorescent proteins, 208
flywheel energy storage (FES), 28
foodborne illnesses, 192
food safety testing, 204–205
Forest Stewardship Council (FSC), 9
fossil energy, 341
free prostate-specific antigen (fPSA), 192
FRET. *See* fluorescence resonance energy transfer (FRET)
fuel cell electric vehicles (FCEVs), 304
fuel cells (FCs), 299–300, 361–362
architectural aspects of, 83–84
basics of, 68–72
economic consideration, 85
efficiency and performance, 84
energy devices, 65–66
environmental impact, 84–85
future trends and opportunities, 86
eco-friendly materials, 86–87
economic benefits, 86
emerging technologies, 86
importance of, 66
low-carbon materials in various fuel cells, 72
alkaline fuel cells, 77–79
biofuel cells, 79–82

molten carbonate fuel cells, 76–77
other fuel cell technologies, 82–83
proton exchange membrane fuel cells, 72–74
solid oxide fuel cells, 74–76
significance of low-carbon materials, 67–68, 72
types of, 68
fullerenes, 155, 157, 214
fabrication and properties of, 156

gallium nitride (GaN), 298
gasification, 338–339
geopolymers, 9
GHG emissions. *See* greenhouse gas (GHG) emissions
global warming potentials (GWPs), 303
GO. *See* graphene oxide (GO)
gold nanoparticles (AuNPs), 161, 234
GQDs. *See* graphene quantum dots (GQDs)
graphene (GR), 13, 51, 79, 151, 212, 229, 232, 323
band structure and properties of, 151–152
fabrication process, 152–153
nanoparticles, 364
graphene oxide (GO), 210, 213, 323
graphene quantum dots (GQDs), 156, 211, 273, 280, 281
graphite, 147, 274
fabrication routes, 148
gravimetric biosensors, 165, 182
green building market, 5–6
green building projects, 9–10
green building technologies, 121–122
green chemistry, 277
greenhouse gas (GHG) emissions, 6, 34, 226, 255, 257, 271, 280, 302, 316, 372
carbon footprint analysis, 6
comparison with traditional materials, 6–7
green supercapacitors, 47
green synthesis techniques, 47, 56, 231
grid integration, 124

Hamiltonian equation, 152
Hartman interferometers, 160
healthcare monitoring, 144
applications of nanocarbon-based materials for biosensing, 166–170
nanocarbon-based materials, 146
carbon black, 153–154
carbon dots and fabrication, 156–157
carbon fibers and properties, 155
carbon nanotubes, 148–150
fullerene, 155–156
graphene, 151–153
graphite, 147–148
sensor types used in low-carbon materials, 157
electrochemical biosensors, 162–164
gravimetric/piezoelectric/mass-sensitive biosensors, 165
optical biosensors, 157–162
thermal sensors, 164–165
heating cycle, 148
heating, ventilation, and air conditioning (HVAC) systems, 122
heavy metals, 54
HER. *See* hydrogen evolution reaction (HER)
heterostructures, 112
hexagonal honeycomb, 151
high-performance thermoelectric modules, 113
high-resolution transmission electron microscopy (HR-TEM), 239
high-temperature PEMFC (HT-PEMFC), 82

human chorionic gonadotropin (hCG), 185
human T-lymphotropic virus type 1 (HTLV-1), 213
hybrid composites, 116, 119
hybrid electric vehicles (HEVs), 304
hybrid energy-storage systems, 23
hybrid nanocomposites, 112
hybrid systems, 124
hydrogen, 131, 299
  fuel cell vehicles, 369
  production and application, 132
hydrogen evolution reaction (HER), 66, 131, 132
hydrogen oxidation reaction (HOR), 66
hydrometallurgical process, 36
hydrothermal carbonization method, 231, 275
hydrothermal methods, 50, 278
hydrothermal treatment technique, 281

immobilizing enzymes, 184
impedance spectroscopy, 58
impedimetric biosensors, 164
implantable devices, 80
impregnation method, 50
improved public health outcomes, 9
in situ synthesis method, 50
intercalation, 148
internal combustion engine vehicles (ICEVs), 303, 314
Internet of Things (IoT), 121, 216
interpolymer interactions, 139
inverse piezoelectric effect, 114
in vitro cell compatibility, 210
in vivo biocompatibility, 210
ion-selective electrodes (ISEs), 163
ion-sensitive field effect transistors (ISFETs), 163
iridium dioxide (IrO_2), 132

Kyoto Protocol, 25

land use change, 372
laser ablation, 278
laser-induced breakdown spectroscopy (LIBS), 321, 338
LCA. *See* life cycle assessment (LCA)
LCMs. *See* low-carbon materials (LCMs)
leadership in energy and environmental design (LEED), 256, 297
lead zirconate titanate (PZT), 115
LIBs. *See* lithium-ion batteries (LIBs)
life cycle assessment (LCA), 25, 96, 279, 297, 309, 315, 341
limit of detection (LOD), 239
liquefied natural gas (LNG), 306
liquid-phase exfoliation, 278
Li–S batteries (LiSBs), 27
lithium batteries, 26
lithium bis(trifluoromethanesulfonyl)imide (LiTFSI), 33
lithium-ion batteries (LIBs), 26, 38, 54
lithium metal phosphate (LFP), 297
lithography techniques, 196
localized surface plasmon resonance (LSPR), 158–159
low-carbon building, 263
low-carbon cement-based composites (LCCs), 365
low-carbon materials (LCMs), 1, 226, 270, 316, 335
  advancements in energy-storage materials, 297
    electric vehicle battery materials, 297–298
    fuel cell and hydrogen storage materials, 299–300
    high-efficiency engine components, 298–299
  applications of, 10
  batteries with, 360–361
  challenges and barriers, 13
    economic and market barriers, 13–14
    regulatory and policy challenges, 14

social and cultural obstacles, 14
technical and material limitations, 13
challenges and future prospects, 283–288
challenges of utilizing materials with lower CO_2 footprint in automotive industry, 303
cost and regulatory certification challenges, 303–304
performance and durability, 304–305
technological barrier, 305–307
and CO_2 footprint, 325–326
community and urban planning, 12
low-carbon urban infrastructure, 12
sustainable cities and communities, 12
criteria and types of, 2–3
drive and uncertainty, 317–318
economic benefits of, 3
cost reduction, 3–4
innovation and market opportunities, 4–5
regulatory compliance, 5–6
environmental benefits of, 6
enhancement of air and water quality, 8–9
impact on ecosystems and biodiversity, 7–8
reduction in greenhouse gas emissions, 6–7
resource efficiency, 7
in environmental monitoring, 280–283
environmental pollution, 319
carbon dioxide, 323–324
carbon monoxide, 324–325
particulate matter, 320–321
volatile organic compounds, 321–323
for environmental sensors, 275–277
essential characteristics of, 317
fuel cells, 361–362
future directions and innovations, 15
emerging technologies and trends, 15
policy and regulatory innovations, 15
sustainability and circular economy, 15–16
future perspectives, 326–327
global examples and best practices, 13
for greener construction, 365
addition of recycled aggregates, 368
adopting recyclable or low-embodied carbon, 368
considering composite window frames, 368
discovering new concrete, 368
embracing recycling, 368
for greener energy, 360
for greener environment, 364
conversion of CO_2 into fuel, 365
pollution remediation, 365
water treatment, 365
for greener healthcare, 364
impact on transportation sector, 307
reduction of greenhouse gas emissions, 307–309
societal and economic benefits, 309–310
importance of, 24–25

low-carbon materials (LCMs) (*continued*)
industrial applications, 10
manufacturing processes, 10–11
product design and innovation, 11
in interior and exterior components, 300
comparison of CO_2 emissions between ICEVs and EVs for better insights into low carbon emissions, 303
sustainable interior materials, 300–302
transportation infrastructure carbon emission reduction strategies, 302–303
manufacturing, 318–319
principles of, 271–275
steps toward greener world with carbon neutrality, 370–372
successful implementations in construction, 9
green building projects, 9–10
infrastructure developments, 10
supercapacitors, 362
activated carbon, 362
carbon nanotubes, 362–363
energy-harvesting materials, 363
sustainable development goals and carbon neutrality approach, 16–17
synthesis and fabrication methods of, 277–280
in transportation and automotive industry, 368–369
in vehicle design, 296–297
Mach–Zehnder interferometers, 160
magnetic nanoparticles, 232
manufacturing industry, 254
manufacturing processes, 10–11
material decisions, 259
material flow analysis, 280
MCFCs. *See* molten carbonate fuel cells (MCFCs)
mechanical exfoliation method, 153, 278
mechanical recycling techniques, 337
agglomeration and pelletizing, 338
shredding and size reduction, 337–338
sorting and separation technologies, 338
mediated e^- transfer (MET), 80
mesoporous carbon materials, 274
metal–air fuel cells (MAFCs), 82, 83
metal–carbon covalent bond, 135
metal hydrides, 300
metal nitrides, 134
metal–organic framework (MOF), 51
metal oxides, 55, 141, 275
metal sulfides, 136
methane (CH_4), 365
methanol (CH_3OH), 365
microbes, 56
microbial degradation, 339–340
microbial fuel cells (MFCs), 82
Micrococcus lysodeikticus, 167
microelectrode array (MEA) technology, 196
micro-electro-mechanical systems (MEMS), 160
microfluidics, 196
microorganisms, 79, 80
molecular design, 111
molecularly imprinted polymers (MIPs), 234
molten carbonate fuel cells (MCFCs), 66, 76–77, 84
molybdenite concentrate, 138
molybdenum carbides, 135
monoclonal antibodies, 188

multi-actor multi-criteria analysis (MAMCA) method, 307, 308
multi-criteria analysis (MCA), 307
multi-criteria decision analysis (MCDA), 307
multielectrode arrays (MEAs), 198
multi-heteroatoms doped carbon (MHC), 139
multiwalled carbon nanotubes (MWCNTs), 149, 166, 213, 274, 324

Nafion, 82, 86, 168
Na-ion batteries (NIBs), 35, 38
nanocellulose (NC), 80, 118, 238
nanocomposites, 111–112
    thermoelectric generators, 113
nanomaterials, 56, 82, 141, 226
nanoparticles, 273, 315
nanotechnology, 15, 125, 148, 157, 315
nanowires, 111
Na–S batteries (NaSBs), 27
native materials, 230
natural fibers, 118–119, 300, 301
natural graphite (NG), 32
natural materials, 116
natural minerals, 138
natural polymers, 57, 170
NC. *See* nanocellulose (NC)
Ni–Cd batteries, 26
nickel cobalt aluminum (NCA), 297
nickel–cobalt–phosphide (NiCoP) electrocatalyst, 141
nickel manganese cobalt oxide (NMC), 297
Ni–metal hydride (Ni–MH) batteries, 26
nitrogen (N), 134
non-renewable resources, 25
N-type semiconductors, 110
nucleic acid aptamer-based biosensors, 189
    applications of, 192–196
    preparation process of aptamers, 190–191
    types of aptamer biosensors, 191

OER. *See* oxygen evolution reaction (OER)
off-grid power systems, 122
one-dimensional (1D) carbon nanotubes, 321
one-dimensional (1D) nano-materials, 229
one-pot green method, 54
o-phenylenediamine, 166
optical aptamer sensors, 191
optical biosensors, 157, 180–182
    fluorescent biosensors, 161–162
    optical interferometry, 159–160
    surface-enhanced Raman scattering spectroscopy, 160–161
    surface plasmon resonance and localized surface plasmon resonance, 158–159
optical interferometry, 159–160
organic electrolytes, 54
organic solar cells (OSCs), 105
    advantages of, 105
    challenges and limitations, 105
organic thermoelectric materials, 110–111
Oslo Fjord City project, 10
Ostwald ripening, 73, 82
oxygen evolution reaction (OER), 66, 131, 132, 136
oxygen reduction reaction (ORR), 66

paper-based biosensors, 200
    advantages of paper as substrate material for, 201
    applications of, 201
        disease diagnosis, 201–203
        environmental monitoring, 205–206
        food safety testing, 204–205
        paper-based wearable sensors for monitoring human physiological and behavioral signals, 203–204

Paris Agreement, 25, 145
particulate matter (PM), 320–321
Pb–acid battery (Pb-A), 28
Peltier effect, 109, 110
PEMFCs. *See* proton exchange membrane fuel cells (PEMFCs)
Perform, Achieve and Trade (PAT) scheme, 6
permanent magnet (PMSYM) materials, 298
perovskite solar cells (PSCs), 105
  advantages of, 106
  challenges and limitations, 106–107
persistent organic pollutants (POPs), 315
petrochemical-based materials, 35
phonon scattering, 112
phosphoric acid fuel cells (PAFCs), 66, 87
photocatalysts, 364
photoluminescence (PL), 156
photovoltaic effect, 97
photovoltaic materials, 103
  applications in energy harvesting from mechanical vibrations, 117
  low-carbon piezoelectric materials, 116
    biodegradable polymers, 116
    innovative composites, 116
    natural materials, 116
  overview of photovoltaic energy conversion, 103
    energy conversion efficiency, 104
    photovoltaic cell structure, 104
    principle of operation, 103–104
  types of low-carbon photovoltaic materials, 105
    device architecture, 107
    environmental and health safety, 108
    environmental conditions, 107
    material properties, 107
    organic solar cells, 105
    perovskite solar cells, 105–107
    research and development, 108
piezoelectric biosensors, 165, 183
piezoelectric coefficients, 114–115
piezoelectric effect, 97
piezoelectricity, 165
piezoelectric materials, 114, 115–116
  introduction to piezoelectricity, 114
    piezoelectric coefficients, 114–115
    piezoelectric materials, 115–116
PLA. *See* polylactic acid (PLA)
plastics, 140
platinum (Pt), 132
platinum-group metals (PGMs), 299
pollution remediation, 365
poly(3,4-ethylenedioxythiophene) (PEDOT), 110
poly(3-hexylthiophene) (P3HT), 110
poly(ε-caprolactone-*co*-lactide) (PLCL), 210
polyacrylonitrile (PAN), 155
polyaniline (PANI), 110
polyaspartamide (PAA), 324
polybenzimidazole (PBI), 82
polycaprolactone (PCL), 119
polychlorinated biphenyls (PCBs), 316
polyclonal antibodies (pAb), 188
polydimethylsiloxane (PDMS), 208
polyether-*block*-polyamide (PEO-*b*-PA_6), 33
polyethylene oxide (PEO), 208
polyethylene terephthalate (PET), 51, 140, 208
polyhydroxyalkanoates (PHAs), 116

polylactic acid (PLA), 116, 119, 325
polymeric waste, 47
polymer matrix, 238
polymers, 116, 228
polypropylene (PP), 177
poly succinimide (PSI), 324
polyvinyl alcohol, 86
polyvinyl chloride (PVC), 163, 177
polyvinylidene fluoride (PVDF), 116
porous membranes, 58
portable energy harvesters, 122–123
portable health monitors, 117
Portland cement, 6, 8
post-carbon economy, 319
potentiometric biosensors, 163–164
power factor, 113
primary batteries, 26
procyanin, 235
prostate-specific antigen (PSA), 192
proteins, 188, 315
proton exchange membrane fuel cells (PEMFCs), 66, 72–74, 84, 85, 299, 362
pseudocapacitors, 47
P-type semiconductors, 110
pumped hydro storage (PHS), 28
pyrolysis, 53, 238, 275, 338–339
pyrometallurgical process, 35

QDs. *See* quantum dots (QDs)
quality control, 344
quantum dots (QDs), 111, 229, 280
quantum yield (QY), 157

Raman spectroscopy, 160
random library, 190
reactive oxygen species (ROS), 209
recycled aggregates (RAs), 341, 368
recycled carbon materials, 53
recycling low-carbon materials
    benefits and impacts of, 340
        circular economy and waste management solutions, 342
        conservation of natural resources, 341
        energy savings and carbon footprint reduction, 341–342
        reduction of greenhouse gas emissions, 340–341
    challenges in, 342
        collection and sorting of low-carbon materials, 343–344
        contamination issues and quality control, 344
        policy and regulatory considerations, 345
        technological and economic barriers, 344–345
    future trends and innovations in, 345
        advancements in recycling technologies, 345–346
        emerging circular business models, 347
        integration of digitalization and artificial intelligence, 346–347
        policy and industry initiatives, 347–348
    for low-carbon materials, 336
        biological recycling methods, 339–340
        chemical recycling approaches, 338–339
        mechanical recycling techniques, 337–338
redox flow batteries (RFBs), 27
reduced graphene oxide (rGO), 213, 229, 241
reference electrodes, 163, 164
refractive index units (RIU), 159
regenerative braking systems, 299
regenerative fuel cells (RFCs), 82
remote sensors, 121
renewable energy sources, 25
reproducibility, 206, 279
research and development, 102

rGO. *See* reduced graphene oxide (rGO)
ruthenium dioxide (RuO_2), 132

Salmonella typhimurium, 205
scanning electron microscopy (SEM), 233
Schwann cell migration, 210
Scotch tape method, 278
SDGs. *See* sustainable development goals (SDGs)
secondary batteries, 26
Seebeck effect, 109, 110
self-powered sensors, 123
sensors, 117, 225
service infrastructure, 372
silicon carbide (SiC), 298, 321
silk, 207, 241
silk-based biosensors, 207
  implantable silk-based biosensors, 210
  traditional biosensors based on silk substrates, 207
    colorimetric biosensors, 208
    electrochemical biosensors, 207–208
    fluorescent biosensors, 208–210
silk fibroin (SF), 210
silver nanoparticles (AgNPs), 161
single-stranded DNA (ss-DNA), 167
single-walled carbon nanotubes (SWCNTs), 149, 213, 274
skin patches, 120
smart agriculture, 121
smart cities, 121
smart grids, 125
smart homes, 121
smart materials, 15
smart textiles, 117, 120
social network analysis (SNA), 308
sodium carboxymethyl cellulose (Na-CMC), 238
SOFCs. *See* solid oxide fuel cells (SOFCs)
solar energy, 95
solar panels, 144
sol–gel method, 49–50
solid oxide fuel cells (SOFCs), 66, 74–76, 87, 299, 362
solid polymer electrolytes (SPEs), 33, 34
solid-state batteries, 38, 41
solid-state electrolytes, 26
solvent-based processes, 339
solvothermal methods, 278
specific surface area (SSA), 274
SPR. *See* surface plasmon resonance (SPR)
square wave voltammetry (SWV), 191, 239
Stockholm Convention, 316
structural engineering, 119
structural health monitoring, 117
sulfonated polyether ether ketones (SPEEK), 82, 86
supercapacitor energy storage (SCES), 28
supercapacitors, 23, 29, 46, 65, 230, 362
  activated carbon, 362
  carbon nanotubes, 362–363
  energy-harvesting materials, 363
  low-carbon electrode materials, 47
    biowaste to, 47–51
    convert plastic waste into, 51–52
    metal and metal oxides to, 55–57
    recycling electronic waste into, 53–55
  other low-carbon and green components of, 57–58
superconducting magnetic energy storage (SMES), 28
superlattices, 112
supply chains, 102, 261
surface engineering, 107
surface-enhanced Raman scattering (SERS) spectroscopy, 160–161, 181

surface functionalization, 211
surface modification, 119
surface plasmon resonance (SPR), 158–159, 181, 182, 212
surface roughness, 119
sustainability, 15–16, 125
sustainable agriculture, 123
sustainable aviation fuel (SAF), 306
sustainable development goals (SDGs), 16–17, 25, 52, 144, 170, 178, 288
synthesis methods, 135
synthetic adsorbents, 325
synthetic graphite (SG), 32
Systematic Evolution of Ligands by Exponential Enrichment (SELEX), 190

Tafel slope, 135
tandem cells, 107
technological barrier, 305
cost and supply chain considerations, 305–306
regulatory and safety standard compliance, 306–307
TEM. *See* transmission electron microscopy (TEM)
template-assisted synthesis, 279
tetraethylenepentamine (TEPA), 324
thermal biosensors, 182–183
thermal conductivity, 112, 148
thermal energy, 95
thermal-hydrothermal method, 51
thermal sensors, 164–165
thermoelectric building systems, 122
thermoelectric effect, 97
thermoelectric generators, 122
thermoelectric materials, 108
applications, 113
future directions, 113–114
low-carbon thermoelectric materials, 110
advantages of organic thermoelectric materials, 111
challenges and limitations, 111
nanocomposites, 111–112
organic thermoelectric materials, 110–111
power factor, 113
principles of thermoelectric energy conversion, 108
challenges and limitations, 112
fundamental concepts, 109
thermoelectric figure of merit, 109–110
thermoelectric modules, 110
thermal conductivity, 112
Thomson effect, 109
three-dimensional (3D) materials, 230
three-dimensional polypyrrole (3DPPY), 238
tight bonding band (TBM) theory, 151
tire pressure-monitoring systems (TPMS), 117
top-down method, 152, 156
toxicity, 96, 107
transducer, 179
transition metal hydrides, 300
transition metal phosphides (TMPs), 137
transmission electron microscopy (TEM), 150, 241, 282
transportation infrastructure, 257
triboelectricity, 118
triboelectric materials, 118–120
triboelectric nanogenerators (TENGs), 118
two-dimensional (2D) nanomaterials, 229–230

urokinase plasminogen activator (uPA), 213

valence bands, 152
van der Waals forces, 274

vascular endothelial growth factor (VEGF), 207
Vibrio parahaemolyticus, 192, 193
volatile organic compounds (VOCs), 9, 321–323
Volmer–Heyrovsky mechanism, 135
voltammetric biosensors, 162–163
V redox flow battery (VRFB), 28

waste management solutions, 342
water treatment, 365
wearable electronics, 117, 120
wearable sensors, 120
wind turbines, 145
wood–plastic composites, 273

X-ray diffraction (XRD), 233

yeast biomass-derived materials, 139
Young interferometers, 160
Young's modulus, 150

zero-dimensional (0D) nanomaterials, 155, 229
zero-emission process, 35
zinc-ion (Zn-ion) batteries, 363
zinc oxide (ZnO), 365
zirconia-based materials, 75
Zn–C batteries, 26, 54
Zn-metal–organic frameworks, 166